Living with the Earth

Living with the Earth

Concepts in Environmental Health Science

Fourth Edition

Gary S. Moore
Kathleen A. Bell

CRC Press
Taylor & Francis Group
Boca Raton London New York

CRC Press is an imprint of the
Taylor & Francis Group, an **informa** business

CRC Press
Taylor & Francis Group
6000 Broken Sound Parkway NW, Suite 300
Boca Raton, FL 33487-2742

© 2019 by Taylor & Francis Group, LLC
CRC Press is an imprint of Taylor & Francis Group, an Informa business

No claim to original U.S. Government works

Printed on acid-free paper

International Standard Book Number-13: 978-1-1385-9056-4 (Hardback)

Visit the Taylor & Francis Web site at
http://www.taylorandfrancis.com

and the CRC Press Web site at
http://www.crcpress.com

Contents

Preface

Intent of the Book

This book is a college-level textbook for introductory courses in environmental health and environmental sciences. It is suitable for both undergraduate- and graduate-level courses for students of Environmental Health, Environmental Sciences, Community Health Education, Epidemiologists, Civil Engineers, Community Medicine, General Medicine, and Public Health. It is also highly recommended for members of Boards of Health, Health Officers and Health Inspectors, and citizens who are members of proactive environmental groups such as Sierra Club, Public Interest Research Groups, Clean Water Action, Environmental Defense Fund, and many others.

Synopsis of the Book

This book contains 14 carefully illustrated and narratively exciting chapters on the subject of environmental health sciences. The contents are derived from a core course Dr. Moore instructed at the University of Massachusetts, Amherst titled "Environmental Health Practices," and this book incorporates both the traditional concepts associated with environmental health and new, emerging, and controversial issues associated with environmental threats to human health. Emphasis is placed on biological, chemical, and physical sources of pollution, and the methods of controlling or limiting those exposures. Careful attention is paid to presenting a balanced view representing opposing scientific views on major issues ranging from "The Greenhouse Effect" to reproductive problems associated with endocrine disruptors.

Key Features

- A wide offering of instructor support materials including PowerPoint™ presentation slides in digital format, a web-enhanced course site, and an examination (test) bank.
- Numerous author-created, detailed technical illustrations and photographs throughout the text and the web-enhanced course site.
- New chapters on emerging diseases and emergency preparedness including the most significant reasons for their emergence, and the major etiological agents associated with emerging diseases and bioterrorism.
- Measures of population dynamics including the contrasts of various cultural philosophies regarding "overpopulation" and the new problems of "underpopulation" in the developed nations.

- New developments in areas of:
  - The emerging problems of asthma and associated air pollutants
  - The genetic basis of cancer and the roles of behavior and pollution in cancer risk
  - The biology of genetics, mutation, and birth defects, and much more

Using the Website

The website has been constructed to provide maximum usefulness for the student and the instructor. The site has been actively tested by student use and modified to meet student and instructor expectations. This is an ongoing and dynamic effort, and you are encouraged to suggest improvements in the website that would make it more useful and interesting. The site has been designed to work with most popular browsers and is fast-loading with most images linked to text and selected for viewing by the user. This design is intended to speed up the use of the site and let users get to where they want quickly without being overwhelmed by graphics that automatically load and slow down the process.

Students

The website address is **https://people.umass.edu/~envhl565/**

> **Username = ulearn**
> **Password = mar99**

Once you connect with the site and enter the correct ID and password, the possible use of the site will be immediately obvious. There are sample test questions for each chapter with separately printable answers.

Associated with each chapter are links to **websites** that will take you to some of the most useful sites maintained on that subject matter by the U.S. government, international and national organizations, academic institutions, and some private groups. These and other sites offer a wealth of information that the student is encouraged to explore. There are also a number of searchable databases where you will be able to enter and find most any environmental topic by keyword.

Instructor Materials

The instructor for this course will have all of the features available to the student and will also have access to chapter-by-chapter **PowerPoint** files that they may download directly from the website. There are nearly 1100 slides with more than 250 images in full color taken directly from the book. The slides are based on the detailed course notes condensed from each chapter. The PowerPoint files may be shown using an LCD projection system in combination with a computer, printed on transparencies, or sent to a service bureau for digital imaging to 35 mm slides. The digital files may be modified by the instructor in **PowerPoint** to meet his/her needs in the classroom. I am also including in the instructor materials section a **test bank** of exam questions on a chapter-by-chapter basis. The questions include multiple choice, true and false, and short essay questions. The answers are supplied along with the page numbers in the text where the answers may be found. The test bank materials are in rich text format (RTF) that may be opened with most any common popular word-processor. The instructor will also benefit from the **searchable databases** provided as web-links to obtain the latest information on environmental health topics. This website is password protected. Instructors may acquire the username and password by submitting a request with your departmental phone number to gmoore9324@gmail.com. Put the word "password" in the subject area.

Acknowledgments

To my wife Lucille, for her unwavering support during my long periods of absence from her side. She accepted my dedication to this project and kept me fed and clothed in the interim. I extend sincere thanks to students who wrote some of the text in selected chapters and provided much-needed proofreading for most of the book. These students all benefited enormously from the experience and grew in confidence and capabilities. I provided instruction in the areas of writing and organization, page layout, and graphic design. Each student took something of importance with them. John Edman, an entomology professor at the University of Massachusetts, provided slides for most of the insects used in Chapter 6. The pictures featuring African children and scenes were provided by Marie Krause Cote who spent several months in Africa assisting as an educator to tribeswomen in producing cloth goods for market. Many of the pictures on wildlife, wetlands, and biomes came from U.S. Government departments including the U.S. Fish and Wildlife Services (USFWS), the National Aeronautics and Space Administration (NASA), and The Environmental Protection Agency (USEPA). Thanks is also given to Charlie Kaniecki, who is the District Health Officer for the Department of Public Health in Western Massachusetts.

Gary S. Moore

Authors

Gary S. Moore is an Emeritus Faculty member in the Department of Environmental Health Sciences in the School of Public Health and Health Sciences at the University of Massachusetts where he has taught several graduate-level environmental courses including courses on Air Quality Assessment (EnvHl 562), Environmental Regulations and Compliance (EnvHl 567), and Environmental Health Practices (EnvHl 565). He has many publications (over 75) in the area of health effects from environmental insults.

Dr. Moore has received several awards for his efforts in the area of providing training and certification programs to Boards of Health in Massachusetts including recognition from the Massachusetts Senate, the Massachusetts House of Representatives, The Massachusetts Association of Health Boards (MAHB), and the Robert C. Huestis/Eric W. Mood Award from the New England Public Health Association (NEPHA). The certification and training programs for Boards of Health have been very successful. This project is a strong foundation for further collaboration between the University and MAHB, providing practical educational opportunities to people serving every community in the Commonwealth.

Dr. Moore is committed to providing the maximum learning experience and highest possible profile for his graduate students. Some of those graduate students have contributed significantly to the preparation of the first edition and are listed below.

Kathleen A. Bell is a consultant and writer who facilitates consulting skills programs across a variety of industries. Kathleen's consultancy began at a small environmental science firm drafting health documents for the United States Environmental Protection Agency Office of Water. Early in her career, Kathleen spent several years at AIG and The Travelers, where she worked on environmental exposure and toxic tort litigation.

Kathleen later joined the Corporate Executive Board (now Gartner), where she served as a Director of Research in both the Washington DC and London offices, advising heads of corporate strategy on process improvements. She has designed and delivered training programs to teams of major multinationals in North America, Europe, South Africa, Australia, and SE Asia.

Kathleen holds a BA in Biology from Cornell University, a Master's in Public Health from the University of Massachusetts-Amherst, and an MBA from the Darden School of Business at the University of Virginia.

She lives in Virginia and enjoys hiking, writing, and spending time with her family.

1

Ecosystems and Energy Flow

OBJECTIVES

A student reading this chapter will be able to:

1. Define the concepts of biosphere and climate.
2. List and explain the factors influencing climate.
3. Define the term "biome." List the major global biomes and discuss their primary features.
4. Show the flow of energy through ecosystems.
5. List the various trophic levels.
6. Explain the various nutrient cycles including the carbon, nitrogen, and phosphorous cycles.
7. Define the term "succession," explain the mechanisms of succession, and discuss the types of human intervention that interfere with succession.

Introduction

We are on a living planet and are surrounded, immersed, and enveloped by life. It is both outside and inside of us. We breath it in, we swallow it, we walk on it, we touch it. Each footstep on a fertile lawn or forest floor will send tremors to trillions of bacteria; millions of algae, fungi, and protozoa; and hundreds of insects and worms. The skin on our bodies, when viewed microscopically, is a teaming matrix of tiny caverns filled with bacteria, viruses, and mites. So dense are the unseen life forms on our bodies that if our own tissues were to instantly disappear, we would still be recognizable by the remaining shell of microscopic life enveloping our body. Every breath draws in untold numbers of fungal spores, bacteria, viruses, and other microbes and plant spores. Life abounds most everywhere inhabited by humans. Life thrives on the nutrients in soil and water, the oxygen and carbon dioxide in the air, and the sunlight that ultimately powers most life. In those areas of earth where nutrients are depleted, oxygen is rare, sunlight is extinguished, or moisture is diminished, life becomes reduced or absent. This is true for most breathing creatures except birds. They exist from the coldest regions of Antarctica to Atacama, the hottest and driest deserts in the world. Conditions for most life are found in a layer about the globe that extends from approximately 8 kilometers in the atmosphere (where some microbial spores and insects may be found) to 8 kilometers below the ocean surface where some unusual life forms adapted to darkness and high pressure survive. This theoretical "layer of life" is called a biosphere because life is thought not to exist outside this area. Most life occurs in a much narrower layer extending from about a 182-meter depth in the ocean, where sunlight is able to penetrate, to the summer snow line of high mountain peaks, where a thin layer of soil supports plant life such as lichens and mosses. Within this biosphere, the forms and quantities of life vary dramatically. Surface or land-based life may be categorized into major regions known as biomes. Biomes are based on the dominant types of vegetation that are strongly correlated with regional climate patterns. We'll explore the various types of biomes in short order, but first let's look at the driving force of these regional differences—the climate.

Climate

What Is It?

Climate varies in latitudinal bands around the globe. Relatively constant warm and humid weather affects most equatorial countries, and as you travel north or south, the number of warm months diminishes and freezing months escalate. Climate can be viewed as average weather within a geographical area viewed over years, or even centuries. Climate, like weather, includes temperature, precipitation, humidity, wind velocity and direction, cloud cover, and associated solar radiation. Climate does change, but over centuries or millennia. Some of these climatic changes are dramatic. There have been many ice ages with the formation of glaciers resulting in major climatic changes globally. These climatic changes have occurred from different phenomena that may include (1) changes in ocean temperatures; (2) changes in the earth's orbital geometry; (3) volcanic activity with increased atmospheric dust and reduced sunlight penetration; (4) variations in solar radiation; or (5) increases in atmospheric gases that absorb heat energy.[1]

Volcanic eruptions have been thought to be a major factor in climate change based on a range of studies that examine the changes in the environment. Volcanic eruptions can cause sudden and severe environmental changes. Some eruptions emit large volumes of sulfur and particulate matter that can change climate and produce depressed temperatures around the world. The impact of nine volcanic eruptions between 1883 and 1982 was analyzed and the study discovered that volcanic eruptions are associated with surface temperature anomalies that usually fall within the normal seasonal statistical variation and that volcanic eruptions may be related to temperature and circulation anomalies. However, questions are still unanswered as to the scale and patterns of the cooling.[2] Many of these gases are produced by human activities such as the burning of fossil fuels releasing large amounts of heat-absorbing carbon dioxide.

Scientists from the U.S. and Australia are reporting that during the last 100 million years there may have been a sudden release of greenhouse gases from the ocean floor. It is believed that gradual warming caused a change in ocean currents, which caused warm water from the surface to plunge deeper into the ocean. The solid methane that was trapped in the ocean sediments became gaseous. The methane would then explode from the sediments that resulted in mudslides. These mudslides allowed the methane to escape into water and then into the atmosphere. The methane reacted with the oxygen and formed CO_2, causing heat to be trapped in the atmosphere. The end result was an increase in ocean temperature by $-15°$ to $-12°C$. This warming was so severe that it wiped out many species of marine life. Even though these conditions happened without humans around, it may be possible that we are creating the same conditions that caused this climate change. Today, a huge amount of heat-trapping greenhouse gases is being released into the atmosphere at a relatively fast rate. Based on work by the National Academy of Sciences, they conclude that the earth's surface temperature has risen by about .555°C in the past century, with accelerated warming during the past two decades. They also report that there is new and stronger evidence that most of the warming over the last 50 years is attributable to human activities.[3] The United States Environmental Protection Agency (USEPA) reports that human activities have altered the chemical composition of the atmosphere through the buildup of greenhouse gases— primarily carbon dioxide, methane, and nitrous oxide.[3] The main greenhouse gas that caused this was CO_2. Since the Industrial Revolution, CO_2 has been released, causing the atmospheric content to rise by 30 percent since the eighteenth century.[4] These climate-changing gases will be discussed in more detail in Chapter 10. Do you want to learn more about climate change and policies? Point your browser to (https://people.umass.edu/~envhl565/weblinks.html) and click on any of the Chapter 1 Web links associated with climate change including:

- The Climate Diagnostics Center
- Climate Prediction Center
- Northeast Regional Climate Center
- United Nations Framework Convention on Climate Change, UNFCCC

How Is Climate Affected?

Climate is most affected by temperature. By the end of the next century, some scientists predict that atmospheric CO_2 levels will be double those of the pre-industrial era. They also suggest that this increase could have been much greater if not for a continuous evolution of energy sources away from carbon producers such as wood and coal to more efficient sources like natural gas and burgeoning hydrogen. Regardless of our decreasing rate of CO_2-to-energy ratio, population explosions and economy increases have started a continuous rise in CO_2 production. As a result, we can expect extreme changes in the world's climate, weather, and lifestyle.[5]

The power that modifies, influences, and controls climate originates with the sun. The amount of sunlight striking a particular area of the globe determines the level of warmth and ultimately the movement of air and the amount of precipitation. Because the earth is a globe, which rotates about an axis that is tilted, the amount of sunlight striking the earth varies by region and time. The intensity or energy from the sun diminishes with the distance from the sun. More specifically, the intensity declines inversely with the square of the distance from the sun. Consequently, our distance of 149.6 million kilometers from the sun means we receive a very small fraction of the sun's intensity (about one 2 billionth of the sun's energy output pointed in this direction).[6]

Earth's distance from the sun has little to do with the seasons. The seasons are caused by the tilt of the earth on its axis as it revolves around the sun. The earth is tilted at a 23.5-degree angle from a vertical axis drawn perpendicular to the plane of the earth's orbit around the sun. This tilt causes some parts of the earth to get more slanting rays of sunlight some of the year and more vertical rays of sunlight at other times. When a hemisphere of the earth is tilted toward the sun, it is summer in that hemisphere. Moreover, because the earth is tilted, seasonal differences will occur as the earth rotates about the sun. As seen in Figure 1.1, the equator receives vertical rays of the sun so that the largest amount of solar energy is absorbed in a band circumventing the globe at the equator. Further, the angle at the equator doesn't vary much as the earth rotates about the sun so that there are no significant seasonal changes, and temperatures remain fairly warm and constant throughout. As an example, New England is tilted more towards the sun in the summer months and receives vertical rays but receives slanted rays in the winter because of the earth's tilted axis (Figure 1.1). Conversely, Australia, which is in the southern hemisphere, experiences just the opposite seasons. The sun impacts the earth in bands of decreasing energy extending north and south from the equator. Warmer

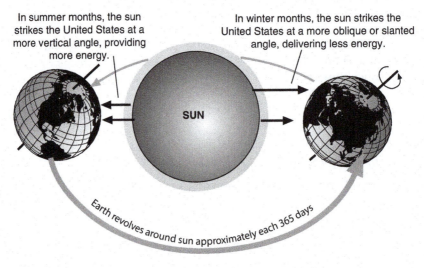

FIGURE 1.1
Variations in the angle of the sunlight striking the earth influence the seasons.

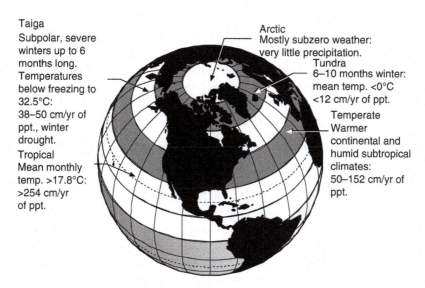

FIGURE 1.2
Variations in global regional temperatures and precipitation.

FIGURE 1.3
Hadley and three-zone models of global circulation.

air can hold more moisture, and therefore may deposit that moisture as precipitation. Therefore, as you move further from the equator, the seasons develop and months of colder and drier air increase, producing major climatic regions following approximate latitudinal bands around the earth (Figure 1.2). Sir George Hadley provided the first written explanation of global air circulation in 1735.[1] Although modified in recent years, the basic concept still holds. The sun warms air at the equator. As the molecules of air absorb heat energy, they spread apart, causing the gases to expand and become less dense. This expanded air rises, traveling away from the equator. The warm air loses its heat by radioactive and convective cooling and subsides back to the earth where it flows towards the equator and completes the Hadley cell (Figure 1.3a). More recent models show that there are multiple Hadley cells known as the three-zone model.[1] In this model, Hadley-type cells are formed on both sides of the equator (Figure 1.3b). Thermal energy is lost and subsidence of cool dry air occurs at about 30° north and south latitudes. The falling dry air compresses and warms. The warmer dry air absorbs moisture, creating some of the driest land areas on Earth at these latitudes. The Sahara Desert in North Africa and the Chilean coastal deserts are representative of these areas. There are other reasons for desert formation and these are discussed later in this chapter under the section Desert Biomes. Moving north or south of these latitudes brings you into the temperate latitude including the bulk of the United States, Central Asia, and Europe. The atmospheric flows in these middle latitudes are highly variable because there is a convergence of polar and tropical fronts with many migrating high- and low-pressure systems that cause frequent weather changes. The collision of warm and cold fronts often produces storm systems resulting in precipitation and ultimately highly productive agricultural land. As can be seen in Figure 1.3b, the large-scale wind flow patterns in the temperate climates are not directly north or south but tend to be deflected to the east in the northern hemisphere and to the west in the southern hemisphere. The deflection of air masses to the east or west is a result of the earth's rotation, causing the deflection of air from its northerly or southerly path, and this is known as the Coriolis effect (Figure 1.4).[1] The effect seen in the temperate North American zone is to produce prevailing winds that flow from the west to the east, and these are known as westerlies. Because of this effect, there is a tendency for storm systems to travel from the west coast to the east coast of the United States. We saw hurricanes like Floyd throughout the 1940s, 1950s, and 1960s. Scientists are now questioning why they seem to be returning. Many meteorologists agree that the climate of the earth shifts naturally and that this trend is normal. Others, however, believe that there is another factor involved, a non-natural factor called global warming. The National Climate Assessment panel reports that there has been a marked increase in most measures of Atlantic hurricane activity since the early 1980s, when we have high-quality satellite data. Measurements from satellite data include intensity, frequency, and duration as well as the number of strongest (Category 4 and 5) storms. The contributions of human versus

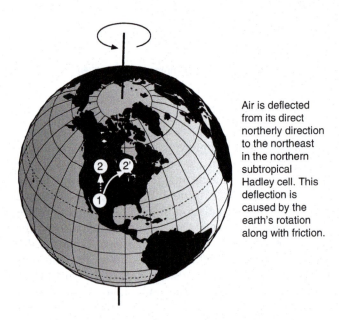

Air is deflected from its direct northerly direction to the northeast in the northern subtropical Hadley cell. This deflection is caused by the earth's rotation along with friction.

FIGURE 1.4
The Coriolis effect produces an easterly or westerly deflection in the expected trajectory of the wind. (Adapted from Godish, T., *Air Quality*, 3rd ed., Lewis Publishers, New York, 1997.)

natural causes associated with these increases are still uncertain because the ability to assess longer-term trends in hurricane activity is limited by the quality of available data.[7]

As we release more and more greenhouse gases into the atmosphere, both the air and ocean temperatures will rise significantly. Hurricanes gain their strength from thermal energy found in air and water; the effect of these increasing temperatures creates the perfect conditions for what scientists refer to as "hypercanes." Storms of this size and power would make hurricane Floyd seem like nothing. Strong winds that approach 241–257 kph stir up cold waters from deep below the ocean's surface. This causes a cooling effect that will eventually destroy the storm. For this reason, we almost never see a Category 5 hurricane. However, warming of the ocean temperatures is suspected in the series of hurricanes that pounded the southeastern American coastline in 2005 including the Category 5 hurricane Katrina that became a Category 3 storm just as it overwhelmed New Orleans and surrounding areas, producing one of the deadliest natural disasters in U.S. history. By August 30, 2005, one day after the Category 3 storm made landfall, three quarters of the city of New Orleans became flooded with as much as 6 meters of water in some locations.[7,8] Several levees breached, causing a flood that was called "the largest civil engineering disaster in the history of the United States."[8,9] We hope not to see frequent "hypercanes," but there have been five hurricanes in the Atlantic since 2005 that had winds exceeding 265 kph. The month of September 2017 saw both Hurricane Irma and Hurricane Maria. Irma, with 297 kph winds, tracked through the Caribbean, British and U.S. Virgin Islands, Cuba, and Florida, causing $66 billion in damage. Maria had 281 kph winds causing massive damage to the Virgin Islands, Puerto Rico, Dominican Republic, and Turks and Caicos.[10] Since most of the U.S. population resides in coastal communities, there is concern for life and property with such storms.[9] Ultimately, climates are determined by the position of the earth with respect to the sun. This influences how much solar energy is absorbed, what wind patterns are established, and how much moisture is deposited. Temperature and moisture combine to become very important factors in establishing climates, which determine the major regional grouping of plants and animals known as biomes.

Ecosystems and Biomes

Ecosystems

Ecosystems are often a component of a biome. The relationship of biosphere, biomes, ecosystems, and populations is shown in Figure 1.5. A system may be viewed as a collection of components that are interdependent or "working together." For example, a collection of gears and springs as you might find in a wall clock or non-digital wrist watch may be viewed as a system that functions in a very precise way to monitor time. Changes or alterations to any component gear may alter or halt this function. The concept of system can be made very inclusive, such as a society with a system of government, or an individual with a system of organs or tissues such as the immune system. The level at which you wish to focus depends on what you wish to study. Ecosystems refer to identifiable areas within nature where the organisms interact among themselves and their physical environment and exchange nutrients largely within that system. An ant farm inside a glass aquarium can be an ecosystem. Maggots and microbes causing decay of garbage in a collection container can be an ecosystem. A pond, tidal pool, river valley, or the planet itself can be considered

FIGURE 1.5
Relationship of biosphere, biome, ecosystem, and populations.

an ecosystem. All of these ecosystems consist of biotic and abiotic components. The biotic components include living organisms and the products of these organisms include waste products and decay such as urine, feces, decaying leaves and twigs, bones, and flesh. The physical or abiotic components of the ecosystem include such things as water, air, sunlight, minerals, and their interaction to produce climate, salinity, turbulence, and other conditions that influence the physical conditions under which organisms survive. The interactions of abiotic and biotic components of the environment are so closely intertwined that they cannot be easily separated for study. A kilogram of soil, for instance, consists of fragments of stone and pulverized rock, water, decaying plants and animals, and numerous living organisms such as bacteria, fungi, protozoa, insects, and worms. Without the biotic components of soil, it could not support the growth of plants and would be lifeless dust.

Biomes

Biomes may be seen as groupings of plants and animals on a regional scale whose distribution patterns depend heavily on patterns of climate. The biome is identified by the climax vegetation or community. A climax community is identified as one that forms in an undisturbed environment and continues to grow and perpetuate itself in the absence of further disturbance. For instance, a mature climax forest of hardwoods (oak, beech, elm, maple) may take a century or more to develop from an abandoned farm field as a series of different plants succeed each other to this final stage. The biome is a community of interacting vegetation, soil types, and animal populations that are adapted to the physical environment of the region. The major biomes of the earth include tundra, taiga, temperate forest and grassland, deserts, and tropical biomes (Table 1.1).[6–12] These major biomes are described briefly in Chapter 3 under the topic of soils with greater attention paid to conditions of soil depth and fertility. In this section we will discuss the biomes in more detail with regard to climate conditions, typical climax vegetation, and predominant mammal populations.

TABLE 1.1

Major Earth Biomes

Biome and Location	Rainfall	Temperature	Climax Vegetation	Animals
Tundra High latitudes of N. hemisphere in a belt around Arctic Ocean.	Less than 12 centimeters per year, strong drying winds.	6–10 weeks growing season. Winters are dark and long (6–10 months). Mean monthly temp. below freezing.	Small shrubs, mosses, sedges, lichens, and some grasses.	Muskox, arctic hare, arctic fox, migratory waterfowl, and caribou.
Taiga Boreal forest in N. hemisphere forming a continuous belt of coniferous trees across N. America through Eurasia. Overlying formerly glaciated area	38–50 centimeters annually. Low evaporation rates make area humid in summer. Winter drought.	Subpolar. Severe winter up to 6 months. Extreme temperatures of below freezing to +32°C.	Needleleaf, coniferous trees (spruce, fir, pine).	Fur-bearing predators such as lynx, wolverine, fisher, and sable. Mammalian herbivores such as snowshoe hares, lemmings, and voles. Large herbivores such as elk, moose, and beaver.
Temperate Forest Western and central Europe, eastern Asia, eastern North America.	50–152 centimeters per year.	Warmer, continental, and humid subtropical climates. Drought during cold winters.	Temperate broadleaf deciduous forest. Oak, maple, beech, hickory, and elm. Autumn leaf fall provides for abundant, rich humus.	Herbivores such as white-tail deer, gray squirrels, and chipmunks. Omnivores such as raccoons, skunks, and black bears. Carnivores largely eliminated such as timber wolves, mountain lions, and bobcats.
Temperate Grasslands N. America in central lowlands and high plains. Steppes of Ukraine, eastward through Russia and Mongolia. Pampas of S. America. Veldt of Republic of S. Africa.	25–50 centimeters per year. Much falls as snow.	Semiarid, warm to hot summer.	Rich mix of perennial grass underlain by fertile soils. Members of sunflower and pea families. Scattered woodlands in wetter areas.	Low in diversity. Bison and pronghorn in N. America. Rodent herbivores including ground squirrels and gophers. Carnivores/omni- vores include coyotes, badgers, and black-footed ferrets.
Tropical Savannas E. Africa, Venezuela, Columbia, Brazil, and Central America.	Dry and wet seasons. 76–127 centimeters with dry season less than 10 centimeters a month.	Mean monthly temperature above 17.8°C.	Perennial grasses, dry zones, with sporadic woods in wetter areas.	Elephants, zebras, and wildebeests. Hoofed mammals such as antelopes, gazelles, and giraffes. Carnivores such as lions, leopards, jackals, and hyenas.

(Continued)

TABLE 1.1 (*Continued*)

Major Earth Biomes

Biome and Location	Rainfall	Temperature	Climax Vegetation	Animals
Tropical Forest Broadleaf forest or rainforest includes areas in Central and S. America, Central Africa, South and Southeast Asia.	In excess of 254 centimeters a year.	Mean monthly temp. above 17.8°C.	Widely spaced emergent trees of 30–36 meters. A closed canopy of 18- to 24-meter trees. Air plants, woody vines and climbers	Most complex biome with numerous birds, insects, and primates.
Deserts West coast of continents between 20° and 30° latitudes including Baja, Ca., Atacama in S. America, and Namib in Southwest Africa. Interior positions of continents such as Gobi of Mongolia, Australian desert, and the Great Basin desert in the U.S. Rainshadow areas of high mountain ranges such as Death Valley, CA, and Peruvian desert of S. America.	Less than 25 centimeters precipitation per year, arid climate in which evaporation exceeds precipitation annually.	May exceed 37.8°C in summer afternoons but drop to below freezing at night. Winters may be cool to cold.	Evergreen or small-leaved deciduous plants, typically with thorns or spines and shallow, extensive root systems. Some plants with long tap roots (6–9 meters) to reach groundwater.	Animals with small body sizes and long appendages. Bright plumage to reflect sunlight. Reptiles have waterproof skin, produce uric acid instead of urine, and hard-shelled eggs.

Based on latest values available for each country.

Tundra

The tundra is limited to the upper latitudes of the northern hemisphere and forms a belt around the Arctic Ocean (Figure 1.6). The continents don't extend southward sufficiently to create a similar biome in the southern latitudes. The tundra appears as a barren and treeless land consisting mostly of mosses, lichens, sedges, dwarf shrubs, and perennial forbs. The scattered shrubs and woody bushes lie close to the ground where they can preserve some warmth. The growing season is very short, lasting 6–10 weeks. The winters are long, dark, and cold with mean monthly temperatures below freezing (<0°C), and precipitation is less than 12 centimeters per year. Dryness is further encouraged by strong, drying winds. Beneath the thin layer of topsoil lies a permanent layer of ice or frozen soil known as permafrost. The permafrost, along with the cold weather, prevents the growth of larger plants and trees, prevents drainage, and leads to the formation of bogs and marshes.

The soils are shallow and poor in fertility because of the limited growing season combined with the slow decomposition of organic matter in extreme cold temperatures. Consequently, the soils and the associated plant life tend to be very fragile, and damage done to the tundra is usually long-lasting.

In 1998, a team of researchers at Vostok, Antarctica drilled 3,263 meters into the ice and brought up ancient ice samples. From these samples it was possible to gather information such as age, temperature at the time the ice formed, and composition of the atmosphere. The composition of ancient atmospheres is shown through the analysis of tiny air bubbles trapped in the ice. Glaciologist Claude Lorius found a strong positive relation between CO_2 and methane. Both of these chemicals are known as greenhouse gases. They have shown the rapid temperature changes of the last ice age. It also relates to the current increase in global temperature. This could be due to the increased levels of greenhouse gases in the atmosphere. This subglacial lake is buried under more than 3 kilometers of ice. It's been

FIGURE 1.6
Typical tundra biome and location(s).

ice covered for more than 15 million years with no sunlight reaching it. Yet, recently thousands of microbes have been discovered thriving in this most extreme environment. Most are single-celled organisms, but a number of multi-celled microscopic organisms have been identified.[13] Computers have been very useful when making paleoclimate models and finding climate mechanisms.[9]

Scientists were able to date the ice core. This was done by using an ice flow model using electrical conductivity measurements, ice accumulation changes, and correlation with other paleoclimate records. The data on greenhouse gases led Lorius to conclude that the greenhouse gases and the ice sheets "played a significant part in the glacial/interglacial climate changes." There is also a problem of which came first, the greenhouse gases or the climate changes. At this time, it is thought that the climate change is a result of the increased greenhouse gas concentration.[9]

The Arctic supports great amounts of wildlife, which appear as large numbers but very few species. These species have adapted to the extremely harsh conditions by morphological, physiological, and behavioral adaptations. Many animals have developed large, heavy bodies with a thick layer of fur (muskox) or feathers. Some species accumulate fat during a shortened growing season (bear) while others adapt by experiencing periodic population crashes and explosions as in the lemming population. Some species such as caribou and waterfowl migrate southward from the tundra during the harsh winter conditions. Some of the better-known Arctic inhabitants include muskox, caribou, grizzly bears, polar bears, wolves, foxes, snowshoe hares, weasels, and mink. A few species of birds (about 6 or 7) such as the snowy owl and ptarmigan also survive here. A number of tundra species can also be found at higher elevations in the mountainous regions of the northern hemisphere where thin soils and cold temperatures prevent the growth of most middle-latitude trees and allow tundra species to grow above the tree line. Here the alpine tundra and alpine grasses flourish with sedges, perennial grasses, and dwarf shrubs.

Taiga

The taiga, or boreal forest, is an almost continuous band of coniferous (cone-bearing) trees extending in a giant arc from Alaska and Canada, through Europe and Siberia.[9] It exists over areas that were once glaciated (Figure 1.7). Such areas tend to have long, severe winters lasting up to six months with temperatures below freezing, and growing seasons between 50 and 100 days. The rainfall tends to be 38–50 centimeters annually, but poor evaporation and poor drainage result in the formation of bogs often overgrown with sphagnum moss. The cone-bearing trees include fir, spruce, and pine. These needle leaf trees are adapted to the taiga through their conical shape that sheds snow, and the waxy, thin needles, which reduce water loss and protect against drying winds. The dark green needles absorb heat and are not shed so that photosynthesis can start immediately with the beginning of the shortened growing season.[12]

Fur-bearing predators including wolverines, sable, fisher, mink, and lynx seek after mammalian herbivores such as snowshoe hares, red squirrels, lemmings, and voles. Many of these species develop white coats during the winter to blend with the snow. A number of larger herbivores may also be found in or near the taiga, including moose, elk, and red deer. Year-round birds of the taiga include finches, sparrows, and ravens. Some birds such as the grosbeak and crossbill become migratory during years when pinecones are reduced in numbers.

FIGURE 1.7
Typical taiga biome and location(s).

Temperate Areas

Temperate Broadleaf Deciduous Forest

The temperate broadleaf deciduous forests (Figure 1.8) are located in western and central Europe, eastern Asia, and eastern North America. Climax deciduous forests include oak, maple, hickory, chestnut, elm, linden, and walnut trees. The temperate forests receive 50–152 centimeters of precipitation distributed evenly throughout the year. Snow and ice conditions during winter months make water unavailable and create droughts during the cold season. The broadleaf trees lose their leaves (deciduous) during the winter to conserve moisture and resist the cold. As the cold fall weather settles in, the leaves begin the deciduous process (leaf-fall) and the leaves turn to brilliant reds, oranges, and yellows as chlorophyll formation is halted and the leaves finally die and fall to the ground, forming a litter that decays further to a rich organic mat called humus.

Mammals in this biome include acorn and nut feeders such as squirrels and chipmunks; omnivores such as raccoons, skunks, black bear, and opossum; and rare carnivores such as mountain lions, bobcats, and wolves. Except for the recent emigration of coyotes from the grasslands, the carnivores have been mostly eliminated by habitat destruction and hunting. Birds include insect-eaters such as hummingbirds, thrushes, warblers, and wrens. Omnivores such as ravens and cavity-nesters such as woodpeckers and chickadees also abound.[11,12] Most of the bird species are migratory and fly south for the winter while a number of mammalian species enter into hibernation.

FIGURE 1.8
Typical broadleaf temperate forest biome and location(s).

Temperate Evergreen Forests

In temperate areas where soil is poor and droughts and fires are frequent, the predominant species tend to be evergreens including coniferous, needle-leafed trees with some broad-leafed evergreens. In such areas, spruce, fir, and ponderosa pine proliferate. These are known as temperate evergreen forests. In cool coastal climates where there is considerable rainfall or frequent heavy fogs, one is likely to see forests of giant trees such as redwoods, which characteristically reach heights of 60 meters to 90 meters. These are temperate rainforests and such areas exist along the Pacific coast of North America stretching from northern California to the Sitka Spruce in Alaska. Chile and New Zealand also have these temperate forests.

Chaparrals

Areas of moderately dry climate and limited summer precipitation are further characterized by small (0.9–4.5 meters) shrubs with leathery leaves that contain aromatic and flammable substances. Consequently, these chaparrals often catch on fire allowing shrubs to re-grow from parts of the plant that survive near the ground. Unfortunately, when such growth is allowed to proliferate near houses that are rapidly expanding into these areas, they become likely fuel for wildfires.

Temperate Grasslands

Temperate grasslands are known by a number of names: North American prairie, Asian steppes, South African veldt, and the South American pampas (Figure 1.9).[9,10] These semiarid regions tend to be located inland along the middle latitudes and experience only

FIGURE 1.9
Typical temperate grassland biome and location(s).

25–50 centimeters of precipitation a year, much of which falls as snow. The predominant plant forms are perennial grasses, forbs, and members of the sunflower and pea families.[10] The perennial grasses grow from buds just below the ground surface and are adapted to fire, drought, and cold, while the well-developed root systems are able to trap moisture and nutrients. The more humid grasslands are populated by sod-forming grasses that have rhizomes or stems under the ground from which new plants are formed. Areas that are drier tend to favor bunch grasses that reproduce by seed. Biodiversity in temperate grasslands is limited. Most commonly seen in temperate grasslands are the herbivore rodents including ground squirrels, prairie dogs, and pocket gophers. Large animals such as bison, pronghorn, grizzly bears, and wolves have largely been confined to isolated preserves or the mountains as humans destroy more habitats.[11]

The Tropical Rainforest

The tropical rainforest is complex in terms of species diversity and structure. The combination of constant warmth, with average monthly temperatures above 17.8°C and precipitation greater than 254 centimeters per year, encourages rapid plant growth throughout the year. The high humidity and continued warmth promote rapid growth and great diversity. There are relatively few of any particular species, but a myriad of species exists in the tropical rainforest. These species account for nearly three quarters of the total number of species on Earth. Further, more than 40 percent of the world's plants and animals grow in the tropical rainforest, which accounts for less than 7 percent of Earth's surface.[11] There are no seasons in the rainforest, so there is no annual rhythm, and each species flowers or fruits at different times of the year on their own cycle. The majority of plants are tall trees with thin trunks that produce branches and leaves near the top. Many of these leaves are elongated at the tips to encourage dripping or shedding of water ("drip-tip" leaves) which reduces mold and insect attack.

FIGURE 1.10
Typical tropical rainforest biome and location(s).

The life of the forest occurs in the canopy. The canopy is an almost unbroken sea of green produced by leaves of trees at 21–30 meters above the forest floor (Figure 1.10). Most of the sunlight is absorbed in the canopy, and it is where the bulk of photosynthesis occurs. resulting in more than 9 Megagrams of new growth per hectare per year. Much of the animal life also lives in the canopy. The leathery green leaves sprouting from branches at the top of trees help to shield the forest floor from the direct impact of intense showers. Occasionally, a taller tree penetrates the lower canopy to form an umbrella-like canopy over the trees below. These trees reach heights of 40 meters and are called emergent trees. Temperatures in the canopy often reach 35.5°C with a humidity of 50–60 percent while the temperature on the floor of the forest is normally lower at 27.7°C with a humidity of 90 percent.

Researchers in Costa Rica have discovered evidence that increasing temperatures have slowed the growth of tropical forests. It has been suggested that the tropical rainforests have actually added billions of megagrams of greenhouse gases into the atmosphere during the 1990s. Although trees take in CO_2 and then release O_2 during photosynthesis, they also release some CO_2 as a byproduct of respiration. When growing rapidly plants take up more than they can produce. However, if growth is slow then the balance will shift. In 1984, researchers Deborah Clark and David Clark began researching the growth rates of adult tropical rainforest trees. The samples included six different tree species with both fast- and slow-growing tree types. They found that the growth of the trees changed each year. In years with high levels of CO_2, the trees had grown slowly, while in years when the level was low the trees grew rapidly.

Less than 10 percent of the sunlight reaches the forest floor and there is intense competition for available energy at this level. There is sparse vegetation in the inner sections of the undisturbed rainforest, and these sections have to compete intensely in an area with little sunlight and poor soil. This part of the rainforest is known as the understory (Figure 1.10). The soils are often soggy and quite shallow because the tremendous biological activity quickly decomposes organic matter and converts it into new life. The soils may be only a few centimeters deep covering red clay. This has caused many trees to adapt by forming shallow root systems and bases that spread at the bottom of the tree much like a buttress. This growth provides support in thin soil and may enhance the ability of the tree to secure nutrients over a broader area.

The reduced light beneath the canopy has encouraged a number of plant strategies to reach the top of the rainforest. Air-plants or epiphytes grow on branches high in the trees. Being supported by the tree limbs, epiphytes extract moisture from the air and trap leaf-fall and wind-blown dust for nutrients. Bromeliads (pineapple family), orchids, cactuses, and ferns are all represented in the canopy as epiphytes. Many plants such as members of the fig family begin life as epiphytes in the canopy and send their root systems downward to the forest floor. Such plants are then called stranglers. Many woody vines grow quickly up the tree trunks when a temporary gap opens in the canopy. Here they flower and fruit in the treetops. These woody vines are called lianas and are often deciduous. Few animals exist on the forest floor. Beetles, butter-flies, and other insects abound in the canopy. A rich array of birds, monkeys, and frogs also inhabit the canopy.

Tropical rainforests have been linked to possible cures for diseases and viruses.[15,16]

Typical rainforests circumvent the globe in a band from 20° south to 30° north and are found in Central and South America, Eastern and Central Africa, Malaysia, Indonesia, the Philippines, Papua New Guinea, and the Solomon Islands (Figure 1.10).

Deserts

Deserts are defined by their arid climates, which average less than 25 centimeters of precipitation a year and where evaporation tends to exceed this precipitation causing a negative annual water budget. When rain falls, it tends to be very localized and unpre-dictable as to when it will occur. Further, some years may have much more rainfall than others. Deserts are not always hot. They can easily reach temperatures higher than 37.8°C on summer days while plummeting to −6.7°C at night. Some deserts may not experience frost while others may have extended periods of subzero temperatures and even snow.[9–12]

Deserts are created by specific geographic and climate conditions.

1. Within the latitudes of 20° south to 30° north and along the western coasts of continents, the winds tend to be easterly (coming from the east) (see Figure 1.3b) and consequently keep moist air rising off the oceans from reaching the coast. This causes some of the world's driest deserts such as the Sahara in Africa, Baja California in North America, and the Atacama in South America (Figure 1.11).

2. Near the 30° latitude, subtropical air descends in association with the Hadley cell (Figure 1.3b) and as it descends it compresses, causing the formation of heat and dry, warm air. This air then absorbs moisture, resulting in little precipitation, and produces such areas as the Sahara and Australian deserts.

FIGURE 1.11
Typical desert biome and location(s).

3. Temperate deserts are generally located in areas known as rainshadows. Rainshadows occur on the lee side of high coastal mountain ranges (Figure 1.12). As moisture-laden air approaches the windward side of the mountains, the air rises, cools, and creates a low-pressure condition causing precipitation. The windward side receives large amounts of rain and snow. The air falls on the leeward side, compresses and warms, and so absorbs moisture producing an arid environment and creating such deserts as Death Valley in California or the Peruvian desert.

4. The Gobi Desert of Mongolia and the Great Basin Desert of the United States are located in the interiors of continents where they are distant from sources of moist air and where the rainshadow effect may also be important.[11,12]

The deserts are not completely lacking in water, and many life forms have adapted to the arid conditions. The major plant life in deserts are evergreen shrubs that have small

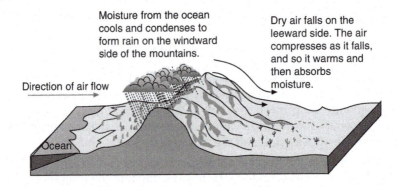

FIGURE 1.12
Rainshadow deserts are created on the lee side of coastal ranges. (Adapted from Turk, J., and Turk, A., *Environmental Science*, Saunders College Publishing, Philadelphia, PA, 1988.)

FIGURE 1.13
Desert Lizard, Utah, Arches National Park, JPEG lizard-60691_1920.

drought-resistant leaves, often spines or thorns, and sweet-smelling oils. The root systems usually extend outward by great distances to capture whatever moisture falls. Other plants have taproots that extend to 9 meters below the ground to reach underground water. Still other plants store water in their tissues to be used during dry spells. These plants are known as succulents. Depending on the species, water may be stored in different parts of the plant including stem, leaf, fruit, or roots. The familiar American cactus is a stem succulent. The aloe plant is an African leaf succulent.

Desert fauna have also adapted to aridity. Body sizes are small with long appendages that maximize heat radiation. Animals may have skin without sweat glands. They may concentrate urine into uric acid, have deposits of fat in tails or humps, or glands that secrete salt without moisture loss. They may be nocturnal or may even be dormant during summer months. Here's an example of a desert lizard from Arches National Park in Utah (Figure 1.13).

Energy Flow

Energy Source

All life requires energy, and ultimately the energy comes from the sun. Sunlight travels across the galaxy at 300,000 kilometers per second to reach us, and in a period of less than 10 minutes it loses all but the smallest fraction of its radiation intensity. Close to 99.9 percent of the sun's energy reaching the earth is reflected into space, absorbed as heat, or evaporated as water. Plants absorb a portion (about 0.1 percent) of this incoming light energy, particularly in the blue and red wavelengths of the visible spectrum. This energy is used by the plants for photosynthesis to create simple sugars from carbon dioxide and water with the release of oxygen (Figure 1.14). The simple sugars are used to store chemical energy and the amount of energy stored by the plant is normally expressed as

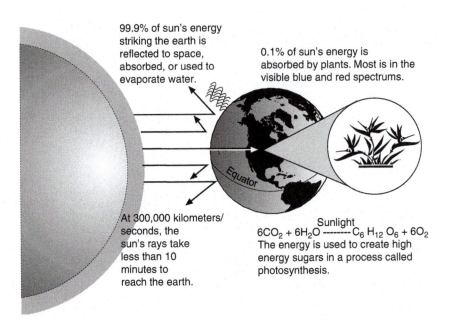

FIGURE 1.14
Distribution of sun's energy striking the earth.

kilocalories of chemical energy stored per square meter of plant tissue per year (kcal m^2/yr). Some 20 percent of this stored energy is used by the plant(s) for respiration and it therefore escapes in the form of heat to the environment. The remainder (from 50 to 80 percent) is converted to growth of the plant, thereby increasing the mass of the plant. The flow of energy and the conservation of matter must obey certain fundamental physical laws (Table 1.2). The accumulation of organic material in an ecosystem is referred to as biomass. Some ecosystems, such as tropical rainforests or marshes, are highly efficient in producing new plant growth because of the availability of nutrients, sunlight, warmth, and moisture. Consequently, the biomass in these areas is much greater than in a desert.

Within the biosphere, energy flows through the interconnecting web of life from organism to organism. The plant produced may be available to a variety of insects and animals, including humans, who must consume the plant to obtain the energy-rich molecules since animals cannot photosynthesize. These consumers, or heterotrophs, convert about 10 percent of the consumed kilocalories into animal flesh or organic matter (Figure 1.15). The 90 percent of the remaining consumed energy is used by heterotrophs in respiration necessary for the energy of motion and the maintenance of body functions, or it is excreted.

As energy is transferred through the food chain, about 90 percent of that available energy is lost with each transfer while the consumer uses one tenth. Consequently, an animal such as a wolf, which consumes deer or rabbits that eat grass, would be a secondary consumer and would receive ($1/10 \times 1/10 = 1/100$) of the available energy in the plant (Figure 1.15). In an example that demonstrates the efficiency of eating at a lower trophic level, 1,360 kilograms of corn and grain would support the energy requirements of one 680-kilogram steer, which in turn would support the energy requirements of one 68-kilogram human (Figure 1.16). If the grain were eaten directly by humans, it could likely support up to 20 people or more in the same amount of time. Only the

TABLE 1.2

Some Natural Physical Laws

Law of Conservation of Matter: While undergoing normal changes in physical or chemical reactions, matter cannot be created or destroyed. Practically, this means the earth has finite amounts of chemicals and nutrients that must be recycled if life is to continue. Oxygen can exist as iron oxides, carbon dioxide, nitrogen oxides, molecular oxygen, water, alcohols, within organic compounds and numerous other forms, but the total amount of oxygen available on the planet remains the same. The Law of Conservation of Matter means that everything is accounted for and that there is a mass balance. As we breathe in oxygen and catabolize plant sugars in the process of respiration, we give off carbon dioxide and water. The plants are able to use the carbon dioxide and water in the process of photosynthesis, which results in the production of sugars and oxygen.

Photosynthesis: $6CO_2 + 6H_2O$ $C6H_{12}O6 + 6O_2$

Respiration: $C_6H_{12}O_6 + 6O_2$ $6CO_2 + 6H_2O + energy$

Without this exchange or recycling of oxygen, the oxygen would soon be used up and life as we know it would be gone. As another example, if we burn 1,000 megagrams per day of solid waste in an incinerator, we will get 1,000 megagrams per day out of the incinerator. The materials discharged will be ash, some unburnable solids, particulates, and gases-but the total matter will weigh the same as that which entered.

The First Law of Thermodynamics: Energy is defined as the capacity or potential to perform work. There are many forms of energy including kinetic, gravitational, chemical, and thermal energy. Energy can be transformed from one form to another but cannot be created or destroyed. In practical terms, the energy of the sun is converted by plants to chemical energy in the form of sugars and plant tissues that can be used by herbivores as a source of energy. Plants use about 0.1 percent of the sun's energy reaching the earth. The rest is radiated back to space, absorbed as heat, or used to evaporate water. Animals that eat the plants convert about 10 percent for their use, with about 1 percent of this converted to animal tissue and the remainder is used in respiration to maintain organ systems, provide motion, and produce heat. The energy used in respiration escapes as thermal or heat energy. Thus, energy was converted from light energy to chemical energy to thermal energy, so the total amount of energy remains the same. However, the energy so produced is not useful to plants or animals for growth in its final form. This relates to the Second Law of Thermodynamics.

The Second Law of Thermodynamics: This law is based on the concept of **entropy**, which means that matter or energy will disperse toward maximum disorder in an undisturbed environment. Just as dye particles will evenly distribute in a cup of water, so will energy move toward a greater state of dispersion that is less useful or unavailable. As light energy is converted to chemical energy and then to heat energy, the heat is radiated ultimately to space where the heat is distributed uniformly throughout. Consequently, there must be a constant supply of energy from the sun to drive the life forces on Earth. Although the total energy has not declined, it moves unidirectionally through the world to become transformed and dispersed into unuseful forms.

wealthy countries can support a human population where meat is one of the main foods consumed. A primarily vegetarian society would be able to feed many times the number of humans with the same amount of grain that is used to feed cattle. Countries such as India, China, and most of the lesser-developed countries (LDCs) such as in Africa are predominantly vegetarian for this reason.

Consumption Types and Trophic Levels

Animals that eat only plants are *herbivores*. Other animals that eat primarily animal flesh often consume herbivores. These animals are called carnivores and may include lions, tigers, hyenas, frogs, or even ladybugs (they feed on aphids). Animals that eat plants and animals are termed *omnivores* and include rats, bears, humans, hogs, and foxes.

All energy comes from the sun and it flows in one direction from plants, which are *producers* and belong to the *first trophic level*. *Primary consumers* or herbivores belong to the *second trophic level*. Caterpillars, grasshoppers, cattle, and elephants are primary

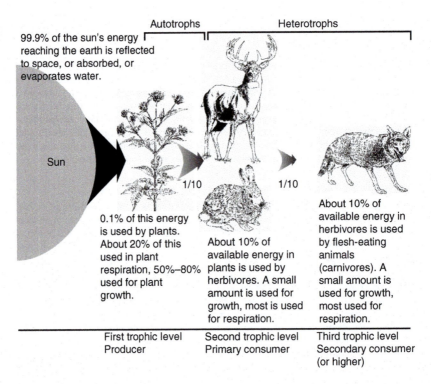

FIGURE 1.15
Relationships of energy flow in an ecosystem.

FIGURE 1.16
Comparison in efficiencies of feeding at various trophic levels.

consumers because they are herbivores that feed primarily on plants. *Secondary consumers* (carnivores) belong to the *third trophic level* (or higher). Each stage through which this energy travels is called a trophic level. Humans, along with many other animal species, are omnivores and may consume plants or animals. This means that many species may belong to several trophic levels. Bears, for instance, may belong to trophic levels two, three, four, and five. Bears will consume berries (herbivore, trophic level two), mice (trophic level three), frogs (trophic level four), or shrews (trophic level five). An example of this complex interrelationship is given in Figure 1.17. These patterns of consumption tend to be complicated and the term "food web" has been used to refer to these complex patterns of consumption among many trophic levels. This is important from a survival point of view since those animals capable of feeding at many trophic levels will be able to survive if portions of their food supply are decreased or eliminated. Because the amount of available energy (kcal) is reduced by a factor of 10 at each trophic level, there is a practical limit to the number of trophic levels that can be sustained. This number seems to be limited to five. As you move from primary consumers to higher trophic levels, the numbers and biomass (dry weight per unit area) of land organisms decreases with each successive level. The plants or producers far outnumber the next trophic level, or herbivores, which in turn far outnumber and have greater biomass than carnivores. This is because at each higher trophic level there is less useful energy available. Consider that there are fewer grizzly bears than caribou, or fewer foxes than rabbits. As previously mentioned, more animals (including humans) can survive as herbivores than as carnivores.

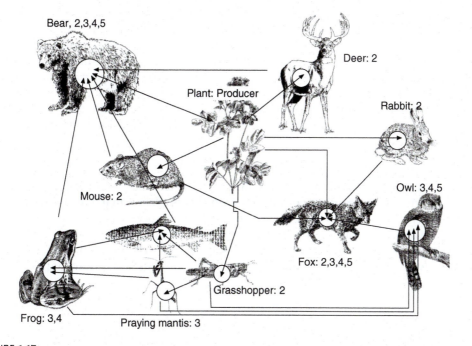

FIGURE 1.17
Simplified food web showing some of the possible trophic levels (represented by numbers) for various species. The arrows show direction of consumption. (Adapted from Turk, J., and Turk, A., *Environmental Science*, Saunders College Publishing, Philadelphia, PA, 1988.)

Nutrients

Recycling

Although there is a constant supply of sunlight, the nutrients tied up in these trophic levels need to be recycled. Some animals may eat, and even prefer, dead or decaying animal matter. Many of us have seen crows or ravens settle down on a road to consume the remains of a small, killed animal. We know that vultures are also primarily scavengers because they prefer to feed upon the dead remains of animals. Some animals that are predominantly omnivores or carnivores may also become opportunistic scavengers. Hyenas and foxes will feed upon the remains of animals that have died or have been killed. The author has witnessed hogs consuming dead rats, which also blurs the separation of scavenger from omnivores or carnivores. Unlike energy, which is rapidly dissipated through trophic levels, nutrients consist of 40 of the 92 known chemical elements that are essential to most forms of life. When nutrients are recycled, the process is called *biogeochemical cycling*. Nearly all life consists of large percentages of sulfur, phosphorous, carbon, oxygen, hydrogen, and nitrogen. These are known as macronutrients. There are also many elements required in tiny amounts such as zinc, manganese, chlorine, iron, and copper that are termed *trace elements*.[6] The cycling of these nutrients is essential because they would soon be depleted by accumulation in dead matter that prohibits the development of new life. Remember from the Law of Conservation of Matter (Table 1.2) that matter cannot be created or destroyed. Another group of organisms essential to the process of biogeochemical cycling are the *decomposers*. Decomposers are insects, bacteria, fungi, and protozoans that sequentially break down complex organic materials into low-energy mineral nutrients that once again may be reabsorbed and used by plants. This process is normally sequential in those specific types of saprophytic organisms, which feed primarily on dead organic matter and first attack dead plants or animals. As these organisms process components of the dead tissues, they leave behind undigested materials that are in turn processed by the next group of decomposers until finally there is nothing left but energy-poor mineral nutrients. This sequence may be so predictable that the science of forensic entomology has developed, and practitioners can judge the time of death of a human from the types of decomposers evident at the time of discovery of the body. Remember the television show *CSI*? Check out (http://www.forensicentomology.com/).

Nutrient Cycles

The recycling of nutrients from the inorganic to organic and back again in a living system was previously defined as biogeochemical cycling. This cycling may take two forms. The nutrients may be sedimentary and originate from soil or rocks, or gaseous and originate from air or water. The gaseous cycle is very efficient compared to the sedimentary cycle and this may be why the major constituents of most life consist of carbon, nitrogen, hydrogen, and oxygen. Representative substances from the gaseous and sedimentary cycles are discussed in the following section.

Carbon Cycle

Humans, along with most animals and plants, are carbon-based life forms. It's not surprising then that carbon is available in such a wide spectrum of forms including the physical states of gas, liquid, or solid, and the chemical forms including organic and inorganic (Figure 1.18). Inorganic sources of carbon include gaseous carbon dioxide, which

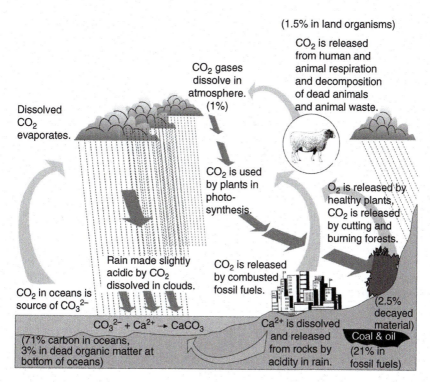

FIGURE 1.18
A simplified carbon cycle (percentages refer to total global carbon).

appears in the atmosphere at about 0.03 percent (300 ppm). Carbon dioxide is released to the atmosphere from (1) the respiratory process of animals and plants, which consume oxygen and release carbon dioxide and water; (2) the combustion of fossil or organic fuels such as coal, wood, oil, and gas; and (3) the decomposition of organic matter (leaves, carcasses, animal waste) by insects and microbes. Carbon dioxide is also dissolved in water where it forms a weakly acidic solution. Small amounts of carbonate and bicarbonate ions are produced, while more than 99 percent of the carbon dioxide remains dissolved in the water without conversion. Therefore, the water and precipitation are only slightly acidified by carbon dioxide. However, the amount of carbon dioxide in the water is enormous and estimates place the level of carbon dissolved in the oceans at 70–100 times greater than present in the atmosphere.[6,9,11] The carbonate ions formed in the water can react with calcium to form calcium carbonate, which mineralizes to form limestone. Limestone can also be formed from shellfish that create an insoluble calcium.

Plants convert inorganic carbon and water to useful energy-rich sugars through the process of photosynthesis. The plants then become food for many forms of animal life, which create new and more complex organic carbon compounds. The carbon is released back to the atmosphere through respiration and the decomposition of plant or animal matter or animal waste products. Although much of the decay of plant and organic matter may occur in weeks or months, decay may sometimes be incomplete because the materials are buried and lying in peat bogs or moorlands, on the bottom of oceans, or beneath the ground where geological processes convert this material into the fossil fuels of coal, oil, or natural gas. These latter materials may be recovered and burned to release carbon dioxide back into the atmosphere. The combination of burning fossil fuels and forests is releasing greater and greater quantities of carbon dioxide into the atmosphere. Many scientists and world leaders think the increase in atmospheric carbon dioxide at the global level is contributing to global warming. This is a topic to be covered in more detail in Chapter 10 on air pollution.

Nitrogen Cycle

Although nitrogen is the major component of the earth's atmosphere, comprising 78 percent of the air we breathe, it cannot be used directly by animals or most plants. Useable forms of nitrogen are in the forms of nitrates (NO_3^-), nitrites (NO_2^-), and ammonia ions (NH_3^+). Therefore, the natural mechanisms must exist that convert atmospheric nitrogen to useable forms. There are two known major natural mechanisms (Figure 1.19). One method involves the conversion of atmospheric nitrogen to nitrates (fixation) by (1) bacteria such as *Rhizobium* sp., which live symbiotically in the root nodules of leguminous plants; or (2) some species of organobacteria that can convert gaseous atmospheric nitrogen to useable nitrate forms. Another natural method occurs when lightning causes atmospheric nitrogen to react with oxygen, converting it though a number of steps to nitrates. Nitrates are also produced in the atmosphere from nitrogen oxides created within the high-temperature confines of boilers or internal combustion engines. Nitrates are also released from the

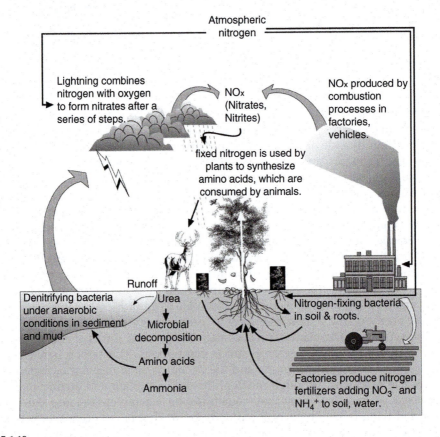

FIGURE 1.19
The basic nitrogen cycle. (Adapted from Turk, J., and Turk, A., *Environmental Science*, Saunders College Publishing, Philadelphia, PA, 1988.)

erosion of nitrate-rich rocks into soil and water. Nitrogenous compounds are continually being recycled from animal wastes and the decomposition of organisms, which are acted upon by soil microbes to produce ammonia and then converted to nitrates by nitrifying bacteria, which is a form usable by plants. Nitrates and ammonia are made by the oxidation of ammonia from nitrogen and hydrogen to create agricultural fertilizers and added to soils to increase fertility. This process is energy consuming and consequently expensive. Once nitrogen is formed into complex organic molecules, it may be recycled back to atmospheric nitrogen by denitrifying bacteria, which live in the mud and sediment of lakes, ponds, streams, and estuaries where they carry out these activities anaerobically (in the absence of oxygen).

Inland waterways, estuaries, and coastal ecosystems around the world have been subjected to a long-term increase in organic matter that has been washed off the land. A major consequence of this is an oxygen-depleted zone off the coast of Louisiana. Many experts say that over-fertilization from organic runoff has passed pollutants released from industries as a threat to marine ecosystems.[14,17]

Phosphorous Cycle

Calcium, iron, and phosphorous are examples of nutrients that flow through living systems in a pattern of recycling that is measured in eons of time or on a geological scale. These substances are gradually leached from sedimentary rock by the actions of rain or erosion from water in a process referred to as the sedimentary cycle. The phosphorous cycle is described here as being representative of this type of cycle. Phosphorous is the main element in compounds such as adenosine triphosphate (ATP). ATP is important in the transfer of chemical energy in animals. Phosphates are also a major component of deoxyribonucleic acid (DNA) and ribonucleic acids (RNA) and may be found in large amounts in shells, bones, and even teeth. Phosphates do not occur as gas but are bound in sedimentary rocks where they are released in small amounts from weathering. Plants take up the released phosphates in the roots, which may then be consumed by animals. Animal wastes and decomposing animals release phosphorous back to the soil for reuse by plants in a fairly efficient manner when not disturbed. However, the processes of mining and agriculture can erode soil and carry phosphorous into streams, estuaries, and oceans where it settles in a mostly insoluble form to the bottom and becomes generally unavailable. Consequently, phosphate fertilizers, which are produced by mining phosphate rock, are used to enrich most agricultural soils. Phosphate rocks are a non-renewable resource that were created millions of years ago. Therefore, sources of phosphates are being depleted rapidly and we are facing the potential of infertile soils where it may take thousands of years for phosphorous to be replaced through geological processes of weathering sedimentary rock.

The Process of Succession and Retrogression

Succession

In Chapter 3, you will read about the degrading of land by overgrazing or overuse leading to retrogression. Retrogression is a negative process that reduces the diversity of an ecosystem leading to fewer and fewer organisms and so weakening the food web and placing greater numbers of species at risk. The normal undisturbed process of nature is one of succession. Succession refers to the predictable and gradual progressive changes of biotic communities toward the establishment of a climax community. A *climax community* is one that perpetuates itself with no further succession within an undisturbed ecosystem. *Succession* usually begins when a disturbance has occurred that kills many or all of the life forms within a region. Complete destruction of areas can occur in association with volcanic eruptions, retreat of glaciers, or impacts from large meteorites. In these events, all life including that in the topsoil is destroyed and primary succession must take place by first creating soil on the barren lava or exposed rock surfaces. This is accomplished initially through the weathering action of rain and dissolved carbon dioxide, which create fragments and dissolved minerals. Dust is captured in cracks and crevices along with microscopic organisms and seeds carried by the wind or deposited by small animals and birds. Mixtures of fungi and algae grow together and are known as lichens. These are highly resilient life forms capable of surviving under sparse nutrient conditions. They gain a foothold and excrete organic acids that further deteriorate the rock surface to release more minerals. This is seen in the lichen growing on the surface of a subarctic rock (Figure 1.20). The organic material provided by decaying lichens and the inorganic minerals released from the rocks serve as nutrients for mosses that gradually replace the lichens, creating a denser growth that supports insects and other invertebrates while adding more organic matter to the mass. Eventually, enough organic matter is present to grow the seeds of some rooted plants. After many years, perhaps hundreds, greater size and varieties of plants and animals move into the once barren area and eventually a climax community may develop. The reason for the succession of plants is that each stage of plants fails to compete successfully with the next stage. The tall grasses have more extensive root systems that draw moisture and other nutrients away from the wildflowers while the grasses grow taller and reduce the available sunlight to these same flowers. Ultimately the grasses are in turn overshadowed by the new growth of trees and so on.[6,9,11]

FIGURE 1.20
Lichen growing on rock surface. (Courtesy of subarcticmike (Mike Beauregard), https://www.flickr.com /photos/31856336@N03/4927975791/, https://commons.wikimedia.org/w/index.php?curid=12833758.)

When soil conditions are disrupted but there remains topsoil and some limited vegetation, succession can take place much more quickly. This process has been termed secondary succession. A common example of this is the return of an abandoned farm field to a climax forest. Since the fertile soil is already present, seeds and rhizomes present in the soil quickly grow. Plants rapidly move in, sprouting from the present seeds or rhizomes, or carried to the site by wind, birds, and grazing animals. These early plants are also known as pioneer plants, and may include wildflowers, followed by tall grasses and compact woody bushes. These grasses will eventually crowd out the wildflowers but new-growth trees such as dogwood and sumac eventually replace themselves. After a time, perhaps a century or two, the climax forest trees will move into a typical temperate forest and include such trees as oak, maple, hickory, and elm. These trees will replace the new forest trees.[6,11]

Retrogression

Not all disrupted ecosystems undergo a progressive return to stability with climax vegetation. Human intervention, particularly, has threatened the stability of ecosystems throughout the world. Activities that cause changes in natural cycles or population such as destroying species or adding new ones can be a threat to environmental stability. Stable ecosystems are ones in which materials are constantly recycled within the system through growth, consumption, and decomposition. These processes tend to balance each other so that there is little net loss over long periods of time in a process called *dynamic equilibrium*. Although minor disruptions occur to most ecosystems, and a true steady state is seldom achieved, there is a long-term balance within stable ecosystems whereby nutrients revolve within that system with little overall loss. However, when the climate changes, soils are poisoned or eroded, or species are eliminated or introduced, the internal compensating mechanisms of the ecosystem may not be sufficient to return the ecosystem to its previous stable state, and a new combination of plant and animal life may be established.

Hurricanes that often strike the east coast released almost 1 meter of rain on eastern North Carolina during two September hurricanes. This rain turned the region into polluted floodwaters. This was caused by decomposing chickens and hogs, rotting farm fields, and ruined neighborhoods. This was one of the most disastrous floods in this region's history. Scientists now feel that the flood has also started an ecological situation that could change the biologically rich waters of the Outer Banks from the mainland of Carolina. These bays and inlets are the second largest estuaries in the country. The main problem is that the flood picked up huge amounts of organic matter. This matter included decomposing vegetation, topsoil, fertilizers, raw sewage, and waste created from hogs and farms. This water went directly into the estuary, which turned the water into a coffee color. This flood may have already killed many marine creatures like fish, worms, crustaceans, and crabs. No matter what happens in years to come, this situation may take years to go back to normal, if the estuary has this ability.[14] As plots of land are overgrazed or intensively farmed, tropical forests burned and then farmed, water resources depleted, or other poor land management techniques employed, complex ecosystems are often replaced. This new ecosystem state may result in fewer overall species in a process called retrogression. Further, the species remaining may be less desirable from a human point of view. Simple systems in which fewer species exist or diversity reduction takes place are at greater risk of failure. The lack of cross-links in an ecosystem food-chain means fewer alternatives for feeding, and when a species is reduced in number or dies off, its predators also die off if other alternatives do not exist.

The earth's climate has been changing rapidly and the temperature has been increasing. As the climate changes, due to human impact, there is an increased danger to human health. There are direct and indirect effects on human health. Direct effects are those like heat stroke, hypothermia, and deaths from hurricanes, floods, and tidal waves. Some indirect effects are the result of changing environments and ecosystems such as crop failure, algae blooms in the world's seas, and mental health disorders. Diseases such as malaria, dengue fever, and cholera are moving into northern countries due to the warming of climates and ecosystems. These threats are mostly due to human actions. Even with immediate action much of what is happening will continue to occur into the future.[15,18]

The human population is exerting enormous pressure upon ecosystems throughout the world as it continues to multiply in logarithmic proportions and develop energy-intensive technologies resulting in the discharge of dramatic levels of toxic substances into the air, water, and land. Most biotic communities are proving unable to respond to the unrelenting pressures of disruption causing major losses in species, soil degradation, desertification, contaminated water, possible climate changes, and other changes in global ecosystems that are not in the best interests of human survival or quality of life.

References

1. Godish, T., *Air Quality*, 3rd ed., Lewis Publishers, New York, 1997, Chap. 4.
2. Sadler, J.P. and Grattan, J.P., Volcanoes as agents of past environmental change, *Global and Planetary Change*, 21, 181, 1999.
3. U.S. Environmental Protection Agency, *Global Warming: Climate*. http://yosemite.epa.gov/oar .globalwarming/content/Climate.html (accessed January 7, 2000).
4. Stevens, W., Lessons from ancient heat surge, *The New York Times*, November 23, 1999.
5. Stevens, W.K., Carbon fuel use down globally, *Sunday Republican*, October 31, 1999, p. A-24.
6. Nadakavukaren, A., *Our Global Environment*, 4th ed., Waveland Press, Prospect Heights, IL, 1995, Chap. 1.
7. U.S. Global Change Research Program, *Changes in Hurricanes*, National Climate Assessment, Washington, DC, 2014. Retrieved July 5, 2018, https://nca2014.globalchange.gov/report/ our-changing-climate/changes-hurricanes.
8. Lush, T., Times staff writer, For forecasting chief, no joy in being right: Max Mayfield strives for accuracy, but worries about complacency, August 30, 2005. http://www.sptimes .com/2005/08/30/State/For_forecasting_chief.shtml.
9. Bradshaw, L. and Slonsky, L.B., A first person account of New Orleans after the hurricane struck, *Socialist Worker*, September 11, 2005. Retrieved July 5, 2018, http://Dissidentvoice.org/ Sept%2005/Bradshaw.Slonsky.
10. *Hurricane Research Division (Database)*, National Hurricane Center, Miami, FL, April 11, 2017 (accessed January 16, 2018).
11. Turk, J. and Turk, A., *Environmental Science*, Saunders College Publishing, Philadelphia, PA, 1988, Chap. 5.
12. Woodward, S., Introduction to biomes, *Geography*, 235, June 1998. Retrieved May 15, 2017, http://www.runet.edu/~swoodwar/classes/geog235/biomes.
13. Oskin, B., Vostok: Lake under Antarctic Ice, *LiveScience*, December 20, 2017. Retrieved July 5, 2018, https://www.livescience.com/38652-what-is-lake-vostok.html.
14. Schoen, D., Learning from polar ice core research, *Environmental Science and Technology News*, April 1, 1999.

15. Sargent, W., *The Year of the Crab, Marine Animals in Modern Medicine*, Norton & Company, New York, 1998.
16. Grouse, L., Turning molecules into medicine: The role of the national Cancer Institute's Developmental Therapeutics Program in Drug Development, National Cancer Institute, U.S. National Institutes of Health, February 2006. Retrieved March 12, 2015, http://www.cancer.gov/newscenter.benchmarks-vol6-issue1/page1.
17. Glanz, J., After the storm, an ecological bomb, *New York Times*, November 30, 1999.
18. Kingsnorth, P., Human health on the line, *The Ecologists*, vol. 29, March/April 1998.

2

Human Populations

OBJECTIVES

A student reading this chapter will be able to:

1. Define the attributes of populations including birth and death rates, growth rate, density, and mobility (immigration and emigration).
2. Calculate rate of natural increase from birth and death rates, and mathematically demonstrate the effects of age-sex composition on a population.
3. Define biotic potential and maximum growth rate, and list the various limits to growth.
4. Identify the population growth forms.
5. Recognize the concept of population explosion with respect to complete and incomplete demographic transition.
6. Define population implosion and discuss the conditions that lead to this phenomenon.
7. Explain the role of urbanization in influencing sustainability of populations.
8. Explain global population projections and differentiate between developed and lesser developed countries with respect to these projections.
9. Discuss the various options for fertility control methods, while contrasting the effectiveness, risks, and benefits of each type.

Introduction

This is a text on environmental health. Why is the subject of human population of any concern here? The answer is brief and startlingly cold. Human numbers have risen with such ferocity that we have overwhelmed the carrying capacity of sections of the earth and threaten the socio-political structures of civilization in some parts of the world. We must come to understand "the environment" and realize that environmental destruction is the national-security issue of the early twenty-first century. The world's billions of impoverished people are surging in an enormous tidal wave into the urban centers around the globe as they desperately seek relief and opportunity from the crumbling ecosystems of rural life. These hordes of billions are fleeing from deforestation, soil erosion, water depletion, natural disasters, hunger, emerging diseases, and hopelessness to find themselves entrenched in uncontrolled settlements of never-ending shanty towns built with corrugated metal shacks or cardboard. Armies of undernourished and pot-bellied children run through the puddled streets of floating garbage where malarial-infested mosquitoes swarm in great black clouds. Electricity, sewage systems, and running water are rare. It is little wonder that in these environs that criminal anarchy is the endemic illness. Overpopulation, infectious disease, unprovoked crime, few resources, and the influx of more refugees, increases the erosion of nation-states leading to the empowerment of private armies, security firms and international drug cartels. This is a vision of the early twenty-first century in many parts of the lesser-developed countries (LDCs), and it threatens to expand along with the growth of human populations.[1]

Since you are reading this text, you are probably sharing in the educational resources of the well-fed and technology-rich western society, where recent studies indicate that overpopulation may no longer be a problem. In much of Western Europe and Japan, declining populations are causing major concerns about the economy resulting from a diminishing work force, falling land and house prices, and a dwindling economy. Where will the industrial workers come from? Who will care for the aged? In the United States, demographic shifts are reshaping the racial landscape. There is an increasing divergence of racial generations creating a gap that pits the "grays" against the "skateboarders." In 2025, the number of elderly white Americans will increase by 67 percent to 47 million while less than half the children will be white. During the end of the same period, 38 percent of the working American population will be non-white (Asian, Black, Hispanic). This mostly white elderly population will be expecting retirement and medical benefits to be paid by a racially diverse group who must yield a greater proportion of their limited income to pay for sharply increasing costs in caring for the elderly. According to the Economic Policy Institute, people of color will be a majority of the American working class in 2032. This estimate is based on long-term labor force projections from the Bureau of Labor Statistics and trends in college completion by race and ethnicity. It is 11 years sooner than the Census Bureau projection for the overall U.S. population, which becomes "majority-minority" in 2043. This is important because those who make up the working class are without a college degree and represent more than two thirds of the U.S. civilian labor force between the ages of 18 and 64. If the non-white population continues to experience wage stagnation and economic inequality the sharply increasing costs for elderly care of mostly whites will become a matter of potential racial and ethnic divides. Further, the combined populations of all nonwhite racial and ethnic groups will make up more than half of the U.S. population by 2043. In 2043, 60.7 percent of people under age 18 will be people of color, while 64.9 percent of those over age 65 will be non-Hispanic white. This shift to a nonwhite majority has already taken place in the grade schools, where children of color outnumber whites in America's public elementary and secondary schools.[2,3]

The world is becoming increasingly bifurcated with the "haves" and "have-nots," and there is growing realization that surging populations, environmental degradation and ethnic conflict are strongly intertwined.[1] Therefore, understanding the dynamics of human populations is a first order of business in beginning the study of environmental health.

The Characteristics of Populations

It's usually easy to recognize a population. They have similar physical characteristics that identify them as a species and tend to inhabit specific areas. A species is normally considered to be a group of organisms that can breed together with the production of a viable and fertile offspring. However, just because members of species can breed together doesn't mean they do. Normally, organisms breed in much smaller collections of like organisms, which are called populations. A population is considered to be the breeding group for an organism. Different species not only have differing physical attributes, but they also differ in the population characteristics. Most species of rats will give birth to several litters a year and have a relatively short life-span of 1–2 years. Elephants have long gestation periods lasting nearly two years, may live more than 70 years, and rarely give birth to more than one at a time. Each population has characteristics that help to identify it. Some of these characteristics are birth rate, death rate, age distribution, and sex ratio. Birth rate refers to the number of individuals added to a population through reproduction (live births) and is normally expressed as the number of live births per 1,000 population (counting the population at the midpoint of the year). The example provided in Figure 2.1 shows a birth rate of 16/1000. Death rate is also similarly calculated using total deaths divided by the

Birth rate equals the number of live children born in a year per 1000 total population

$$\text{Birth rate in year Y} = \frac{\text{Number of live children born in year Y}}{\text{Midyear population in year Y}}$$

$$\text{Birth rate in year 1998} = \frac{4,345,600 \text{ (children born in 1998)}}{271,600,000 \text{ (population in mid-1998)}} = 16/1000$$

$$\text{Death rate in year Y} = \frac{\text{Number of deaths in year Y}}{\text{Midyear population in year Y}}$$

$$\text{Death rate in year 1998} = \frac{2,172,800 \text{ (deaths in 1998)}}{271,600,000 \text{ (population in mid-1998)}} = 8/1000$$

Rate of natural increase in year 1998 = (Birth rate – Death rate) = 8/1000 or 0.8%*
*These are approximate numbers for the U.S. used only for example.

FIGURE 2.1
Calculations of birth rate, death rate, and rate of natural increase.

mid-year total population. The rate of natural increase is determined by subtracting the death rate from the birth rate. The rate of natural increase reflects the growth rate in which migration is not considered. The growth of a population in the absence of migration must depend on the birth rate being higher than the death rate. The calculation of the rate of natural increase depends on other factors beyond this simple calculation. A more accurate estimation of population growth may be made if the sex ratio and age distribution of the population is known. The age-sex composition of the population has a profound effect on the birth and death rates of a country. This is so because the probability of dying or giving birth within any given year depends upon the age and sex of the population members. Clearly, an aging population is going to have a higher death rate and lower birth rate than a population with many young women of reproductive age. The sex ratio refers to the number of males relative to the number of females in the population, while the age distribution describes the percent of the population in various age categories from birth to death. In most societies, it is useful to separate the age groups into pre-reproductive juveniles, reproductive adults, and post-reproductive adults. Some approximate differences in age-sex composition for different countries are shown in Figure 2.2. These graphs are created by plotting the number of males and females for age groups in 5- or 10-year increments within a population. The lesser-developed countries (LDCs) of the world such as Mexico and Africa, and much of Asia, more closely approach the broad-based pyramid-shaped graph in Figure 2.2. The distribution for these countries has a very broad base indicating that birth rates have been very high in recent years. Further, there are many young children in the population who will likely reach reproductive age in the next few years and so rapidly expand the population of these countries in the next generation. It has been estimated that the group of young women entering childbearing years will be the largest ever. Additionally, there will be over 1 billion young people between the ages of 15–24 and 3 billion under age 25.[2] Those countries with more stable populations such as the United States or France have approximately an equal number of people in the age groups from 0 to 35. As the younger groups mature, there will be approximately the same number entering the reproductive years so that the population will remain fairly constant. This is complicated in the United States by the more than one million people immigrating into

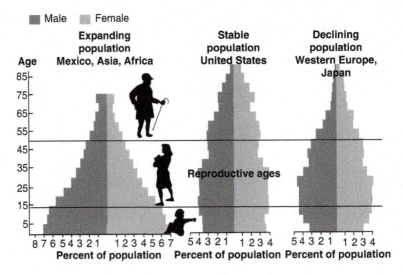

FIGURE 2.2
Age distribution types in human populations.

the U.S. each year. It is the only industrialized country where large population increases are projected.[4] Many Western European countries and Japan are experiencing a decline in growth rates. These declining populations are creating economic crises in these and other countries, reducing their national power. The populations of most European countries will decline in the next generation, and the decline will be dramatic in Germany and Russia. There is mounting evidence that the global population explosion is ending. Population stability is maintained when the birthrate stays at 2.1 births per woman. If the birth rate is above 2.1, the population rises; below that, it falls. The birthrate is already well below 2.1 in the advanced industrial world. In countries such as Mexico or Turkey, the birthrate is falling but will not reach 2.1 until between 2040 and 2050. In Bangladesh and similar poor countries, the birthrate is also falling, but it will take most of this century to reach 2.1.[5]

This is characterized by an age distribution where there are fewer infants and children than young adults. In the years to follow, there will be fewer people in the reproductive age groups, and a population decline will occur in these countries if current trends continue.

Even with this information, predicting what will happen to future populations is a difficult task. Imagine that a large population of youngsters enters the pyramid at the base and travels upward in a group of diminishing size (because of deaths) into reproductive age and then even smaller numbers into post-reproductive age. If this were a constant behavior of the population, predictions would be easy. However, wars, disease, famine, and reproductive intervention may cause the next generation to be reduced in number. This creates a constrictive formation or pinching at the base as seen in Western Europe (Figure 2.2). Since it is not easy to predict war, disease, or even human behavior with regard to reproduction, then estimating future populations is a matter of some difficulty. Many countries are laboring with issues of unexpected population explosion or population implosion right now.

Although it is apparent from this discussion that accurate population predictions may be difficult, there are other indicators useful in estimating future population changes. These include emigration, immigration, infant mortality, and total fertility rate.

In nature, when the density of organisms becomes too great, the intense competition for food, water and other resources damages the entire population. Some species have the ability to disperse or migrate out of the area and in doing so, temporarily relieve the overcrowding. This process is called *emigration*. The pressure for resources may be the consequence of rapid population growth, and/or the disappearance of essential resources brought on by natural phenomena such as drought. When species emigrate from an area, they must *immigrate* or enter into another area. Driven by natural disasters, war, disease, and disappearing resources, the numbers of refugees worldwide may exceed 15 million, with about 880,000–1.4 million immigrants entering the United States each year, including more than 200,000 who enter illegally. The number of immigrants is likely to remain high for many years, and this is contrasted with an American fertility rate that has been below replacement level for the 20 years. Immigration has, therefore, had a disproportionate impact on population growth in the United States. The consequences of immigration to the United States will be discussed later in this chapter.

As discussed earlier, death rates are calculated by dividing the total number of deaths by the total mid-year population. However, deaths of persons in the pre-and reproductive age groups will have a greater effect on future population growth than somebody dying in the post-reproductive ages. Deaths of the young will mean a loss of potential births. Because of this impact, infant mortality is often a measured population characteristic that is used in conjunction with death rates. Infant mortality can be determined by the following relationship.

$$\text{Infant mortality } (\%) = \left(\frac{\text{No. infant deaths}}{\text{No. live births}} \right) \times 100\%$$

The birth rate is useful for determining how many births are occurring in the total population but does not tell what the current fertility patterns are. A useful predictor of reproductive behavior in a population is total fertility rate. Total fertility rates (TFR) represent the number of children a woman in a given population is likely to bear during her reproductive lifetime providing that birth rates remain constant for at least a generation. Recent figures show that the world TFR is 2.7 with the more developed nations having a TFR of 1.6 and the less developed countries excluding China having a TFR of 3.5.[6-8] The TFR is useful in determining the replacement level which is defined as a TFR that corresponds to a population exactly replacing itself. The replacement TFR level for most countries is accepted as being 2.1, although some LDCs have such a high infant mortality because of poor nutrition, disease, and medical services that the TFR may be as high as 2.7. Almost all European countries are experiencing below-replacement fertility rates with levels as low as 0.8 in China, Macao SAR (Special Administrative Region), while many of the LDCs have reduced the TFRs, most are still well above the replacement level of 2.1, with the highest recorded TFR of 7.3 for the year 2017 is Niger.[8] The world population will reach 9.8 billion in 2050, up 31 percent from an estimated 7.5 billion now, according to projections included in the *2017 World Population Data Sheet* from the Population Reference Bureau (PRB).[9]

If a person knows how many females of reproductive age are in a population based on the age-sex distribution and also knows the average number of births per female from the TFR, then it is easy to predict the next generation population providing the birth rates do not change. Of course, birth rates often do change as a consequence of changing social behavior, wars, famine, disease, or migration. The inability to predict these phenomena or their effects on the TFR contributes to inaccurate demographic predictions.

Population Dynamics

There are periodic upsurges in many populations that lead to overwhelming numbers. Whether these population explosions occur in rabbits, lemmings, soldier ants, or locusts, there are always some natural pressures that bring the population back into balance with their natural surroundings. Bacteria reproducing by binary fission can grow to millions of organisms within a few hours given the right conditions, while many fungi grow cubically and could easily grow to fill an entire room overnight if pressures didn't exist to limit their growth. Fish and amphibians produce thousands of eggs, plants release millions of seeds, and many animal species produce multiple litters of offspring in a year. Every species has an inherent reproductive capacity to produce offspring that is generally many times greater than needed to replace the parents that die.

Clearly, if these organisms and offspring all survived, we would soon be overwhelmed by the sheer numbers of these species. In fact, even the human population growing at a rate of 1.6 percent for the next few 100 years would completely cover every square inch of the earth and its oceans. The unrestricted growth of populations resulting in the maximum growth rate for a particular population is called its biotic potential. The biotic potential of species differs markedly and is influenced by: (1) the frequency of reproduction;

(2) the total number of times the organism reproduces; (3) the number of offspring from each reproductive cycle; and (4) the age at which reproduction starts. Some bacteria reproduce every 20 minutes by splitting in two (binary fission), while polar bears give birth only once in every 3 or 4 years, which gives them a much lower biotic potential. This biotic potential is never actually reached for any species because the growth is limited by one or more factors that are collectively referred to as environmental resistance. Environmental resistance refers to those pressures that limit population and may include such factors as disease, wars, predatory behavior, toxic waste accumulation, or species competition. The actual rate of increase is then described by the biotic potential minus the environmental resistance as shown in Figure 2.3.[9] The sum of these limiting factors serve to control the numbers of a species that can survive in defined area over time, and this represents the carrying capacity of that area for a particular species. The carrying capacity of a defined system can be demonstrated in the laboratory by cultivating a bacterial organism in a nutrient broth at 37°C and removing organisms hourly in order to count them. This is accomplished by making serial dilutions of the original broth culture and plating 1.0 ml of each dilution onto separate nutrient agar plates. Incubation of these plates results in the growth of discrete colonies that can be counted to determine the population at hourly intervals (Figure 2.4). By plotting these numbers, a typical sigmoid or "S" growth curve can be demonstrated (Figure 2.5). The growth curve initially shows a lag phase in which the organisms show no increase in growth rate but are undergoing changes that ready them for rapid proliferation. This is followed by a rapid increase in growth where the population doubles every few hours. The population increases at a rate proportional to its size, or exponentially. This is known as the *exponential growth phase*.

To demonstrate logarithmic growth, consider that a single bacteria splits every 30 minutes. If a single bacteria starting at time "0" divides in 30 minutes to form two bacteria, then in 30 minutes, each of the two divides to form four bacteria. If this were to continue unabated for 24 hours, there would be 48 generations for a population of over 280 trillion organisms. Can you determine the number that would be produced if allowed to grow uninterrupted another 4 hours?

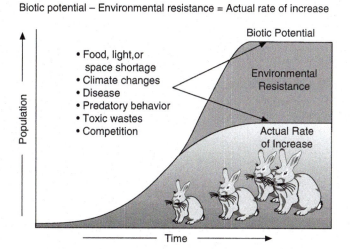

Biotic potential – Environmental resistance = Actual rate of increase

FIGURE 2.3
Biotic potential – Environmental resistance = Actual rate of increase.

Calculation: Number of colonies on plate x reciprocal of dilution of sample = bacteria/ml. In this example, there are 20 colonies on the plate of 1:100,000 dilution = 2 million bacteria/ml. A growth curve can be constructed if the original inoculum is counted by this process hourly for 24 to 48 hours.

FIGURE 2.4
The plate counting technique is one counting process that may be used to count bacteria for generating a bacterial growth curve.

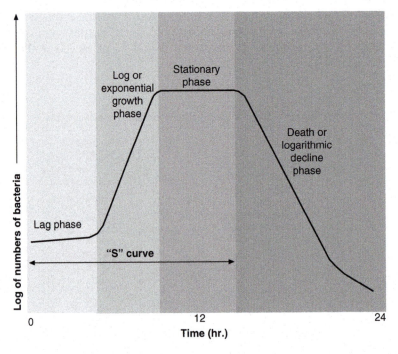

FIGURE 2.5
Bacterial growth curve showing various phases of growth including lag, exponential, stationary, and death phases.

Eventually, the limiting environment of the culture tube restricts the growth of the bacteria by an accumulation of toxic wastes, and lowered pH. The number of organisms dying eventually comes to equal the number of living organisms so that a stationary level is reached. This also represents the carrying capacity of this culture tube environment. If left to grow without removing wastes, or adding nutrients, the bacterial population will die off rapidly in what is known as the *death phase*. This "S" type growth curve was described in 1932 by the Russian ecologist G.F. Gausse when he studied the growth of experimental populations of *Paramecium caudatum*, a freshwater protist.

If an organism grows too rapidly and the population escalates beyond the carrying capacity of the environment in which it is located, a "J" type growth curve may develop (Figure 2.6). This curve is somewhat misleading, because it is truncated before the outcome of the event is revealed and may more accurately be referred to as geometric growth. There are many examples in nature where insects, mammals or other life forms have an explosive growth in population that far exceeds the carrying capacity of their habitat. The consequence of this is a precipitous decline in the population once food supplies are gone, disease intervenes, or the habitat is changed in some other way to create hostile conditions. This behavior sometimes oscillates every few years as in the case of lemmings that inhabit the arctic tundra north of the Canadian forest. Every 3 to 4 years the population explodes, then crashes the following year, followed by a 2-year cycle of slow recovery (Figure 2.7).[11] There are other patterns of growth as well. Predator-prey interactions such as lynx and rabbit is likely to produce an oscillating pattern for both species with the predator showing increases and declines following such changes in the prey. As the rabbit population increases, the lynx will have more prey to feed upon and its population will grow. Eventually, there will be sufficient predators to cause a reduction in the rabbit population and its decline will cause a decline in the predator lynx population soon after as seen in Figure 2.8a. Another type of growth curve is seen in Figure 2.8b in which

FIGURE 2.6
"J" type growth curve showing growth extending beyond the carrying capacity of environment.

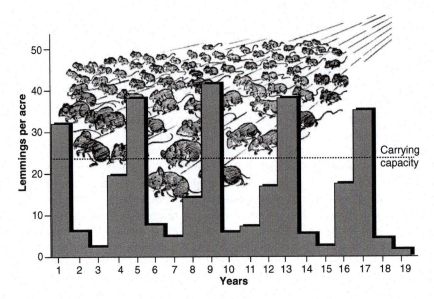

FIGURE 2.7
Typical population crashes seen in lemming populations every 3 to 4 years.

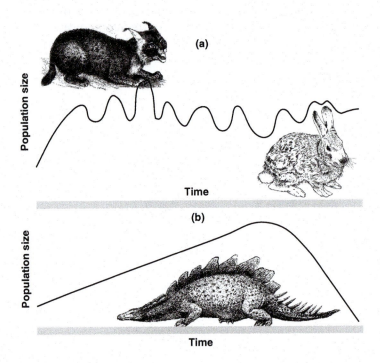

FIGURE 2.8
(a) Predator-prey and (b) extinction type growth curves.

a population becomes extinct. This type of curve can be drawn for more and more species everyday as human intervention eliminates species and habitats, spreads pesticides, and hunts down larger species for human uses. The disappearance of wildlife and species extinction will be discussed in more detail in Chapter 3.

The science of demography has categorized two different reproductive strategies for living organisms. One strategy is known as a *k-strategy* in which large organisms with relatively long life spans have only a few offspring, but devote their energies to protecting and nurturing the offspring to enhance their individual survival until they can reproduce. Examples falling into this group would be elephants, deer, large cats, swans, and humans. These populations normally stabilize at the carrying capacity of their environments where density dependent factors limit growth. Density dependent factors include such items as food supply, which becomes more limiting as the size of the population grows. The pattern for mortality of these species can be demonstrated by plotting a survivorship curve in which the fate of young individuals is followed to the point of death in order to describe mortality at different ages.[12] This curve can be used to predict the probability that a newborn will reach a specific age which is useful in determining the life expectancy within a population The "type I" curve typical for the k-strategy survival population shows that the majority of infants reach sexual maturity and many survive to maturity eventually dying of diseases from old age (Figure 2.9). This is in sharp contrast to *r-strategy* populations that are typically small, short-lived organisms, which produce large numbers of offspring and receive little or no parental care. This is characteristic of species of invertebrates, fish, mollusks, plants, fungi, gypsy moths, locusts, grasshoppers, and some rodents. Large numbers are produced, but few offspring survive. There is high mortality among the offspring as depicted by the Type III survivor curve (Figure 2.9). These organisms are limited by density-independent factors such as a drought that dries up a pond, or sudden climactic changes such as El Niño, which alters the temperature of the water, making it uninhabitable for certain species. Such populations often demonstrate wildly fluctuating numbers with population explosions and crashes.

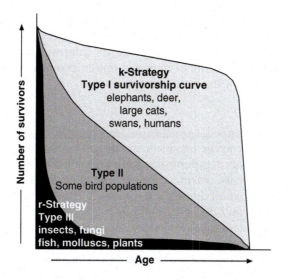

FIGURE 2.9
Characteristic survivor curves showing Type I (prosperous human population), Type II (impoverished human population and certain bird populations), and Type III (fish, plants, fungi, etc.). (Adapted from Turk, J., and Turk, A., *Environmental Science*, Saunders College Publishing, Philadelphia, PA, 1988.)

A type II survival curve that lies between I and III is characterized by a high mortality in the infant or beginning stages of life from disease, genetic defects, or predation. When this initial die-off is passed, the survivors die at a smaller but constant rate over the life-span of the population.[12] This is characteristic for some bird species and for human populations under the constant stress of poor sanitation, infectious diseases, and inadequate food supplies. In such populations, there is a tendency to produce more offspring because of the higher mortality in infants or offspring.

Population Trends in the World

Demographers use the information on population size, fertility rates, migration, birth and death rates, growth rates, infant mortality, density, age-sex composition and other factors to statistically characterize human populations. Their purpose is to predict what will happen to that population over time. You may question why this is so important. What is the value of this information? The statistical knowledge of human populations certainly dates back thousands of years to the beginnings of civilization when support systems for maintaining large groups of interdependent people became necessary. Government structures developed and there was a need to raise taxes, maintain armies, harvest and store foodstuffs, provide education or training, build roads, and promote commerce. These activities provide the infrastructure on which society depends to this day. Even in the smallest towns, the local officials must have census data to know how to plan for possible expansions in schools, sewer and septic systems, road construction, power lines, building lots, and land conservation. Humans require significant infrastructure to maintain a healthy, educated, and physically comfortable lifestyle. This takes significant planning, and the knowledge of how many people to account for. There is little that goes on in society today that doesn't depend on a knowledge of human numbers. Marketing products, predicting enrollments, estimating tax incomes, analyzing insurance risks, determining future costs for elderly care are just a bare mention of the uses for population numbers by age, by race, by sex, by interests, by disease, and so forth. Understanding the growth patterns of human populations is vital to a nation's interest and perhaps to the survival of the human race.

Historical Trends

Core samples taken from the arctic ice, and from deep-sea beds reveal a chaotic history of extremes in the earth's surface temperature over several 100,000 years. Then, as if almost a fluke, the earth's temperature warmed and stabilized about 10,000 years ago. Humans began to develop agriculture and domesticate animals. The ability to feed larger groups of people from planted crops and domesticated herds allowed civilizations to develop. The inefficiency of nomadic existence was no longer necessary, and soon the world's population began to increase as population groups could sustain more children. Although accurate predictions of human populations are questionable before circa 1700, estimates of primitive civilizations around 8,000 BCE (before Christian era) place the number of humans at about 5 million (Figure 2.10). The advent of agriculture is thought to have promoted the growth of human populations to about 500 million by 1650. The world's population then doubled in the next 200 years to about 1 billion people. During this time, the industrial

FIGURE 2.10

History of human population from early man, *Homo habilis*, to the present showing exponential growth beginning around the Industrial Revolution. (Adapted from Turk, J., and Turk, A., *Environmental Science*, Saunders College Publishing, Philadelphia, PA, 1988.)

revolution began, introducing mechanization into agriculture with increased production and delivery of foodstuffs; the production of fabrics and hardware; and the construction of domiciles. The ability to house and feed more people caused an eruption in the birth rates and numbers of people. By 1930, there were nearly 2 billion on the planet and it took less than 80 years to double the world's population. In the late 1800s, the beginning concepts of sanitation emerged with the chlorination of public water supplies. The introduction of public sewers, pasteurization of milk and food protection programs soon followed. Certainly by the early 1900s, most major cities in the United States and Europe understood these concepts and employed the principles of sanitation. Consequently, the spread of communicable diseases through milk, water, and food was dramatically reduced. Death rates plummeted and so growth rates exploded into the twentieth century. New discoveries in antibiotic therapy, vaccines, and medical treatment saved many more lives and increased life span. The consequence was alarming in its effect on the growth rate. In 1975, the world's population had increased to 4 billion people, and it took less than 45 years for the population to double from 1930. In the fastest population spurt in history, fewer than 40 years for the population to double, the world reached 6 billion people on October 12, 1999. Today the world's population is estimated to be greater than 7,449,698,000.[13] Consider that the world's population in 1943 was a round 2 billion people and therefore in my lifetime (so far) it has more than tripled.

The rate of growth of the world population from the mid-1700s until the present is shown in Figure 2.11. The growth rate peaked in 1970 at 2.06 percent and has declined since then to a level of 1.2 percent.[7,8] This decline in growth rate is directly related to declines in births prompted by international efforts beginning during the 1960s to stem the tide of the growing global population. This concern escalated from private groups and then spread to national governments and then to international agencies. The majority of the population increases were being seen in the lesser-developed countries (LDCs), and it became clear that the population increases were so rapid and so large that they were passing the carrying capacity of these countries and threatening their stability. The majority of these nations are racing to reverse this trend and stabilize the populations

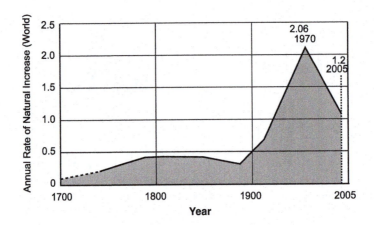

FIGURE 2.11
The world population growth rate from 1700 to 1997.

before major calamities occur. However, the efforts at achieving population stability should have occurred decades sooner. The combination of *population momentum* and a large population base in pre- and reproductive ages makes this effort very difficult. PRB's 2050 projections show Africa's population will more than double to 2.6 billion by 2050 and account for 57 percent of the global population increase by that date. The Asian population will increase by about 750 million to 5.2 billion, while Europe (including all of Russia) is headed for a population decline from 745 to 736 million. The Americas' population is seen increasing to 1.2 billion from 1 billion, and Oceania (which includes Australia and New Zealand) would rise from 42 to 63 million. In January 2017, the United States experienced one birth every 8 seconds and one death every 11 seconds. Meanwhile, net international migration is expected to add one person to the U.S. population every 33 seconds. The combination of births, deaths, and net international migration will increase the U.S. population by one person every 17 seconds.

In order to understand how we reached this point, it is helpful to look at the relationships of growth rate, doubling time, demographic transition, and incomplete demographic transition.

Growth Rate

A rate is normally expressed as one quantity in relation to another. Runners like to keep track of the time it takes to run a 10-kilometer course in minutes per mile. This is a rate measurement, and runners can study their progress by decreasing the minutes per mile that it takes to run the course. In a similar way, growth rate can be used to assess additions or deletions to the population. People enter a population by births or immigration and leave a population by deaths or emigration. The rate of births is the ratio of births to the population, and death rates represent the ratio of deaths to the population. Growth rate is then determined by the birth rate minus the death rate. If the growth rate is positive, there will be an increase in population. Even if the rate of growth is slowed (declines), the population will accumulate. Similarly, even if the runner's pace is slowed, he/she will still accumulate miles. Consequently, we are still contributing large numbers of people to the population despite a declining growth rate. This addition becomes exaggerated when the base of people is larger. In 1975, the growth rate was close to 1.75 percent with a global

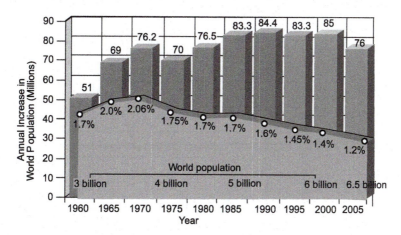

FIGURE 2.12
Annual increase in world population versus global growth rate.

population of 4 billion people leading to 4 billion × 0.0175 = 70 million new infants each year. On October 12, 1999 the population reached 6 billion people with a growth rate of 1.4 percent leading to 6 billion × 0.014 = 84 million new infants per year. Today, world population is growing by 76 million people per year (about 240,000 people per day) and we have passed 7.45 billion people. The Latest UN mid-range estimate suggests that there will be 9.8 billion people on the earth by 2050 with total population shrinking over the next 100 years.[9,14] Despite the successes in reducing the worldwide growth rates, the population has grown so much that even the smaller growth rates lead to additions of larger numbers of people to the global population (Figure 2.12). This is referred to as population momentum. For example, in Botswana, a country experiencing a drop in life expectancy, the population is expected to double by 2050 due to population momentum.[10] Added to the problem of population momentum are the differences in age-sex distribution in LDCs versus the developed countries. Figure 2.2 compares the age-sex distribution for the lesser and more developed countries. The very broad base for LDCs indicates that birth rates have been high and that young children represent the largest percent of the population. As these children reach reproductive age and bear their own children, the population will rapidly expand. The combination of huge populations that have a large percent of young pre-reproductive people have a population momentum that will foster significant population increases through the next generation.

Doubling Time

Another useful way to demonstrate growth rate is to present it as doubling time, or the number of years for a human population to double its size. The doubling time can be calculated according to the following relationship:

$$\text{Doubling time} = \frac{0.70}{\text{growth rate}} \times 100$$

This relationship is derived from the formula for calculating compound interest at the bank and is an approximation. It is useful to remember that a growth rate in any population compounds itself since the individuals entering the population will ultimately reproduce

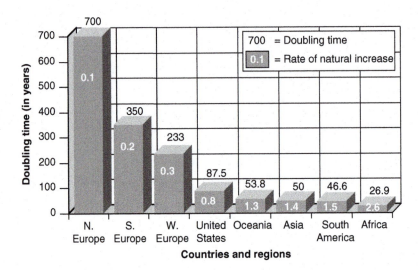

FIGURE 2.13
Population doubling times and rates of natural increase comparing certain developed and lesser-developed regions and countries.

and add more people to that population. As the growth rate increases, the doubling time decreases. Populations from 8,000 BCE to the 1650 doubled about every 1,500 years as evidenced by the relatively slow rate of population increase (Figure 2.9). The population next doubled in 200 years, then 80 years, then 45 years, and 36 years when the world reached 4 billion people in 1975; it is now over 7,449,698,000 people in 2017. Viewing population growth in this way is useful because most everybody can understand what life would be like if the number of people in our particular location were doubled within a generation. A graphical presentation of doubling time versus the rate of natural increase is shown in Figure 2.13. A rate of 4.0 percent annual increase will double the population for a country in 18 years. The World Bank estimated Kenya to have such a rate in 1979, dropping to 3.7 percent in 1988 with a population of 23.3 million.[15] The population in Niger in 2018 is 22, 311, 375 with an annual rate of increase of 3.88 percent and a TFR of 7.35, and its doubling rate has been increased to 20.5 years.[16] Although this may be viewed as improvement, such an increase in so short a time will likely surpass the carrying capacity of that country and it will be incredibly challenged to build roads, houses, schools, and sanitation facilities to accommodate this many people in so short a time.

Demographic Transition

Understanding population trends in the world requires an explanation of the differences in growth rates occurring between the developed countries of Western Europe, North America, and Japan versus the lesser-developed countries in Latin America, Asia, and Africa. The developed countries have exhibited slowly declining death rates over the last century with the gradual introduction of sanitary principles (chlorinated water supplies, pasteurized milk); increased agricultural production; and improved medical care (vaccines, antibiotics). During this time, the industrialization of society and mechanization of agriculture reduced the economic benefit of more children, while those

FIGURE 2.14

Demographic transition to developed countries showing declining death rates and birth rates resulting in a low rate on natural increase. (Adapted from United Nations Population Fund, *The State of the World Population 1999: 6 Billion, A Time for Choices*, United Nations Population Fund, New York, 1999.)

children that were born had a better chance at survival. Therefore, as the death rate declined, so did the birth rate. This has resulted in a diminishing difference between birth rates and death rates and a very low rate of natural increase, resulting in a stable population with very long doubling times (Figure 2.14).

Incomplete Demographic Transition

The lesser-developed countries had a much different experience. Most of these countries are/were agrarian societies without government-sponsored programs for social security. These are subsistence economies where rural inhabitants extract a living directly from plants and animals requiring much human labor to accomplish simple tasks without the aid of electricity or tap water. Children in the household are required to cultivate crops, care for livestock, cook food, and produce simple products for market. Such activities may require up to 6 hours a day collecting wood, water, and fodder. Children in India as young as 6 are expected to perform these duties. As resources diminish, there is a need for even more hands to collect water and firewood that are further from the home.[14] Such children, especially males, are also viewed as a source of security in old age because the governments in LDCs don't have the resources to institute social security or have unstable policies that fail to capture the trust of its citizens.[15] The populations in such countries had remained stable for perhaps thousands of years with high birth rates and high death rates. It was not uncommon for half of the children in a family to die, thereby maintaining a relatively low growth rate.[17] Then suddenly, the developed nations introduced into these developing countries better nutrition, modern medicine, insecticides, improved water quality, and methods to isolate human wastes. The result has been a startling decrease in deaths, but without the time for adjustment in birth rates. The number of births has remained high for many of the reasons discussed above. The combination of high birth rates and lowered death rates has resulted in very high rates of natural increase (Figure 2.15).

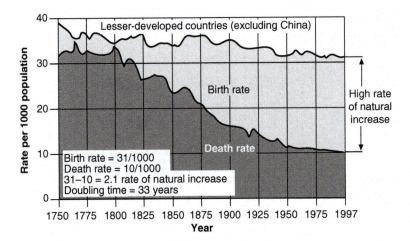

FIGURE 2.15
Incomplete demographic transition for lesser-developed countries showing declining death rates with birth rates remaining high resulting in a high rate of natural increase. (Adapted from United Nations Population Fund, *The State of the World Population 1999: 6 Billion, A Time for Choices*, United Nations Population Fund, New York, 1999.)

Current Population Trends

The world's population is growing at a rate of 1.09 percent annually and reached 7,449,698,000 people in 2017. This growth rate is a remarkable decline from the peak of 2.06 percent growth rate reported in 1970 and we entered the year 2000 with a declining rate. According to the United Nations rates are expected to continue to fall gradually for the next 20 years, and more rapidly thereafter.[4] However, we may have already set in motion the potential for destructive impacts to global ecosystems and some socio-political structures as the population momentum sweeps around the world. These problems are not equally balanced, nor are the population increases equally distributed. It is expected that the human population will reach 9.8 billion by the year 2050.[19,14] The U.S. media has announced in bold headlines that the "population explosion is over," and that the real crisis ahead is a "population implosion" as many countries now face fertility rates that are below the replacement level of 2.01.[20] This may be true in the developed nations where the United Nations predicts that during the next century, industrialized nations will have fewer people than it has today.[2] From a different view, over 95 percent of the annual increase in the world's population is occurring in the LDCs, and more than 74 of the less developed countries are expected to double their population in less than 30 years adding nearly 80 million people to earth annually. A clear division has developed between the developed and less developed countries that will likely have enormous geopolitical consequences. We need to examine the current population trends based on these differences. Are you interested in learning more about current population trends and policies around the world? Visit https://people.umass.edu/~envhl565/ and click on one or more of the following sites:

- U.S. Census Information
- World Health Organization
- World Watch Organization
- Zero Population Growth

- U.S. Population Clock
- International Programs

Population Decreases in the Developed Countries

Declines in Fertility

In 1970 there were 19 countries reporting declining fertility rates while more recently 68 countries have reported below-replacement fertility rates.[4,8] Nearly all of the European countries have below-replacement fertility rates as do the Asian countries of China (1.6), Japan (1.3), Thailand (1.7), and South Korea (1.2), and Hong Kong (1.0).[8,19] In North America, the countries of Canada (1.5) and America (2.0) are also at below-replacement levels. Those countries with the lowest fertility rates include China, Macao (0.8), China, Hong Kong (1.0), Taiwan (1.2) and Slovenia (1.2).[8] The European birth rates of the 1980s were already remarkably low, and then declined precipitously by another 20 percent to 1.4 in the early 90s and plummeted to 0.0 on the eve of the year 2000. These rates cannot sustain a population, and even if the European rate was to return to the replacement rate of 2.1 by the year 2060, Europe will have lost almost 25 percent of its population. Nearly 50 years ago about 32 percent of the world's population lived in modern nations of Europe, North America, and Japan. Today, 20 percent of the world's population lives in these countries and by 2020 it is projected to be 16 percent. In 2050 that figure will have dropped to 12 percent of the world's population.[20] The population in these regions fell so rapidly and so far that it was unprecedented in the world's history. The potential for major economic, geopolitical, and environmental changes remains to be discovered, but are likely to be dramatic. These low-fertility countries will have to look to active older people and immigrants to supply needed services and contribute to the economy.[3] There was such concern that for the first time ever, the U.N. population convened experts and leaders from around the world in December of 1997 to consider aspects of the low and rapidly declining fertility rates.[20]

The global fertility rate in the years 1950–1955 was 5.0 and on course to explode the world's population beyond its carrying capacity. The rate remained steady for another 15 years until the period 1975–1980 when the global TFR had fallen to four children. In 1995 the rate had fallen to slightly less than 3, and it fell below 2.79 in 1997. Demographers had earlier made projections of human populations reaching nearly 10 billion people by the year 2050. Now their projections would show a population that is 650 million short of the original figure because there are people that will never be born. Now the United Nations predicts that the population will reach 8.9 billion people by 2050, and 10 billion by the year 2100, following a twenty-first-century medium fertility scenario.[21]

Concerns About Decline

The industrialized countries are deeply concerned about the static or declining work force. By the year 2025, Japan will lose 12.2 million people in the age group 15–64, while Germany will lose 4.5 million and Italy will see a loss of 4 million. These developed countries are experiencing a rapidly aging population and a significant decline in the ratio of working-age people. In Western Europe there is an average of less than 4.3 workers in the age of 15–64 years for each elderly person. The ratio of workers to elderly is projected to decline further by the year 2025 with Britain at 2.9–1, followed by France (2.7), Germany (2.5), and Japan (2.3) where almost a third of the population will be elderly.[18] If the present trend of a tumbling population in Japan continues, its population in 2100 will be 55 million people, down nearly 50 percent from the 125 million people today.[22] In developed countries, populations are projected to slowly rise until 2025, and then decline so that the total number in 2050 will be about the same as today.[19] There is a concern throughout Europe and Japan that the declining population will result in decreasing house and land prices as the demand declines along with the population. A smaller work force and declining economy are also viewed as creating a less important country. Further, a high proportion of elderly non–wage earners in countries where governments support the elderly place an ever-increasing burden on the diminishing ratio of wage earners. There is likely to be economic pressure on a country that exhibits a high dependency ratio of many aged versus fewer young to alleviate the imbalance by delaying retirements and reducing retirement benefits. We see some of that already occurring in the United States.[23] The world economies are becoming more globalized and those countries experiencing a declining population with a high percentage of non-working people will be unable to compete against nations of young, lower wage, productive workers such as China and India. Consequently, there is a growing effort among some governments to reverse this population trend.[23] Vastogirardi, a small village in southern Italy, has experienced a decline in fertility rate for many reasons, including many women choosing to have a career and men enjoying their bachelorhood.[21] The people of Vastogirardi fear that they will lose their culture, traditions, and schools. Although birth rates are low all over Europe, Italy and Spain have the lowest of all: an average of 1.2 bpw, slightly lower than Greece and considerably lower than the United States' rate of 2.1 bpw.[24] One of the main reasons Italy's birth rate is so low is because state aid to families with children is lower in Italy than other countries, such as Germany or France. Since Italian women who work are often still responsible for domestic chores, it is not uncommon to avoid having a second child in order to fulfill their obligations to work and the home. This village has also witnessed a variety of businesses close due to people moving to bigger towns. There was only one child born in Vastogirardi last year.[24]

Officials in Kyokushi, a small town in the southern island of Kyushu, Japan, are offered a gift of $5,000 to parents who have a fourth or subsequent child. Other Japanese towns offered cash rewards and also considering reduced land prices and cash incentives for newlyweds.[22] It seems that very few couples are impressed, and they feel that the amounts offered are too small. Many feel that the children are costly to raise and interfere with their free time. Others feel pressured to raise perfect children and do not want the burden of having to be perfect parents. Many children that feel this burden of being raised with such stringent expectations often rebel. Countries facing declining birth rates such as Japan are offering matchmaking events and robot babies (to inspire couples to have their own), while Denmark and Singapore have saucy ad campaigns that remind couples of their civic duty to procreate. Some offer baby supplies. Finland gives away a maternity package of warm weather clothes. Russia has offed SUVs, refrigerators, and cash for those conceiving on a National Day of Conception. Many countries are offering "baby bounties" such as stipends, tax breaks, and other financial incentives to have children.[25,26] The declining fertility may also be attributed to the growing number of women who marry later in life, or not at all. The total fertility rate dropped from 1.57 in 1989 to 1.39 in 1997, and is now 1.3.[8,27] In Japan, the average woman marries at 27, and nearly 17 percent of women in their early 30s are unmarried.[28] Makoto Atoh, director of the General Institute of Population Problems in Tokyo, points to the advancement of women as the major factor in fertility decline. In Japan, the failure of husbands to participate in child care and housework is also considered a significant factor in influencing women to have fewer children.[22] Higher education for women, with new aspirations and higher incomes is considered to be a factor for declining fertility rates in many countries.[20] In fact, as illiteracy among women decreases in a country, the average number of children born to those women declines, and conversely,

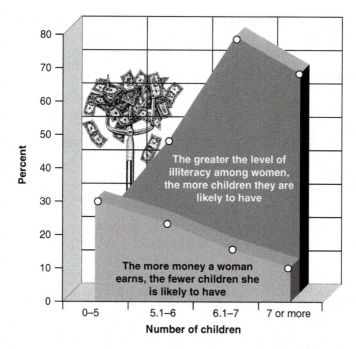

FIGURE 2.16
The number of children a woman is likely to have based on her level of literacy or the earned income she contributes to the household. (Adapted from Kristof, N.D., *New York Times International*, Sunday, October 6, 1, 1996.)

the more income a women brings into the family, the fewer children she is likely to bear (Figure 2.16).[14,19] Unfortunately, women's rights as well as their voices are often silenced around the world with over 100 million girls in the next decade to be married before their 18th birthday and some before they turn 10 years old.[14] Also, pay is often reduced during a woman's maternity leave making it even less likely that she will have many children, if any.[27] In Italy, unmarried children make a practice of continuing to live at home with their parents contributing to one of the lowest birth rates in the world.[23] Another factor that may decrease fertility rates is increased urbanization. As more people move to cities, children become more expensive to maintain and are unnecessary for farm work. Improved methods of contraception, later marriages, more divorces, and substantially reduced infant mortality (giving parents confidence their children will survive) may all play a role in decreasing fertility rates.[20,23] Is the concern about declining birth rates realistic? A large population is not a necessary condition for global power and influence as illustrated by the most rapidly growing nations in Africa, Asia, and Latin America. These nations are growing the most rapidly, but are also the most poverty-stricken. The population increases are expected for three reasons: declines in mortality, population momentum, and a fertility rate that is still above the replacement level (despite a dropped rate from 6 bpw to 3 bpw).[19] Certainly if a population were the answer to global power, India and China would be the world's two superpowers, with Israel and Great Britain being essentially without power or influence.[16] Dr. Nafis Sadik illustrates that Brazil and Japan had similar growth in economy between 1960 and 1984. Brazil's per capita GNP (gross national product) doubled in that time while Japan escalated by 7.4 times. The high fertility rate consumed much of the expanding economy to provide for the larger population while Japan's economic and political clout rose in the pressure of a declining fertility.[28] The concerns about declining populations may not be warranted, but the concerns are apparent and as countries act on those concerns, the future may see upward shifts in fertility for some of those countries as governments look for successful ways to influence fertility behavior.

Increasing Death Rate

Some countries in Eastern Europe and the former Soviet Union have experienced mortality increases. The trends influencing the rise in adult mortality have been cited as increased poverty, inadequate nutrition, increase stress and overcrowding, and a greater incidence of infectious diseases.[4] These factors have been most likely brought about due to the current political changes and instability in the region, and if continue could be a source for concern.

Fertility Rates in the United States

The United States has a demographic history of rapidly changing patterns of fertility while mortality remained stable (Figure 2.17). Subtle changes in social attitude appeared to produce rather significant changes in fertility rates and subsequently cause marked changes in the rate of natural increase. It is not surprising that the "Great Depression" resulted in major decline in fertility reaching a low of 2.2 in 1936, and then showed a rapid increase during and after WWII. The post-war demographers however, predicted a fertility decrease, but were surprised by the continued and rapid increase in population peaking in the early 1960s. The population then rapidly declined to a below-replacement level in 1971. The demographers had predicted that the birth rates would remain high after the 1960s, but once again were unable to predict the decrease. There are explanations given retrospectively that attribute the decline to a more educated female population seeking greater career opportunities and challenges. The ideal American family size grew smaller as couples realized that large families were an economic disadvantage. The availability of inexpensive and effective birth control methods made it easier for couples to reach this goal. Unpredictable changes in social attitudes had produced dramatic changes in fertility and the rate of natural increase. The TFR continued to drop from 1971 through 1976. It rose again in the 1975–1980 period to 2.1 for unexplained reasons and then fell again in 1986 to a low of 1.7. Since then the TFR has continued to rise to around 2.0 and again falling to an estimated 1.87 in 2017. It has remained below the 2.1 replacement level for the past 35 years.[29]

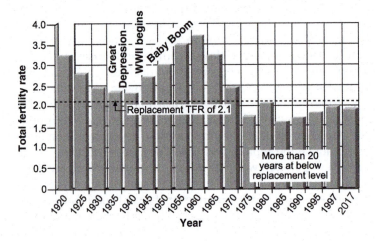

FIGURE 2.17
Total fertility rate (TFR) for U.S. from 1920 to 1997. (From U.S. Bureau of the Census, Suitland, MD.)

Although the TFR has remained below replacement levels, the United States has one of the most liberal immigration policies in the world, and immigration adds at least another 850,000 to 1.2 million people each year to the county's population with an estimated 12 million unauthorized foreign-born people in the United States in 2006.[13,24] There were 43.3 million foreign-born residents in the Unites States in 2015 that included both unauthorized and legal immigrants.[30,14] If federal immigration policies do not change, these numbers will likely go unchecked for the next 50 years. Since TFR has been at or below replacement level for 27 years, immigration is considered to be among the most important demographic factors in the growth rate of the country. About 82 million of the projected 390 million people living in the United States in 2050 will either have immigrated since 1991 or be children of those immigrants. This group will then represent 21 percent of the nation's population, and total growth due to natural increase (0.8 percent) and net migration (0.2 percent) would be expected to double the nation's population in less than 70 years.

The majority of immigration to the U.S. for the last 30 years has originated from Latin America and Asia. Hispanic women have been averaging more than three children each, black women more than two, and white women less than two. Within the next 25 years, the under-65 population of Hispanic and Asian Americans is expected to double as the number of white Americans under 65 will be fewer than in 1995.[20] The number of Black Americans under 65 is expected to increase by a third during this period, and the white elderly population will balloon by 67 percent to over 47 million people so that in the next 25 years, more than 75 percent of the elderly but less than half of the children in America will be white. This divergence in race and age is leading to a situation in which an expanding population of elderly whites will be expecting support from a working population of tremendous diversity. Another and coinciding event is that the aging Baby Boomers (those born between 1946 and 1964) are becoming the largest and most politically potent group of the elderly in history. This generation of retirees will depend on a proportionally smaller group of racially diverse workers to support them through mandated government programs such as Social Security and Medicare.[20]

The Social Security system was established on a pay-as-you-go basis such that the money a person pays each year is not saved for their pension, but is used to pay the pension of an older person alive at the time. The system was established assuming the ratio of workers to retirees would remain high, but this is not happening. There were 16 workers per retiree in 1950, and by the year 2020 there will be only 2.2 workers per retiree (Figure 2.18).

FIGURE 2.18
Number of workers per retiree in the U.S. from 1950 to 2025.

Consequently, the percent of worker income going to the Social Security system has increased and will likely increase more drastically, with a probability of one of the supporting workers being white and the other being non-white. Such a trend may create a gap between racial generations on issues of school funding and social security. The powerful elderly white electorate is demonstrating its unwillingness to fund an increasingly minority school-age population across America while consuming greater amounts of limited resources for expensive health care.[32] In the state of Florida, Medicaid nursing home costs multiplied more than five fold, from $200 million to $1.2 billion in the 11 years from 1985 to 1996. It is expected to reach a budget destroying $8 billion in the next few years.[32] The education and support of our increasingly racially diverse children is critical if we are to maintain a position of economic competitiveness in the world and if we are to expect ungrudging support of the elderly white.

There are competing bill and competing solutions in the U.S. Congress with regard to immigration. Steep fines and prison would face those who support or shield illegal immigrants. The Bill would also authorize the construction of a multi-billion-dollar, several-hundred-mile fence along the Mexico–U.S. border. The Senate is still grappling with this issue and there are many who support the President's approach that would provide a mechanism for unauthorized foreign-born people to earn legal immigrant status after several years and U.S. citizenship after that. There is division among Democrats and Republicans on this issue.[31]

Current Population Trends in Lesser-Developed Countries

The Population Reference Bureau projects the 2050 world population will be 9.8 billion. That will be an increase of 31 percent from the current estimated population of 7.5 billion. The average worldwide TFR (average lifetime births per woman) is now around 2.5. But this number varies dramatically from country to country. The less developed nations have the highest fertility rates. The three countries with the highest fertility rates are Niger, 7.30; Chad, 6.40; and Somalia, 6.4. Africa's population is expected to more than double to 2.6 billion by 2050. This will be nearly 60 percent of the global population increase by that date. India will become the world's most populous country with 1.7 billion people, overtaking China, which will decline from 1.39 billion to 1.3 billion. India will show the largest numerical increase in population of any country, adding 323 million people by 2050, followed by Nigeria with the second largest increase of 220 million. What's number three for numerical increase? That will be the Democratic Republic of the Congo at 134 million.[9]

The contention by the U.S. news media that the "population explosion is over" may be premature. There remain a number of reasons to be concerned about the world's expanding population. [6]

- The world's population is expected to increase by 85 to 100 million people per year for many years because of population momentum and continued high fertility rates among the LDCs.
- Although fertility rates have declined globally, 33 countries still have fertility rates in excess of six children per woman, accounting for almost a half billion people and continues to grow geometrically.

- In the last 30 years, more than 2 billion people were added to the global population despite the declining fertility rates in many countries. Ninety-eight percent of these annual increases in populations were in LDCs.

- Nearly a quarter of the world's population are so impoverished that they must survive on less than a dollar a day. Nearly a quarter lack drinking water or sanitation, while 20 percent lack adequate food. More than a half billion are homeless or without adequate shelter, and 86 countries are presently classified by the U.N. Food and Agriculture Organization as low-income and food-deficient because they lack the resources to purchase sufficient food.[6]

- There is 4.8 billion people in developing countries and three fifths of them lack basic sanitation.[3]

- Death rates are increasing because of aging populations in developed countries. In developing countries, a return of infectious and parasitic diseases, AIDs, war and other violence is increasing the death rate. Ninety seven percent of the deaths from emerging and re-emerging infectious diseases (malaria, dengue fever, tuberculosis, cholera) have occurred in the LDCs, causing more than 25 percent of the 68 million deaths in these countries since 1995.

- 95 percent of HIV-infected persons live in LDCs.[33]

- AIDs has killed more than 4.5 million people in Sub-Saharan Africa alone since 1985, and HIV infects more than 10 percent of the adult population in these countries and has left more than 10 million children orphaned. It is now the leading cause of death in Africa.[8,33,34]

- As the world's billions scramble for resources to feed, clothe, and house themselves, they are devastating natural resources causing: pollution of the oceans, lakes, streams and air; deforestation; desertification; the extinction of numerous plant and animal species; the elimination of wetlands; the decline in arable land resulting in diminished food supplies; and mass migration to cities creating the phenomena of urbanization.

- Many areas with high fertility rates are beginning to see rising death rates due to the spread of HIV, depletion of aquifers, and a shrinking amount of cropland available to support each person. For example, in Botswana life expectancy has fallen from 61 years in 1990 to 44 years in 1999, and is expected to drop even further.[31]

Many of these concerns are covered in the following chapter, but since populations are likely to increase the problems in these areas, this seems the right place to discuss future population trends.

Predicted Future Trends in Populations

We have already seen how difficult it is to predict population growth. There is, however, general agreement that the world's population will continue to grow substantially over the next 50 years with most of the growth occurring in the lesser-developed regions of Latin America, Africa, and parts of Asia. By 2050 the world's total population is expected to be around 9.07 billion (Table 2.1). Despite birth rates falling in many developing countries,

TABLE 2.1

Demographic Indicators of World and Regional Population Development

	Total Pop. (millions) 2005	Project Pop. (millions) 2050	Average Growth Rate 2005	Total Fertility Rate	Doubling Time in Years
WORLD TOTAL	6,464	9,075	1.2	2.6	58.3
Developed Regions[a]	1,211	1,236	0.3	1.57	233.0
Lesser-developed Regions[b]	5,253	7,839	1.4	2.82	50
AFRICA	905.9	1,937	2.1	4.83	33
Eastern	287	678	2.4	5.41	29.1
Middle	109	303	2.7	6.18	25.9
Northern	190	311	1.7	3.04	41.1
Southern	54	56	0.4	2.81	175
Western	263	587	2.3	5.59	30.4
ASIA (excluding China)	2,617	3,888	1.6	3.0	23.3
China	1,303	1,437	0.6	1.6	116
EUROPE	728	653	0.0	1.42	–
Eastern	297	223	–0.5	1.29	–
Northern	96	105	0.3	1.66	233.0
Southern	149	139	0.3	1.37	233.0
Western	186	186	0.2	1.56	350.0
LATIN AMERICA & CARIBBEAN	561	782	1.4	2.47	50
Caribbean	39	46	0.8	2.42	87.5
Central America	147	210	1.5	2.55	46.6
South America	375	527	1.4	2.44	50.0
NORTH AMERICA	331	438	1.0	1.98	70.0
Canada	32	43	0.9	1.49	77.7
United States	298	395	0.9	2.04	77.5
OCEANIA	33	47.6	1.2	2.27	58.3
Australia/New Zealand	24	33	1.0	1.78	70.0

Source: United Nations Population Fund, *State of World Population 2005: The Promise of Equality, Gender Equity, Reproductive Health and the Millennium Development Goals,* 118 p, 2005. http://www.unfpa.org/swp /swpmain.htm).

[a] More developed regions comprise North America, Japan, Europe, Australia and New Zealand.

[b] Less-developed regions comprise all regions of Africa, Latin America and the Caribbean, Asia (excluding Japan), and Melanesia, Micronesia, and Polynesia.

they are expected to remain above replacement levels for a considerable time in many regions. Almost all growth will take place in the less-developed regions: today's 6.1 billion population in underdeveloped countries is expected to increase to 8.4 billion in 2050. By contrast, the population of the more developed regions will remain broadly unchanged, at around 1.3 billion (Figures 2.19). The exception is that the population of the United States is expected to increase by 21 percent from 322 million in 2015 to 389 million in 2050. In contrast, 39 countries will have smaller populations in 2050 compared to 2017. China will have the largest numerical population decrease of 44 million with Japan showing a reduction of 25 million and Ukraine losing 9 million.[35-36.] The median or best estimate by the United Nations is that the world population will stabilize at 11 billion people around the year 2150 if the world fertility rate drops to 2.06 and life expectancy is 85 years (Figure 2.20).

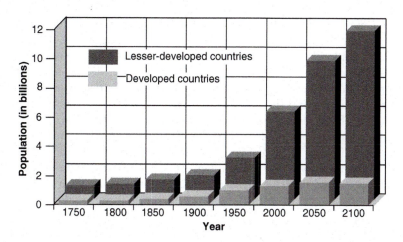

FIGURE 2.19

Population comparisons between developed and lesser-developed countries from 1750 to 2100. (Adapted from Population Reference Bureau, 1997 World Population Data Sheet, 1997, and the United Nations Population Fund, *The Right to Choose Reproductive Rights and Reproductive Health*, The State of the World Population, 1997, New York, 1997.)

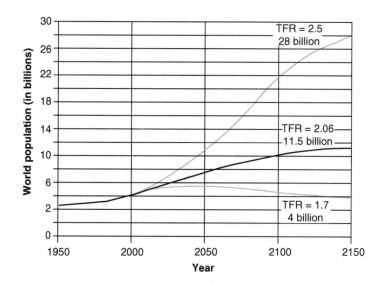

FIGURE 2.20

Projected world populations to 2150 based on various total fertility rate possibilities. (Adapted from Doyle, R., *Sci. Am.*, 276, 1997; Montavali, J., *E Mag.*, November/December 28, 1996.)

Urbanization

What Is Urbanization?

There is another dynamic to the world's increasing population. The mass migration of people to the cities is known as urbanization. In 1950 when black-and-white television was beginning to become commonplace in American homes, 30 percent or 750 million of the world's 2.2 billion people lived in cities. There was only one city in the developing world that had more than 5 million people. Forty years later, in 1990, there were nearly 2.3 billion living in cities or nearly 43 percent of the global population. In 2017, the degree of urbanization worldwide was at around 54 percent. North America was the most urbanized continent worldwide, with 81 percent of the population living in cities. Additional statistics from the World Bank estimate America's degree of urbanization at 81.62 percent as of 2015. Latin America and the Caribbean were also reported as having a high degree of urbanization, with 80 percent of the population living in cities. Europe ranked third in terms of degree of urbanization. Meanwhile, as of 2010, the global share of people living in urban areas was estimated at 51 percent. This share is projected to increase to 70 percent of the global population by 2050. The movement of masses of people into urban centers is prompted by the rapid destruction of rural ecosystems as increasing hordes of people attempt to eke out an agricultural life on land of diminishing quality. The depletion of resources, and the lack jobs and opportunities drive countless millions of desperately impoverished people to the cities.[37]

In many of these countries, the rate of urbanization exceeds the ability of local and national governments to provide necessary employment and services such as sewage disposal and clean water supplies. Uncontrolled settlements erupt as unwelcome growths around nearly all of the major developing cities and continue to grow even faster than the core urban centers. These shanty towns often are dense clusters of corrugated metal shacks or cardboard walls covered with plastic wrap.[1] According to a World Resources report, an estimated 20–50 percent of urban inhabitants in developing countries live in impoverished slums or squatter settlements without access to adequate water, sanitation, and refuse collection.[39] Nine of ten of the world's largest cities are in developing countries (Figure 2.21).

Few residents have electricity, proper sewage disposal, or clean water supplies. People urinate and defecate in streams that are filled with garbage and which incubate clouds of swarming mosquitoes.[1] Whether viewed by satellite or from windows of a car, internal African borders are disintegrating into huge megalopoli as evidenced by the countries of Ghana, Togo, Benin, Nigeria, and the Cote d'Ivoire (Figure 2.22). Borders no longer clearly exist as shanty towns merge into each other creating an endless obscene vista of abandoned cars, wire mesh, metal shacks, and stores built of shipping containers facing into puddled streets of floating garbage.[1] These urban growths are breeding grounds for disease, violence, and environmental degradation.

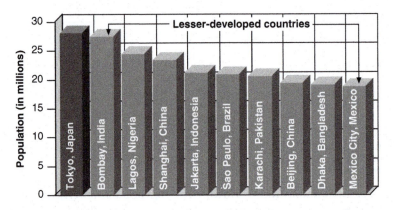

FIGURE 2.21
World's largest cities projected for 2015. (Adapted from The Environment, a new positive image, *The Weekly Review*, Kenya, November 10, 24, 1995.)

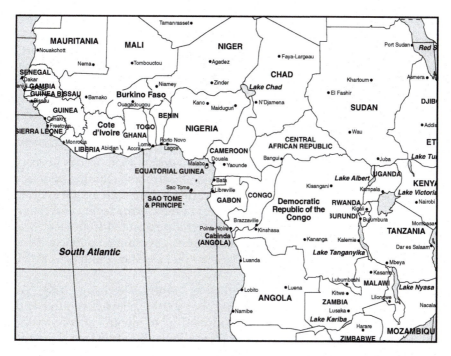

FIGURE 2.22
Areas of Western Africa where traditional borders have broken down by merging megacities.

The Role of Urbanization in the Spread of Disease

Three quarters or 8 million of the 12 million HIV positive people worldwide are located in Africa where the disease is accelerated by war, refugee movements, and uncontrolled settlements linked by improved road systems. Tuberculosis, cholera, and malaria have returned with a vengeance. Substandard housing, inadequate sanitation, and unsafe water in the densely populated cities account for 10 million deaths globally each year, with 4 million infants and children dying from waterborne diseases alone.[38] Such disease outbreaks produce dramatic economic losses. As an example, a 1994 outbreak of the plague in the Indian city of Surat killing 54 and affecting 5,000 others caused economic damages approaching $1.5 billion in U.S. dollars and caused nearly half a billion people to leave the country in a mass exodus.[38]

Violence in Developing Urban Centers

Violence in densely populated urban centers of developing countries is not a certainty. However, there are a number of countries where violence appears to be on the increase, consistent with significant population increases. In the African tribal community, it is customary and natural to eat at any table with any family. An entire village or community is considered to be the family. However, as exodus to urban areas occurs, traditions and customs often erode and communal existence evaporates as young men seek survival through joining gangs and descending into a criminal, violent process and making cities such as West Africa some of the unsafe places in the world. Gunmen roamed the streets with AK-47s in Somalia in 1992 and cruised the streets of Mogadishu in their armed "technical" rusting pick-up trucks, terrorizing the country. Somalia had become a country with no working economy, no police force, and no government.[40] Internal clan disputes warring over limited natural resources have kept stability very remote despite the intervention of American forces under Operation Restore Hope. This effort erupted into a disastrous armed conflict on October 3, 1993 when U.S. rangers raided a site near the Olympic Hotel where followers of the warlord Mohammed Farrah Aidid were meeting.[41] In 1994, gunfire and mortar attacks triggered terror and stamp stampedes at the Zaire–Rwanda border in one of the world's most vicious war zones. The million refugees of Rwanda's civil war stretched for miles as they accumulated in these death camps and succumbed by the thousands to cholera, dysentery, measles, and other diseases. The Prime Minister of Rwanda, Faustin Twageramunga, tried in desperation to coax his people back into the capital city where violence had disrupted all services including phones, stopped the flow of water, electricity, and turned it into an armed camp with military checkpoints.[42]

There was an ongoing conflict dating from 1993 in the Darfur region of western Sudan, predominantly between the Janjaweed, a militia group recruited from local Arab tribes, and the non-Arab peoples of the region. The Sudanese government publicly denies that it supports the Janjaweed but allegedly supplies arms and assistance and has participated in joint attacks with the group.

The Western media described this conflict as "ethnic cleansing" and "genocide." A British Parliamentary Report estimated that over 300,000 people had already died and the United Nations estimates that 180,000 had died in 18 months of conflict More than 1.8 million people had been displaced from their homes. Two hundred thousand fled to neighboring Chad.[43,44]

Urban poverty is a socially destabilizing force where hordes of the impoverished people fleeing decaying rural ecosystems to urban centers are threatened by the loss of native customs in expanding cities. Here, their values are undermined, physical necessities are absent, and the spread of diseases is imminent.[1] Many countries in Africa, and others outside, see the rise of criminal anarchy, unprovoked crime, and the erosion of nation-states and international borders as disease, overpopulation and disappearing resources become prominent.

The Control of Population

Empowerment or Force?

The International Conference on Population Development (ICPD) was held in Cairo beginning September 5, 1994. The major idea or topic behind the conference was women's rights. The draft program of action proposed that a full range of reproductive and health-care services, including contraception and sex education should be made available to all.[42] One of the most crucial points of the conference was the need for participation of women in the global work force at different levels. Education and employment were discussed as a tool to make women more powerful. The stabilizing of human population is no longer a mystery. It is best accomplished by improving women's education, providing them with paid employment, and making effective and inexpensive contraception available.[46] The United States continues to this day to support the Program of Action of the 1994 International Conference on Population and Development (ICPD), which was negotiated and accepted by 179 governments. The conference set far-reaching goals linking global health, human rights, and development. A key principle of the ICPD is that couples and individuals have the right to decide freely and responsibly the number, spacing, and timing of their children. Couples also have the right to information, education, and means to do so. Reproductive health care programs should not include any form of coercion.[47] It has been estimated that if women had the power to make decisions about sexual activity and its consequences 80 million unwanted pregnancies, 20 million unsafe abortions, 50,000 maternal deaths, and 333 million sexually transmitted infections could be avoided.[33] Unfortunately, this is not an easy solution in many countries where poverty, cultural traditions, and lack of educational opportunities exist and contribute to illiteracy of nearly 1 billion people worldwide. The largest gathering ever of world leaders reaffirmed the need to maintain gender equality, reproductive health, and HIV/AIDS at the top of the development agenda at the 10-year review of ICPD known as IPCD+10.[41] Two thirds of the illiterate population are women, due to young girls who are denied this education. In most of the Asia and in Muslim and Catholic countries, a woman's body may not be considered her own, nor the child in her womb. Collective decisions are often made about a woman and it is accepted because the extended family is the institution that works; the State is often not trusted.[46] Countries attempting to bring population growth under control without first empowering women and providing effective birth control have often resorted to oppressive population control policies.

Population Policies in Some Countries

India

India was the first country to introduce family planning in 1951. The programs focused on the use of the unreliable "rhythm" methods, followed 10 years later by the use of intrauterine devices (IUDs) as the TFR continued to climb. The use of IUDs led to a number of complications in largely undernourished and anemic population, which had little access to proper medical care. Consequently, there was little acceptance, and populations continued to rise. Efforts shifted to male sterilization in the 1970s with significant government abuses as attempts were made to achieve regional sterilization quotas.[48] More recently, the aspirations of the people to lead a better life after being exposed to widely accepted electronic media have led more people to realize the virtue of a small family. In October of 1997, the focus shifted to a broader approach to address the need to limit family size.[28] Literacy campaigns particularly among women combined with education about the advantages of smaller families are playing a greater role in causing this trend. India conducts a census every ten years. The most recent reports available suggest that India has succeeded in slowing its rate of growth from 2.1 percent annually to less than 1.8 percent. However, this rate is still sufficient to double the population in less than 30 years. Some parts of India continue to have a high fertility rate. Son preference has led couples that have girls to continue to have children until a son is born.[42] Girls are often seen as an economic liability. There have also been sex-selective abortions and unreported female infanticide.[42] This continued oppression of females in India has led to high rates of illiteracy and less health care.[49]

If a stabilized population growth is to be achieved, India must improve the status of Indian women by further expanding access to education, increasing opportunity for employment, and support programs that encourage later marriage.[33] All states in India have made notable progress in improving education and literacy among women over time, but equality has yet to be achieved.[42,49] The average age of marriage has increased considerably and there has also been an increased use of effective family planning methods.[49] Government efforts to achieve birth reduction quotas through sterilization, abortion, or heavy socio-economic penalties may result in regional and/or temporary successes. However, these efforts are contrary to the popular will and may lead to abuses and failures. In 1977, the Prime Minister Indira Gandhi led her Congress Party to a crushing electoral defeat that was associated with these efforts to force a sterilization program on men during her previous 2 years of emergency rule.[50]

Peru

Critics of a sterilization program that began in Peru in 1995 charge that government health workers are taking advantage of poor rural women who speak only local dialect. They report that the health workers are rewarded by promotions and money and so vigorously pursue government-imposed sterilization quotas by offering clothes and food to women in exchange for agreeing to become sterilized.[45] The TFR for Peruvian women who are uneducated and live in rural areas is 7.0, while the average TFR for Peruvian women with some college education is 1.7. The nation's poor women are, therefore, likely targets for sterilization. Critics further argue that the women are not being told about alternative methods of contraception or that tubal ligation is nearly always reversible. The opportunity for making informed decision by these women is often minimized. An official of the U.S. Congressional Subcommittee on International and Human Rights Operations questioned whether the Peruvian Government is more interested in doing family planning or population control.[51]

China

China continues to enforce a one-child policy in the nation's largest cities such as Beijing and Shanghai. The one-child policy has successfully reduced China's population growth by 250 million.[52] There are severe penalties in much of China for having a second child that may include job loss and fines equivalent to 3 years' salary for each parent.[45] It wasn't long ago (1980s) that family planning goals were accomplished through forced abortions and sterilization. The one-child policy, which was announced in 1979, is gradually yielding to a healthier economic growth. The improved economy has permitted couples in some medium-sized cities to have additional children by simply paying the steep fines without concern for retribution. However, economic growth also appears to be reducing the number of births as families find new economic prosperity. Also, there are financial incentives to encourage couples to abide by the one-child policy.[52] The emerging government policy of China is moving away from enforcement to education and encouragement. There is an emphasis on later marriages, longer intervals between children, and smaller families.[52]

Some officials have come to realize that persuasion is more effective than threatening in effective long-term family planning goals.[52-54] Improvements in quality and choice of contraceptives in recent years has increased the effectiveness of family planning. But the one-child policy has led to increased sex discrimination against women. Threats have led to underreported births of both males and females, with the birth of a girl being twice as likely to be ignored.[52] It is believed that one half to one third of the difference in infant sex ratios is due to underreporting. Unreported daughters may be adopted out, left with relatives, or abandoned in orphanages. Sex ratios are further skewed by illegal use of ultrasound to determine fetal sex, which has led to widespread abortion. The girls who survive often have less health care and education.[52]

Mexico

The population of Mexico is currently 119 million, more than 8.91 million of whom live in Mexico City alone.[8,55] If it weren't for the success of sex education and family planning practices, Mexico's population could have been over 134 million today. The population growth rate has dropped from 3.5 percent in 1974 to 1.9 today.[8,55] Twenty-five years ago, the average number of children in a Mexican family was seven; today it is down to 2.6. The emphasis has been on promoting smaller families and better relationships. Some areas are even offering free birth control. Many families are realizing that large families aren't really possible with Mexico's poor economy. Today, more than one in three women work outside the home, whereas 20 years ago less than one in five did. Also, educators are handing out leaflets about medical care, offering courses on sexuality, and providing free HIV testing.

Iran

It is estimated that Iran will double its population from 65 to 115 million by 2050.[2] However, Iran's family planning operation reaches almost the entire population and is one of the most successful operations worldwide.[2] There are advocates that disseminate information, and there are family planning messages advertised on television. Thirty-five thousand health volunteers operate mobile clinics, offering free contraceptives. There is one catch: condoms are only offered to married couples, since premarital sex isn't acknowledged. Iranians that plan on getting married must first have a day of counseling, where family planning is discussed. Many people in Iran are waiting until they are 22 or 23 years of age before they start a family, and most of them only want two children.[2] Contraceptive use by Iranian women has increased from 26 percent to 75 percent since 1979.[2] Although fertility rates are still high, they continue to drop rapidly.

Family Planning versus Population Control

The government-directed programs that set a policy for establishing an optimum population size are referred to as population control. In the countries described above, the policies were directed at reducing population. In a number of developed countries such as in Western Europe, policies may be directed at promoting a higher TFR by providing State financial support for hospital expenses related to childbirth; for monthly support for the parent(s) during the first year or two of the child's life as in Austria; or even tax deductions or special bonuses. Population control is in contrast to family planning programs that are directed at assisting couples in having the number of children they desire regardless of how many. Family planning is parent-centered and places the desires of the individual or parent for children before that of a state or national policy. Some policy makers argue that this style of family planning may not be enough to stem the tide of population growth in these countries where many children are considered an important resource for gathering wood and water, harvesting crops, or performing other forms of manual labor. The simple act of getting water may take several hours a day, and it takes longer and longer as resources diminish. Children are also viewed as a source of security in old age in those countries without a financial support system for the elderly.[17] The persistent desires for large families in many LDCs is estimated to add nearly 660 million people in the next 50 years.[8]

There is the additional problem of funding needs for family planning services in foreign countries. The ICPD in Cairo (1994) agreed that it would take at least $17 billion annually to provide services for population and reproductive health programs, and the prevention of STDs including HIV. Unfortunately, annual world expenditures are much less than half of this amount with the dollars contributed appearing to have leveled off.[8] One issue contributing to the decrease in funds supporting family planning efforts has been the freezing of much of the promised United States aid due to the anti-abortion lobby in the republican controlled congress.[14,56] Such shortfalls in LDCs are expected to result in as many as 220 million unintended pregnancies. As many as 88 million of these will likely be terminated by abortions. The pregnancies and abortions may cause as many as 117,000 additional maternal deaths with up to 1.5 million women with some form of chronic health problem. This unmet demand for family planning services may likely cause an increase of 1.2 billion in global population.[6] The major elements to offering a better quality of life for the world's billions may depend on: (1) the willingness of governments to set reasonable policies with adequate funding; (2) develop greater motivation for small families through improved economies; and (3) universal access to knowledge and use of family planning services.

In July 1999, delegates from 180 countries approved proposals to slow world population growth, including offering women greater access to abortions as well as sex education to adolescents.[50] This was a significant advance from what was agreed to at the ICPD in Cairo because it forces governments to increase and improve access to reproductive health programs. The belief is that women who are educated and have access to reproductive health care have fewer children. There are new goals in place to increase access to universal health care, increase education for women, and reduce maternal mortality and HIV infection.[57] These goals have been reaffirmed at world meetings and were at the top of the development agenda at the 10-year review of ICPD known as IPCD+10.[46] USAID is the world's largest family planning bilateral donor and is committed to helping countries meet the family planning and reproductive health needs of their people. USAID supports **Family Planning 2020's** goal to reach 120 million more women and girls in the world's poorest countries with access to voluntary family planning information, contraceptives, and services by 2020. In 1965, fewer than 10 percent of women in the developing world (excluding China) were using a modern contraceptive method, and the average family size was over six. Since then, in the 31 countries where USAID focuses its support, modern contraceptive prevalence has increased to 30 percent, and average family size has dropped to 4.4. USAID's work in reproductive health also focuses on ending child marriage, female genital mutilation/cutting, gender-based violence, and more.[58]

Methods of Fertility Control

Introduction

The process of conception requires that a viable sperm have access to a healthy egg. When the sperm enters the egg, the tail is released and the ovum secretes a protective chemical barrier that prohibits other sperm from entering. Methods that prevent fertilization of the egg are called contraception, and there are hundreds of different procedures and formulations available today. The enormous variety of contraceptive tools attests to the interest of society in preventing unwanted pregnancies. These methods vary in their risks to health, their efficacy in preventing pregnancies, ease of use, acceptance, and costs. Some of the methods and characteristics are described below.

Contraceptive Methods That Are Reversible

Natural Birth Control and Family Planning

This process is sometimes referred to as the rhythm method and relies on voluntary changes in sexual behavior. The method requires that the female predict her fertile periods by examining cervical mucous and/or tracking basal temperature or charting her menstrual cycle and ovulation times on a calendar. She would then abstain from sexual intercourse during times of the month when she is fertile. This method has proven to be very unreliable because menstrual cycles vary, and the narrow window for sexual opportunity is not always observed.

Abstinence

The term abstinence has come to mean the exclusion of sexual intimacy including oral-genital kissing and mutual masturbation. These latter activities are often referred to as "outercourse." Either of these methods can be 100 percent effective in prevention of pregnancies as long as there is not an opportunity for sperm getting near the vagina. The problems with "outercourse" are the increases in the probability of spreading sexually transmitted diseases (STDs) through oral-genital contact, and the motivation and opportunity for intercourse is increased.

Hormonal

Oral Contraceptives

Since the introduction of the Pill in 1960, it has become widely accepted and is now used by nearly 40 percent of women who wish to avoid pregnancy. The combined effects of the synthetic hormones estrogen and progesterone in the Pill prohibits the maturation of ova, prevents the development of the uterine lining, and forms cervical mucous creating a barrier against sperm. The amount of estrogen supplied in the Pills has been decreased to reduce health risks. There remains a limited risk of increased blood pressure, the formation of blood clots, strokes and heart attacks for some women. This risk is low in most healthy women. The risk is further reduced for women less than 35 years of age and for those who do not smoke. A physician's prescription is required in most Western countries, including a complete gynecological exam. When taken according to directions, the Pill approaches a 97–99 percent effectiveness in preventing pregnancy (Figure 2.23). Oral contraception offers no protection against the transmission of sexually transmitted diseases.[59]

FIGURE 2.23
Various forms of the oral contraceptive pill designed to be taken daily.

A progestin-only pill known as a mini-pill is also available for women who don't want to use estrogen. The mini-pill is 96 percent effective and has few side effects except for the possibility of irregular menstrual bleeding.

Depo-Provera

This is a long-acting synthetic hormone (progesterone) that is injected into muscles of the arm or buttocks every three months. It has an efficacy of 99 percent and is especially desirable for women who don't want to keep track of taking daily Pill or inserting a device or chemical into the vagina. Most women lose their menstrual period within a year on this hormone. The adverse side effects are minimal and usually limited to a slight weight gain, occasional dizziness and headaches.[59]

Norplant

Norplant consists of five or six silicon rubber capsules (Figure 2.24) containing progestin that are inserted under the skin of the upper arm of the female using minor surgery. Small amounts of progestin are time-released over a 5 year period and works by suppressing ovulation, thickening the cervical mucous, and inhibiting the growth of the uterine lining. Norplant has been successfully tested in more than 40 countries demonstrating minimal health risks with nearly 100 percent effectiveness in preventing pregnancies. It has been used more frequently in the U.S. since it entered the market in 1991, but the upfront costs that may exceed $550 are sometimes a deterrent even though the cost of the Pill over the same 5 year period is more than twice that amount.

Spermicides

Chemicals that are designed to kill sperm are called spermicides. The substance most commonly used is nonoxynol-9 and it comes in a variety of forms including jellies, creams, foams, and waxy suppositories (Figure 2.25). These are all meant to be inserted far enough to cover the cervix and so providing a chemical and physical obstacle to sperm. Spermicides are most useful when used together with a barrier method such as a condom. Together they provide a 98 percent effectiveness.

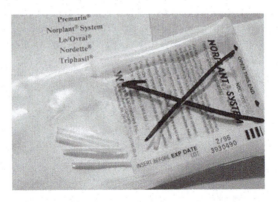

FIGURE 2.24
Norplant® showing elongated capsules.

FIGURE 2.25
Samples of spermicidal jellies, creams, and lubricants.

Barrier Methods of Contraception

Male Condom

Condoms are typically elastic material of latex rubber or polyurethane designed to be rolled onto the erect penis before intercourse (Figure 2.26). They appear in a variety of colors and textures; with and without lubricants; and with and without a reservoir that is designed to capture ejaculate. The effectiveness of condoms to prevent pregnancies is reduced to 88 percent because of improper use or storage, which permits sperm to enter the vagina through tears, rips, or spillage. Formerly it was believed a greater effectiveness could be achieved if the condom is used along with a spermicidal lubricant. Studies have found that the spermicide nonoxynol-9 might actually irritate the vaginal walls making infection in users more likely. Due to this study the CDC has changed its recommendations, advocating for use of a lubricated condom alone until this finding can be researched further.[61] Condoms are one of the top three methods used in the United States and provide the additional benefit of protecting against the transfer of HIV.

FIGURE 2.26
Male condoms appear in a variety of colors, shapes, and textures.

Female Condom, Diaphragm, and Cervical Caps

The female condom is a soft elastic polyurethane conical bag with an elastic ring (Figure 2.27). The narrow end is inserted into the vagina leaving the wider end to extend out over the labia. It has a reported failure rate of around 12 percent based on use in the United States.

The diaphragm is a thin latex rubber cup with a flexible rubber-coated ring intended to cover the cervix. Diaphragms come in a variety of sizes and must be fitted to the female by a trained health professional. Moreover, the user must be familiar with the correct method to insert and position the diaphragm. It is intended to be used with a spermicidal cream or jelly on the concave or inside surface before insertion, and then left in place for up to 8 hours after intercourse. This permits sufficient contact time to kill any remaining sperm.

The cervical cap is similar in design but smaller than the diaphragm (Figure 2.28) and is also fitted by a health professional. It is designed to be used with a spermicide and left in place for several hours after intercourse. Its effectiveness is somewhat less than a diaphragm because it is smaller and more difficult to properly insert (Figure 2.29).

Intrauterine Devices (IUDs)

IUDs are plastic T-shaped devices that are folded with a special instrument and inserted by a trained health professional into the uterus such that the arms of the "T" open up and cross the uterus where it is braced. Some brands come with a slow releasing synthetic

FIGURE 2.27
Female condom is inserted into the vagina with the wider open end extending over the labia.

FIGURE 2.28
Sponge (left) and cervical cap (right) designed to be placed over the cervix with spermicide.

FIGURE 2.29
Intrauterine device showing approximate placement in the uterus.

progesterone, while others are copper wrapped. The specific action of IUDs is unknown, but it is believed they interfere with egg fertilization by sperm. The IUDs must be removed or replaced periodically every 1–4 years depending on the type used. The effectiveness in preventing pregnancies can reach 95 percent. Risks include pelvic inflammatory disease, severe cramps, heavy menstrual flow, uterine perforation, tubal infections, and infertility. These are among some of the reasons for negative publicity that prompted the manufacturer to remove the Dalkon shield from the market in the 1970s.

Contraceptive Methods That Are Permanent

Sterilization has become one of the most popular methods for contraception in the United States among married couples who have achieved their desired level of parenthood. Male sterilization by vasectomy is usually an outpatient procedure accomplished by making an incision on either side of the scrotum and snipping out a piece of the vas deferens. The two ends of the vas are tied off or cauterized. There is some discomfort in this area for about a week. The risks are minimal and effectiveness in preventing unwanted pregnancies in partners is better than 99 percent. Women may be sterilized by tubal ligation, which is a slightly more involved surgical procedure requiring small incisions near the navel and just above the pubic area. The abdomen is inflated with a gas so the physician can easily locate the fallopian tubes with the fiber-optic light on the laparoscope and tie off or cauterize the tubes. This procedure blocks the entry of eggs into uterus, and eggs released from the ovaries dissolve and are reabsorbed into the body. Tubal ligations are largely reversible while vasectomies have generally not met with similar success at being reversed. However, developments in microsurgical techniques have resulted in a reversal approaching 50 percent depending on the elapsed time since the vasectomy and the skill of the surgeon.[60]

Abortion

Abortion refers to the medical means of terminating a pregnancy. In the United States this is considered to be a decision that a woman and her physician may make within the first trimester without legal restriction. This right was given under the 1973 Roe versus

Wade Supreme Court decision.[62] However, individual states were given the right under later court actions to impose restrictions on abortions. Despite these restrictions, about 1.6 million abortions have been legally performed each year in the United States since 1980.[60] Nearly 60 million abortions occur annually on a worldwide basis. Three quarters of the world's population live in countries where it is legal to induce abortion medically. There are different forms of abortion depending on how many weeks pregnant the woman is at the time of the abortion. The number of weeks of pregnancy is determined from the first day of a woman's last menstrual period. Most pregnancies are terminated in the first trimester when legal and health consequences are minimal. These are usually referred to as therapeutic abortions and are normally accomplished by vacuum aspiration. Abortion can also be safely induced within the first nine weeks of pregnancy by administering the drug RU-486. This steroid blocks the action of progesterone and causes the uterine lining to break down and so expelling the lining and the fetus, which terminates the pregnancy. The drug RU-486 was developed in the late 1970s by a French researcher consulting for Roussel-Uclaf, a French pharmaceutical company. There continues to be fears by many, including antiabortion groups, that RU486 would popularize abortion by making it available within a doctor's office and therefore remove the need for abortion clinics, which are frequent targets of protestors. Despite many setbacks including halted productions and distribution, RU-486 is now being widely used in France and on September 28, 2000 was approved by the FDA for sale in the United States. The approval of this drug gives women an alternative to surgical abortion for the cost of a standard abortion.[63]

References

1. Kaplan, R., The coming anarchy, *The Atlantic Monthly*, 273(2), 44, 1994.
2. Valerie, W., *People of Color will be a Majority of the American Working Class in 2032: What This Means for the Effort to Grow Wages and Reduce Inequality*, Economic Policy Institute, June 9, 2016. Accessed March 2018, http://www.epi.org/publication/the-changing-demographics-of-americas -working-class/.
3. Hussar, W.J., and Tabitha, M. B. *Projections of Education Statistics to 2022* (NCES 2014–2051). U.S. Department of Education, National Center for Education Statistics. U.S. Government Printing Office, Washington, DC, 2014.
4. United Nations, *The State of the World Population 1999: 6 Billion, A Time for Choices*, United Nations, New York, 1999.
5. Freidman, G., *Population Decline and the Great Economic Reversal*, Stratfor Worldview, February 17, 2015. Accessed March 2018, https://worldview.stratfor.com/article/population-decline-and-great -economic-reversal.
6. Population Reference Bureau, World Population Data Sheet, and the United Nations Population Fund, *The Right to Choose Reproductive Rights and Reproductive Health*, The State of the World Population, New York, 1997.
7. United Nations Population Fund, *The State of the World Population 1997, The Right to Choose Reproductive Rights and Reproductive Health*. New York, 1997. Accessed December 2006, http:// www.unfpa.org/swp97e/toc.html.
8. Population Reference Bureau, *World Population Data Sheet*, August 2005. Accessed December 2008, http://www.prb.org.
9. World Population Data Sheet, Population reference bureau, 2017. Accessed February 2018, http://www.prb.org/Publications/Datasheets/2017/2017-world-population-data-sheet.aspx.
10. Odum, E., *Fundamentals of Ecology*, W.B. Saunders Company, Philadelphia, PA, 1959.

11. Meyers, J. and Krebs, C., *Population Cycles in Rodents*, Scientific American, New York, 38, June, 1974.
12. Turk, J. and Turk, A., *Environmental Science*, Saunders College Publishing, Philadelphia, PA, 1988, Chap. 7.
13. U.S. Census Bureau, U, S. and Word Population Clocks, January 25, 2018. Accessed January 2018, https://www.census.gov/popclock/.
14. Nierenberg, D., and MacDonald, M. *The Populations Story...So Far*, World_Watch, September/October, 2004. pp. 14–17.
15. Perlez, J., Birth control making inroads in populous Kenya, *The New York Times*, Sunday, September 10, 1989.
16. Worldometer, Elaboration of data by United Nations, Department of Economic and Social Affairs, Population Division. World Population Prospects: The 2017 Revision. (Medium-fertility variant). Accessed February 2018, http://www.worldometers.info/world-population/niger-population.
17. Dasgupta, P.S., Population, poverty and the local environment, *Scientific American*, February 42, 1995.
18. Freed, S.A. and Freed, R.S., One son is no sons, *Natural History*, 37(1), 1–31, 1985.
19. Bongaarts, J., Demographic consequences of declining fertility, *Science*, 282, 419, 1998.
20. Wattenburg, B.J., The Population explosion is over, *The New York Tomes Magazine*, November 23rd, Section 6, 60, 1997.
21. Deen, T., *Development: Population Explosion Not Over*, Warns U.N., Inter Press Service, March 1998. Accessed February 2006, http://www.oneworld.org/ips2/mar98/population.html.
22. Kristof, N. D., Baby makes 3, but in Japan that's not enough, *New York Times International*, Sunday, October 6, 1, 1996.
23. Ayres, R., *The Economic Conundrum of an Aging Population*, World-Watch, September/October 2004, pp. 45–49.
24. Stanley, A., Blissful bachelorhood and the shrinking village, *New York Times*, November 16, 1999. Accessed November 2006, http://www.nytimes.com/news/world.
25. Lily, H., What countries around the world give their citizens for having children, Quartz, April 19. https://qz.com/200728/what-countries-around-the-world-give-their-citizens-for-having-children/ 2014 (accessed February 2018).
26. Adam, T., It's official: Japan's population is dramatically shrinking, The Washington Post, February 26, 2016. https://www.washingtonpost.com/news/worldviews/wp/2016/02/26/its-official-japans-population-is-drastically-shrinking/?utm_term=.087d88266f20 (accessed March 2018).
27. Funabashi, K., Reassessing the value of children, *Japan Echo*, 26, February 1999. http://www.japanecho.co.jp/docs/html/260111.html (accessed April 2017).
28. Sadik, N., United Nations Fund for Population Activities, 1985 Report.
29. CIA World Factbook, United States Total Fertility Rate, Source, CIA World Factbook. Accessed October 2017, https://www.indexmundi.com/united_states/total_fertility_rate.html.
30. Jie, Z. and Jeanne, B., *Frequently Requested Statistics on Immigrants and Immigration in the United States*, Migration Policy Institute, March, 2017. https://www.migrationpolicy.org/article/frequently-requested-statistics-immigrants-and-immigration-united-states (accessed March 2018).
31. Martin, P., *The Battle Over Unauthorized Immigration to the United States*, Population Reference Bureau, April 11, 2006. http://www.prb.org (accessed May 2017).
32. Tilove, J., Racial generation gap growing, *Sunday Republican*, February 22, B2, 1998.
33. United Nations Population Fund, *The State of the World Population* 2000. Accessed December 2006, http://www.unfpa.org.
34. Newsweek, 10 Million Orphans, January 17, 2000.
35. Population Matters, *Current Population Trends*, March 2016. Accessed November 2017, https://www.populationmatters.org/wp-content/uploads/2016/07/SI_12_current_population_trends.pdf.
36. Population Reference Bureau, World Population Data Sheet. http://www.prb.org/Publications/Datasheets/2017/2017-world-population-data-sheet.aspx (accessed March 2018).

37. Statista, Degree of urbanization (percentage of urban population in total population) by continent in 2017. Accessed March 2018, https://www.statista.com/statistics/270860/urbanization-by-continent.
38. The Environment, A new positive image, The Weekly Review, Kenya, November 10–24, 1995.
39. Lynn, W., Megacities: Sweet dreams or environmental nightmares?, *Environmental Science and Technology*, 33, 238A–240A, 1999.
40. Nelan, B.W., Taking on the thugs, *Time*, December 14, 27, 1992.
41. Church, G., Anatomy of disaster, *Time*, October 18, 40, 1993.
42. Gibbs, N., Cry the forsaken country, *Time*, August 1, 28, 1994.
43. Smith. R., How many have died in Darfur, *BBC News*, February 16, 2005. http://bews.bbc.co.uk/world/africa/4268733.stm.
44. Egeland, J. UN's Darfur death estimates soars, *BBC News*, March 14, 2005. Accessed December 2006, http://bews.bbc.co.uk/world/africa/4349063.stm.
45. Xin, W., Population vs. development: Challenge of the new century, *Beijing Review*, May 1–7, 12, 1995.
46. UNFPA, Consolidating the achievements of the ICPD, May 2006.
47. U.S. Department of State, Population. Accessed March 2018, https://www.state.gov/j/prm/population.
48. Conly, S.B. and Camp, S.L., India's family-planning challenge: From rhetoric to action, Country Study Series #2, The Population Crises Committee, Washington, DC, 1992.
49. Arnold, F., Choe, M.K., and Roy, T.K., Son preference, the family-preference process and child in India, *Population Studies*, 52, 301, 1998.
50. Kataria, S., India succeeding in slowing population growth, *Reuters*, February 02, 1997. Accessed November 2006, http://www3.elibrary.com/id/2525/getdoc.c...Q002D029&Form+RL&pubname=Reuters&pubur=0.
51. Sims, C., Using gifts as bait, Peru sterilizes poor women, *New York Times*, February. 14, 1, 1998.
52. Choi, C.Y. and Kane, P., China's one child family policy, *British Medical Journal*, 319 i7215, 992, October 9, 1999.
53. Arthur, E.D., *Assistant Secretary for Population*, Refugees and Migration Testimony before the House International Relations Committee Washington, DC, December 14, 2004 *"Archived copy." Archived from the original on 2011-07-21. Retrieved 2009-07-31.*
54. Faison, R., China's strict 1-baby policy giving way to economic changes, *Springfield Sunday Republican*, August 17, 1, 1997.
55. McConahay, M.J., A smaller but better future, *Sierra*, July/August, 1999.
56. Roach, J., *Family Planning urged for Developing Nations*, October 6, 1998. Accessed October 2006, http://www.enn.com/enn-news-archive/1998/10/1000689/famplan.asp.
57. U.N. Moves to Slow Population Growth, *Weekend Gazette*, A5, July 3, 1999.
58. USAID Family Planning and Reproductive Health, November 30, 2017. Accessed March 2018, https://www.usaid.gov/what-we-do/global-health/family-planning.
59. Donatelle, R. and Davis, L., *Pregancy, Childbirth, and Birth Control, Access to Health*, Third edition, Prentice Hall, Englewood Cliffs, N.J., 1988, Chapt 8.
60. Doyle, R., Global fertility and population, *Scientific American*, March 26, 1997.
61. Mayo Clinic staff, spermicide, Mayo Clinic, Mayo Foundation for Medical Education and Research (MFMER). Accessed July 2018, http://www.unaids.org/whatsnew/press/eng/durban120700.html.
62. Boston Women's Health Collective, *The New Our Bodies, Ourselves*, Simon & Schuster, New York, 1984, p. 218.
63. Kaufman, M., Abortion pill wins FDA OK, *Union News*, September 29, 2000.
64. Montavali, J., Contents under pressure, *E Magazine*, November/December 28, 1996.

3

Environmental Degradation and Food Security

OBJECTIVES

A student reading this chapter will be able to:

1. Discuss the impact of population on resources and ecosystems.
2. Define the following terms: retrogression, soil erosion, desertification, deforestation, wetlands destruction, and wildlife destruction.
3. Define the term "food security."
4. Discuss the main causes of food insecurity among nations worldwide.
5. List the suggested steps that might be taken to minimize global food insecurity.
6. Explain the most likely reasons for growing food insecurity in the United States.

Introduction: The Debate

The ability of our planet to sustain and feed the dramatic increases in human population growth has been an on-going debate stretching back over 200 years. An English clergyman and economist by the name of Thomas Malthus published an article in 1789 titled "An Essay on the Principle of Population." He concluded that human populations grow exponentially (2,4,8,16...) and that plants grow arithmetically (1,2,3,4...) so that eventually uncontrolled human growth would lead to "misery and death" as growth surpasses the capability of the land to feed the expanding population. The predictions of Malthus were not widely accepted initially and were even scorned as the opportunities for migration relieved the population pressures and technological advances in agriculture increased the available food supply.

It wasn't until the post–World War II period when growth rates sharply escalated that world leaders and demographic scientists pressed the population alarm, proclaiming that overpopulation was the greatest threat to continued habitation of the planet by humans. Those holding these views are often considered to be neo-Malthusians. Government reports and published papers exclaimed that the world in the year 2050 will become more polluted, greatly crowded, ecologically unstable, and vulnerable to economic and political disruption. The population by 2050 is predicted to be 10 billion people. By that time, there would likely be more regional water shortages and major reductions in forested land. Significant deterioration in agricultural soils has been reported through the processes of erosion, desertification, waterlogging, and salinization. World climate is predicted to change from the influence of greenhouse gases such as carbon dioxide that are produced in ever greater amounts by human activities such as combustion of fossil fuels. These gases are then thought to absorb greater amounts of heat energy from the sun, producing global warming. The combustion of fossil fuels is also thought to increase the formation of acid compounds in the air leading to increased acid deposition in lakes, streams, and rivers that become intolerant to many forms of aquatic life. There are predictions of increasing extinctions of animal and plant species as pollutants increase and habitats are destroyed for agriculture and development. There is no shortage of professionals and laypeople who predict worldwide calamities as a direct consequence of unbridled human growth.[1-5]

Technology and Policy Will Save the Day

There are many who counter these views and argue that overpopulation is not the central cause of a degraded environment, but rather that such degradation develops from poverty and an unequal distribution of wealth. It is suggested by some that the earth has resources capable of feeding numbers as great as 16 billion.[1] Still others predict the earth has an enormous productive potential capable of supplying food for 400–1,000 billion people.[6] Such wide deviation in the predicted carrying capacity of the earth highlights the uncertainty in making such predictions. Additionally, some fear that the human population may have already become so large and the momentum for further growth so great that any technologies that might mitigate these growth pressures cannot be developed or applied quickly enough to prevent disasters from occurring.

The counter view to "population as a problem" is that the real threat to global stability is the failure of nations to pursue economic trade and research policies that increase food production, more evenly distribute food and resources, and limit environmental pollution. These views usually recognize that more people are not a problem but an asset that drives the need for greater technology and innovation to sustain larger numbers of people. People who adhere to these views are sometimes called "Cornucopians" after the mythical horn-of-plenty. The development of genetically superior crops with dramatically higher yields became known as the "Green Revolution." Preceding the Green Revolution was the development of more efficient and economical energy sources.

The Green Revolution

In the mid-1960s, an interdisciplinary team of scientists from a worldwide network of agricultural centers began to operate under the Consultative Group on International Agricultural Research (CGIAR). Research from these groups has led to the development of new varieties of important food crops such as rice, wheat, and corn.[7] The International Rice Research Institute in the Philippines was responsible for developing genetically improved rice strains that doubled the yields of rice per hectare. Further, strains were developed that resist diseases, pests, drought, and flooding. Corn production per hectare has continually risen since the 1960s, with the world average rising from 2.24–3.92 megagrams/hectare. Corn is one crop that has shown increasing yields per hectare of one to two percent even without the contribution of genetic engineering or other technologies. Using the latest technologies and best seeds, the state of Iowa now averages in excess of 9 megagrams/ hectare of corn, or five times the world average, and Iowa expert farmers now average nearly 18 megagrams/hectare.[7] The new varieties have been shortened with thicker stalks to support the heavier weights while the leaves were also shorter and more upright to reduce the problem of shading and crowding of nearby plants. These plants respond to increased fertilization by producing much greater yields of corn, rice, or wheat. Wheat crops in Mexico exploded by a factor of 4.3 times the yield per hectare when genetically enhanced seeds were used.[7] Dramatic increases in yields per hectare of these crops ranging from 50 to 200 or even 400 percent have been seen in many areas around the globe. The International Rice Research Institute in the Philippines developed genetically improved strains of rice that have increased yields by twofold.[6] There are now varieties being developed that include resistance to drought, flood, salt, and pests in order to further boost yields. The striking increased production brought by the incorporation of these new varieties of seeds and processes has earned the name Green Revolution (Figure 3.1).

FIGURE 3.1
The world markets and the Green Revolution may promote monoculture technology that could prove to be ecologically unstable. (Courtesy of Discimagery.)

Historically, improving crop yield was a slow process that involved identifying and selecting seeds that were better performers within existing crops. These seeds varied slightly in genetic makeup from the others, in that they possessed the characteristics necessary to produce somewhat greater yields per hectare. Investment in international agricultural research has allowed scientists to rapidly accelerate this process through crossbreeding, use of bacteria to introduce new genetic material, and mutation by ultraviolet or gamma radiation. Crossbreeding involves the selection of genetic and morphologic plant variants that possess one or more of the desired characteristics and then crossing the plants in the hope that the most desired characteristics of the different plants will appear in the new crossbred plants. As an example, corn with a heavy stalk but few ears might be crossed with corn that has many ears but a thin stalk that won't ordinarily support the additional ears of corn. The desired product would be corn that has a thick stalk and many ears (Figure 3.2a). Bacteria that infect plants may be altered to add new DNA rather than disease DNA to plant cells, thereby transforming the plant's genetic makeup. Last, a plant's genes may also be altered by using ultraviolet or gamma radiation or chemicals such as ethyl methane sulfonate to produce mutations.[8] These mutations produce many variations of the plants, some of which may be beneficial, with the desired characteristics of many ears and a thick stalk, as described in this example (Figure 3.2b). Over 2,300 varieties of crops have been created using this technique. This method may produce mutations beyond those desired, including some that have never been identified, documented, or studied.

More recently, these time-consuming processes have been improved with the advent of genetic engineering, which involves tailored alteration of genetic information to produce

FIGURE 3.2
Historically, plant improvements were achieved with crossbreeding and artificial mutation. (a) Hybridization; (b) induced mutation. (Adapted from Turk, J. and Turk, A., *Environmental Science*, 4th ed., Saunders College Publishing, Philadelphia, PA, 1988.)

new and unique plants. As scientists have improved their knowledge of plant genomes, they have been able to improve the specificity of changes to that genome, down to the level of changes to a unique gene. Whereas the technique of radiation-induced mutations may or may not lead to a beneficial plant mutation that increases yield or confers other plant benefit, genetic engineering involves precise genetic changes, reducing uncertainty.

One of these genetic techniques is gene transfer. Gene transfer techniques were initially limited to transferring specific bacterial genes that might be beneficial to certain plants. Most plants are difficult to manipulate with gene therapy. However, newer technology uses a "gene gun" that incorporates the desirable foreign DNA fragments into a plastic bullet. The bullet is fired by the gun into a barrier that shatters the bullet, creating a genetic spray that pierces the plant target at over 1,600 kilometer per hour, and via this "particle bombardment" the foreign DNA enters thousands of cells in a single shot.[7,9] Plant cells are then cultured with growth hormones to produce plant shoots in a few weeks that exhibit characteristics associated with the newly incorporated genetic materials (Figure 3.3). Viral-resistant barley has been created this way, and these techniques are being used to develop corn that can play host to nitrogen-fixing bacteria and so require less fertilizer.[6]

On March 3, 1988 the USDA and Delta & Pine Land Company announced that they had a patent on an agricultural genetic technology that made plants make sterile seeds. This Terminator™ Technology is also known as Genetic Use Restriction Technology (GURT). A terminator seed has the code for an embryo-killing toxin genetically spliced into the gene sequence before a gene promoter sequence. To allow the parent plant to produce seeds a genetic spacer is inserted between the toxin and promoter sequences. This way, in early germination, seeds are produced because the toxin and promoter are not touching. When a protein recombinase is activated that recognizes the spacer as unnecessary, it removes the spacer sequence. The toxin code and promoter are now adjacent and the resulting toxic protein kills the embryo in late embryogenesis. If planted, the seed sits in the ground producing nothing, forcing the farmers who use these seeds to buy new ones every year from multinational corporations.[10] At present, 1.4 billion farmers globally rely on farm-saved seeds for their survival, with 1 billion of those farmers in Asia. In fact, farming via seed saving alone is responsible for 15–20 percent of the world's food supply, which suggests that this method of farming remains an important component of global food production.[11]

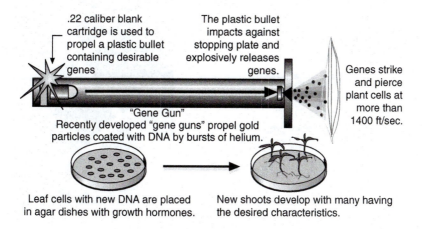

FIGURE 3.3
Methods of gene transfer in plants may use the "gene gun" that blasts the desirable genetic fragments into the plant. (Adapted from Budiansky, S., 10 billion for dinner please, *U.S. News and World Report*, September 12, 1997, p. 57.)

If implemented, the terminator seed technology could lead to a loss of farmers who farm via seed-saving, who are not able to afford to buy new seed each year. Further losses would be seen in the loss of genetic diversity through local crop varieties and even a decrease in food sovereignty for many countries.[12] Additional concerns exist that the mutated seeds might cross with wild plants, thereby introducing this uncertain and untested technology into the broader ecosystem. Given an overwhelming amount of concern, the United Nations enacted a ban on this technology. Although not binding, it sent a strong message.

Many thought that this Terminator technology controversy ended in 2000 when the agricultural giant Monsanto, which had purchased the Delta & Pine Land Company, agreed not to commercialize the genetically modified plants. While Monsanto continues to stand by its pledge not to commercialize the terminator seed technology, many think that may change. Additionally, smaller biotech companies are researching the commercialization of GURT seeds.[12]

Other agricultural techniques being developed involve the improved application of fertilizers using global positioning satellite (GPS) systems and new soil-tilling methods that reduce erosion (Figure 3.4). "Precision farming" depends on the use of GPS satellites that exactly match the rates of application of seeds and fertilizer to existing soil conditions. Signals from the satellite are sent to the computers on board the tractors, which determine the coordinates and adjust the amount of fertilizer dispensed in each area of the field. This process reduces waste and increases yields. There are now "no-till" planters that cut a narrow trough or slit through the crop residues into the sod, drop the seed in place, and then cover the seed. This substantially reduces soil erosion.[6] New techniques are being developed all of the time.

However, there remain concerns about the application of new agricultural technologies. The new breeds of plants that are selected or genetically altered to produce higher yields often require productive soils, along with extensive application of fertilizers, water, and insecticides to promote the growth of the plants and protect against insect invasion.[13] The higher yields that may be achieved often prompt farmers to plant single crops or monocultures because they are profitable in the short term, but which could result in extensive ecological damage through depletion of soil nutrients, erosion, and the attraction of pests.

Computers on-board tractor receive signal from GPS satellite and determine field coordinates, then adjust fertilizer dispersion.

Global positioning satellite (GPS) sends specific signals...

..on location and local soil conditions to receiving systems on tractors

Some artwork based on Metro Creative Graphics, Inc

FIGURE 3.4
"Precision farming" techniques use satellite maps and onboard computers to adjust control rate of seed and fertilizer spread. (Adapted from Budiansky, S., 10 billion for dinner please, *U.S. News and World Report*, September 12, 1997, p. 57.)

The high-yield crops' need for more water, pesticides, fertilizers, and production equipment creates an ever-increasing energy demand, as these components all require fuel to be produced or delivered. Many areas of the world that experience drought or poor water control (necessary for rice production) have not widely adopted these crops. Diffusion of high-yield crops is slow in areas that have inadequate infrastructure or market access such as sub-Saharan Africa or the hillside systems of Latin America and Asia.[13]

There is little debate that these advances in agricultural technologies have contributed significantly to reducing hunger in millions of people. Many lesser-developed countries (LDCs) have adopted the seeds and technologies produced via the Green Revolution. China's rice and maize crops are almost entirely planted in modern varieties, and although some exceptions exist, most of the growth in production in developing countries is a result of the higher yields generated by the Green Revolution.[13] Nearly three fourths of rice and wheat crop areas and more than half of all maize croplands were planted in modern varieties.[14] However, the growth of the human population in many of these LDCs has exceeded the capacity of even these technological wonders in agricultural production,[6] while the energy to support high-yield cropping techniques is becoming more expensive.[13]

Agricultural researchers claim that we could expect more increases in crop yields providing that money continues to flow into agricultural research and into credit and technical assistance for farmers who wish to apply the technologies. As populations continue to expand, especially in LDCs, the pressure is on governments and scientists to increase crop yields, to maintain soil quality, to maximize energy sources, and to ensure access to clean water for crop irrigation. Although global cereal production is at a high right now, with production exceeding expectation, we cannot assume this will continue indefinitely.[15]

Energy

Although there have been major technological advances in fuel use, LDCs continue to use wood as a primary fuel for cooking and heat. Wood is being used at such a rapid pace in some LDCs that forested regions have been decimated and the collection of wood for fuel may require several hours each day or as much as 25 percent of average income.[16] Across the globe, 2.4 billion people use wood fuel each day to cook, boil water, and stay warm. In Africa alone, wood is the only source of fuel in 2/3 of households. Interestingly, wood energy has been increasing in popularity in developed countries too, as people are more often choosing to use wood pellets in pellet stoves as an affordable means to heat a home. To meet this increasing demand, the production of wood pellets increased over 40 percent between 2012 and 2015.[17]

More commonly, the history of fuel use in developed nations has involved a move from wood to more energy-efficient fuels such as oil and gas. As the Industrial Revolution entered the seventeenth and eighteenth centuries, a process for making charcoal from wood was developed; charcoal became the main fuel for smelting of iron and virtually depleted much of the forested land of the United States. Technology again entered the picture as wood resources became a limiting factor, and a process was developed that converted coal to a useable coke form necessary for powering "iron refineries." As the need developed for economical fuels to power automobiles, trucks, tractors, and eventually planes, petroleum was refined into gasoline and oil. The requirements for petroleum increased, and new technologies were developed that permitted the detection and extraction of previously unrecoverable sources, driving the cost of oil down for many years so that it has become an inexpensive fuel for much of the world when compared to total wages. Prices escalated sharply in 2006 due strong petroleum demand, limited refining capacity and surplus oil production, and concerns about political tensions with oil-producing countries such as Iran.[18] Although the price has dropped since that time, it is on an upward trend again. Unfortunately, high oil prices place significant hardships on billions of people around the globe.

A few developed nations appear on the verge of entering yet another energy transition as the use of petroleum is reduced in favor of less polluting alternatives. Humankind continues to respond with the development of new, more efficient, energy sources. As technologies expand, they create the possibility of feeding, clothing, and housing even greater numbers of people. Authors such as Julian Simon argue that human adaptation and change will avert disaster as more technologies are developed in response to growing population pressures.[19,20] Although such authors argue that technology will support ever-increasing populations, there is a practical limit to the maximum number of people this planet can sustain with a quality of life.

The tragedy of the human condition is already evident in many parts of the world. Tens of thousands die each day from malnutrition and diseases in countries where resources have been exhausted and the land laid to waste. This is in contrast with countries that have successfully brought population growth under control, have enormous wealth, and a very high standard of living, such as the United States, Western Europe, Japan, Australia, New Zealand, and other developed countries. Poor developing countries are sometimes saddled with blame for the world crises because of high growth rates and related environmental damage, though perhaps the blame best sits with the leadership of those countries, who often do little to improve the living conditions of their citizens. Paul Ehrlich, professor of biological science at Stanford University, thinks the impact of humanity on the earth's life supporting systems is not merely decided by the numbers of people, but also by how people behave. The impact of human activity on environments can be summarized by the following relationship:[21]

$$I = P*A*T$$

where:
  I is the impact of human energy-related activity on the globe
  P is the population size
  A is the affluence in terms of per capita consumption
  T represents the technologies to supply each unit of consumption

According to Ehrlich, it is energy-intensive activities that cause most of the environmental degradation. The average energy consumption per person in the developed "rich" countries is 7.4 kilowatt while the average person in the LDCs uses less than 1 kilowatt. Let's estimate that about 2 billion people live in the world's "rich" nations. At 7.4 kilowatt, their total environmental impact as measured by energy use would be 14.8 trillion watts. More than 6 billion people live in poor nations, so hence their total impact was about 6 trillion watts. Measured by energy use impact, the 3X-smaller population of developed countries accounts for nearly three quarters of the global environmental destruction. About 5 percent of the world's population lives in the United States, where 35 percent of all raw materials are consumed.

The Role of Conflict

A United Nations report shows a widening economic gulf occurring among the world's poorest countries, driven by wars and political infighting.[22] Such unrest is thought

FIGURE 3.5
Agricultural lands in fragile, conflict-ridden countries, 1980–2016. (Adapted from World Bank, World Development Indicators, environment-agricultural products. http://wdi.worldbank.org/tables.)

responsible for reductions in per capita gross domestic product in the Congo, Burundi, Somalia, Sudan, and the Central African Republic, as examples. As can be seen in Figure 3.5, conflict has a negative effect on food production. The area of land planted to grow crops remains the same in these countries over a 10-year period, suggesting a flat food supply, especially considering that a population continues to grow. Conflict has this unfortunate but predictable negative impact on food security. Unfortunately, conflicts are increasing, especially between non-state forces and in the Near East and North Africa. A recent FAO report indicates that due to ongoing conflicts in Syria, 50 percent of the population requires food assistance; in Yemen, 50 percent of the population requires food assistance; and in Iraq, 9 percent of the population requires food assistance.[23]

Attitude and Behavior

The question posed by many is does human society have the capacity to adjust to the dizzying pace of population increases without major political upheavals, where war and environmental destruction result in human calamity. We have seen in this chapter that neo-Malthusian arguments predict the inevitable misery of humankind as the population momentum outstrips the carrying capacity of the earth.[22] We have also presented the optimistic "Cornucopian" outlook that more people will drive the technology to support the world's growing billions. Which of these views will ultimately be realized? Will we progress in a smooth transition to a world of global stability and health through carefully planned and thoughtful agricultural, environmental, and governmental processes, or will regional, national, and personal interests prevail at the expense of the larger global community, resulting in the destruction of our environment and depletion of world resources? What are the attitudes and behaviors that may impact this outcome?

The Tragedy of the Commons

If we are to believe the arguments proposed by Garrett Harden, the individual or nation that follows a healthy environmental protection ethic may be doomed unless everybody else follows the same path.[24] Where there is common land, air, or water, some people will take advantage of this situation and overgraze, overfish, or pollute that environment to their own advantage and at the expense of others. The deforestation and soil erosion in ancient Greece,[25] the overgrazing in the Sahel of northern Africa,[16] and the polluting of air in New England by plumes of sulfur dioxide emissions from Midwestern states are all examples of this behavior. As one person or one segment of society emits their toxic brew into the environment while avoiding costly controls, the profit margin is enhanced in the short term. Seeing the advantage, others are likely to follow this course, providing no adverse consequences are immediately applied. It is likely that many members of any society will pass on the consequences of their destructive actions if they will benefit in the short term and receive little or no negative consequences from that action. Harden refers to this as the tragedy of the commons.[24]

The Pioneer

In the past, the consequences of laying waste to a land were minimized by the ability of the population to move on or even to emigrate, just as the pioneers in America could move west. There were enormous resources that permitted the early pioneers to cut trees, harvest the land, and trap or shoot wildlife with ecological impunity. Unfortunately, this pioneer mentality cannot be continued indefinitely in the presence of massive population increases and limited resources. The opportunity of mass emigration has been largely eliminated throughout the world as most habitable areas are now populated, bound by country borders, or privately owned, restricting access. Yet societies continue to exploit the world's resources. The consequences of unrestricted use of resources beyond their replacement level will eventually result in ecological and human disaster. We must seek a sustainable development where resource use comes into balance with replacement and the discharge of pollutants is minimized.

Declining Investment in Technologies

The optimistic attitude of the "Cornucopians" towards population increases demands an investment in new technologies that will help to sustain these populations. Unfortunately, the major gains in food crops experienced as part of the Green Revolution are unlikely to continue in the absence of investment in research and development. Further, these technologies, along with financial support, must reach the farmers in the developing countries. Just as it would be impossible to feed today's population with 50-year-old technology, it will be impossible to feed the population of the year 2050 with today's technology. However, government funding for such organizations as the CGIAR centers, which are largely responsible for the Green Revolution, has been falling. Despite contributions from the World Bank to shore up the losses, more than 100 scientists left the organization in 1993–1994 alone.[6] Further, the United States aid for agricultural programs has sharply declined by more than 75 percent since 1980.[6]

Family Planning Cuts

The United States contributes significantly to assist with family planning (FP) and reproductive health (RH) in developing countries. The World Health Organization estimates that 214 million women in developing countries lack access to a modern method of contraception.[27] Every year, more than 300,000 women die from complications during birth, with most of these deaths in developing countries. In 2017, the U.S. donated $608 million to FP/RH, which was up from $425 million in 2006.[28] The U.S. contribution levels to global FP are intertwined with the country's ongoing debate over abortion. The

United States has historically withheld funds from the U.N. Population Fund (UNFPA), as it did during the Bush Administration because of disputes over abortion. The "global gag rule" originated with the Reagan presidency and continues by banning discussion of abortion in RH clinics, even when considered legal in many countries.[26] There continues to be severe shortfalls in planning assistance in LDCs resulting in up to (1) 220 million unintended pregnancies, (2) 117,000 additional maternal deaths and 1.5 million women who experience permanent impairment, and (3) 9.3 million additional deaths of infants and young children.[4] The loss of U.S. population funds has scaled back or eliminated RH services in some of the world's poorest countries where fertility rates are the highest, such as Kenya and Ethiopia.[26] The consequence of failing to meet the contraceptive needs of women in these countries is that there will likely be an additional 23 million unplanned births, 22 million induced abortions, more than 300,000 pregnancy-related deaths, and 1.4 million infant deaths.[26]

What Now?

We are now faced with two very different views on the effect of world population on the human condition and global ecosystems. The pessimistic view warns that population momentum will hurl humankind beyond the carrying capacity of the earth, creating enormous human misery and suffering and dramatic destruction of environmental support systems. Others view people as an asset that will finally prevail with the development of new technologies and policies that will avoid a cataclysm. As with most polarities in life, the reality is probably somewhere in between and is already being played out in regions around the world. Many of the LDCs are showing a downward trend in birth rates associated with policies that empower women through education and economic opportunity. Overall, world growth rates are declining and the public news media boldly broadcast that the population problem is over. Certainly, these are very substantial reasons to be optimistic about the future. However, just as there are positive examples, so are there dramatic failures that give reason to be apprehensive about the future.[29] Just as we can believe in technology as rescuing humankind, we can also point to declining investments in agricultural research, FP, and the preservation of ecosystems. Social and political will may not be capable of making the necessary adjustments, and ecological damage may already be sufficiently severe that it is irreversible in many parts of the world.[30] The ability of people to predict the consequences of increasing populations depends on too many variables to be accurate. There are no simple answers to these issues. We can, however, examine the current state of human affairs and global ecosystems to capture a glimpse of where we have been and where we are now. This may give students of environmental health enough information to develop their own vision of the future world.

Impacts on the Environment

Imagine a spider web, glistening with morning dew in the dawn light. Touch one strand of the web and all strands move and vibrate. Our environment is so intertwined that each human action ripples across the environment causing imperceptible changes throughout the system that become magnified by the billions of people. As the population increases, the need for food increases. Land is cleared or grassland is cut and uprooted. Topsoils are lost, nutrients depleted, and balanced ecologies of insects, birds, and mammals are upset as habitats are destroyed. Fertilizers and pesticides are applied that wash into streams and seep into groundwater, causing soil and water contamination that biomagnify up the food chain causing reproductive abnormalities and disease in wildlife and humans. As insects develop resistance and natural soil nutrients are depleted, more chemicals are applied to croplands. These chemicals require energy in their manufacture that is drawn from the combustion of fossil fuels. These combusted fuels release pollutants into the air that produce acidic deposits and climate-warming gases. Climate change and acidification of soil and water may cause further ecological damage contributing to a downward spiral from a sustainable environment. The circle of influence moves from more people, greater deforestation, poorer soil, more deserts, fewer species of animals and plants, and less food. This is the sequence of events we will explore in the following pages.

Deforestation

Although life is highly adaptable, most species require a source of energy such as sunlight (required by most plants), nutrients, and moisture. Organisms are also affected by the presence of oxygen and the availability of warmth. Life exists in a relatively narrow shell about the globe extending a few kilometers into the air and below the sea. The richest and most diverse forms of life tend to be in the tropical rainforests while harsher conditions at the top of the Himalayan Mountains or in the frozen, bleak landscape of the polar ice caps harbor very few life forms. Just as conditions of temperature, moisture, and nutrients vary in regions around the world, so do the types and diversity of vegetation and animal life change. Several major regions exist around the world identified by the major types of vegetation. These are called biomes and include tropical rainforests, temperate forests, prairies, deserts, and arctic tundra. Since our present topic is deforestation, we'll focus on tropical rainforests because the majority of them occur in areas of the world at risk from overpopulation and many are being threatened with slash-and-burn techniques to make room for croplands.

Rainforests

The tropical rainforests (Figure 3.6) are located predominantly in the LDCs in a latitudinal band forming a belt around the globe from 20¼° south to 30¼° north. Rainforests are located in many countries along this band including Mexico, Belize, Salvador, Guatemala, Nicaragua, Columbia, Ecuador, and Brazil, eastward to Cote d'Ivoire, Nigeria, Congo, and the Central African Republic, and further east to Malaysia, Indonesia, Philippines, New Guinea, and the Solomon Islands (Figure 3.7).

The benefits of the tropical rainforests include being (1) a major producer of oxygen for the global atmosphere, (2) the major carbon dioxide sink, (3) a potential source of new pharmaceuticals useful in the treatment of human disease, and (4) an important source of species diversity. Rainforests cover less than 10 percent of the earth's surface but contain more than a third of the earth's biomass, which is instrumental in absorbing large amounts of carbon dioxide and releasing great quantities of oxygen and so serving as the "lungs of the planet."[31] In 1995, a team of British, Australian, and Brazilian ecologists found proof that the Amazon rainforest absorbs tons of carbon dioxide produced by burning fuels. The globe's 404 million hectares of rainforests may absorb one-sixth of the amount of carbon dioxide produced by burning fuels each year.[32]

Some of the most important and historical drugs isolated from plants include those used for treating circulatory disorders and high blood pressure. Some of these come from the Indian Smoke Root, Rauwolfia serpentine. Caloniea, from the Malaysian tropical rainforest tree is presently being developed and tested for the treatment of HIV-infected persons

FIGURE 3.6
Typical lowland tropical rainforest. (From NASA, The Earth Observing System Educator's Visual Materials, 1998, http://eospso.esfc.nasa.gov/cos-edu.pack/p24.html.)

in the United States.[33] Plants from the Peruvian rainforests could help doctors in their fight against deadly tuberculosis. In a study of 1,250 plant extracts from Peru, about 46 percent showed an ability to inhibit the growth of *Mycobacterium tuberculosis*, the bacterium that causes tuberculosis. There is a myriad of other pharmaceuticals being used and investigated in the treatments of cancer, blood pressure, HIV, and other disorders. There are extensive efforts underway to investigate the possible pharmaceutical benefits from tropic rainforest plants before these plants disappear as a result of ecosystem destruction.

In spite of the numerous benefits from rainforests, they are disappearing at an alarming rate. This process is leading to deforestation, which has been defined as the permanent decline in crown cover of trees to a level that is less than 10 percent of the original cover. Prior to the human population explosion, the tropical rainforests are estimated to have covered 16 million kilometers or nearly 11 percent of the earth's land area. By 1987, tropical rainforests were disappearing at the rate of 17 million hectares each year, representing a loss of 46,500 hectares each day. Thankfully, these rates have slowed, due to increased conservation awareness. Now the rainforests are being lost at the rate of 7.5 million hectares of forest per year.[34] This number is still too large of a loss—equivalent to losing nearly 27 football fields of forest/minute. WRI reports that half of the world's original forests have been lost and only one-fifth of the remaining forests are "frontier forests"—pristine areas that have not been disturbed or degraded by human activities.[35] The Food and Agricultural Organization (FAO) estimates the rate of South American tropical deforestation at 54,000 square kilometers annually or about the size of the United Kingdom.[36]

Tropical deforestation contributes significantly to the net movement of carbon (carbon dioxide) to the atmosphere, as the concentration of carbon in forests is higher than in the non-forested areas that replace them. Additionally, the tropical rainforests occupy less than 7 percent of the land surface but contain more than half of all plant and animal species. The greatest problem associated with tropical deforestation may be related to the massive extinction of species. The diversity of plant and animal species is affected by (1) destruction of habitat, (2) the formation of fragments of forested area less than 100 square kilometers, and (3) edge effects occurring as far as 1 kilometer into the forest from adjacent deforested areas.[37]

FIGURE 3.7
World map showing regions of major tropical rainforests. (From NASA, The Earth Observing System Educator's Visual Materials, 1998, http://eospso.esfc.nasa.gov /cos-edu.pack/p24.html.)

This boundary zone has greater exposure to winds, easier access by livestock and humans, and many other effects that result in a loss of plant and animal species at the edge. Edge effects greatly magnify the adverse effects of deforested areas.

The Brazilian Amazon is the largest region of tropical forest in the world containing 31 percent of the world's total.[37] The National Aeronautics and Space Administration (NASA)'s Global Inventory Monitoring and Modeling Study showed that in 1987, fires burned 199,000 square kilometers of the Brazilian part of the Amazon rainforest, or an area greater than New York by one-and-a-half times.[38] During the months between July and October 1987, four percent of Brazil's Amazon jungle was destroyed, and most of this destruction followed the route of the World Bank-financed road through Rondonia known as BR-363. Brazil's poor and destitute have flocked to this area in search of arable land and they clear the land by setting fire to it. NASA's heat-sensitive images from satellites show nearly 350,000 fires were started in 1987. NASA can now track fires globally via satellite on a daily basis and those images are available online.

Despite worldwide attention, the deforestation of the Amazon continues. Although the rate of deforestation slowed for about a decade, between 2004 and 2014, the number of deforested hectares in the Brazilian Amazon increased from 2015 to 2016, to nearly two 809,000 hectares of rainforest lost per year.[39] For the years 2014–2015, the figure was lower, 607,000 hectares lost/year, but this too was an increase from the 485,000 hectares lost the year before that. This is an unfortunate trend, mostly driven by small farmers looking to grow soy crops to sell to large corporations like Cargill. This destruction is ongoing in the nearby Bolivian rainforest as well, where annually over 323,000 hectares of rainforest are lost to deforestation, an increase from an average of 121,000 hectares lost annually in the 1990s.

Unfortunately, too, the soil is so thin and of such poor quality that the farmers growing crops have to clear new plots every year or two. With no leaf canopy to break the intense downpours, the thin topsoil is easily washed away. The birds and other natural predators to insects are eliminated through habitat destruction, and mosquitoes often abound creating greater opportunities for vector-borne disease transmission. Colonization and subsequent subsistence farming is among the leading reasons for clearing the rainforest, but clearing for cattle pasture, commercial agriculture, and logging also remain very high.[40]

The consequences of tropical deforestation are well documented and well understood globally. There are, in fact, 219 international groups concerned with saving the Amazon forests.[32] Unfortunately, reforestation efforts have met with dismal failure in the wake of the continued rapid growth of human populations. These fires are not limited to Brazil but are spanning over a number of countries including Uruguay, Bolivia, and Paraguay producing a plume of smoke covering over 6 million square kilometers that causes respiratory irritation and closes down airports on many days.[33] The NASA satellite imagery allows the tracking of forest fires across the globe, which is helpful to compare the changes year by year. In 2006 nearly 28 percent of the Brazilian Amazon was subjected to incipient human pressure associated with fire activity where fire zones are defined as the 10-kilometers radius around a forest fire. These were detected by satellite between 1996 and 2002. Human pressure may vary within fire zones. More than 60 percent of fire zones are concentrated near deforested or urban zones that have been subjected to intensive pressures such as logging. The remaining 30–40 percent is found in more remote locations, suggesting that the fires are for shifting agriculture, set by traditional *mestizo* communities or indigenous populations. Forests in these areas may be subjected to less intensive pressures such as hunting, harvesting of non-timber forest products, and selective logging. The number of annual forest fires has nearly tripled from 16,000 per year in 2000 to almost 42,000 per year showing an acceleration of incipient human activity.[41]

Central America has been experiencing tremendous losses of its forests to industrial fell-ing and the practice of slash-and-burn agriculture by the rural, landless poor. The felling of native forests has been leading to serious soil erosion in these parts. Panama, El Salvador, and Guatemala have been found to have the worst soil erosion in the world according to a study conducted by the Center of Studies and Social Action of Panama. Nearly 98 per-cent of El Salvador's natural forest cover has vanished with Panama and Guatemala close behind.[42] Dramatic price increases in oil by OPEC in the early 1970s placed kerosene and other petroleum fuels out of reach for millions of the worlds' poorest and left them no alternative but wood. Poor families in El Salvador and Guatemala are spending an entire day every week looking for firewood and traveling up to 12 kilometers from their homes to find it.[42] Increases in world oil prices may likely worsen this situation.

The tropical forest of Africa is 18 percent of the world total and covers 3.6 million square kilometers of land in West, East, and Central Africa. Annual rates of deforestation in the region vary from 15,000 hectares in Gabon to nearly 300,000 hectares in Cote d'Ivoire. West African forests have been reduced to patches. African forests are the most depleted of all the tropical regions with only 30 percent of historical stands remaining.[43] Just as in part of South America, many Africans must travel kilometers away from home to gather firewood and/or expend 25–30 percent of their income for wood to burn.

A study published by the WorldWatch Institute, a nonprofit research organization dedi-cated to the analysis of global issues, concludes that nearly three quarters of the middle mountain range in Nepal has been deforested in the last 40 years. India has lost 40 percent of its forest cover within the last 30 years, and Pakistan has only a fraction of its original forested land. The deforestation contributes to extensive flooding costing Pakistan about a $1 billion each year in crop, building, and infrastructure losses.[44]

The rate of deforestation in the Pacific and Asia could decimate some 182 million hect-ares. The greatest rates of deforestation are occurring in Indonesia, followed by Thailand, Malaysia, India, and Laos.[44] Hundreds of thousands of hectares are also being lost in Nepal, Cambodia, and Vietnam. Some of this is due to illegal logging activities.

The rates of deforestation are slowing in some parts of the world and increasing in others (Table 3.1). There is significant room for improvement. Although 70 tropical countries have agreed to support the International Tropical Forests Action Program (TFAP) initiated in 1993, the full implementation of this program will require substantial investments of time and resources.[45] According to the Global Environmental Outlook Report released by the United

TABLE 3.1

Regional Forest Area and Deforestation Rates over Time

	Forest Area sq. km Thousands		% Change Forest Area
	1990	**2015**	**1990–2015**
World	41,283	39,991	−0.03
East Asia & Pacific	6,280	6,413	0.02
Europe & Central Asia	10,200	10,434	0.02
Latin America & Caribbean	10,242	9,272	−0.09
Middle East & North Africa	199	232	0.17
North America	6,507	6,572	0.01
South Asia	789	834	0.06
Sub-Saharan Africa	6,990	6,144	−0.12

Source: World Development Indicators: Deforestation and Biodiversity, The World Bank, Washington, DC, 2017.

Nations Environment Program (UNEP) at Nairobi, Kenya, "International and national funds and political will remain insufficient to halt further global environmental degradation."[30]

Forests in Developed Countries

The area under forest cover in developed countries has remained fairly unchanged during the last decade.[30] The forested areas in these countries are stable and increasing in the temperate regions where alternative fuels and forest management programs have sustained timber harvests. The major issues surrounding temperate forests include (1) damage by air pollution, (2) the loss of biological variety, and (3) fragmentation with the loss of forest edge. Many Western European countries have actually experienced an increase in forest cover. Since the mid-1960s, Sweden, France, and the United Kingdom have experienced increases of forested area of 22, 34, and 37 percent, respectively. The United States has seen a significant return of forests compared to the eighteenth and nineteenth centuries when farmers cut down all but 2 percent of the trees in the northeast.[46] However, in several important ways the forests are fundamentally different from those of previous centuries. At ninety years old, the forest ecosystems are not yet mature. Ironically, the oldest trees are preserved in cities and suburbs. Exotic species from other continents have taken hold. Many native species are threatened by diseases and pests. Expanding suburbs have partitioned or fragmented the forest into patches creating lots of forest "edges." Maine, Vermont, and New Hampshire as well as northern New York contain 10.5 million hectares of forest known as the Northern Forest. This precious resource is in serious danger according to the Northern Forest Alliance, a coalition of about 25 local and national environmental groups. The danger stems from the process of clear-cutting by lumber companies, recreational pressures from millions of campers,[46] and the constant threat of acid deposition from air pollution.

Nevertheless, developed countries also bear inescapable responsibility for the rapid loss of forest sources. North America, Europe, and Japan, with just 16 percent of the global population, consume two-thirds of the world's paper and paperboard and half of its industrial wood. The World Bank has invested in sustainable forest development and offered incentives to companies to harvest a variety of tree species and plant replacement trees. The Brazilian government has begun talks with representatives of the major industrialized nations on the progress of an ambitious program to save the Amazon rainforest. The project, which was set up in 1991 and is funded mainly by the G-7 countries, aims to reduce deforestation, protect the Amazon's unique biodiversity, and encourage sustainable development. Brazil has lifted a ban on new logging permits in the Amazon rainforest after landowners and loggers agreed to slow their rates of forest destruction. The total area of selectively logged forests in the Amazon is unknown, although estimates indicate this activity may affect 10,000–20,000 square kilometers of forest per year in the Brazilian Amazon. Some of these forests are converted to agricultural and pasture land soon after timber is harvested, while other areas remain as logged forest. A large part of the timber harvested in the Brazilian Amazon, or approximately 47 percent in 2001 and 43 percent in 2004, is thought to be illegal. Illegal logging often happens when the authorities willingly look the other way. Despite signing 2017 deals worth $123 million from the UK and Germany for anti-deforestation projects, both Brazil's federal government and its states, including Mato Grosso and Acre, continue to allow extensive logging, legal and illegal, and other activities that lead to deforestation.[47] Additional problems arise when licensed loggers fail to follow forest management plans or they harvest illegally in public unclaimed lands, causing significant damage to the forest often associated with wildfires.[41]

Soil Degradation

What Is Soil?

Don't make the mistake of telling a farmer that his "dirt" must be really good to grow such excellent crops. Normally, the farmer will acidly reply that this is "soil," not dirt. What's the difference? The difference is that you can grow things in soil. Soil consists of small particles of rock and minerals mixed with a major proportion of plant and animal matter in various stages of decay. This organic matter is filled with living organisms including bacteria, algae, fungi, protozoa, insects, mites, and worms. A handful of soil is a complex, living ecosystem that provides the nutrients and conditions for plant growth. As plants grow and ultimately die, they serve as the nutrient source for a living mass of microorganisms that effectively recycle the organic materials into fertile soils. This organic material supports the growth of plants, which may in turn become the nutrients for higher forms of animal life.[7] Plants are called autotrophic because they synthesize their own food from inorganic substances. Plants derive carbon, oxygen, and hydrogen from carbon dioxide and water. Plants also require nutrients they derive from the soil including nitrogen, phosphorous, calcium, magnesium, and potassium. These are called macronutrients because plants use significant amounts of them. There are a number of minerals that plants require in only trace amounts. These are known as micronutrients and include such substances such as zinc and molybdenum. Soils vary in their physical nature. Some soils are sandy and so contain larger "gritty" particles that allow water to flow through rapidly and carry away or leach nutrients. Clay soils consist of extremely small or "greasy" particles that are bound together very tightly by chemical attraction. These soils tend to be impermeable to water allowing rain to puddle on the saturated surface. Silt tends to be composed of fine particles that are larger than those producing clay but smaller than sand particles. Soils best suited to agriculture consist of sand, silt, and some clay in a homogeneous mixture referred to as loam. Fertile soils contain complex organic matter that has been biologically broken down so that original plant and animal matter is unrecognizable. This organic matter is humus and it serves to (1) retain moisture much as a sponge, (2) serve as an insulator to heat and cold, and (3) bind and release nutrients to plants in useable forms. Soils that lack humus have low or poor fertility.[7]

Soil Biomes

Not all soils are equally fertile, and soils vary substantially across the globe in their ability to sustain intensive agriculture. The most fertile soils for the growth of food crops such as corn, wheat, and rice are located predominantly in the global temperate zones of northern America, Europe, and Asia. A smaller belt of the temperate zone climate may be found in Argentina, southern Africa, Australia, and New Zealand (Figure 3.8). Such temperate zones of forest, woods, and grassland tend to have deep fertile soils of 1–2 meters because the moisture and temperature of these regions permits the growth of substantial plant cover while permitting organic matter to accumulate more rapidly as it decomposes. In higher latitudes the growing seasons become very short and substantially less organic matter is produced. The boreal forest or taiga is an example of this biome and is represented by a great circle of spruce and pine trees stretching from Alaska through Canada and Russia. Such organic matter is decomposed very slowly, resulting in thin soils. In even higher latitudes is the tundra. This is an area where there are no trees, but great expanses of mosses, lichens, sedges, and occasional low-lying bushes. The soils here are very thin because the cold weather restricts plant growth and also inhibits plant matter decomposition. Further, the thin soil lies atop a permanent layer of ice or frozen soil known as permafrost. Any disruption to the

FIGURE 3.8
World map showing regions of major soil biomes including temperate, taiga, tundra, deserts, and tropics.

sensitive tundra takes many years to recover. Southern or "horse" latitudes at 20°–30° north and south latitudes will exhibit some of the earth's driest and largest deserts such as the Sahara because the regions receive a constant supply of warm, high pressure.

Soil Erosion

Consider that the soils that sustain life around the globe are really extraordinarily thin and vary from 1 to 2 meters deep to just a few centimeters. This critical layer of soil is in constant flux as new organic matter decomposes and is added to the soil, and some of it is carried or swept away by a process called erosion. In most natural conditions, the replacement of soil is equal to or greater than erosion. This allows soil depth to be maintained or to increase. However, the practice of poorly managed agriculture has caused erosion to proceed more rapidly, resulting in impoverished soils. As woods are cut down and fields are plowed to plant crops, soils are lost to the effects of wind and runoff water (Figure 3.9). Techniques to prevent soil erosion have been practiced in many parts of the world for centuries. Traditionally, such practices have included the planting of nitrogen-using (corn) and nitrogen-fixing (beans) crops in alternative years in the same space or planting them in alternating bands in the same year. This is called rotation. When fields remain unplanted every few years, this resting period permits moisture and nutrients to return in a process called fallowing. When farmers plant on hillsides, they must create level fields along the contour of the slope in a process called terracing. Otherwise erosion will occur at an enormous rate and wash away the topsoil in a generation or two.[7] However, these techniques are not widely used, and soil erosion looms as one of the greatest potentially devastating problems facing the world today even as most countries fail to recognize it as such.[41]

According to the United States Department of Agriculture (USDA), soil surveys show that more than 30 percent of natural top soils in North America have been lost since farming began in the United States.[48] The average rate of topsoil erosion in the United States in the early 1980s was 10.5 megagrams per hectare with many of the agriculturally intensive states showing levels of 19.5 megagrams per hectare. At this higher rate, and assuming an average topsoil of 30.5 centimeters, there would be a complete loss of topsoil in less than 150 years.[48]

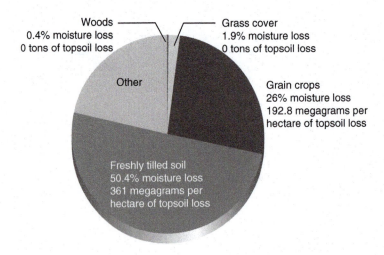

FIGURE 3.9
Loss of soil and moisture in a season per hectare under various ground cover conditions. (Adapted from Turk, J., and Turk, A., *Environmental Science*, 4th ed., Saunders College Publishing, Philadelphia, PA, 1988.)

Nearly half of the cropland in the United States is losing soil to erosion with about 1.2 million hectares being destroyed every year. In areas of moderate rainfall and temperate climate, it takes 30 years to form 2.5 centimeters of topsoil, which can be lost in 10 years in the presence of excessive grazing, mono-cropping, and ground cover destruction.[49] The devastation of croplands in the Midwest in the 1930s by wind erosion led to the term "dustbowl" and sub-sequently the creation of the Soil Conservation Service, which was renamed in the 1990s to the Natural Resources Conservation Service (NRCS). The NRCS has inventoried soil erosion in the major croplands of the United States and shown that through crop rotation, improved tilling methods, and the Conservation Reserve Program (CRP) there has been either an improvement or no increase in average erosion rates since the 1980s.[49] The CRP was started under the Food Security Act of 1985 and offers farmers financial incentives to remove erod-ing land from production for periods of 10 years so they may recover. Despite these improve-ments, excess erosion still occurs in southern Iowa, northern Missouri, parts of western and southern Texas, and eastern Tennessee. A large section of the Great Plains stretching from Nebraska to northeastern Texas is unstable and a prolonged drought could cause the entire area to become a Sahara-like desert.[49]

Even as soil is pushed to its capacity for growing crops, global demand for food is increasing. Driven by logarithmically expanding populations, the developing nations of the world are forced to continuously cultivate their best farmland while moving agricul-ture into areas with decreasing agricultural sustainability. Many of these countries look to the United States for food, which increases both demand and commodity prices. Farmlands in the United States have little time to lie fallow or be rotated as the land is pressed into service to grow food for the world's billions. Food exports continue to rise. North America now exports more than 100 million tons of grain per year, which is a fourfold increase over exports in the 1950s, and three times the combined exports of Australia, New Zealand, and Western Europe.[7] A significant reduction in the United States food production would have a devastating effect on the world food supply.

When viewed globally, soil erosion claims over 404 million hectares every year, rendering them incapable of crop growth. Further, 485 million hectares of global cropland are losing topsoil so rapidly that these hectares are expected to become unproductive in the next few decades.[50] India is averaging topsoil losses at the rate of 20.2 meggrams per hectare and is losing about 11.2 billion megagrams of topsoil a year at a rate double that of the United States. Food security is a major issue in Africa, West Asia, and Latin America. The major concern in these areas is the availability of land and the prevention of further degradation. More than 404 million hectares of land in Africa is moderately or severely degraded. Nearly 47 percent of the grazing lands in Latin America have lost their fertility because of erosion, overgrazing, and salinization.[30] As LDCs move toward adding another 5 billion people in the next decades, continued soil erosion in these countries could result in harvest reductions of 30 percent or more during the same period.

The Process of Desertification

Fences pieced together by plywood, corrugated cardboard, and rusted scrap metal are the futile efforts of the residents in Komsomolky to slow the inevitable advancing desert from finally consuming their town. The Kalmyks who inhabited the region for nearly 400 years understood the fragility of the Black Lands in Southern Russia, which were once the bed of a larger Caspian Sea. They never tilled the soil. They raised sheep with flat-bottomed feet and did not let cattle graze in the area. These careful custodians of the grassy steppes of central Kalmykia were expelled by Josef Stalin in 1943 and in their place were brought irrigation projects to develop cropland and sharp-hooved sheep. Because the topsoil lay over the former extensive bed of the Caspian Sea, the irrigated land became a salt marsh and the sheep destroyed the thin topsoil causing the desert to begin advancing in the 1950s.[51] The encroaching sands buried everything in sight and forced 25 towns to be abandoned while threatening 40 others. State-financed plans to reverse the trend have been halted as money ran out in the mid-1990s.[51] The agricultural mismanagement of this land has promoted the desertification of this delicate ecosystem and it will take generations for the land to be restored even if the practices of mismanagement are halted.

What Is Desertification?

The previously described scenario is being repeated in a number of areas throughout the world, but it wasn't until the late 1960s that desertification became a matter of international concern. Worldwide, with more than 4 million hectares of farm land becoming unproductive each year, "dust bowls" are increasing in number, raising legitimate concerns about our planet's capacity to feed a rapidly growing population. During a period of severe drought in the Sahelian region of Africa from 1968 to 1973, the people suffered enormous tragedy through starvation. This drew worldwide attention to the problems of trying to survive on arid and semi-arid land, especially on the perimeters of deserts. As a consequence, there have been a number of initiatives including a United Nations Conference on Environment and Development that was held in Rio de Janiero in 1992 (Rio Earth Summit). Here, more than 100 countries began the process of developing a legally binding international contract to combat desertification. This resulted in the United Nations Convention to Combat Desertification (known as the Convention). According to the Convention, desertification was defined as "land degradation in arid, semi-arid and dry sub-humid areas resulting from various factors, including climactic variations and human activities."[52] The definition was adopted by the UNEP and areas susceptible to desertification were defined as those subjected to arid climates (arid, semi-arid, and dry sub-humid). In 2007, the UN announced that 2010–2020 would be the decade devoted to combat global desertification via the United Nations Decade for Deserts and the Fight Against Desertification. Activities promoted include raising awareness in countries at risk and fostering activities to promote sustainable lands, such as planting trees and seedlings. June 17 is an annual desertification day, a day to promote net degradation neutrality.[53]

  About 6 billion hectares or one third of the earth is dry land, and 1 billion hectares (or 16 percent of the earth's surface) of these dryland hectares are hyper-arid deserts where there is little or no growth. This leaves 5 billion hectares or one third of the earth's surface that falls into the categories of arid, semi-arid, and dry sub-humid climates (Table 3.2). These areas include 51,720,000 square kilometers in nearly 100 countries that may be directly or indirectly affected by desertification. Some 900 million people live in areas susceptible to drought and desertification, and 200 million people, or about 22 percent, are presently affected by this process.[43] Even more disturbing from a socioeconomic view is that the majority of areas at high risk of desertification are also home to some of the world's most impoverished people in LDCs, especially in Africa and Asia (Figure 3.10). Poverty is being globally recognized as among the major reasons associated with the degradation of arid, semi-arid, and dry sub-humid regions.[52] Expanding populations force people to cultivate and overuse marginal lands that become increasingly less resilient to human and climactic disturbances. Growth conditions fluctuate from season to season and even within seasons. Dryland areas may receive an entire season's precipitation in

TABLE 3.2

Types of Land Areas at Risk of Desertification by World Region in Square Kilometers × 1,000

Climate Type	Africa	Asia	Australia	Europe	North America	South America	Total
Hyperarid	6,720	2,777	0	0	30	260	9,780
Arid	5,040	6,260	3,030	110	820	450	15,710
Semiarid	5,140	6,930	3,090	1,050	4,190	2,650	23,050
Dry Subhumid	2,690	3,530	510	1,840	2,320	2,070	61,500
TOTAL AREA	30,335	43,508	8,923	10,498	25,349	17,611	136,224

Source: Brazil Embassy, What is desertification? Science and Technology Section, National Research Network's Brazilian Regional Center, February 30, 1997, http://brasil.emb.nw.dc.us/nuna/deserts/ingles /se/~edesert/desertmi.html#dm; Moeller, D.W., *Environmental Health*, Harvard University Press, Cambridge, MA, 1997, Chapter 19.

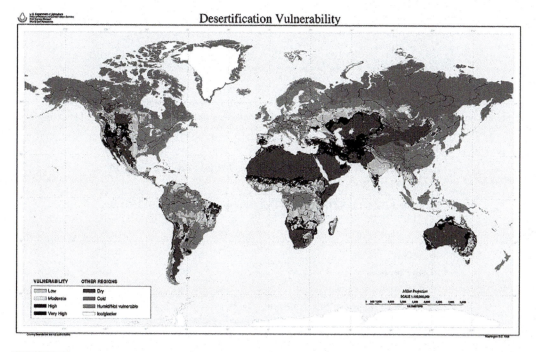

FIGURE 3.10
Global desertification map. (From National Resource Conservation Service, https://www.nrcs.usda.gov/wps /portal/nrcs/detail/soils/use/worldsoils/?cid=nrcs142p2_054003.)

a few days, or experience prolonged droughts, even if the average rainfall over a decade is normal. The people who inhabit these regions must adjust to the conditions, since poverty and the need for food are enormous pressures that defy a flexible land use response[54] (Figure 3.11).

The Costs of Desertification

The process of desertification is ongoing, and about 60,000 square kilometers are being removed from production annually because of poor land management including overgrazing, improper irrigation, and intensive use. Economic losses are calculated to be $40 billion

FIGURE 3.11
Land degradation in arid, semiarid, and dry, sub-humid areas resulting from various factors, including climatic variations and human activities, defines the process of desertification. This was once completely forested land in Africa.

TABLE 3.3

Estimated On-Site or Immediate Losses from Land Degradation
with Loss in Productivity

Estimated Direct Loss from Land Degradation (in Billion US Dollars)	
Africa	9.3
Asia	21.0
Australia	3.1
North America	4.8
South America	2.7

Source: Brazil Embassy, What is desertification? Science and Technology Section,
National Research Network's Brazilian Regional Center, February 30,
1997, http://brasil.emb.nw.dc.us/nuna/deserts/ingles/se/~edesert
/desertmi.html#dm.

while the cost of recovering these lands worldwide is estimated at $10 billion annually.[52] Some $40 billion in losses from land degradation are considered to be on-site or immediate losses from reduction in productivity and are distributed according to Table 3.3. Costs may be many times higher if you were to include off-site or indirect costs involving the loss of wells from silt formation, illiteracy among the young as they are forced to scavenge instead of attend school, and the costs of environmental refugees and emergency aid.[52] There are significant and increasing costs in human misery, since the world's largest population increases are occurring in regions at greatest risk of desertification. For instance, the cost from the plight of environmental refugees is difficult to measure. There is no 100 percent accurate way to estimate the numbers of environmental refugees since they are not officially protected through the United Nations.[52] The flight of refugees from degraded land has caused extensive disruption in developing countries, resulting in economic, social, and political difficulties. These problems may worsen as populations increase in LDCs causing greater poverty, increased desertification, and reduced food production.

Another loss difficult to measure is the reduction in biodiversity of important food crops. Many of our important food crops such as wheat, barley, and millet originate in drylands. The species that have adapted to drylands are few in number and their loss to non-sustainable agriculture methods is a great threat to biodiversity. This is further exaggerated by the Green Revolution, which has developed high-yield crops but at the cost of fewer varieties. Overall, we are seeing a reduction in the genetic pool of important food crops. Consequently, the lack of genetic variability in the important food crops increases food security risks from changes in climate, emerging or re-emerging pests, changing soil conditions, or other unpredictable events.

Wetlands

What Are They?

Wetlands are found everywhere across the globe, from tropical forests to the furthest reaches of the tundra (Figure 3.12). Wetlands appear virtually in every biome except the frozen expanses of Antarctica. They can exist in both high and lowlands. Wetlands are those areas of land where water saturation is the major factor influencing the nature of soil development and the communities of plants and animals that live in the soil and on the surface.[55]

The National Research Council defines a wetland as "an ecosystem that depends on constant or recurrent shallow inundation or saturation at or near the surface of the substrate."[56] There is a somewhat different definition of wetlands used for regulatory purposes and written under the Clean Water Act (40 CFR 230,3[t]) as "those areas that are inundated or saturated by surface or ground water at a frequency and duration sufficient to support, and that under normal circumstances do support, a prevalence of vegetation typically adapted for life in saturated soil conditions." Because land that is classified as wetlands falls within the guidelines of a protected environment under federal law, the ultimate decision in this matter is affected by USEPA's definition of a wetland. Some ambiguity in delineating wetlands still exists. For example, the number of consecutive days that are required for land area to have surface water can vary from 14–21 days depending on the scientific or legislative source. Other factors such as growing season, temperature changes, or the existence of records of environmental activity for the specified land site all must be taken into account. The key wetland features that are in question include the hydrology, anaerobic substrate, and biota. Policy makers and scientists dispute the number of wetland features that must be proven present to classify an area as a wetland.[57,58]

Wetlands are zones that fall in between areas of land and water and that both separate and join the land and open water. Wetlands are covered or soaked by water for long periods during the year. They provide habitat for numerous species of plants and animals including some that have specifically adapted to wetlands and appear nowhere else such as cattails, mangrove trees, or wood ducks.[57] Wetlands appear in a variety of forms including swamps, bogs, prairie potholes, bottomland, hardwood forests, and estuaries. Swamps are often transitional environments between large open water and drier land. They are home to species from both extremes and are marked by flood tolerant trees and shrubs.[55] Bogs are usually a stagnant, acidic, nutrient-poor ecosystem filled with peat and sphagnum moss. Its abundance of decaying material consumes enormous amounts of carbon dioxide. Prairie potholes are freshwater marshes found primarily in the upper Midwestern United States. This glacially formed landscape, pockmarked with potholes

FIGURE 3.12
Global Distribution of Wetlands. (From National Resource Conservation Service. https://www.nrcs.usda.gov/wps/portal/nrcs/detail/soils/use/worldsoils/?cid=n rcs142p2_054021.)

FIGURE 3.13
Various types of wetlands. (From USEPA, Office of Wetlands, Oceans, and Watersheds, *Facts about Wetlands*, August 25, 1997, http://www.epa.gov/owow/wetlands/facts/types.html.)

that fill with snowmelt and spring rains, forms one of the most important wetlands in the world, supporting more than 50 percent of North American migratory waterfowl and wide varieties of invertebrates and plants (Figure 3.13).[56] River swamps or bottomland hardwood forests are found along streams and rivers in the Southeast and South-Central flood plains of the United States. These are deciduous forested wetlands consisting of gum and oak trees adapted to areas of seasonal flooding or covered with water most of the year. The trees in these wetland systems often have fluted or flaring trunks and knees or aerial roots as seen in the Bald Cypress.[55] Estuaries are inland marine waterways found world-wide. They are dominated by grass or grass-like plants and lie mostly between barrier islands and beaches. Nutrient-rich tides make estuaries among the most naturally fertile habitats in the world.[56] The USEPA lists a number of wetland categories including bogs, bottomland hardwoods, fens, mangrove swamps, marshes (tidal, saltwater, freshwater), playa lakes, prairie potholes, swamps, vernal ponds, and wet meadows.[55]

Benefits of Wetlands

Because they are often inaccessible muddy landscapes, associated with insects and snakes, wetlands have not historically received much positive press. However, wetland areas are very beneficial. For example:[55–57]

- Wetlands purify and replenish water supplies. They serve as natural recharge areas to groundwater. They act as natural water filters because the rich biomass of plants and microscopic animals strain pollutants from water flowing towards streams or soaking into the ground.
- Wetlands are extremely rich in biomass (the amount of plant and animal life). They are equivalent in biomass hectare for hectare to a tropical rainforest. Wetlands in the United States are habitat for 5,000 species of plants, 190 kinds of amphibians, and one third of all bird species.
- Wetlands are an important source of food. Many species of fish and shellfish such as shrimp, crabs, and crayfish spawn in salt marsh wetlands.

- Wetlands absorb large amounts of carbon dioxide from the air taken in by plants or stored in peat moss.
- Wetlands control flooding in low-lying areas, since they work like a sponge, soaking up floodwaters from overflowing rivers. Vegetation slows the water flow thereby sparing farms and communities downstream.
- Wetlands protect coastal areas from storms. Fierce coastal storms can create huge waves that wetlands are able to absorb and so buffer communities further inland.
- Wetlands provide recreation and even beauty. Hunters and sport fishers find enjoyment and often dinner in wetlands.

Wetland Losses

Despite their value, wetlands are being threatened in many parts of the world. As an example, despite concern for the destruction of the Amazon rainforest, another part of the Amazon is also suffering from rapid population and development. Brazil's Panatanal, the largest wetlands in the world at 138,000 square kilometers, is also being threatened by human development, jeopardizing many of the 2,000 rare species of animals and birds that live there.[62] It was recently named a World Heritage site. In the Netherlands, the Dutch Wadden Sea is a valuable and extensive wetland environment of salt marshes, wet meadows, sand banks and dunes.[63] It covers 2,700 square kilometers, playing a significant role in flood prevention in this mostly flat country. As a major tourist site, it also faces pressure from tourism while balancing the need for awareness and conservation. Near Sri Lanka's capital city Colombo, an environmentally important wetland called the Muthurajawela Wetland is threatened by urban, agricultural, and industry developments, despite its useful role in flood prevention and water filtration.[63] Thankfully, countries are increasingly aware of the value of protecting unique biomes. Table 3.4 details the changes in protected wetland regions and terrestrial areas across the globe.

The United States has lost more than half of its wetlands since the early days of colonization. Wetlands were considered a nuisance that made farming difficult and as settlements moved westward, these areas were filled in. The great farm areas of Ohio, Iowa, and California have seen 90 percent of the wetlands disappear. There were no federal laws to protect wetlands until 1972 when Congress enacted amendments to the Clean Water Act,

TABLE 3.4

Trends in Protected Areas by Geography

	Terrestrial Protected Areas % of Total Land Area		Marine Protected Areas % of Territorial Waters	
	1990	2014	1990	2014
World	8.2	14.8	5.1	12.2
East Asia & Pacific	10	15.6	8.8	17
Europe & Central Asia	5.7	12.6	3.5	13
Latin America & Caribbean	8.8	23.3	3.7	15.5
Middle East & North Africa	3.8	11.7	0.5	3.8
North America	9.7	11.6	11.5	16.4
South Asia	5.3	6.6	1.2	2.3
Sub-Saharan Africa	10.3	15.3	2.2	6.1

Source: World Development Indicators: Deforestation and Biodiversity, The World Bank, Washington, DC, 2017.

which prohibited discharge of fill material into "waters" of the United States without a permit from the U.S. Army Corps of Engineers. A federal district court ruled that this included wetlands and unleashed a bitter controversy over issues of property rights and the role of government.[57] Unfortunately, the Army Corps of Engineers has been granting quick building permits in environmentally sensitive areas under the Clean Water Act. Permit 26, one of the 30 different quick permits under this Act, allows alteration of up to 4 hectares of wetlands if they are isolated from other bodies of water. The regulations allow for small lots to be developed without forcing extensive documentation about the environmental impact of the project. Acquiring such permits may take as little time as 1 month.[59] Florida has lost an estimated 800 hectares of wetlands in a 4-year period from this process. Nationally, the estimate is tens of thousands of wetland hectares lost since the program started. From the mid-1970s to the mid-1980s, 1.05 million hectares of U.S. wetlands were destroyed. The most recent and greatest losses have been in Louisiana, Mississippi, Arkansas, North Dakota, South Dakota, Nebraska, and Florida. An estimated 120,000 hectares of wetlands are drained or filled every year. About 87 percent is lost to agriculture, with urban development taking the rest.[57] More than three quarters of U.S. wetlands are privately owned and so preservation attempts are difficult to mandate. One successful approach incorporated a provision in the 1985 Farm Bill by removing farm program benefits from farmers who drain wetlands for agriculture. Enacted under the same legislation was the CRP, which provides financial assistance to farmers who remove lands from cultivation and so promote the restoration of wetlands habitat. With wetlands in the United States rapidly vanishing, ecologists are seeking methods of restoring or replacing them. A symposium at the annual meeting of the American Association for the Advancement of Science in February 2000 offered evidence that human-made wetlands may be just as good at water filtration as untouched wetlands, but they may not offer the same rich resources for wildlife. These efforts have seen a return of thousands of hectares to wetlands along with significant increases in waterfowl populations.[60] A report released by the U.S. Fish and Wildlife Service (FWS) revealed that between 1998 and 2004, there was a net gain of 77,598 hectares of wetlands and other waters—a gain of 12,949 hectares annually. This is the first time ever that a study has reported a net gain in water resources. The problem, however, is that the increase is mostly in new pond hectarage (283,000 hectares, an 11 percent increase). A closer evaluation of the report reveals a net loss of more than 200,000 hectares of naturally occurring wetlands such as estuarine emergent wetlands (salt marshes), freshwater emergent, and freshwater shrub wetlands. Natural wetlands provide for waterfowl and shorebird habitat, floodwater reduction, shoreline stabilization, fish habitat, and other functions whereas special-purpose ponds and ornamental lakes offer little vegetation and provide limited wetland functions with the most use as a storm-water retention device.[61]

The Loss of Biodiversity and Extinction of Species

Background

The extinction of species is not a new phenomenon. There have been many catastrophes on Earth, but none so destructive and overwhelming as the "mother of mass extinctions" some 250 million years ago at the end the Permian period. This mass extinction was far more severe than those events that occurred 65 million years ago and caused the extinction of the dinosaurs and nearly half of all species.[64] The Permian extinction caused 90 percent of all species in the oceans to disappear and two thirds of reptiles and amphibian families to perish. Thirty percent of insect orders were lost. The events leading to this mass extinction are hypothesized to involve a huge and prolonged series of volcanic eruptions that released large volumes of carbon dioxide into the air and increased climactic instability.[64] The return of life may have taken 5 million years and resulted in dramatic changes in the forms of life that emerged. Very likely, most of the species (perhaps 99 percent) that have ever existed on this planet are now extinct. The diversity of species has ultimately recovered from each of these cataclysmic events as the rate of new species through evolution was greater than the extinction rate. Today, more than 1.8 million currently living species have been identified. There is general consensus, however, that the planet is now losing more species than are being created, and that the activities of humans are the reason for a rapidly growing species extinction and loss in biodiversity. Biodiversity refers to the range of animal and plant species and the genetic variability among those species. The greater the genetic variation, the more likely there will be survivor species in the event of major catastrophes.

Since the planet has experienced a series of cataclysms over several million years and has recovered each time with abundant and diverse life, what is the major concern about the loss of a relatively few species? In order to respond to this question, we need to remember that the forms of life that emerged after cataclysms were much different, and that humans are a relatively recent phenomenon on the history of the earth. A reduction of species diversity will often result in an unstable ecosystem that is in danger of collapse if key species in that ecosystem disappear. Since genetic variation is the basis for evolution, the diversity of genes and species "affects the ability of ecological communities to resist or recover from environmental change, including long-term climactic change."[65] A human-induced cataclysm resulting in major losses of species could lead to a retrogression in the global ecosystem incapable of supporting large populations of complex creatures such as humans.

Previous extinctions have been associated with major climactic changes arising from such forces as volcanoes, or geologic changes that forced continents to clash and seas to rise or fall. Records of fossils show that entire groups of organisms including fish, reptiles, birds, and mammals have replaced one another over long periods of time. Birds and mammals, better able to respond to changes, began to proliferate as the numbers and species of reptiles declined (Figure 3.14). These changes are measured in millions of years. Even the cataclysm of volcanic eruptions in the Permian period is thought to have occurred over several hundred thousand years. This is in sharp contrast to the sudden and precipitous global changes being caused by humans. The massive increase in human population with corresponding changes to and pollution of land, air, and water is resulting in such rapid ecosystem changes that many species are unable to adapt. The extinction of species is thought to be occurring more rapidly than at any other time in recorded history.

More recently, a loss in biodiversity has also become a public health concern. One of the examples of this is the potential loss of medicines derived from plants. It has been estimated that there is one useful pharmaceutical out of every 125 plant species. Unfortunately, 3,000 of the world's 300,000 catalogued plant species may become extinct in the next 25 years.[58] Disease epidemics may also be associated with a loss in biodiversity as ecosystems are disturbed and disease-producing organisms or conditions emerge. As an example, the deforestation of tropical rainforests eliminates the natural predators to mosquitoes, which may then proliferate along with many vector-borne diseases including malaria (see Chapter 6).

Loss in Biodiversity

A 1992 report by UNEP titled "The Global Biodiversity Assessment" was among the first comprehensive, peer-reviewed reports ever to examine the earth's biodiversity. The report concludes that (1) 4,000 plants and 5,400 animal species are threatened with extinction,

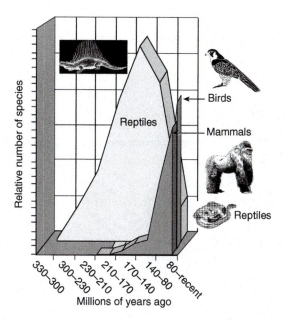

FIGURE 3.14

Comparative evolution of birds, mammals, and reptiles over geologic time. (Adapted from Turk, J., and Turk, A., *Environmental Science*, 4th ed., Saunders College Publishing, Philadelphia, PA, 1988.)

(2) species have recently become extinct at 50–100 times the average expected natural rate, and (3) 1.75 million or about 13 percent of the 13–14 million species on Earth have been scientifically identified.[66] Background extinction rates through geological time have been roughly estimated at the rate of one mammal and two bird species every 400 years. Documented extinctions for mammals and birds during the last 400 years are already at 50 times that rate, and this figure is probably underestimated. The highest levels of biodiversity are in the tropical forests where current rates of species extinction have been estimated at 1–11 percent per decade.[67] This report was one of the first efforts of a group of 168 United Nations countries that had formed the UN Convention on Biological Diversity in the earlier 1990s.

A study by the World Conservation Union in October 1996 showed that the world is entering a period of major species extinction rivaling five other periods across the past half billion years. Of the 4,327 known mammal species, 1,096 are at risk, and 169 are in extremely high risk of extinction in the wild in the immediate future.[68] Besides mammals, 11 percent of birds, 20 percent of reptiles, 25 percent of amphibians, and 34 percent of fish are threatened. Of the 26 orders of mammals, 24 are threatened, with the most at risk being certain species of elephants, primates, rhinoceros, and tapirs. The pattern of species extinction and threatened species follows closely the greatest population densities. The countries with the most threatened mammals are Indonesia (128 species), China (75 species), and India (75 species), where 43 percent of the world's population lives and which are the most densely populated countries (Figure 3.15). A 2006 United Nations report, the Global Biodiversity Outlook 2, claimed that we would need to make unprecedented efforts to slow down the enormous losses of species if the 1992 UN biodiversity goals were to be achieved by 2010.[69,70] This report revealed that the estimated pace of extinctions was 1,000 times faster than at any time in the past and that the major causes such as habitat change, over-exploitation, invasive alien species, climate change, and nutrient loading were not likely to change in the near future.[69]

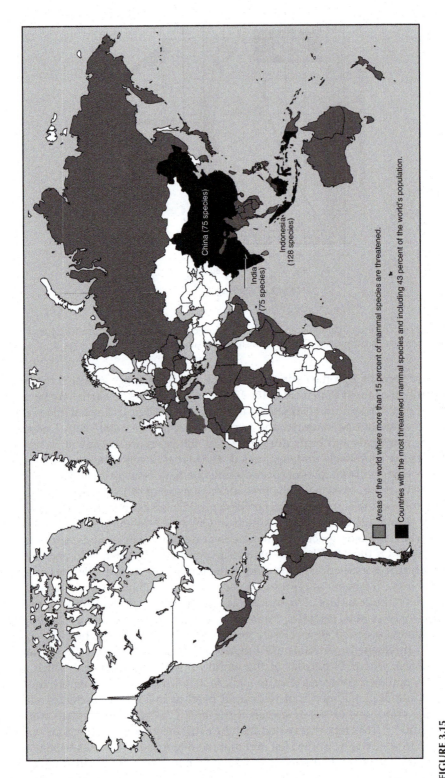

FIGURE 3.15
The majority of threatened mammal species occur in lesser-developed countries. (Adapted from Doyle, R., *Sci. Am.*, 276, 32, 1997.)

In 2010, the Parties to the Convention on Biological Diversity agreed to adopt and extend a further biodiversity plan, spanning the years 2011–2020. It was titled The Strategic Plan for Biodiversity, which included several goals and biodiversity targets.[71] By example, Target 12 states: "By 2020 the extinction of known threatened species has been prevented and their conservation status, particularly of those most in decline, has been improved and sustained."[71] The plan includes indicators to track annual progress against the targets. For Target 12, the indicator is the "Wildlife Picture Index," which uses "camera trap" data to measure over time the presence or absence of specific animals and the evenness of their distribution in the areas under study. There are five strategic goals and twenty targets across them. In this way, there is a quantitative assessment of global efforts to prevent further loss of rich biodiversity. Even still, the risks of continued losses are there. The United States is the country that has the potential for losing more plant variety than any other country. More than 4,500 of all existing plant species are threatened with extinction. Australia is next, with approximately 2,245 species threatened, followed by South Africa with 2,215.[5] Wisely, the UN Convention is already planning to extend its biodiversity work well beyond 2020.

Threats to Biodiversity

Some of the threats to biodiversity include (1) the destruction and fragmentation of habitat; (2) the harvesting of wild species for food, lumber, and other products; (3) the introduction of non-native animals and plants; (4) the pollution of the air, water, and soil with toxic chemicals and waste products; and (5) the increased ultraviolet radiation and global warming resulting from atmospheric changes and possible human activities.[54]

Loss of Habitat

The most significant threat to biodiversity today is the elimination of habitat for agricultural and habitat purposes. As expanding populations assimilate land by cutting down forests, filling in wetlands, and plowing fields, most species are unable to adjust to these rapid changes, and so they perish. The greatest biological diversity occurs in the rainforests where much of the habitat destruction is occurring. However, habitat destruction occurs on many levels and even threatens the mass extinction of freshwater mussels. More than half of the 300 species of mussels in the United States are in trouble to some degree, and most of this is brought on by human intervention such as the building of dams, gravel mining, sewage, and polluted runoff.[72] For a lesson in how quickly ecosystems fragment across international borders, consider Lake Victoria. Lake Victoria is the world's second largest freshwater lake, covering an area of 67,850 square kilometers. This vast expanse, about the size of the Republic of Ireland, forms the headwaters of the River Nile. Prior to 1970, Lake Victoria had more than 350 species of fish from the cichlid family. Ninety percent of these were unique to the lake. Fishing for tilapia and Nile perch provides a living for many of the Luo people who live along the lake. Unfortunately, the overfishing of Nile perch and tilapia caused a collapse in the lake's biodiversity and deforestation in the three countries that border the lake: Uganda, Tanzania, and Kenya. Wood is needed to dry the oily perch, but resulting deforestation is jeopardizing the lake and its fish. Forest clearing has increased siltation and eutrophication (the process by which a water body is overwhelmed by plant life and nutrients at the expense of oxygen) in the lake, jeopardizing the Nile perch and tilapia fishery. The cause of the problem with Lake Victoria illustrates how profound and unpredictable tradeoffs can be when management decisions do not take into account how the ecosystem will respond to stressors.[4]

Over-Harvesting

Many species have been harvested to the point of extinction. For example, many fish stocks such as cod in the North Sea off England are heavily exploited, with as much 60 percent of the fishable stock being removed annually. This occurs, despite warnings by the International Council for the Exploration of the Sea (ICES) and government action to reduce allowable catches or closing fisheries completely. The data suggest that management efforts have not been successful.[73] The Northern "right" whales are estimated to be fewer than 300 in number. Their decline is attributed to devastating over-harvesting by turn-of-century whalers who preferred to hunt these whales because of their slow-moving and docile nature. Hence, they were called the "right" whale to hunt. Deaths are attributed today to the slowness of the beasts in which collision with ships results in one third of the whale deaths per year.[74] The small number of remaining whales leads to a loss of genetic variability through inbreeding, which further jeopardizes the species.

In other harmful practices, people along the coast of the Philippines use dynamite and cyanide to stun fish and then pound the reefs with crowbars to scare them out into waiting nets. Some fish are eaten while others are sold as exotic aquarium fish. The natives see this as necessary for day-to-day survival. The process, however, damages the coastline coral reefs. Although efforts are under way through the Coastal Resources Management Project to empower local communities to act in their own self-interest, it is difficult to protect the environment from impoverished people simply trying to survive.[75]

Organized crime in countries such as the United States, Japan, Russia, and many other countries has discovered a new profitable market—wildlife. The species they trade include parrots, elephants, whales, bear, tigers, snakes, and rhinos. Body parts from many of these animals are used in some cultures to stimulate sexual prowess. The shift in interest for criminals is that they used to use Boa Constrictors to carry cocaine across borders. Now they sell the snakes for faster and easier money.[76]

Non-native Species

In the aftermath of World War II, no one paid much attention to the invasion of the tiny Pacific Island of Guam by a seemingly harmless but non-native tree snake. The opportunistic brown tree snake, which grows up to 2 meters in length, hitched rides from the island of Manus to Guam, slithering into the machinery used by the United States military. Without natural enemies on Guam, the population of brown snakes grew quickly within 30 years and eliminated eight of the Island's natural bird species and sharply reduced several others.[77] The freshwater mussels previously mentioned, already under attack by changing habitat and pollution, are also being displaced by Zebra mussels inadvertently transferred from the Black, Caspian, and Azov Seas. Many non-native species of fish including large-mouth bass and carp have been implicated in the extinction of as many as two thirds of the native species, and now threaten many others.[78] Rainbow trout had never encountered "whirling disease" before the parasite was unknowingly transplanted here from Europe. In some reaches of the Colorado River, the disease has wiped out all but one percent of the population of young rainbow trout and has also infected trout in many streams in the Rocky Mountains on both sides of the Continental Divide.[79]

Pollution

Acid deposition, promoted by the release of nitrogen oxides and sulfur oxides from combustive processes, has caused dramatic ecologic changes in the United States, Europe, and China, as well as in many cities of LDCs. The acidification of lakes and streams has led to juvenile recruitment failure among fish resulting in the disappearance of many species in a number of industrialized countries. The acidification may also be leading to forest decline in many parts of the world, and it is thought to contribute to the loss of 40–50 percent of the mycorrhizal fungi in parts of Western Europe. These fungi are considered "key" to many ecosystems because they live on plant roots symbiotically and facilitate the uptake of nutrients to the plants.[80]

A significant decrease in some species of fish and birds on the Southern California coast has been attributed to the 80 percent decline in the population of zooplankton. This decline is explained by some researchers as a consequence of rising ocean surface temperatures of 1.2°C–1.6°C. Some scientists suggest this warming is a result of the "greenhouse" effect. A steeper density gradient, resulting from ocean surface warming, has slowed the upwelling to the surface of nitrates, phosphates, and other nutrients used by phytoplankton. The phytoplankton have diminished sharply from the lack of these nutrients. Because zooplankton feed on the phytoplankton, their numbers have also been markedly reduced and consequently this effect is carried up the food chain to larger fish. If the ocean surfaces continue to warm over the next few decades, the biological impacts could be devastating.[81]

Deformed frogs have been reported in Minnesota, Wisconsin, South Dakota, Vermont, and even Quebec. The frogs being found have limb abnormalities. Some have stumps, others have four front limbs, sunken or missing eyes, and small or missing sex organs. Aquatic frogs appear to have the most severe abnormalities. Amphibians without these deformities are difficult to find.[82] The cause of the abnormalities is uncertain but may be caused by increased ultraviolet radiation resulting from stratospheric ozone depletion, heavy metals, pesticides, or a combination of these factors. Studies are under way to determine the cause, and there is concern that these effects may signal potential health effects to human populations.[82] Based on key findings of the 2006 Global Amphibian Assessment (GAA), there are 5,918 known amphibian species. About 32 percent of 1,896 species are threatened compared to 23 percent of all mammal species.[83] The GAA reports that at least 34 amphibian species are known to be extinct and that as many as 165 species may be extinct, while up to 43 percent are declining. Habitat loss and fungal diseases are major contributors to losses while many species are declining for unknown reasons.[83]

Protecting Endangered and Threatened Species

Legislation first aimed at protecting wildlife in the United States was introduced as a bill in 1926. It had little authority to protect. In 1969, the Department of the Interior added species to the list that were threatened with extinction worldwide. This international initiative helped to form the Convention on International Trade in Endangered Species (CITES), which attempts to reduce the trafficking of rare and exotic animals and/or plants.

In 1973, the Endangered Species Act (ESA) was promulgated in the United States. A species' status is regularly evaluated by the FWS and the National Marine Fisheries (NMF) to determine if it should be listed as threatened or endangered based on scientific data alone. Consideration is not given to economic or political concerns and no branch of the federal government may proceed with a project that might threaten an endangered species or its habitat without first contacting the FWS.[84] Because projects may be stalled or blocked when threatened species are encountered, there are many critics of the ESA. However, 99.5 percent of the 100,000 projects going to consultation proceeded, usually with some modification. Less than 0.5 percent of the projects were prohibited. Even so, some isolated cases received national attention and thrust the ESA into the spotlight. The snail darter, a 7.6-centimeter fish, became a celebrity in 1978 when it was listed under ESA as endangered and so temporarily halted the construction of the multi-million-dollar Tellico Dam project on the Little Tennessee River. Thousands of jobs were expected to be lost in the Pacific Northwest when the spotted owl was listed as threatened in 1990, causing a fury among the loggers in the area.[75] The ESA seeks to protect the presently listed 2,300 species and subspecies of mammals, mollusks, and plants classified as endangered in the 50 states.[85] The ESA also attempts to protect some 200 additional species classified as threatened, and using trade sanctions, intends to protect 523 foreign and oceanic species outside the United States. Many environmentalists praise the ESA for reducing the extinction rate of some animal species in the United States, and even increasing numbers in as many as 65 species. Numbers have increased for the American bald eagle, whooping cranes, grizzly bears, red wolf, and peregrine falcons and many other species (Figure 3.16). More than 29 species have been delisted from the government's endangered species list, including the bald eagle, the peregrine falcon and the little-known Pahrump pool fish.[86] This marks the first time in the 30-plus year history of the ESA that such a large number of species have been targeted for removal from the list. Some groups, such as the Environmental Defense Fund (EDF), hope that politics does not drive the process, as efforts are made to show Congress that the ESA is working and deserves less criticism.[86] However, many opponents of ESA argue that its powers are excessive, that it overrides individual property, threatens jobs,

FIGURE 3.16
Endangered or threatened species whose numbers have increased include the whooping crane, peregrine falcon, American bald eagle, grizzly bear, and red wolf. (From U.S. Fish and Wildlife Service (USFWS), http://fws.gov/; Courtesy of Whooping crane by Steve Hillenbrand; grizzly bear by Don Redfern; bald eagle by Robert Felds; gray wolf by USFWS.)

and interferes with economic progress. ESA is the only United States federal law that addresses biodiversity. Many organizations are calling for Congress to consider reforming the ESA to include protections for private property owners. Bills submitted before Congress propose major changes in ESA that include (1) compensation to landowners if federal actions lower property values, (2) better scientific justification for listing species, (3) tax incentives to property owners for habitat conservation planning, and (4) removal of protection of endangered species habitat.[65,87] The notion of providing property owners with incentives has gained support based on a study that determined that most populations of endangered species in the United States live in a relatively few critical "hot spots" covering less than two percent of the land area, and most of these critical tracts are on private land.[88] These "hot spots" contain endemic species. They are limited to a small area and they are being disturbed by urban development and agriculture. These hot spots are concentrated mostly in Hawaii, California, and the Southeast, particularly Florida. Spending money on these limited areas would be very cost-effective.[88]

Experts now generally agree that it is important to protect large areas of landscape to protect species and keep them off the endangered species list to begin with, while permitting greater flexibility in balancing development and species protection.[88] Scientists and policy makers are learning that absolute bans on wildlife destruction has led to disaster in LDCs because (1) large areas of land have been destroyed by the uncontrolled growth of wildlife, (2) the wildlife is eliminated as a natural resource for the native human populations, and (3) these policies require very costly enforcement. The solution, as seen in some wildlife markets in the United States, is to allow and regulate markets for use of these species in ways that will protect the native peoples and balance the use of natural resources.[90] In some programs, like one called "campfire" in Zimbabwe, the villagers actually are the owners of the wildlife and are allowed to sell permits, under some governmental control, to hunters that allow the villagers to profit while maintaining a sustainable population.[91]

Scientists are constantly working on new ways of protecting and rebuilding populations of endangered species. However, these techniques have greatly varying degrees of success. The cloning of Dolly the sheep presented scientists with new options for the repopulating of endangered species. At China's National Academy of Sciences, zoologist Chen Dayuan and his team have already used adult panda cells to produce a cloned embryo. Some teams are already even working on cloning cells taken decades ago from species that are now extinct.[89]

Food Security

One of the biggest debates for the twenty-first century concerns whether or not the world can produce enough food to feed another few billion people. Population activists, as well as some social scientists, believe the answer is no, and that the only solution is to limit the number of mouths to feed. However, many agronomists and other scientists say yes, provided there are ample funds for research into improved plant varieties and agricultural technologies to increase yields.[92] Depending on whether you live in an impoverished nation or a wealthy one, there is ample room for pessimism or optimism.

Are you interested in learning more about environmental degradation and food security around the world? Visit <http://www-unix.oit.umass.edu/~envhl565/weblinks.html> and click on one or more of the following sites under Chapter Three.

- General Links to Environmental Degradation
- UNEP
- Rachel Carlson web site
- Rainforest
- NASA Landsat Pathfinder Humid Tropical Forest Inventory Project
- Search.spacelink.nasa.gov
- The Amazon
- World Conservation Monitoring Centre—Home Page
- World Resources Institute
- Earthwatch Institute International
- NASA Spacelink—An Aeronautics & Space Resource for Educators
- Soil erosion
- https://www.unenvironment.org/
- Food and Agriculture Organization of the United Nations (FAO)
- Wetlands
- Office of Wetlands, Oceans, and Watersheds
- EPA: Facts About Wetlands
- Biodiversity
- U.S. FWS: The ESA
- Endangered Species Program, U.S. FWS
- Food Security
- United Nations Feed the Hungry Site
- Second Harvest

Food Production

For nearly 40 years, the world production of grain rose by more than 2 percent a year but declined to scarcely 1 percent a year in the 1990s. This level is markedly below the 1.5 percent annual increase in the world's population.[4] Twenty years ago, the story was grim. The 1997 world carryover of grain stocks was at a low level of only 55 days of consumption, which is only 4 days more than the record low of 51 days of consumption experienced the year before, in 1996. The "carryover" of a grain is the ratio of production to use in a given year. A 70-day supply in carryover stocks of grain is considered important to maintain a minimal level of food security in the world.[4] This 1996 low surplus in grain drove prices of wheat and corn up, resulting in 2.5 percent increase in grain harvest to 692 million hectares and a record 1997 harvest of 4.12 megagrams per hectare. Even so, the world was unable to rebuild depleted stocks to a secure level.[4]

Fast forward to the present day, when thankfully it is a different story, with bumper crops across all major grains. This ensures greater food security by ensuring adequate carryover amounts. In particular, the global wheat carryover forecast is up 16 percent from 2017's carryover.[93] Global wheat production in 2017 was 744,155 thousand tons. The 2018 global wheat production is predicted to be 755,512 thousand tons, a 2 percent increase over 2017's production and a 10 percent increase from 2009 production of 687,154.09 thousand tons.[96]

Food security is said to occur when all people have physical and economic access to the basic food they need to work and function normally.[83] While we see good years and lean years in crop production, malnourishment continues to affect almost a billion people without adequate access to food. In 2016, there were an estimated 815 million undernourished people in the world. Although this number has decreased from the 2003 high of 947.2 million undernourished, the trend toward hunger is again on the rise, since 2013's number of 775.4 million.[94]

The World Health Organization estimated in 2016 that nearly 11 percent of the world's population was undernourished and did not receive a minimum required level of dietary energy.[95] Country-specific reports reveal troublingly higher numbers. For instance, for the period from 2001 to 2016, in Afghanistan, 26.8 percent of the population was unnourished while in neighboring Pakistan, 22 percent of people were undernourished. In India, 19 percent of the population is undernourished. In Iraq, 22 percent of the population is undernourished. In many African countries, the reality is most stark. In Ethiopia, a staggering 40 percent of the population does not receive adequate daily calories, while in Zambia, the percentage of undernourished is 44 percent of the population. The countries with the most critical or low food security are shown in Table 3.5. Most of these are in Africa, where food crises are ongoing, often exacerbated by conflict and natural disasters such as drought. Children are especially vulnerable from malnutrition and can suffer from both stunted growth and wasting, leaving them vulnerable to infection and at an increased risk of dying. Despite these concerns, trends in recent cereal yields are expected to meet the demand for much of the world except Sub-Saharan Africa, based on projections through the 2025.[96]

Cereals are central to the human diet and represent more than 63 percent of calorie intake when indirect consumption of grain fed livestock is considered. Cereal production

TABLE 3.5

Most Food-Insecure Countries, Based on the Global Hunger Index

Most Food-Insecure Countries[a]
1. Central Africa republic
2. Chad
3. Sierra Leone
4. Madagascar
5. Zambia
6. Yemen
7. Sudan
8. Liberia
9. Niger
10. Timor-Leste

Source: *The World's Ten Hungriest Countries,* October 12, 2017, https://www. concernusa. org/ story/ worlds-ten-hungriest-countries/.

[a] For several at-risk countries, no food-insecurity data exists, including South Sudan, Somalia, and Syria. This is often due to ongoing internal conflict, which makes data collection impossible.

has been lagging behind the growth rate of the human population since 1984, but some areas of the world are faring better than others.[96] If one looks at regional trends, Africa and the Middle East including West Asia and Northern Africa are doing poorly in terms of food production. Africa continues to experience widespread political instability, government neglect of agriculture, an AIDs epidemic, and rapid population growth.[96] South Asia (India, Pakistan, Bangladesh) are experiencing a rise in per-capita cereal output to around 225 kg, while Southeast Asia (China, Japan, Indonesia) has seen a marked upward shift following the late 1980s. The U.S., Canada, and Australia provided more than 20 percent of the global cereal harvest, although these countries collectively represented less than 6 percent of the world's population. The main reason for the decline in production in these countries is linked to the European Union (EU), which flooded the world market with their government-subsidized cereal surpluses and drove the market price down.[96]

Reasons for Regional Food Shortages

Distribution of Population and Food

Today there is a large food gap between developed countries and many developing nations. Already this gap poses moral and political as well as economic problems of major dimensions, and if it widens, these problems will become increasingly serious. The majority of the global population lives in developing countries, which face the greatest food security problems. Approximately 1 billion lived in the poorest 30–40 countries, roughly the same number as in the well-fed developed nations. Moreover, the proportion of the world total living in the developing nations will continue to grow, as 98 percent of the annual increase in the world's population is occurring in the developing countries. The growing food gap between rich and poor nations has been—and will continue to be—in large part due to the differences in rates of population growth. Food output has been increasing somewhat more rapidly in lesser developed nations. However, because of a more rapid rate of population increase in the poor countries, their average rate of *per capita* increase in food output was less than one third of the rate in the developed countries. Although there has generally been steady growth in the world production of most crops from the 1960s forward, when viewed as per-capita production, there are unequal results among countries. The food production in Africa has shown a steady increase since the 1960s, with a doubling in food production. However, the daily calories per capita trend has not kept pace with the increased food production, due to the enormous population increases. A view of regional daily per capita calories demonstrates this, where a person living in Sub-Saharan Africa receives 1,000 *fewer* calories per day than a person living in an industrialized country (Table 3.6). Other regions also show significant per capita daily calorie deficits, including South Asia. Overall, if the world hopes to successfully feed the 10 billion people expected by the year 2050, there would need to be 50 percent increase in food production globally.[94] In a fundamental sense, the world food problem, viewed globally, is less a problem of absolute shortages than a problem of unequal distribution between rich and poor countries. Moreover, the inequality of food distribution is also a basic problem within many developing countries.[96]

Global food production of major cereals (wheat, maize, rice) has increased across the past 25 years. While production in agriculture and cereals may have increased overall, the rate of these increases has slowed, especially in developed countries (Figure 3.17a–c). As the graphs show, the recent annual rates of increase in cereal production in developing countries is significantly greater than the rate of increase in developed countries. It is due primarily

TABLE 3.6

Trends in Global and Regional per Capita Daily Calorie Consumption: kcal per Capita per Day

	Trends in Global and Regional per Capita Food Consumption (kcal per capita per day)					
Region	1964–1966	1974–1976	1984–1986	1997–1999	2015	2015 per Capita Daily kcal Difference versus Industrialized Countries
Industrialized countries	2,947	3,065	3,206	3,380	3,440	0
Developing countries	2,054	2,152	2,450	2,681	2,850	−590
Near East and North Africa	2,290	2,591	2,953	3,006	3,090	−350
Sub-Saharan Africa	2,058	2,079	2,057	2,195	2,360	−1,080
Latin America and the Caribbean	2,393	2,546	2,689	2,824	2,980	−460
East Asia	1,957	2,105	2,559	2,921	3,060	−380
South Asia	2,017	1,986	2,205	2,403	2,700	−740

Source: Global and regional food consumption patterns and trends, World health organization, Geneva, Switzerland, www.who.int/nutrition/topics/3_foodconsumption/en.

FIGURE 3.17

(a–c) Trends in global cereal production, in thousand tons of Wheat, Maize, and Rice. (Adapted from *OECD-FAO Agricultural Outlook*, 2018, stats.oecd.org/.)

to production increases in developing countries that global cereal production has increased year by year, from 1,575 million tons in 2000 to 1,973, million tons in 2010 to 2,293 million tons in 2017. The growth rate of agricultural production worldwide slowed from a 3.0 percent growth rate in the 1960s to 2.3 percent in the 1970s and less than 2.0 percent in the mid-1990s.[96]

Pessimistic View

Some authors such as Lester Brown, President of the Worldwatch Institute, express dire consequences for future food production. He maintains that agricultural technologies are going unused, fisheries and rangelands are approaching the limits of biological productivity, fresh water is becoming increasingly scarce, fertilizer applications are becoming less effective, and social disintegration is occurring in many LDCs.[97] There is also concern that the tropical and subtropical environments are so fragile in many LDCs that they will be unable to sustain significant food production increases. About 45 percent of the potential cropland in sub-Saharan Africa and Latin America is under forest or in protected areas.[98] Nearly three quarters of the potential cropland is so fragile that development would result in dramatic losses in biodiversity, increased carbon dioxide emissions, decreased carbon storage capacity, and high financial and ecological costs.[99] Many experts agree that the potential to expand cropland area is quickly disappearing in most areas of the world.[100] China represents a startling example of this problem. More than 21 percent of the world's population lives in China, while its arable land covers less than 7 percent, resulting in a per person share of 0.086 hectares or only 25 percent of the world average.[101] The hectarage for agricultural development is diminishing while existing farmland is being over-cultivated, polluted, and nearly one third of the land is threatened by soil erosion.[85] Overgrazing and reclamation of farmland has turned a quarter of the country's grassland into desert with more than 300 cities or towns having a shortage of water.[101] In many of the LDCs, agriculture is hampered by unfavorable growing conditions. There is a continuing threat of civil strife, the impoverished cannot afford to pay for food, and governments cannot provide the policies or infrastructure to the rural poor that promote domestic agriculture production.

Optimistic View

A number of organizations—including the FAO of the United Nations; The World Bank; and the International Food Policy Research Institute (IFPRI), which is a component of the Consultative Group of International Agricultural Research (CGIAR)—have published studies looking at food production estimates from 2010 to 2030.[96,102–104] The studies and projections point to a number of hurdles with respect to worldwide food production but are similar to the FAO prediction that expected production increases will be able to meet the effective demands of a rising world population providing there are substantial investments in agricultural research.[102] Effective demand refers to that "demand level" corresponding to purchasing power without regard to food requirements. Unfortunately, a sharp rise in the number of food-related crises has shifted priorities. In 1986, the World Food Program directed 75 percent of its available monies to development projects. By the mid-1990s, more than 85 percent of its resources were going to meet emergency and refugee needs as those affected by natural and political disasters rose from 44 to 175 million people in the same years.[105] This need continues today, with famine in areas of the world like South Sudan and war-torn countries like Syria. This last revelation is not optimistic but must be included to realistically balance the view that we can meet the world's effective demand for food.

Sources

Where will the food come from to feed the world's billions? Increases in food supply must come from one or more of the following sources: (1) increases in yield (tons per hectare), (2) increases in arable land placed under cultivation, and (3) cropping intensity (fewer fallow periods or more than one crop per year or field). The FAO projects that 66 percent of food production will come from increased yields, about 21 percent from arable land expansion, and the remaining 13 percent from increasing cropping intensity (Figure 3.18).[94]

Worldwide gross agricultural production is expected to increase in the next 20 years, but its growth will likely be at a slower rate of 1.8 percent per year compared to the previous decades because population growth rates and food demand have slowed in developing countries, and there is inadequate income growth in many countries to purchase large supplies of food.[106]

There are scientists who believe that ability to expand cropland is limited, and that it is disappearing in many areas of the world. The potential for increasing agricultural land is limited by (1) the significant costs of developing an infrastructure in remote areas, (2) the lesser productivity of these alternative areas, and (3) the trade-offs in environmental destruction of sensitive ecosystems.[100]

The FAO projection of increased yields is based on their assessments that there is room for improvement for farmers who are not currently achieving peak yields. Many countries are showing only half the yield (3.7 metric tons) being experienced in the best-performing countries (6.7 metric tons).[100] The FAO projects that South Asia will make the strongest contribution of increased yields in crops of about 82 percent while Latin America and Africa will experience less than 53 percent yield increases.[94]

In addition, countries and regions are sharing best food production practices, not only about farming but also about the entire food value chain. Sharing information across countries and regions is one tactic that may enable additional increases food. At a recent World Bank workshop, representatives from countries in Asia shared information on

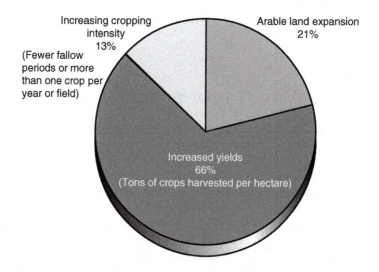

FIGURE 3.18

Projected sources of food increases. (Adapted from FAO, http://www.fao.org [WAICENT/faoinfo/econ-tie /giews/English/fo/fo9804/httoc.htm].)

the Asian rice value chain with West African nations.[113] The rice value chain in Asia is integrated and well-managed. It includes farmers, distributors, sellers, and buyers, both wholesale and retail. Factors such as country commitment, supportive policy, infrastructure investments, and ongoing research have been keys to success. Since rice is a critical source of daily calories in West Africa, any improvements in the yield are useful. Improving how a food value chain is managed can lead to significant reduction of food loss and food waste. Since the loss can occur anywhere, from the initial harvesting to storage, packing, transport, and final point of sale, it takes awareness and education to identify area of improvement.[114] This type of holistic best practice thinking can help regions like West Africa significantly improve crop yields and overall process.

Not everyone thinks there are significant improvements possible in terms of increasing food production and yield. It is argued that increasing numbers of farmers in developing countries are already using crops with the highest genetic yield potential, and that rice crops at experiment stations in Asia have leveled off for a number of years.[100] Further, the high input approach of using genetically programmed, high-yielding varieties have required large amounts of water, pesticides, and synthetic fertilizers. This approach can increase yields but may also cause the loss of genetic stocks and reduction in soil quality. Further, many LDCs do not have the funds necessary to support the increased costs of fertilizers, pesticides, and seeds. Alternative models are being evaluated and promoted that are friendlier to the environment. Integrated farming systems are being proposed that employ sustainable crop intensification strategies. These strategies include (1) improved irrigation systems, (2) structured water pricing to reduce overuse, (3) alternative rotation of crops, (4) selective pesticide use, (5) use of pest-resistant varieties, (6) improved soil testing and fertilizer application, (7) regional crop breeding programs, and (8) more education of farmers.[94] High-input and crop-intensification strategies may not be viable in fragile ecosystems because of drought risk, inadequate support structures, and lower yield response. In these areas, alternative strategies might involve (1) erosion and moisture control, (2) recycling nutrients, (3) and integrating annual and perennial crops in what is known as mixed farming.

Food Security

Worldwide

Temporary deficiencies in food and famine may occur in the presence of droughts, floods, political strife wars, or poor harvests. These events are on the increase and produce significant suffering in the short term. However, chronic undernutrition is a more difficult and pervasive problem resulting in a food security crisis in many LDCs. In such areas, more and more poor are unable to afford food, conditions for agriculture become more unfavorable, and governments have been unable to provide the infrastructure or policies needed to promote domestic production.[94] The lack of food security means people are unable to acquire the basic food they need to function normally because they are restricted physically and economically. Civil strife, inadequate marketing, poor storage facilities, and inadequate delivery capabilities reduce physical access to food. Protein-energy malnutrition (PEM) is the most devastating form of malnutrition and is the one most referred to when world hunger is considered.[107] The world production of food calories is 20 percent greater than 1975 even with an increase of population by 70 percent. This means that on average there are at least 2,940 kilocalories (kcal) of food per person per day and sufficient to meet the world's needs. However, great disparities in income and arable land make the distribution of these food calories very uneven. Poverty is the primary cause of hunger in the world today. The combination of unequal income distribution, corrupt governments, conflict, and natural disasters contribute to the burdens of poverty and hunger. There are 815 million food-insecure and malnourished people in the world and more than half of these people, 489 million, live in countries struggling with conflict.[94] Conflict causes a decrease in productive agricultural land. Unfortunately, the number of conflicts is increasing. Since 2010, conflicts between armed groups that are not government sponsored or related increased by 125 percent. Perhaps unsurprisingly, the number of refugees and displaced has also increased, doubling in 10 years to 64 million people today.

It is estimated that 1.08 billion people in developing countries live on less than $1 day.[107] Although global hunger decreased for nearly a decade, it is now increasing again, from 777 million people in 2015 to 815 million in 2016. Unfortunately, this trend is expected to continue. Food insecurity is most pronounced in sub-Saharan Africa, South Eastern Asia, and Western Asia. In these regions, people may be undernourished with a per capita food supply of 2,170 calories per day or about 60 percent of the per-person calories available to people in developed countries. In other areas of the developing world, there is reason to be more optimistic.[94]

Although there is a hypothetical potential to meet the worldwide demand for food, the reality is that food crises in Africa, South Asia, and other countries are likely to persist. The IFPRI has made recommendations that will hopefully lead to sustainable land use, efficient agricultural systems to produce low-cost food, and sufficient food that is physically and economically accessible to every person.[108] In order for these goals to be met, IFPRI recommended that (1) governments of LDCs need to be strengthened; (2) agricultural and extension services need expansion; (3) agricultural yields need to be intensified with methods that sustain natural resources; (4) low-cost, efficient agricultural input and output markets need to be developed; and (5) international assistance needs to be expanded.[108]

As countries develop, a population's diets may change. A trend worth watching is the increase in the production of meat in developing versus developed countries, as compared to the production of cereals (Figure 3.19a–c). As we looked earlier at the deforestation challenge, consider that sometimes the newly deforested land is used for pastures rather than to grow crops for human consumption. As the global demand for meat increases, especially in developing countries, so too will the demand for pasture lands and crops to feed the animals. This will be an ongoing tradeoff of growing crops to feed the increasing human population or raising livestock and growing crops to feed them.

Hunger in America

Hunger threatens millions of Americans. The rate of hunger in America grows steadily worse in the face of prosperity, creating our own food insecurity problems as the gap widens between the rich and the poor, the secure and the threatened.[109] The U.S. Department of Agriculture estimates that in 2015 there were 42 million Americans, including 13 million children, who were regularly "food insecure."[110] Food insecure means that a household lacks consistent access to healthy food and must make trade-offs between paying housing costs or buying healthy food. This number has been annually decreasing, from a high of more than 50 million Americans in 2009 that were classified as food insecure. Food insecurity rates vary widely across the country, from state to state, and even within a state, but are highest in the rural South. Assistance may be available to food insecure people, but not always. More than 20 percent of food insecure people make too much in income to qualify for federal food assistance.[110] Looking at specific financials, in 2015, food insecure

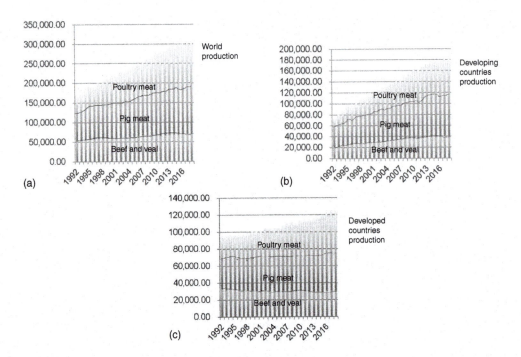

FIGURE 3.19

(a–c) Trends in production of meat from cows, pigs, and poultry. (Adapted from *OECD-FAO Agricultural Outlook,* 2018, stats.oecd.org/.)

individuals reported spending about $62 each week on food, which averages nearly $3.00 per meal. This amount leaves little margin for error in the case of an emergency, if transportation costs increase, or an unexpected medical crisis occurs. Food banks try to meet the need for people unable to afford food but unable to qualify for aid. Two out of five households that receive support from food banks have at least one person employed, and in 50 percent of these cases, the employment is full-time.[111] The number of working families requiring food aid is increasing as a percentage, but the majority of people requiring aid still include those with disabilities, single mothers, and seniors on fixed incomes. The millions of food insecure households in the United States cannot reach this goal and must often make choices between paying for other necessities such as housing, medical care, or fuel for heating or transportation.[112] Food insecurity in America is a consistent problem, and additional government cuts are not likely to result in increased personal giving. Obtaining enough food for its citizens to pursue a healthy and active lifestyle should be a primary objective of any government.

References

1. Motava, W.J., Contents under pressure, *E. Magazine*, November/December 27, 1996.
2. United Nations Environment Program, Global environment outlook, *Environmental Health*, April 22, 1997.
3. Kaplan, R., The coming anarchy, *The Atlantic Monthly*, February 1994, p. 44.

4. Fornos, W., World population overview, *The Population Institute*, December 30, 1997. http://www.population institute.org/overview97.html.

5. Barney, G.O., *The Global 2000 Report*, Pergamon Press, New York, 1980. (Vol. 1, Summary Report; Vol. 11, Technical Report; Vol. III, Government Global Model), 360 p.

6. Budiansky, S., 10 billion for dinner please, *U.S. News & World Report*, September 12, 1997, p. 57.

7. Turk, J. and Turk, A., *Environmental Science*, 4th ed., Saunders College Publishing. Philadelphia, PA, 1988, Chap. 9.

8. Safety of Genetically Engineered Foods, National Research Council (US), *Committee on Identifying and Assessing Unintended Effects of Genetically Engineered Foods on Human Health*, National Academies Press (US), Washington, DC, 2004.

9. Segelken, R., Gene shotgun, *Cornell Chronicle*, 18(33), May 14, 1987, p. 3.

10. The terminator technology: New genetic technology aims to prevent farmers from saving seed, *RAFI (Rural Advancement Foundation International) Communiqué*, March/April 1998.

11. Lombardo, L., Genetic use restriction technologies: A review. *Plant Biotechnology Journal*, 12, 995–1005. doi:10.1111/pbi.12242, 2014.

12. Gordon, B., Terminating the "terminator" seed: Broken promises and sterile seeds could alter the lives of food producers, *Utne*, March 2, 2006. http://www/utne.com.

13. World Resources Institute (WRI), Chapter 10: Food and agriculture, *A Guide to the Global Environment*, World Resources 1996–1997. http://www.wri.org/wri/wr-96-97/fa-txt2.html (accessed March 7, 2018).

14. Alexandratos, N., (Ed.), *World Agriculture: Towards 2010*, An FAO Study, John Wiley and Sons, Chichester, U.K.; *Food and Agriculture Organization of the United States*, Rome, Italy, 1995, p. 38.

15. Market Monitor, Agricultural market information system. http://www.amis-outlook.org/home/en/ (accessed March 10, 2018).

16. Dasgupta, P.S., Population, poverty and the local environment, *Scientific American*, February 1995, p. 42.

17. Forests and Energy Infographic. http://www.fao.org/resources/infographics/infographics-details/en/c/852816/ (accessed March 7, 2018).

18. MSNBC.com, High global oil prices seen through 2007: U.S. government makes prediction on demand outlook, geopolitical worries, *Reuters*, May 3, 2006. http://www.msnbc.msn.com/id/.

19. Simon, J., *The Ultimate Resource*, Princeton University Press, Princeton, NJ, 1981, 495 p.

20. Simon, J. and Kahn, H., *The Resourceful Earth*, Basil Blackwell, Oxford, UK, 1984, p. 585.

21. Ehrlich, P., Too many rich people: Weighting relative burdens on the planet, *Our Planet*, 6, 31, 1994.

22. AP (Associated Press): Geneva, U.N. gap widening between countries, *Daily Hampshire Gazette*, September 25, 1997, p. 1.

23. Conflict causing spikes in food insecurity in the Near East and North Africa, December 21, 2017. http://www.fao.org/news/story/en/item/1073611/icode/ (accessed March 7, 2018).

24. Harden, G., The tragedy of the commons, *Science*, 162, 243, 1968.

25. Runnels, C., Environmental degradation in Ancient Greece, *Scientific American*, March 1995, p. 96.

26. Nierenberg, D. and MacDonald, M., The population story...so far, *World-Watch*, September/October 2004, pp. 14–17.

27. World Health Organization (WHO), Family Planning website. http://www.who.int/topics/family_planning/en/ (accessed March 7, 2018).

28. The U.S. Government and International Family Planning & Reproductive Health Efforts, Henry J. Kaiser Family Foundation, July 10, 2017. http://www.kff.org.

29. Repetto, R. (Ed.), *The Global Possible Resources. Development and the New Century*, Yale University Press, New Haven, CT, 1985, p. 538.

30. United Nations Environmental Program (UNEP), Global environmental outlook, *Environmental Health*, April 22, 1997.

31. Briefings, Satellite provides dim view of Amazon, *Greenpeace*, November/December 1988, p. 22.

32. AP, Scientists hopes are borne out: Rain forests and carbon dioxide, *The New York Times*, November 8, 1995, p. A4.

33. Farnsworth, N.R., New drugs from the rainforest, *World Health*, 2, 49, March/April, 1996.
34. Deforestation. https://www.worldwildlife.org/threats/deforestation (accessed March 7, 2018).
35. Earth's green lungs begin to fade, *International Wildlife*, 25, September/October 1999.
36. Iwokrama, International centre for rain forest conservation and development, *Global Rain Forests*, May 9, 2006. http://www.iwokrama.org/forest/globrainforests.htm.
37. National Aeronautics and Space Administration (NASA), Tropical deforestation and habitat degradation in the Brazilian Amazon Basin, The Earth Observing System Educator's Visual Materials, 1998. http://eospso.esfc.nasa.gov/cos-edu.pack/p24.html.
38. Moeller, D.W., *Environmental Health*, Harvard University Press, Cambridge, MA, 1997, Chap. 19.
39. Tabuchi, H., Rigby, C., and White, J. Amazon Deforestation, Once Tamed, Comes Roaring Back, *The New York Times*, February 24, 2017. https://www.nytimes.com/2017/02/24/business /energy-environment/deforestation-brazil-bolivia-south-america.html.
40. Butler, R., Deforestation in the amazon, 2004, *Mongabay*, May 5, 2006. http://www.mongabay .com/brazil.html.
41. Baretto, P., Jouza, C., Mogueron, R., Anderson, A.M., and Salomao, R., Human pressure on the Brazilian Rainforests, March 2006, World Resources Institute: Amazon, Amazon Institute of People and the Environment, March 2006. http://pdf.wri.org/amazon_glossy_overview.pdf.
42. Hernandez, S., Environment: Afro-Asian trees to the rescue in Central America, *InterPress Service English Newswire*, August 1, 1996.
43. Brevik, J.J., Global environmental outlook, *Environmental Health*, April 22, 1997.
44. Haniffa, A., Degradation hits human security, *India Abroad*, 12, 106, 1996. http://www2.elibrary .com/id/2525/getdoc.c.cos&form=ri.&pubname=lndia-Abroad&puburl=O.
45. Depth News Asia, Action for tropical rainforests, *World Press Review*, August 13, 1993.
46. Brown, L.R. and Woff, E.C., *Soil Erosion: A Quiet Crisis in the World Economy*, Worldwatch Paper, Worldwatch Institute, Washington, DC, 1984, p. 49.
47. Maisonnave, F., *It's War in the Amazon, Says Brazil's Top Environmental Law Enforcer*, November 11, 2017. http://www.climatechangenews.com (accessed March 7, 2018).
48. United States Deptartment of Agriculture (USDA): *Analysis of Policies to Conserve Soil and Reduce Surplus Crop Production*. Washington, DC, USDA, Agricultural Economic Report No. 534, 21, 1985.
49. Doyle, R., Soil erosion of cropland in the U.S., 1982–1992, *Scientific American*, October 34, 1996.
50. Stutz, B., The landscape of hunger, *Audubon*, 2, 95, March/April, 1993.
51. Filipov, D., Desert's drift buries Russian region's dreams, *Boston Globe*, September 27, 1997, p. A26.
52. Brazil Embassy, What is desertification? Science and Technology Section, National Research Network's Brazilian Regional Center, February 30, 1997. http://brasil.emb.nw, dc.us./nuna /deserts/ingles/se/~edesert/desertmi.html#dm.
53. United Nations Decade for Deserts and the Fight Against Desertification. http://www.un.org /en/events/desertification_decade/background.shtml (accessed March 7, 2018).
54. National Aeronautics and Space Administration (NASA), Desertification Atlas (Africa) Maps I-1, Global Change Master Directory, October 24, 1997. http://gcmd.gsfc.gov/cgibin/detdy?=di splay+dif&entry=NBld0270/101.
55. EPA-OWOW (Environmental Protection Agency, Office of Wetlands, Oceans, and Watersheds), Facts About Wetlands, August 25, 1997. http://www.epa.gov/owow/wetlands/facts/types.html.
56. Secraig, B., What is a wetland? *Sierra*, May/June 1996, p. 44.
57. Enteractive, Disappearing wetlands, Earth Explorer, February 1, 1995. http://www3.elibrary .com/id/2525/getdoe.c... lorer&puburl=http*c*//www.enteractive.com (accessed March 7, 2018).
58. Adler, T., Scientist dispute legislator's take on wetlands, *Science News*, 148, 56, 1995.
59. Cushman, J.H., From wetlands to asphalt, a parcel at a time, *The New York Times*, October 27, 1996, p. 22.
60. Stevens, W.K., Prairie ducks return in record numbers, *The New York Times*, October 11, 1994, p. 1.
61. Dahl, T.E., *Status and Trends of Wetlands in the Conterminous United States 1998 to 2004*. U.S. Department of the Interior; Fish and Wildlife Service, Washington, DC, 2006, p. 112.
62. Marina, M., *Progress Threatens Brazil's Vast Wetlands: Foes Say Canal Would Drain Panatanal*, CNN Interactive, Online Internet, April 23, 1997. http://www.cnn.com/earth/9704/23/brazil .wetlands.

63. *The Economic Values of the World's Wetlands*, World Wildlife Fund, Gland/Amsterdam, January 2004.
64. Erwin, D.H., The mother of mass extinctions, *Scientific American*, July 1996, p. 72.
65. Callahan, J.R., Vanishing biodiversity, *Environment Health Perspectives*, 104(4), 386, 1996.
66. Hilchey, T., Biodiversity study sees more species in danger, *The New York Times*, November 14, Science Section 1, 1995.
67. Worldwide Fund for Nature, Climate Change and Biodiversity Conservation: Worldwide Loss of Biodiversity, Executive Summary, 1996, Chap. 3. http://www.panda.org/ida/climate /cIimate.3/Pagel.html.
68. Doyle, R., Threatened mammals, *Scientific American*, 276, 32, January 1997.
69. Doyle, A., Humans spur worst extinction since dinosaurs, *Reuters*, Know Now, March 20, 2006. http://go.reuters.com.
70. Ananthaswamy, A., Earth Faces sixth mass extinction, *NewScientist.com news service*, March 18, 2004. http://www.newscientist.com/article.ns?id=dn4797.
71. UN Convention on Biological Diversity. http://www.cbd.int/2011-2020/about/goals (accessed March 7, 2018).
72. Cushman, J., Freshwater mussels facing mass extinction, *The New York Times*, accessed October 3, 1995, p. C1.
73. Cook, R.M., Sinclair, A., and Stefarmson, O., Potential collapse of North Sea cod stocks, *Nature*, February 6, 1997, p. 358.
74. Allen, S., Trying to do the right thing. *The Boston Globe*, September 26, 1995, p. 27.
75. Butler, S., Rod? Reel? dynamite? A tough-love program takes aim at the devastation of the coral reefs, *US News and World Report*, November 25, 1996, p. 56.
76. Lemonick, M.D., Animal genocide, mob-style, *Time Magazine*, November 14, 1995, p. 77.
77. Jones, A., Stowaway snakes deplete bird populations on Guam, *CNN Interactive*, 1, November 9, 1996.
78. Luoma, J., Boom to anglers turns into disaster for lakes and streams, *The New York Times*, November 17, 1992, p. 3.
79. Stevens, W., Bizarre parasite invades trout streams, devastating young rainbows, *The New York Times*, September 19, 1995, p. C9.
80. Wilson, E.O., *The Diversity of Life*, Harvard University Press, Cambridge, UK, Chap. 1.
81. Kaiser, J., Vanishing zooplankton, *Science News*, March 11, 1995, p. 157.
82. AP, Many reports of deformities among frogs are puzzling, *The New York Times*, October 13, 1996, p. 33.
83. International Union for Conservation of Nature and Natural Resources (IUCN), Global Amphibian Assessment, May 2006. http://www.globalamphibians.org.
84. Chadwick, D.H., Dead or alive: The endangered species Act, *National Geographic*, March 21, 1995.
85. U.S. Fish and Wildlife Service, Threatened and Endangered Species System, Summary of Listed Species, May 8, 2006. http://ecos.fws.gov/tess_public/Boxscore.do.
86. AP (Associated Press), Animals to come off list, *Daily Hampshire Gazette*, May 1998, p. 1.
87. Biddison, J., De-Prioritizing people, *Townhall.com*, March 28, 2006. http://www.townhall.com/.
88. Stevens, W.K., "Hot spots" for American endangered species cover surprisingly little land, *The New York Times*, January 1, 1997, p. 1.
89. Mzurek, R., Back from the dead, *New Scientist*, October 9, 1999. http://www.newscientist.com /article/ng6422071.800.html.
90. Budiansky, S., Killing with kindness, *US News & World Report*, November 25, 1996, p. 48.
91. Satchell, M., Save the elephants: Start shooting them, *US New and World Report*, November 25, 1996, p. 52.
92. AP (Associated Press), Experts debate limits of earth's resources, *CNN Interactive*, October 27, 1996.
93. Sosland, J. Global wheat carryover seen headed to all-time high, *World-Grain.com*, August 25, 2017.
94. State of Food Security 2017 report. http://www.fao.gov, (accessed March 2, 2018).
95. Nutrition Landscape Information System (NLiS), WHO. http://apps.who.int/nutrition/ landscape/report.aspx.

96. Dyson, T., World food trends and prospects to 2025, *National Academy of Sciences Colloquium*, 96(11), 5929–5936, May 25, 1999.
97. Brown, L.R. and Kane, H., *Reassessing the Earth's Population Carrying Capacity*, W.W. Norton, New York, 1994.
98. The World Bank, World Development Report 1992, The World Bank, Washington, DC, 1992, p. 135.
99. Ehrlich, P.R., Ehrlich, A.H., and Daily, G.C., Food security, population and environment, *Population and Development Review*, 19(1), 7, 1993.
100. Pinstrup-Andersen, P. and Pandya-Lorch, R., *Alternating Poverty, Intensifying Agriculture, and Effectively Managing Natural Resources*, Food Agriculture and the Environment Discussion Paper No. 1, International Food Policy Research Institute, Washington, DC, 1994, p. 6.
101. Xin, W., Population versus development: Challenge of the new century, *Beijing Review*, May 1–7, 1995, p. 12.
102. Food and Agriculture Organization of the United States (FAO), *The State of Food and Agriculture 1994*, FAO, Rome, Italy, 1994, p. 11.
103. Mitchell, D.O. and Ingco, M.D., *The World Food Outlook*, The World Bank, Washington, DC, 1993.
104. Rosegrant, M.W., Agcaoili-Sombilla, M., and Perez, N.D., *Global Food Projections to 2020: Implications for Investment*, International Food Policy Research Institute (IFPRI) Food, Agriculture and the Environment, Draft Discussion Paper, IFPRI, Washington, DC, 1.
105. Webb, P., A time of plenty, *A World of Need: The Role of Food Aid in 2020*, International Food Policy Research Institute (IFPRI) 2020, Brief No. 10, Washington, DC, 1995, p. 2.
106. Pimental, D., Harvey, C., Resosudarmo, P. et al., Environmental and economic costs of soil erosion and conservation benefits, *Science*, 267, 5201, 1117–1123, February 24, 1995.
107. Hunger Notes, World Hunger Facts 2006, World Hunger Education Service, Updated January 21, 2006. http://www.worldhunger.org/articles/Learn/world%20hunger%20facts%202002.htm.
108. International Food Policy Research Institute (IFPRI), A 2020 vision for food, agriculture, and the environment, *Draft Paper (117PRI)*, Washington, DC, June 1, 1995.
109. Roberts, L. and Roberts, S.V., Hunger: A startling crisis, *USA Weekend*, March 27–29, 1998, p. 4.
110. Food insecurity in the United States', feeding America. http://map.feedingamerica.org (accessed March 3, 2018).
111. Shapiro, L. and Rosenberg, D. How hungry is America? *Newsweek*, March 14, 1994, p. 59.
112. Food research and Action Center (FRAC), Hunger and food insecurity in the U.S., September 10, 2005. http://www.frac.org/html/hunger_in_the_us/hunger_index.html.
113. Asia shares key strategies for rice value chain success with West Africa, AfricaRice, the World Bank and the West and Central African Council for Agricultural Research and Development (CORAF/WECARD). http://africarice.blogspot.com/2018/02/asia-shares-key-strategies-for-rice.html (accessed February 27, 2018).
114. Food loss and food waste, Food and Agriculture Organization of the United Nations, Rome, Italy. http://www.fao.org/food-loss-and-food-waste/en/ (accessed February 28, 2018).
115. National Resource Conservation Service. https://www.nrcs.usda.gov/wps/portal/nrcs/detail/soils/use/worldsoils/?cid=nrcs142p2_054003.
116. National Resource Conservation Service. https://www.nrcs.usda.gov/wps/portal/nrcs/detail/soils/use/worldsoils/?cid=nrcs142p2_054021.
117. OECD-FAO Agricultural Outlook, March 8, 2018. https://www.stats.oecd.org/.
118. *World Development Indicators: Deforestation and Biodiversity*, The World Bank, Washington, DC, Last updated date September 18, 2017.
119. *The World's Ten Hungriest Countries*, October 12, 2017. https://www.concernusa.org/story/worlds-ten-hungriest-countries/.
120. http://www.who.int/nutrition/nlis/en/.

4

Environmental Disease

OBJECTIVES

A student reading this chapter will be able to:

1. Define "environmental disease."
2. Discuss the separate concepts of environment and disease.
3. Describe the distinct types of human environmental disease including infectious disease, physical and chemical injury, ionizing radiation, developmental disease, neoplastic disease, and nutritional disease.
4. Explain the process of genetic replication in mammalian cells.
5. List the major genetic components and discuss their primary actions and features.
6. Describe the major mechanisms of developmental and genetic disease.
7. Outline a schematic of the new processes in recombinant-DNA and genetic screening techniques.
8. Describe the concepts of the origins of cancer and cancer risk factors.

Introduction

On a stretch of road in a New South Wales town in Australia, there have been more cancers than statistically expected in the town's residents. In fact, there have been ten known cases of breast cancer, eight cases of prostate cancer, five bowel cancers, three stomach cancers, three non-Hodgkin's lymphomas (NHLs), two leukemia diagnoses, and two liver cancers. So far. One resident commented, "There's not actually a lot of properties on this street." Among people living on this stretch of road that is not even 5 kilometers long, 39 people have battled cancer in the past 15 years.

In 2015, it was reported that the town's water supply was badly contaminated with PFAS, a chemical used to fight fires. The cancer cluster is located only a few kilometers from an air force base, which had used the PFAS foam extensively for firefighting training.[1] These chemicals, perfluorooctane sulfonate (PFOS) and perfluorooctanoic acid (PFOA), have been linked to six diseases, including cancer. They aren't found only in firefighting foam, but in many common household products such as non-stick frying pans and waterproof clothing. The bad news is that when the chemicals get into our bodies, they stay for 3–5 years. During that time, they can cause disease.[2] This is environmental disease.

Environmental Disease

Environmental disease refers to any pathologic process that is the consequence of external factors, including exposure to physical agents, such as UV radiation from the sun; exposure to chemical agents such as pesticides or water pollution; poor nutrition; and social or cultural behaviors that cause disease. Factors involved in the development of environmental disease may include polluted air and water, excess noise, UV radiation, ionizing radiation, food contamination, occupational exposure, toxic wastes, poor nutrition or diet, stress, and tobacco use. The World Health Organization (WHO) broadly defines environmental health risks as "all the physical, chemical and biological factors external to a person, and all related behaviors, but excluding those natural environments that cannot reasonably be modified."[71]

Definitions of "environmental disease" may vary slightly. For instance, in a scientific study, the definition might be narrow, confined only to those factors that can be measured, while in a newspaper article, the definition might be broader. In this chapter we will look at several different types of environmental disease and will see that a large percentage of people across the globe are affected by environmental disease.

There is rising public awareness and concern about environmental disease, for several reasons. These include an increased ability to measure human exposure to and the effect of environmental risks, focused research efforts linking environmental exposure to genetic damage and/or disease, and broad global communication about numbers of people impacted by environmental disease, especially children, and how to reduce those numbers. A recent study by the WHO concluded that 23 percent of global deaths and 26 percent of deaths among children fewer than 5 years of age were caused by controllable environmental factors. This suggests that if the disease-causing environment risk factors were improved, whether through better sanitation, cleaner air, or reduced chemical exposure, fewer people would die.[3]

Many factors combine to cause disease. It is the role of scientists to tease apart how much of a disease's incidence—the number of new cases of the disease per year among the population—is caused by the environment. The environmental component is generally the controllable part of a disease. This is why it is important to know which environmental factors contribute to disease. What can be changed to reduce or limit human exposure to that disease-causing factor and therefore reduce the incidence of that disease? For instance, by 2050, urban air pollution, specifically particulate matter (PM) and ground-level ozone, is on track to be the main environmental cause of mortality worldwide, outpacing water pollution. At present, global premature deaths from exposure to PM are approximately one million per year. That number is expected to more than double by 2050, to almost 3.5 million premature deaths per year. Most of these deaths will be in countries and cities where the air quality continues to exceed the recommended PM level of 10 ug/ml.[4] If a person lives in an urban area, these numbers are especially concerning. Is it safe to exercise outside? How serious are the risks? How can cities and countries reduce this risk?

These concerns have driven global research efforts aimed at understanding the environmental aspects of diseases, including cancers and developmental abnormalities. Current scientific efforts are directed at the cellular level where injury to the genetic makeup of the cell can have ominous implications for the individual and possibly for several generations of progeny that will carry the genetic errors as inherited traits. A study of mutations, birth defects, or cancer must start with knowledge of the genetic code.

While environmental disease presents a global health risk, the specific environmental factor may vary. For instance, in the United States, many cities are clogged with smog and ground-level ozone during the summer months. Geography may play a factor. Smog builds up in cities that have mountains nearby, like Los Angeles and Santiago, Chile, creating more significant air pollution than in cities where winds disperse pollutants. Environmental factors causing disease may vary by a country's level of development, wealth, and sometimes, its priorities. Aging infrastructure or lack of infrastructure may impact water quality and sanitation, major environmental factors. Levels of agriculture may also affect the degree of environmental disease-causing factors. In the European Union, there is tremendous concern over the continued use of the chemical glyphosate, the main ingredient in the world's bestselling pesticide. Although health effects of glyphosate remain unclear, the chemical has been found in 45 percent of Europe's topsoil and there are strong anecdotal links and some scientific evidence suggesting a connection to increased incidence of NHL, a type of cancer.[5,6]

Awareness of the impact of environmental disease has developed, albeit unevenly, across the globe over the past 50 years. Environmental disease follows a pattern of cause and effect, exposure and response. For environmental disease to be managed and decreased, there must be an understanding of the link between exposure to the environmental factor and a related deleterious health effect. Establishing these causative linkages takes years of research, information gathering, and analysis.

In the United States, the possibility of environmental disease, specifically, that manufactured and discarded chemicals in our environment could potentially produce harm, erupted into the public consciousness in 1978 when the New York State Commissioner of Health declared Love Canal in New York a health emergency with his document, "Love Canal: Public Health Time Bomb."[9] A firestorm of activity lasting 2 years revolved around this revelation. The resulting emotional intensity accelerated the condition to a point where the federal government was forced in 1980 to relocate more than 700 Love Canal homeowners at a cost exceeding $30 million. The public, now aware and critical of government, wanted answers and immediate resolution in December 1982, when local flooding spread dioxin throughout Times Beach, Missouri. The dioxin was a contaminant in waste oil sprayed as dust control at a horse arena and onto unpaved roads in the mid-1970s.

Newly conscious of these risks, the public responded to news about chemicals and radionuclides that could cause negative health effects. Public advocates such as Ralph Nader, Lester Brown, and Sam Epstein authored books vigorously challenging humanity to stop polluting the environment with synthetic chemicals. These and other books speak of "chemical violence...approaching epidemic proportions," "the merchandising of poison...a huge global growth industry," and "toxic chemicals (that) threaten every man, woman, and child."[10–12] Environmental advocacy has found its way onto the television airwaves in documentaries such as the *The Poisoning of America* narrated by Hugh Downs.[13] Award-winning films like *An Inconvenient Truth* and *Erin Brockovich* and dedicated social media further share useful environmental health information about the environment.

Globally, awareness of environmental risks and related disease often develops after a local disaster such as the Bhopal, India, chemical release in 1984 that killed thousands or the Tianjin, China port explosion in 2013 that killed 50 people and injured 700.[14] Unfortunately, poor and middle-income countries often have fewer resources devoted to maintain good air and water quality, let alone prevent environmental disasters that kill and maim. Further, as the country's population suffers, the country itself loses productivity due to environmental disease and premature death. Even countries with adequate resources to enforce regulation may choose not to, perhaps for political or economic reasons. Just as there is a cost from environmental disease, there is also a cost to ignoring environmental disease—and it is not insignificant. Table 4.1 outlines the cost in lives and years lost to death and disability, from a number of diseases seen around the globe. Helpfully, the research indicates the proportion of the disease related to the environment, which indicates that things can be improved.

TABLE 4.1

The Environmental Component and Cost of Disease

Disease	Proportion of the Disease Attributable to Environmental Factors	Burden of Disease, in DALYS[a] from Preventable Environmental Risks	Global Deaths Attributable to the Environment
ASTHMA	44%	11 M	169,449
CARDIOVASCULAR DISEASES	31%	119 M	5,000,000
CHRONIC OBSTRUCTIVE PULMONARY DISEASE	35%	32 M	1,193,590
MUSCULO-SKELETAL DISEASES	22%	23 M	32,000
LOWER RESPIRATORY INFECTIONS	35%	52 M	566,361
MALARIA	42%	23 M	258,702
CANCERS	20%	49 M	1,600,000
DIARRHOEAL DISEASES	57%	57 M	360,751
NEONATAL CONDITIONS	11%	26 M	270,087

Source: Prüss-Ustün, A. et al., *WHO Report, Preventing Disease Through Healthy Environments: A Global Assessment of the Burden of Disease from Environmental Risks*, World Health Organization, 2016.

[a] Calculated in Disability Adjusted Life-Years, a measure that combines years of life lost to death and years of life living with a disability from the illness.

Ironically, sometimes chemicals that are meant to decrease an environmental health risk may have an opposite effect. An example is methyl tertiary-butyl ether (MTBE). MTBE was slated to be an additive to automobile fuel to foster more efficient combustion. This was meant to not only reduce the amount of toxic chemicals released into the air but also to decrease the amount of fuel burned altogether. However, MTBE has been found in increasing amounts in drinking water supplies throughout the country, including Long Island, New York. Since MTBE has been classified as a possible carcinogen by the U.S. Environmental Protection Agency, this is now a health risk.[15]

Defining the Term "Environment"

"Environment" can be defined as the "sum of all surrounding physical conditions in which an organism or cell lives, including the available energy in living and non-living materials," while defined elsewhere as "the surrounding conditions or influences that affect an organism or the cells within an organism."[16,17] The term environment, in its broadest definition, refers to personal and cultural behavior including smoking, diet, alcohol consumption, sexual and reproductive patterns, workplace, infections, food additives, and pollution along with the strictly physical environment.[7] In this chapter, "environment" as it relates to disease includes not only the natural world, but also the world humans have made.

Therefore, as noted earlier, pollution is a component part of that environment. Pollution may be defined as "the presence of a foreign substance—organic, inorganic, radiological or biological—that tends to degrade the quality of the environment so as to create a health hazard."[16] For instance, exposure to certain pesticides, such as paraquat and rotenone, have been associated with a higher incidence of development of Parkinson's disease, with users 2.5 × more likely to get the disease than people who were not exposed to these pesticides. Organo-chlorine pesticide exposure, such as DDT and Dieldrin, has also been associated with Parkinson's disease.[18] Previously, Parkinson's disease had been thought of as a purely genetic disease.[8] This new association with environmental pollutants may be the first in a series of diseases previously thought of as genetic, but which may have an environmental component.

In another example, consider the impact of pollution on premature death. In 2015, pollution caused 16 percent of all global deaths, which translates to nine million premature deaths. Many of these deaths are concentrated in certain countries with bad air pollution. Of those 9 million deaths, 2.5 million of them were in India. The majority of these, about 1.8 million deaths, are attributed to polluted air. These are deaths from diseases with a large environmental component, including heart disease, respiratory tract diseases, chronic obstructive pulmonary disease, and lung cancer.[19,20]

What Is Disease?

Disease is defined as "a definite pathological process having a characteristic set of signs and symptoms which are detrimental to the well-being of the individual."[16] Diseases may have a number of different causes, some controllable, such as washing hands with soap and water to prevent the spread of germs, and some uncontrollable, such as one's age. The environmental component of a disease is considered controllable, in that if there are

TABLE 4.2

Human Disease Forms

1. Infectious disease
2. Physical and chemical injury
   - Mechanical
   - Thermal
   - Ionizing radiation
   - Chemical
3. Developmental disease
4. Neoplastic disease
5. Nutritional disease

improvements made, the disease incidence will decrease. Studies have shown that many common diseases have a significant environmental component.

Environmental disease may be classified by the type of disease, as seen in Table 4.2. These diseases may be infectious in nature, carried by water or insects. They may occur from chemical injury or physical injury, including radiation or heat. Or these diseases may affect newborns and children via developmental defects occurring during embryogenesis. Neoplastic diseases such as cancer also may develop from multiple genetic defects caused by environmental factors. Nutritional environmental diseases are often the consequence of poverty and its related environmental conditions, such of lack of access to healthy food and clean water. We will look at each of these disease groups now.

Infectious Disease

The greatest mass of humanity suffers from infectious diseases caused by the pathologic processes of a microbial agent or virus in the body. Disease agents may be intracellular, extracellular, or both. The agents of such disease that are always inside the cell or intracellular include viruses (HIV, measles, mumps, human papilloma virus), chlamydia (often associated with nongonococcal urethritis), and rickettsia (epidemic typhus, Rocky Mountain spotted fever). Metazoans are multiple-celled animals and describe organisms that are always extracellular because of their significant size and would include parasitic nematodes and roundworms (Ascaris, Trichinella); the cestodes or segmented flatworms (tapeworms); and the trematodes including the liver, lung, and intestinal flukes. Insects may also be included in the category of extracellular parasites and would include mites, ticks, and lice (Table 4.3). Organisms that are intracellular or extracellular include bacteria (strep throat, botulism, staph infections, gonorrhea), mycobacteria (tuberculosis), fungi (histoplasmosis, ringworm, jock itch [candidiasis]), and protozoa (giardiasis, amebiasis). In order for a disease to perpetuate itself, the host organism must be in constant supply. Such sources can be living organisms or inanimate objects that provide the conditions for the organisms

TABLE 4.3

Agents of Infectious Disease

Always Intracellular	Intracellular or Extracellular	Always Extracellular
Virus	Bacteria/mycobacteria	Mites/ticks
Chlamydia	Fungi	Metazoa: Tapeworms/nematodes
Rickettsia	Protozoa	

to survive and multiply, and also provide the conditions necessary for transmission. These sources are called reservoirs. Reservoirs may include humans, animals, and nonliving objects. As human populations grow and spread into previously unsettled regions, it is likely that they will be exposed to new reservoirs and potentially new diseases never experienced before. Diseases such as yellow fever, ebola, and possibly the HIV virus may have been introduced to human populations in this manner.[23]

Most human diseases are harbored within the living bodies of humans and are transmitted directly or indirectly to others. Diseases may be transmitted from persons who exhibit signs of the disease or may be transmitted by persons known as carriers who exhibit no symptoms but harbor and transmit the disease organisms. Some people may never develop symptoms while others may pass the disease to others during a latent or symptom-free stage. Humans serve as reservoirs for such diseases as AIDS, hepatitis, and tuberculosis. A number of wild (sylvatic) and domestic animals serve as reservoirs of disease agents that infect humans. Diseases that are spread mostly from animals to humans are called zoonoses, and may include Lyme disease, Eastern equine encephalitis, Rocky Mountain spotted fever, rabies, bubonic plague, salmonellosis, toxoplasmosis, trichinosis and more than 140 other known zoonotic diseases. Some disease agents that have animal reservoirs may mutate in these animals such that they are able to change their characteristics enough that they continue to cause epidemics year after year. One such disease is influenza or "the flu." In 1997, a strain of influenza that had been previously thought to only infect birds was linked to deaths in Hong Kong. It was a particularly virulent strain and there was concern that a pandemic similar to the large influenza pandemic of 1918 might ensue. The 1918 pandemic infected nearly every person on Earth and was responsible for an estimated 30 million deaths.[24] The major nonliving reservoirs of most human diseases are soil and water where disease-producing fungi and bacteria such as *Clostridium perfringens* or *Cl. botulinum* occur. Many organisms can survive in water including parasites and bacteria that produce gastrointestinal disease and that may be passed to the water from human feces or animal wastes.

The major routes of infections for most infectious disease agents include (1) direct contact, (2) vehicles, and (3) vectors. The types of contact transmission involved in transmission include (a) direct, (b) indirect, and (c) droplet contact. Direct contact transmission occurs when the disease is passed directly from a source to the host by physical contact without an intervening object. Also known as person-to-person contact, this type of transmission may occur through kissing, touching, or sexual intercourse. HIV, infectious mononucleosis, influenza, the common cold, and measles are examples of diseases transmitted in this way. When the disease agent is transmitted from a reservoir to a host by a nonliving object, the process is called indirect contact transmission. The object on which the disease is transmitted is often referred to as a fomite and may include eating utensils, glassware, shared toothbrushes, tissues, toys, syringes, and bedding. As examples, hepatitis and HIV viruses may be transmitted by contaminated syringes. Organisms such as the viruses involved in the common cold may be transmitted by droplets, whereby the organisms are contained within mucous droplets expelled into the air via sneezing, coughing, or even talking. These droplets travel short distances to be inhaled or ingested by persons nearby (usually less than approximately a meter).

When disease agents are transmitted by air, water, food, bodily fluids, like blood or saliva, or via drug-use, this is referred to as vehicle transmission. Airborne transmission refers to disease agents that travel a distance greater than one meter by air and are usually carried by dust particles or extremely small droplets. Fungal spores, staphylococci, and measles virus can be transmitted by particles in the air. Diseases spread by water usually come from water sources that have been contaminated by sewage or animal waste that contains disease-causing organisms. Foodborne disease usually occurs when food is improperly refrigerated or insufficiently heated, which permits the disease organism to grow to numbers capable of producing toxins or invasive numbers of bacteria. In many cases, the disease organisms are carried in the flesh of a food product that has been previously infected (e.g., tapeworm or trichinosis) or has come into contact with the offending organism and serves to simply transport the organism (some forms of hepatitis). Diseases that are transmitted by air, water, or food are discussed in subsequent chapters.

The last means of transmission, vectors, are living organisms that carry pathogens from one host to another, either from another human that is infected or from an infected animal. Most vectors of significance to humans are arthropods (joint-legged invertebrates) including insects, spiders, and crabs. Insects represent the largest and most significant vectors of human disease. When the arthropods transmit the disease by simply carrying the organism(s) on the feet or other body part to a person's food or an open sore on the person's body, then the transmission is passive and the process is referred to as mechanical transmission. In such diseases, the disease agent goes through some reproductive stage in the vector and is transmitted by a bite, fecal matter, or vomit of the vector in an active process referred to as biological transmission. Lyme disease and diseases such as ehrlichiosis are spread by the hard-bodied genus *Ixodes* ticks such as deer ticks or Lone Star ticks, depending on the region of the United States. In Europe, Lyme disease is also spread by such hard-bodied ticks and is seen most often in people living in Central Europe, particularly in the Czech Republic, Estonia, Lithuania, and Slovenia.[25]

Some of these new vector-borne diseases are not easily detectable, and therefore are more difficult to treat than Lyme disease.[26] The parasite that causes malaria is transmitted by the bite of a female anopheline mosquito. The parasite (*Plasmodium sp.*) undergoes a sexual cycle inside the mosquito before it infects a human by a mosquito bite. Since the sexual cycle occurs in the mosquito, the mosquito is known as the definitive host. The asexual reproduction takes place in the human who is called the intermediate host. Vector-borne diseases like malaria have a huge worldwide importance in that they cause debilitating disease in millions of people while also hurting the social and economic climates of many countries.

Diseases that already exist in a community and persist at a low but constant incidence are said to be *endemic*. When there is marked increase in the incidence of disease within a limited area affecting ever-increasing numbers of people, this is referred to as an *epidemic*. Should the epidemic spread throughout the world, as in the case of various outbreaks of influenza, it is called a *pandemic*. Many diseases such as measles are acute and have a rapid onset, are usually self-limiting, and are of relatively short duration. Diseases having a slow onset and lasting for extended periods are said to be chronic and may include cancer, emphysema, some forms of heart disease, or AIDS.

Physical and Chemical Injury

Physical injury includes mechanical, thermal, and radiation injuries. Mechanical injury results from a transfer of a damaging excess of kinetic injury to tissues resulting in abrasions, lacerations, punctures, contusions, broken bones, and projectile wounds.[27] Automobile accidents, falls, shootings, stabbings, and other forms of trauma to the human body continue to be among the most common forms of death and bodily injury in the world. The leading cause of injury-related death in the United States is motor vehicles, accounting for nearly 40,000 deaths per year with more than 40 percent of those deaths in the 16–19-year age group. Alcohol is a contributing factor in nearly half of all motor vehicle deaths.[28] Falls are the leading causes of accidental deaths in the home, with nearly 32,000 people dying from falls in a single year in the United States alone.[29] Nearly 40,000 children are injured each year from cribs, high chairs, and bunk beds. Another 27,000 are injured in infant walkers, especially from falling downstairs. More than 750,000 falls and injuries occur from accidents on stairs, while falls from ladders account for another 100,000 injuries each year in the United States. Elderly persons above 75 years of age are 12 times as likely to die from falls as all other groups combined.[27] Violence within the home and the community has become a significant public health problem. Firearm-related deaths include homicides, suicides, and unintentional shootings. In the United States, deaths from firearms total nearly 38,000 per year, with more than half of these being suicides.[29] U.S. gun deaths classified as homicides have hovered at around 12,000 for several years. The majority of these deaths are males aged 15–24 years of age. Notably, the gun homicide rate is 25.2 times higher in the U.S. than other developed countries.[30]

Thermal injury occurs when tissues are exposed to extremely low or extremely high temperatures. Low temperatures can produce hypothermia or frostbite. Frostbite occurs when extreme cold freezes body water and forms intracellular crystals that burst cell membranes to kill affected cells. Very high temperatures can produce hyperthermia or burn injuries. Burn injuries kill about 500 people a year in the United States. These deaths are generally from residential fires and most commonly occur among the very young or the very old, the Black or Hispanic populations. Major reasons for the fires include children playing with matches, faults in electrical devices including heating sources, and careless smoking.[28]

Ionizing radiation occurs when radiant energy interacts with matter to form charged particles. This may include electromagnetic radiation (gamma and x-rays) or particle radiation (alpha, beta, and neutron). Exposure to radiation may occur from radon gas in the home, diagnostic use of x-rays, use of radionuclides in medicine, and emissions from catastrophically failed nuclear power plants such as Chernobyl and Fukushima or the testing of nuclear bombs. Bodily injury results from the radiolysis of water molecules in tissue that form free radicals (ion pairs). These free radicals react immediately with components of the cell causing cross-linking of macromolecules and changing the tertiary or three-dimensional structure of the molecule. These changes may impair the function of proteins, deoxyribonucleic acid (DNA), and other molecules, resulting in reactions from subtle interference to cellular death.[27] Ionizing radiation can also cause cancer through DNA damage, which we discuss later in this chapter.

We are all exposed to some level of ionizing radiation, occurring naturally from rocks on the earth and rays from the sun. In addition, there is a level of background radiation globally due to the nuclear tests and explosions that have taken place since the 1940s. Radioactive isotopes released in accidents have included Iodine-131 (I-131), Cesium-137 (Cs-137), strontium-90 (Sr-90), and plutonium-239. Once in the body, Iodine-131 accumulates in the thyroid gland, where it may cause cancer. As with most environmental disease factors, children are more susceptible to disease from ionizing radiation.[31]

Since we cannot see or feel this background radiation, perhaps it can best be understood as it relates to inanimate objects, specifically steel. Since 1945, there have been about 1900 atomic weapons tests. All of these tests released radiation into the Earth's atmosphere. This radiation is pervasive in the air we breathe. It is a small amount but it is there. Now consider steel, an alloy of iron and carbon that is used to construct many things, from buildings, to ships, to medical instruments. Steel is made by passing air through it to "cure" it. Medical instruments that require precise measurements of radiation cannot be made with steel created after the first atomic bomb. All that steel has background radiation in it. The only steel that has no background level of radiation comes from sunken ships made before the 1940s, as water is a good block of radiation.[32]

Chemical injuries result from exposure to chemicals that produce a deleterious effect on organ systems by stimulating or depressing metabolic function(s). Some chemicals may function at trace levels to promote or stimulate organ response while larger doses may slow activity or destroy it entirely. The major mechanisms for chemical injury may be categorized as (1) interfering with enzyme activity, (2) directly combining with a cell component other than enzymes, or (3) producing a secondary action in which a chemical causes the release or formation of a more harmful substance. In this last case, allergens represent a good example in that otherwise harmless substances such as pollen, animal dander, or even certain foods can cause life-threatening allergic reactions in some people.

Some chemicals produce a systemic disease. Systemic refers to conditions or a process that affects the whole body. Many environmental pollutants such as lead, mercury, cadmium, and chlorinated hydrocarbons are systemic poisons in that they affect a broad array of organ systems. Systemic pollutants are normally ingested, though inhalation and skin absorption may also occur. Once in the body, they are distributed through the bloodstream to various organs where common manifestations include central nervous system disorders, stomach and bowel disorders, and disruptions to the blood-forming tissues. We discuss toxins in-depth in the next chapter.

Developmental Disease

Every individual within each species is very much a representation of the information contained in the genome together with the expression of those genes in the development of that individual. Developmental disease occurs when faults or mistakes occur within the genes (or chromosomes) or when stages in fetal development are disturbed. Mistakes occur within the genetic code when the genetic information is disturbed by chemical or radioactive insult, inherited errors, or random (spontaneous) errors in transcription or replication. Major categories of genetic disease include (1) single gene defects or (2) cytogenetic defects (abnormalities in the number or structure of chromosomes).[27] When defects arise during the embryonic period of development, these are said to be developmental or *teratologic* defects. While some teratologic diseases are inherited and others are produced by cytogenetic defects, most are the result of a disruption of normal processes during embryo development in an individual whose genetic makeup would otherwise have produced a disease-free individual. The causative factors in this case are extrinsic or extra-genetic factors. Chemical agents that produce these disruptions during embryo development are referred to as teratogens. This is most often an environmental disease in that external factors are involved including maternal behaviors (tobacco and alcohol use, inadequate nutrition, use of various medications such as Accutane or DES, stress), infectious diseases including toxoplasmosis or German measles, or exposure to chemical and physical agents in the home or workplace environment. The defects occurring during the embryonic stages of development are said to be congenital defects. More detailed information on developmental disorders is given later in this chapter.

Neoplastic Disease

The new and uncontrolled growth of abnormal tissue from the transformation of normal body cells is referred to as neoplasia. Such growths are referred to as tumors or neoplasms. Tumors may be benign or malignant and vary dramatically in form, rates of growth, potential injury to host, and appearance. Malignancies may derive from a number of exogenous factors (arising from the environment) and/or endogenous factors (originating from within body). The exogenous factors may include (1) habits (tobacco use, poor nutrition, alcohol use, sexual activity), (2) ionizing radiation, (3) chemical exposure, (4) environment (socio-economic, geographical, and occupational differences), and (5) oncogenic viruses. Endogenous factors contributing to cancer risk may include (1) gender, (2) age, (3) hormonal imbalance, (4) impaired immune system, and (5) genetic predisposition.[27] This topic is covered in more detail later in this chapter.

Nutritional Disease

The consequences of overpopulation including an inadequate food supply in many parts of the world have been discussed in Chapter 3. Inadequate diets are also seen to occur in developed nations such as the United States. Fast food represents an increasing percentage of the American diet. Critics of this trend point to the over-processing of the food for the purpose of homogeneity. This processing removes much of the nutritional value from the foods, often while increasing the sugar and fat content of the food. Obesity in the United States has risen from 8 percent of the population in 1980 to 20 percent currently.[33] In lesser-developed countries (LDCs), nutritional diseases are often coupled with political, cultural, or emotional problems that make solutions more difficult than simply increasing caloric intake. Malnutrition, as a consequence of inadequate diet, is often associated with ignorance and poverty. In the broadest definition of environment, malnutrition may be considered an environmental disease, but it must be known that in most nutritional diseases there is a social or emotional state that is contributing to the problem. The consequences of an inadequate diet are far more disturbing to the child than to the adult because the child needs food for growing in addition to functioning. Deprivation of nutrients at vital stages in development may result in stunted growth, delayed maturation, permanent organ damage, or even death. In certain LDCs in Africa, many children 1–3 years old display a disease of dietary insufficiency known as kwashiorkor. These children are severely protein deprived and may exhibit anemia, diarrhea, graying or reddening of hair, edema, ascites (distended belly), fatty degeneration of the liver, failure to grow, and apathy with extreme weakness. Nutritional marasmus may appear in children one year of age who lack sufficient food and show signs of wasting, failure to grow, and complications of infections with multiple vitamin deficiencies. Infantile scurvy develops from a lack of vitamin C (ascorbic acid) in the diet and consequently the synthesis of collagen and mucopolysaccharides is disturbed. This results in widespread hemorrhages, anemia, failure to heal, tenderness in limbs, and possible cardiac failure with sudden death.[27]

Are you interested in learning more about environmental disease? Visit <https://people.umass.edu/~envhl565/> and click on one or more of the following sites under Chapter 4.

General Links to Environmental Disease

- OncoLink: A University of Pennsylvania Cancer Center Resource
- www.organogenesis.com
- Center for Children's Environmental Health and Disease Prevention Research
- Welcome to the March of Dimes Home Page
- Your American Cancer Society (ACS)
- e.hormone: Your gateway to the environment and hormones
- NCI's CancerNet Cancer Information
- LBNL ELSI Home Page

- Gene Gateway: Exploring Genes and Genetic Disorders
- Human Mutation
- CDC
- Welcome to PubMed
- Human Genome https://www.genome.gov/10001772/all-about-the-human -genome -project-hgp/. WHO http://www.who.int/phe/en/

The Role of Genetics in Environmental Disease

Structure and Function

DNA is the thread of life that weaves through all living things. It defines us and programs both the commonalities and differences among and between species. DNA is a spiral staircase-shaped molecule over three billion steps long, compressed in the nucleus of living cells. It is the informational macromolecule that is the basis of heredity. The storage of biological data occurs within DNA and this information is expressed via transcription and translation of that data into proteins. Its chemical structure has been known for more than 45 years.

DNA is located within the cell nucleus of each of the human body's 100 trillion cells (except mature red blood cells) (Figure 4.1). Each nucleus contains 46 chromosomes

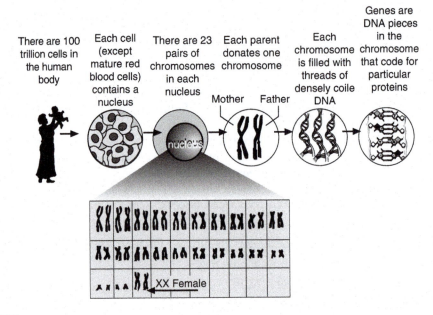

FIGURE 4.1

A perspective of DNA in the human cell. (Adapted from Elmer-Dewitt, P., The genetic revolution, *Time*, January 17, p. 46, 1994.)

arranged in 23 pairs. Each parent contributes one of the chromosomes in every pair. The chromosomes in eukaryotic cells are tightly coiled (condensed) around a group of proteins known as histones. Genes are segments of DNA that code for a particular protein. The DNA of each species or cell type is species-specific and reproduces that species precisely with the exception of mutations. When viewed in a concentrated, pure form outside the cell, it is a clear, thick, viscous liquid that looks like water but has the consistency of mucous. When unfolded, DNA is a very long, thread-like molecule that is formed by four recurring subunits called mononucleotides. The mononucleotides have three components: one purine or pyrimidine base, one molecule of phosphoric acid, and one molecule of a sugar called 2-deoxy-D-ribose. Except for the base component, these nucleotides are identical and attached to each other in a long chain by phosphodiester links between carbon atoms in each molecule (Figure 4.2). The base components are either adenine and guanine (purines) or thymine and cytosine (pyrimidines). These four different recurring mononucleotides are the basic building elements of the DNA coding system from which all proteins are ultimately formed including over 1,000 enzymes, numerous structural proteins, lens protein, myosin in striated muscle, hemoglobin in blood, nucleoproteins, and mucous. The proteins are essential for growth, the building of new tissue, and the repair of injured or broken-down tissue. The genetic message for the production of protein is imparted initially by the specific sequence of the four basic mononucleotides forming the long DNA strand.

FIGURE 4.2
The chemical backbone structure of DNA. (Adapted from Lehninger, A.L., *Bioenergetics: The Molecular Basis of Biological Energy Transformations*, W.A. Benjamin, New York, 1965, Chapter 11.)

As an example, starting along one section of the DNA strand you might find a sequence that reads A_C_G_A_A_C_G_T_. Chromosomes are then gene sequences along strands of DNA compressed and floating in the fluid of the cell nucleus. A thousand genes may be contained in a single chromosome, and it is the precise replication (copying) of this genetic information that ensures genes are duplicated and transferred to daughter cells. The two strands intertwine to form a double-helical structure so that the two molecules run in opposite directions (Figure 4.3). The pair nucleotide bases fit together in a precise way held together by hydrogen bonds between the base pairs and by hydrophobic bonding (Figure 4.3). Each strand of DNA contains the same information because the base pairs are complementary, in that adenine is always paired with thymine, and guanine is always paired with cytosine. The chromosomes do not exist as single structures in higher plants and animals but are homologous pairs such that every chromosome has a near twin counterpart.

Dividing cells provide each daughter cell with one complementary strand of the parent and then synthesize a complementary strand of the parent to produce double-stranded DNA that is a duplicate of the parent DNA. This process is referred to as replication. Consider the human genome of 23 chromosomes pairs as 23 chapters in a book that describes the complete set of instructions for making a person. Within each chapter are sentences (genes) composed of words that are each three letters in length. Each chapter has its own unique sentences and words. There are only four letters available to make these words and they are A (adenine), G (guanine), C (cytosine), and T (thymine). There about 25,000 sentences (genes) in this book of 23 chapters. These sentences instruct the body in how to make proteins. In fact, tens of thousands of different proteins are dictated by the sequences stored in the genetic code, the 23 pairs of human genes.

In this genetic code, a group of three nucleotide bases (three letters) is called a codon and specifies the position of a single amino acid in a protein molecule. Proteins are composed

Double helical DNA

FIGURE 4.3
The double helical structure of DNA and the mechanism of hydrogen bonding.

of specific sequences of amino acids linked by peptide bonds and the sequence of amino acids ultimately determines how the completed protein folds or layers into specific three-dimensional shapes. The shape of the completed proteins is critical to its function. For example, enzymes are proteins that interact with a substrate much like a key in a lock to catalyze the formation of a new molecule or product. Changes in the three-dimensional structure of the enzyme could reduce or eliminate this lock and key interaction and cause outcomes that might range from barely perceptible to severe, including death.

Protein Biosynthesis

The DNA molecule is locked in the nucleus and cannot reach the ribosomes outside the nucleus in the cell cytoplasm where protein synthesis occurs. The DNA must first be transcribed to another molecule called messenger RNA or mRNA. Messenger RNA is quite similar to its DNA counterpart except it is single stranded, contains the nucleotide base uracil instead of thymine, and the sugar D-ribose instead of 2-deoxyribose in the four mononucleotides. The complementary mRNA once formed, diffuses through the nuclear membrane into the cytoplasm and attaches to the surface of cellular ribosomes where protein synthesis occurs. The mRNA now carries the code, which will direct the specific sequence of amino acid molecules to be inserted into the peptide chain(s) that will form the protein. The coding is accomplished through the sequence of three consecutive (triplet) mononucleotide units (codons) in the mRNA that each represents a single, specific amino acid molecule (Figure 4.4). The nucleotide codon triplets (i.e., A-U-G) are read by the ribosome's protein synthesizing machinery. Each codon triplet represents a unique amino acid. Amino acids are first prepared for this process by an activation enzyme for each amino acid that forms a high-energy link between the amino acid and an amino acid carrier molecule called transfer or soluble RNA (sRNA) that is specific to each of the 20 amino acids. The sRNA molecule has a sequence of corresponding nucleotides that recognize and attach to a specific codon triplet. The recognition site for nucleotides on the sRNA is called an anticodon and is a complementary triplet of bases to each mRNA codon. The activated amino acid plus its anticodon forms a specific bond with the corresponding codon of mRNA. Thus, the amino acid tyrosine with its specific anticodon will attach only to the triplet codon sequence of A-U-U appearing as a sequence on the mRNA attached to the surface of the ribosome. Once fixed in position, enzymes form peptide bonds between this amino acid and the peptide under construction, in a specific sequence of amino acids that, when done, will form a complete protein. As the protein synthesis completes, other enzymes peel the peptide chain from the ribosome. A great deal of energy is input into this reaction through the formation and breaking of peptide bonds via adenosine triphosphate (ATP). The accuracy of replication is therefore ensured thermodynamically by the creation of very stable attachments between the triplet codon and its anticodon. However, mistakes do occur, and if a single amino acid is out of sequence, the coding is disrupted, an amino acid may be eliminated or replaced, and the corresponding protein will be altered.

Diseases of Genetics and Development

Genetic Abnormalities

A change in the nucleotide base sequence of DNA is called a mutation. When such changes occur, there is often a resulting change in the protein encoded by that gene. If this protein is an enzyme, that enzyme may become less active or completely inactive as its amino

FIGURE 4.4

A schematic representation of protein synthesis on the ribosome. (Adapted from Lehninger, A.L., *Bioenergetics: The Molecular Basis of Biological Energy Transformations*, W.A. Benjamin, New York, 1965, Chapter 11.)

acid sequence is changed, causing alterations in the active site of the enzyme. Most genotypic changes are harmful or damaging, but some mutations are beneficial. That is how species evolve and change over thousands and millions of years. Many simple mutations are neutral or silent and cause no change in the product of the encoded gene. This occurs when the amino acid that is changed is in a non-vital portion of the protein molecule and the protein function remains. However, other changes cause variations in protein function resulting in a spectrum of decreased or increased activity. These are known as genetic polymorphisms reflecting the existence of several variations of the same gene. As an example, the P450 genes produce enzymes that detoxify chemicals such as certain polyaromatic hydrocarbons (PAH) that occur in tobacco smoke. Variations within the P450 genes can result in decreased activity of the resulting enzymes. This decreased activity may be associated with an increased risk of cancer. Other genes, such as GSTM1, also have been shown to have a natural function of inactivating some carcinogens in tobacco smoke. This gene is deleted in 50 percent of Caucasians, which may increase the risk of lung cancer from secondhand smoke by 2.6–6 times in individuals with the deletion.[34]

Many human diseases are associated with genetic defects. These defects can occur as (1) single-gene defects known as a point mutations or base substitutions or (2) cytogenetic defects in which there are abnormalities in the number or structure of chromosomes. A point mutation occurs when a single nucleotide base is replaced at one point in the DNA molecule with a different base. The DNA replicates with a substituted base pair such that A:T might be substituted for C:G. This base substitution could result in the incorporation of an incorrect amino acid into a synthesized protein, thereby rendering the protein less active or inactive. The resulting effect may be imperceptible or neutral or may result in one of many serious human ailments such as cystic fibrosis (CF), phenylketonuria (PKU), hereditary spherocytosis, alpha-1-antitrypsin deficiency, Gaucher's disease, achondroplastic dwarfism, hemophilia, or certain forms of diabetes (Figure 4.5). Point mutations may occur spontaneously because of occasional mistakes in the process of DNA replication, without any intervention of external factors. There are also agents in the environment including chemicals or radiation that promote or cause such mutations. These agents are called mutagens. When such mutations occur, they become part of all the daughter cells of that cell. If the mutations occur in the cells that lead to reproductive cells, these defective genes may be inherited just as are healthy genes.

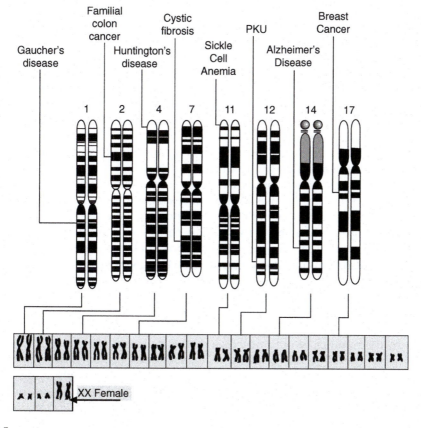

FIGURE 4.5
Chromosome locations of some genetic diseases. (Adapted from Elmer-Dewitt, P., The genetic revolution, *Time*, January 17, p. 46, 1994.)

Chromosomes occur in pairs, so that there are two copies of each gene, one copy from each parent. Gene expression is typically referred to as dominant or recessive. A dominant gene needs only one copy in the pair of chromosomes to be expressed phenotypically, that is, in the physical appearance. Recessive genes require two copies to be expressed. Brown eye color is a trait resulting from a dominant gene while blue eye color is from a recessive gene pair. For blue eyes to appear, both copies of that eye color gene must be for the recessive blue eyes.

With dominant gene mutations, these are usually mild defects since they are always expressed and have been passed down over multiple generations. The hereditary disease, spherocytosis, a form of anemia, is caused by an autosomal dominant point mutation. Dominant genes that are lethal before reproductive age would not persist in a population, as the carriers would die before the mutation could be passed down to progeny. A lethal dominant gene produces Huntington's disease, but this disease is normally expressed in the fourth and fifth decades.[27]

Some genes are co-dominant. Co-dominant genes are partly expressed if present as a single allele (an allele is one of a pair of genes situated at the same site on paired chromosomes) and fully expressed if present on both. If the gene is defective on one chromosome, with a normal gene on its counterpart chromosome, the expression of the gene may be subdued. When defective on both chromosomes, the disease may be fully expressed, and may even be fatal. An example of a co-dominant gene is that which produces Sickle cell disease. The term "sickle cell" describes the distinctive crescent or "sickle"-shaped erythrocytes (red blood cells), which develop in de-oxygenated erythrocytes carrying hemoglobin S. It is a disease seen primarily in the black race. Hemoglobin S differs from normal hemoglobin A by the substitution of a single amino acid in one of the three polypeptide chains that make up the protein portion of the molecule. The misshapen erythrocytes are quite fragile and become "sickled," or curved, and accumulate in the tiny capillaries where oxygen is released to the tissues. This causes vascular obstruction, tiny infarcts, and fibrosis. When the defective gene responsible for sickle cell disease is present on one chromosome, about 40 percent of the hemoglobin will be hemoglobin S. When defective in both genes, the sickle cell disease will be fully expressed leading to severe anemia and probable death in early adulthood, if untreated.

Other genetic diseases are inherited as autosomal recessive, which means that a defective gene must occur in both chromosomes at complementary sites in order for the disease to be expressed. It also means that both parents, if healthy, were each carriers of the defective gene, which they passed along to the child. A person with this type of genetic disease is deprived of an essential protein resulting in a potentially severe disease, such as PKU or CF.[27] There are also dominant and recessive defective genes on the sex chromosomes as well, producing a variety of sex-linked diseases, including hemophilia-A, color blindness, and forms of muscular dystrophy. Since males have one copy of the Y-chromosome, even recessive sex-linked genetic diseases are expressed in men, as there is no other allele to mask the defect.

Changes in the number of chromosomes or the structure of them can also produce genetic diseases. These are referred to as cytogenetic defects. Such defects tend not to be inherited, but result from disorders occurring during meiosis, when eggs or sperm are formed in the parent, or when the embryo is developing. These defects involve large parts of the chromosome and most often appear as deletions or additions of chromosome material. Most of these defects are fatal and fail to survive embryonic development. About 20 percent of fetuses that spontaneously die while in the uterus have this type of genetic defect. There are survivable cytogenetic disorders, however, and one of the more recognizable is trisomy 21. Trisomy 21 or Down's disease is characterized by the addition of an extra chromosome to chromosome number 21, which is expressed phenotypically as various degrees of mental retardation, predisposition to leukemia, congenital heart disease, and characteristic facial features including epicanthal folds of the eyelids causing an oriental appearance of the eyes, flat nose or absent bridge, and low-set ears.

Teratologic Diseases

Diseases or physical defects that arise from exposure to a specific external factor during embryonic development in an otherwise genetically-sound person are called teratogenic diseases and the external factors are teratogens. Teratogens may include ionizing radiation; chemicals such as accutane, dioxin, diethylstilbestrol (DES), thalidomide, organic mercury, and alcohol; pathogenic infections including toxoplasmosis, German measles (rubella), cytomegalovirus, and syphilis; and maternal conditions such as age, stress, nutritional adequacy, smoking, drinking, hormone levels, and previous number of births (Table 4.4). Most teratogens exert their effect during specific critical windows of time in the early stages of tissue and organ formation known as organogenesis. In humans, this is a period occurring from the 18th day to the 60th day after conception with the most sensitive period around 30 days. Exposure prior to this time is very often fatal to the developing fetus while exposure after eight to nine weeks will not produce major organ defects, as the organs will be developed, but can produce blindness, mental retardation, damage to sexual organs, and other physical problems. As an example, the sedative thalidomide, synthesized and marketed in Western Europe in the 1950s, was eventually linked to a birth defect in which the long bones of the arms and legs did not develop so that hands and feet were attached directly to the torso. This condition was known as phocomelia or "seal limb" disease.[24] Most of the women who gave birth to children with this defect had taken thalidomide during the first trimester of pregnancy. Sadly, more than 10,000 such births occurred before the drug was removed from the market.

A similar critical window of exposure was demonstrated for German measles (Rubella). It was discovered in the early 1940s that women who contracted this disease during the first trimester were at high risk of having babies with congenital cataracts, deafness, and heart abnormalities.

TABLE 4.4

Selected Teratogens

Chemicals/Pharmaceuticals	Pathogens
• Accutane (Isotretinoine)	• Cytomegalovirus
• Alcohol	• Rubella virus
• Cigarette smoke	• Syphilis
• Dioxin	• Toxoplasmosis
• Methyl mercury	• Herpes simplex virus
• Thalidomide	• Varicella virus (chicken pox)
• Tetracycline	• Venezuelan equine encephalitis
• Vitamin A (large doses)	

Other
• Ionizing radiation

Source: Chung, W., *Teratogens and their Effects*, Columbia University, New York, 2005.

Just as chemicals ingested by a mother-to-be during critical window of an embryo's development can cause physical abnormalities, so too can the absence of certain important nutritional compounds. The absence of folic acid, a B vitamin, from a pregnant woman's diet is associated with neural tube defects including spina bifida and anencephaly. Spina bifida is characterized by the protrusion of a portion of the spinal cord from the spinal column, while anencephaly is a disorder in which most of the brain is absent. Consequently, the simple addition of this vitamin to the diet of a woman about to conceive can prevent a seriously debilitating and often fatal disease.

There is a growing list of physical and chemical agents that have the ability to produce birth defects, including excessive heat. It should be remembered, however, that 600 to 800 of these agents are teratogenic in laboratory animals, but not necessarily humans. About thirty chemicals are currently known to be responsible for human birth defects.[9,35,36] However, the science of teratology is relatively new and the little that is known rests upon those substances and conditions that produce clear abnormalities. Very little is known about teratogenic substances that produce only mild deficiencies, such as a lack of vigor or slight losses in mental acuity.

New studies have suggested a much stronger link between fetal development and many adult-onset diseases. A study in England found that women who were pregnant during times of severe food shortage gave birth to more children who became obese as adults. It is thought that factors such as food shortages during development may "program" the fetus to retain food, predisposing them to obesity. Other possible diseases that may have a fetal implication may include heart disease, breast cancer, as well as other physiological characteristics such as body size and metabolism.[37] Expectant mothers wishing to protect their fetus from teratogenic effects should be under a physician's care, pursue a healthy diet (including folic acid), avoid smoking and alcohol consumption, reduce stress, and limit exposure to known chemical teratogens and infections known to produce teratogenic effects.

New Approaches in Genetics

Methods of Studying Genes

The artificial techniques for creating recombinant DNA developed rapidly in the 1970s and 1980s permitting the transfer of genes between unrelated species and creating a powerful new science called genetic engineering. These techniques relied upon the discovery of a number of critical tools and techniques including the discovery of restriction enzymes. These are microbial enzymes that are used to cut DNA into several predictable and reportable pieces. These fragments come together to form genetic "maps" that are unique to each individual, as we all contain many slightly different forms of some genes. Most of these polymorphisms are harmless in effect, but invaluable in using DNA as forensic evidence in courtrooms, as well as genetic screening tests.[38] These fragments of DNA from different sources can also be spliced together through the enzymatic action of DNA ligase and so create a recombinant-DNA molecule (Figure 4.6).[39] Once created in a test tube, the recombinant-DNA molecules can be inserted in cells where copies of the gene may be produced, or protein products may be created from the newly inserted gene. The protein product can then be harvested and used for many different purposes. One of the original uses for this process was in the production of insulin. The two small polypeptide chains of insulin are created by inserting the gene for each chain into two different bacteria, which are then cultured in large amounts to produce the polypeptide chains. The polypeptides are linked to another compound normally produced by bacteria and so must be separated from the

The double-stranded DNA is cut with specific restriction enzymes at particular recognition sites.

Once cut, the DNA fragment with "sticky ends" is released to join with a fragment from another DNA source cut by the same restriction enzymes.

The fragments from different sources join to form a linear or circular molecule (as in a plasmid). The final step is to join the backbones of the two DNA fragments with DNA ligase, producing a molecule of recombinant DNA.

FIGURE 4.6

Restriction enzymes help create recombinant DNA. (Adapted from Tortora, G. et al., *Microbiology: An Introduction*, 5th ed., Cummings Publishing, Redwood City, CA, 1995.)

compound and then the insulin polypeptide chains are joined together to form a complete insulin molecule. The genetic manipulation in insulin production required methods for inserting the recombinant-DNA molecules into cells. The method developed relies on a fragment of self-replicating DNA molecule called a cloning vector to which the target gene is attached and transmits the gene from organism to another (Figure 4.7). Once inside the bacterial or yeast cell, the cloning vector will replicate itself including the newly inserted target DNA sequence. The most commonly used vectors are plasmids, which are circular, self-replicating, gene-containing DNA fragments found in most bacteria and some yeasts such as *Saccharomyces cerevisiae*. Restriction enzymes can be used to cut out a section of the plasmid and insert the desired DNA sequence into it. The plasmid must then be inserted into bacterial or yeast cells. One method involves placing bacterial cells in a solution of calcium chloride, which makes them competent to take up the plasmids when placed together in a process called transformation. In addition to human insulin, many useful proteins are developed in this way including erythropoietin, growth hormone, epidermal growth factor, bone morphogenic proteins, taxol, monoclonal antibodies, factors useful in treating hemophilia, and many others.

Another application of gene manipulation is already becoming commonplace in the food that we eat. Techniques are employed to insert DNA from a different species into the genes for crops like corn. These genes encode proteins that may reduce the reliance on fertilizers, make crops more palatable in taste and appearance, and encode defenses against pests. The U.S. in particular has been at the forefront of this technology. Nearly 90 percent of corn planted in the U.S. has genes that confer resistance to herbicides, while 80 percent of corn seeds planted are modified to be resistant to insects. Soybean and cottonseeds may also contain these additions to the genome.[41] China, facing a burgeoning population, has also

FIGURE 4.7

The genetic engineering of insulin. (Adapted from Tortora, G. et al., *Microbiology: An Introduction*, 5th ed., Cummings Publishing, Redwood City, CA, 1995.)

embraced these technologies.[40] These genetic technologies are not without controversy. One possible consequence is the decline of insects and other fauna, including honeybees and the Monarch butterfly.[42]

In addition to making products, recombinant-DNA technology can be used to produce millions of exact copies (cloning), which is useful in genetic analysis. This technology enables researchers and clinicians to determine the exact sequence of nucleotide bases in DNA using techniques that are now largely automated. Such sequencing permits the identification and cloning of mutant genes such as those responsible for CF. This cloned gene can then be used to diagnosis CF and other genetic diseases by way of techniques based on hybridization. Once a gene is located and its sequence of nucleotide bases determined, a hybridization test can be devised to detect the presence of the gene even before the disease is manifested in a process called genetic screening. Such diagnostic tests exist for CF, Huntington's disease, breast cancer, Duchenne muscular dystrophy, retinoblastoma, neurofibromatosis, and many others.

One such hybridization technique is called Southern blotting, and a variation on this has been automated by manufacturers into a probe array (gene chip) that permits rapid genetic screening of a wide array of genetic disorders. In the original Southern blotting technique, DNA is extracted from human cells and fragmented by restriction enzymes (Figure 4.8). The fragments are separated by gel electrophoresis and then transferred to nitrocellulose filter by blotting. The DNA is transferred to the filter as an exact copy of the gel and is exposed to a radioactive probe. If you are screening for CF, the probe consists of cloned DNA fragments with the defective gene and these are radioactively labeled. Once exposed, the probe will pair up with complementary bases (hybridize) on the patients separated DNA on the filter if a close match exists. When exposed to X-ray film, the gene fragment containing the hybridized pair will show up as a band in the developed film and confirm the presence of the abnormal CF disease gene fragment in the patient.[43] The introduction

FIGURE 4.8

The hybridization technique called Southern Blotting. (Adapted from Tortora, G. et al., *Microbiology: An Introduction*, 5th ed., Cummings Publishing, Redwood City, CA, 1995.)

of gene chip technology has greatly simplified and automated this process. In a typical assay, the user extracts and purifies the patient nucleic acid, prepares a fluorescent-labeled target including the genes of interest, and fragments the DNA for hybridization. The material is applied to a single-use probe array (gene chip) that contains hundreds of thousands of different oligonucleotide probes (Figure 4.9). Computer controlled equipment incubates, washes, and processes this material automatically. The probe arrays are then laser scanned to detect any fluorescent-tagged probes indicating a complementary match with the patient's DNA. In this way, genetic disease can be rapidly screened. Such devices are useful in sequence analysis, genotyping, and gene expression monitoring.

Such technology was used in the Human Genome Project, a global effort to map all human genes. The project was completed by April 2003, resulting in a detailed map of human DNA and the sequences of three billion nucleotide bases positioned along about 25,000 genes. Before this project, original human gene number estimates were much higher, between 50,000–140,000 genes.[44]

FIGURE 4.9
A probe array (genechip) used for genetic screening developed by Affymetrix Inc., Santa Clara, California.

The Hunt for Environmental Genes

Similar techniques are also being used in the Environmental Genome Project. The Environmental Genome Project studies the relationships between environmental exposure, gene variations called single nucleotide polymorphisms (SNPs), and disease likelihood. Researchers have identified certain genes and specific mutations on these genes that enhance environmental risk.[45] There is, for example, the gene coding for the enzyme paraoxonase, which produces various versions of the enzyme. These versions differ in detoxifying capability. Ten percent of Caucasians and 25 percent of Asians express a version of the enzyme paraoxonase that converts organophosphates such as the nerve gas sarin into a less toxic form nearly ten times faster than those who do not produce this version.[43] This became evident during the sarin nerve gas attack in Tokyo subway by Aum Shinrikyo cult members in 1995. Genetic variation in cytochrome P450 and NAT families can increase bladder cancer risk by six fold in smokers compared to non-smokers. Every person is suspected of having a unique set of drug or toxin metabolizing enzymes, which leads to differences in response to exposure. Gene variations for environmental exposures may have broad implications for setting environmental policy and regulations. Since the coming trend in environmental risk assessment is to include genetic susceptibility and so determine both distribution and ranges for risk, the Environmental Genome Project may assist in this effort. Substantial technical obstacles exist including: reliability of the gene sequencing process; determining the relevance of polymorphisms to disease; and demonstrating how genotypes relate to phenotypes (physical manifestations).

The project is currently in its second phase, having completed phase one, in which scientists sequenced 647 genes that assist in apoptosis (cell death), DNA repair, drug metabolism, oxidative stress and other cell processes. Phase two involves scanning all the genes from the 95 people involved in the study, broadening the set from only the 647 cell function genes to the entire gene-coding region, called the exome. This information will be used to study further variation across the genome. Part two involves second-generation DNA sequencing technology.[45]

Although the technology is exciting, legal and ethical questions abound. For instance, will workers be obligated to submit to genetic screening before working with toxic chemicals? Will the results of the screening be a basis for job denial? A recent Reddit "Ask Me Anything" talk on January 29, 2018, featured several directors from the National Human Genome Research Institute and sparked additional questions from participants about these ethical and moral issues, especially related to children with a predisposition for a genetic disease.[46]

The Promise of Gene Therapy

Genetic screening will rapidly advance the discovery of genetically based diseases while offering improved capabilities in gene therapy. Gene therapy may involve removing cells with defective genes from a person and replacing those genes with normal or healthy genes and then placing the cells back into the patient where the cells will begin to produce normal product and "cure" the disease. As fantastic as this sounds, these procedures are being attempted by such people as Dr. W. French Anderson, director of gene-therapy programs at the University of Southern California/Norris Comprehensive Cancer Center in Los Angeles. Dr. French and colleagues performed the first federally approved gene therapy procedure in 1990 when they replaced a faulty gene in Ashanthi DeSilva, a 4-year-old girl with severe acquired immune deficiency (SCID or "bubble-boy" disease). The disease results from a deficiency in the enzyme adenosine deaminase because of a rare and deadly mutation. Dr. French et al. extracted some white blood cells from the girl and exposed them to viruses that were genetically engineered to carry the normal gene for the production of adenosine deaminase. When the white blood cells were returned to her bloodstream, her cells began producing the necessary enzyme that restored immunity. The transplanted cells eventually die, requiring follow-up treatments.

Although very promising, the success of this gene therapy story has not been easy to duplicate. Many technological problems in subsequent human trials initially prompted the National Institutes of Health (NIH) Committee on Gene Therapy to suggest NIH funding for such activities be confined to laboratory animals.[47] Two decades of gene therapy experimentation produced a number of adverse reactions and at least one death. It is unclear if these adverse reactions were related to the gene therapy or not.[48] Recently, however, there has been some success. In 2017, the Food and Drug Administration approved three gene therapy products, the most recent of which targets a specific inherited genetic mutation that causes vision loss.[49,50]

Challenges remain. Technical problems revolve around methods of introducing DNA into the right place in human cells, and then having the genes express themselves. Ideally, inserting the "corrected" genes into undifferentiated human stem cells would provide long-term or permanent cures, as the stem cells would serve as an unending source for the differentiated cells throughout the body. Unfortunately, stem cells are few in number and divide infrequently, thereby making the introduction of vector-borne genes difficult. Even so, gene therapy sits on our horizon as a promising technology and a possible solution to curing gene-based diseases. Even infectious diseases, such as AIDS, may be cured via gene therapy. A new treatment under development is to introduce a gene into an HIV-infected person's white blood cells, the cells that HIV targets. This introduced gene would produce a protein that prevents the HIV virus from being able to replicate, thus denying the virus the reservoir that it requires to reproduce.[51]

In the meantime, other practical approaches to using recombinant-DNA technology are in genetic screening where DNA errors can be detected. In some cases, the patient may be advised to follow certain activities and diets that will reduce their risk from the disease. In other cases, it may be possible to counteract the effect of faulty genes by replacing missing or abnormal proteins.

The Ethical Dilemma

The discovery of defective genes is progressing at a much faster pace than the development of treatments for the related diseases, which gives rise to many ethical and legal dilemmas. One such dilemma is whether to inform patients that they carry a gene for an as yet unexpressed genetic disease that is difficult or impossible to treat. As an example, Huntington's disease can now be identified through genetic screening. However, this disease is autosomal dominant and so is invariably expressed, but later in life normally manifesting in the 4th and 5th decades of life as a relentlessly progressive disease leading to mental retardation, emotional disturbances, a loss of self-control, debilitation, and death. There is presently no known cure for the disease. What are the implications of telling someone in his or her 20s that they are destined to develop the disease later in life?

A gene mutation in the gene BRCA1 indicates a 50–85 percent risk of breast cancer when present.[52] The protein produced via this gene is involved in fixing errors in DNA. Knowing they possess a mutated form of this important gene may cause some women to decide to prophylactically have one or both breasts removed before cancer develops. Other women with this mutation have opted to have their breasts removed at the first indications of cancer.[53] Some patients believe that having this knowledge of a defective gene is not liberating and could lead to more immediate and harmful health consequences, including severe depression and even suicide. However, the genetic knowledge gained by a person may be important to other family members. In the event that severe disease or early death is likely, some argue there must be an ethical obligation to communicate it to a spouse or a significant other because it influences their lives, such as the choices of whether or not to have children.[54]

The other ethical and legal questions being asked are whether employers or insurance companies could use the information to identify people that may be insurance or occupational risks. What would a potential employer do if they knew in advance a person was at increased risk to developing cancer or Alzheimer's disease? Prenatal testing could ultimately reveal a comprehensive disease profile of an individual potentially leading to a type of eugenics through abortion, or perhaps stigmatism by society because of revealed inborn genetic deficiencies. The ethical and legal questions are enormous and only beginning to be addressed as the weight of the technology settles upon the social consciousness. Perhaps there is no more pressing need for genetic information than in the area of cancer, and yet the information that comes with it is often fearful.

Cancer

What Is It?

The clinicians and scientists treating and researching cancer now understand that it is a genetic disease caused by multiple mutations in genes responsible for controlling cell division.[55] Most cancers develop from the interaction of genetics with the environment. Genetic factors by themselves (heredity) are said to cause only about 5–10 percent of cancers while the remainder can be attributed to external or environmental factors that act in conjunction with acquired and genetic susceptibility.[56]

The use of the word "environment" in relation to cancer causation is thought to have originated with Dr. John Higginson who was the founder and director of the WHO's International Agency of Research on Cancer. When he reported that nearly two thirds of cancers had an environmental cause, he was "...considering the total environment, cultural as well chemical...mode of life...what surrounds people and impinges on them."[21]

Some of the major factors associated with cancer risk are presented in Figure 4.10, including diet and tobacco usage. Poor diet represents one of the highest risks associated with cancer. Poor diet refers to eating patterns such as a diet high in fat and low in fruits, vegetables, and grains. This particular dietary pattern increases risks of colon, uterus, and breast cancers. In fact, an unhealthy diet may contribute to 35 percent of all cancers. Diet and other risk factors are discussed in more detail later in this chapter under the subtitle "Cancer." The use of tobacco products is thought to contribute to 30 percent of cancers (Figure 4.10). High risks of cervical cancer are associated with certain sexual behaviors such as having sexual intercourse earlier in life, having multiple sexual partners, or contracting specific sexually transmitted diseases.[22] Estimates of cancer risk associated with reproductive behavior suggest that such risks account for 1–13 percent of all cancer deaths with the best estimate being 7 percent. Medicines and some medical procedures are also included as cancer risks such as DES that caused reproductive cancers in the offspring of women who used the drug during pregnancy to prevent premature births. DES is a non-steroid estrogen that was used as a substitute for natural estrogenic hormones.

The majority of cancers result from a combination of exogenous (external environmental) factors and endogenous (internal) factors that facilitate the accumulation of genetic errors on the prolonged pathway to cancer. Many of these factors may enhance susceptibility in this process (Figure 4.11). Endogenous factors are those from within, which include gender and ethnicity, health status, age, and heredity (genetic polymorphism). External, environmental factors include cigarette smoking, inadequate nutrition, exposure to cancer-causing

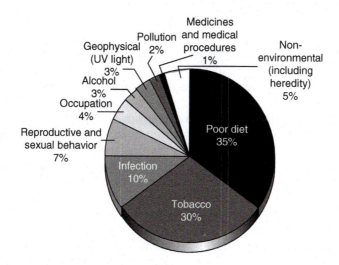

FIGURE 4.10

Proportions of cancer deaths associated with various environmental factors. These are best estimate figures. (Adapted from *Morbidity and Mortality Weekly Report*, June 19, p. 47, 1992.)

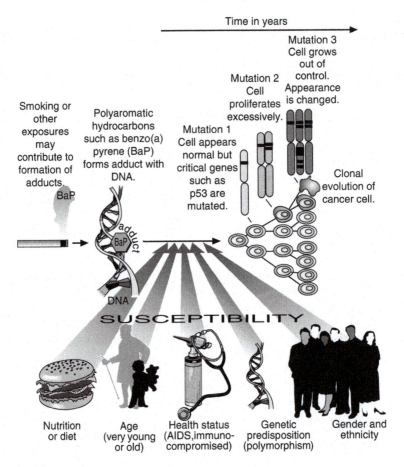

FIGURE 4.11
The evolution of cancer cells and role of susceptibility. (Adapted from Perera, F.P., *Science*, 278, 1068–1073, 1997.)

chemicals and physical agents (ionizing radiation), and oncogenic viruses. It has been hypothesized that frequent cellular telephone use may increase the risk of certain types of brain cancer but studies have been inconclusive.[57] Eventually, a single abnormal cell accumulates enough mutations to bypass all of the regulatory mechanisms that normally control cell growth. Rather than dying, it divides again and again, accumulating further mutations that discard or silence protective genes that ordinarily monitor replicating DNA for errors. The malignant cells grow into a small mass that stimulates the development of nutrient-bearing blood vessels in a process called angiogenesis. The blood vessels allow further growth of the tumor and also establish a pathway for the slippery cancer cells to metastasize to other body sites.[58] Cancer produces its damaging effects through (1) local damaging cell and organ effects, (2) metastasis, and (3) systemic effects. Local damaging effects include the obstruction of airways, urinary tract, and digestive tract, resulting in infections and stasis. Cancer can erode blood vessels causing acute or chronic hemorrhaging or the perforation of intestines or the stomach, causing peritoneal infections. Cancerous adhesions may develop in the abdominal or thoracic organs. When bone marrow is affected, healthy bone marrow is replaced by cancer, leading to anemia and thrombocytopenia. Very often, malignancies will travel, producing tumors at distant sites via a process called metastasis. Tumors may also produce systemic effects causing (1) endocrine tissues to produce either too much or too little of essential hormones, (2) alteration of intrinsic blood coagulation process, (3) autoimmune inflammation of the skin, and (4) wasting away.[27]

How Does Cancer Develop?

Cancer develops from multiple mutations in the DNA, often in the genes producing proteins that control cell division. Some of these genes are tumor suppressor genes, which produce proteins that can repair defective cells or help destroy damaged cells. When these genes undergo mutation, the DNA repair system breaks down. In a typical pathway leading to cancer, a chemical agent such as benzo(a)pyrene (BP), associated with smoking cigarettes, combines with human DNA to form an adduct (carcinogenic residues bound to DNA) that leads to increased mutations that eventually accumulate and lead to cancer. The smoke from cigarettes contains as many as 4,000 compounds with more than 40 of these known to be carcinogenic. Included among these are BP, benzene, and arsenic. The act of smoking (or being exposed to passive smoke) brings these substances into intimate contact with the delicate tissue of the respiratory tract. Other defective genes (such as ras) produce proteins that inappropriately stimulate cell division. Collectively, these cancer-causing genes are called oncogenes.

Cancers are thought to emerge through a process known as clonal evolution whereby a single cell develops a genetic mutation that enables it to divide even when normal cells don't replicate (Figure 4.11).[59] These cells copy their defective DNA to daughter cells that show similar unrestricted growth. At some point, a descendant of this line undergoes additional mutation(s) that removes the mutated cell further from cellular regulation, allowing the progeny to pass through tissue, enter the bloodstream, and travel to other body locations and colonize distant organs (metastasize). A brain cancer known as *astrocytoma* demonstrates this mechanism. It follows a defined sequence of genetic changes involving at least (1) inactivation of the p53 gene, (2) loss of a gene on chromosome 9, (3) enhancement of epidermal growth receptor by another altered gene, and (4) loss of one copy of chromosome 10.[45] In another example, certain colon cancers often result when defects develop (1) in both alleles of a gene called APC (adenomatosis polyposis coli), (2) in certain ras genes that promote cell growth, and (3) in the p53 gene. The genetic changes occur early in APC while subsequent changes in the other genes appear much later. These cumulative genetic errors may be facilitated by defective genes such as hMLH1 and hMSH2, which code for proteins that correct errors within the order of nucleotide bases during DNA replication by scanning for these errors, detecting them, and repairing them much like a "spell-checker" on a computer (Figure 4.12). Defects in these genes result in incomplete or non-functioning proteins that permit replication errors to accumulate within the DNA. Eventually, the most important step in the development of a cancer cell is its failure to die when genetic mutations render them cancerous. This process of cellular self-destruction is called apoptosis and occurs when cell damage accumulates to the point where cell integrity is compromised.[60] At this point, a normal cell produces enzymes that cause the cell's DNA to disintegrate in an orderly way, causing cell death.

One of the most important genes coding for enzymes involved in apoptosis is p53, a tumor suppressor gene. More than half of all human cancers involve a p53 gene either mutated by free radicals or inherited as a defect. The normally functioning enzyme halts cell division until DNA can be repaired or turns on the mechanism for cell destruction if the DNA is severely damaged. Mutations occurring along the p53 gene appear at characteristic sites and with unique base sequences known as biomarkers depending on the damaging environmental agent at fault. Cigarette smoke will cause damage to specific locations along

Genes such as hMLH1 and hMSH2 code for proteins that correct errors in the order of nucleotide bases during DNA replication. These proteins are known as "spell-checker" proteins, and defects in the gene will produce defective "spell-checker" proteins that fail to correct errors. When such errors occur in genes controlling cell growth, a major step forward in the development of a cancer cell occurs.

FIGURE 4.12
The action of "spell-checking" proteins created by hMLH1 AND hMSH2 genes.

p53 whereas UV light will produce damage at different sites. This information may eventually be useful in risk assessment or even legal proceedings.[61] Certain p53 mutations may also reduce the ability of cells to limit blood vessel formation. A tumor stimulates its growth by creating a blood supply via *angiogenesis*. It is this nourishing process that is believed to give cancer cells their enhanced ability to multiply and spread. Dr. Judah Folkman and his colleagues tested anti-angiogenesis compounds for years as a possible mechanism for cancer treatment. They discovered that endostatin and angiostatin may prevent and reverse the formation of blood vessels in a broad array of tumors growing in rodents.[58] However, 20 years later, this area of research continues, as studies have shown that very high doses of these compounds actually increase the number of cancer cells in laboratory animal tumors, by causing an inflammatory and immunosuppressive response within the tumor. Further work is needed to find the therapeutic dose.[62]

Many cancer cells are unstable and discard both copies of the p53 gene while other cells retain mutant copies. It turns out that cells with normal p53 genes are quite sensitive to chemotherapy while those with mutant p53 genes tend to be very resistant. This sensitivity of cells to p53 actually is the cause of many of the difficult side effects of chemo and radiation therapies. Cells that are only slightly mutated are destroyed, which causes the common side effects of nausea, hair and weight loss, and a weakened immune system. Some researchers have found that by suppressing p53 during these therapies the side effects may be lessened.[63] New therapies are being sought that may prove more effective against cancers with mutant p53 genes. Cancer is a complicated disease that occurs in different types of cells within the body, often expressing many different genes, influenced by multiple growth factors in different ways, and variably responsive to pharmaceuticals.

TABLE 4.5

Known Cancer Genes

TP53	When functioning properly, TP53 (P53) helps limits mutations in a cell. It repairs damage in a cell's DNA or starts a cell's death when the damage is too great. TP53 is mutated in more than 50% of all cancers.
RB1	The first tumor suppressor gene to be discovered, RB1 demonstrated the genetic link of cancer and the impact of one or two faulty genes. When it is not working, eye cancer is the common result.
Ras	Ras are a family of genes involved in cell signaling. A defective ras gene causes cells to divide repeatedly. The defective gene plays a role in 90% of pancreatic cancers, 50% of colon cancers, and 25% of lung cancers.
BRCA1/BRCA2	BRCA genes are tumor suppressors. They repair damage to cellular DNA. When BRCA1 or BRCA2 proteins are not functioning properly, the risk of breast cancer increases sharply.

Source: *The Most Famous Genes in Cancer Research*, January 18, 2018, https://www.rapamycinpress.com/blog /science-talk/most-famous-genes-cancer-research/.

Consequently, combination treatments of chemotherapy with multiple drugs, gamma radiation, and other interventions in combination are proving more successful.[64]

With the use of DNA genechips to screen cancer cells by the proteins that they produce, cancers will be named by the genes they express and not by the site such as colon or stomach cancer.[65] Eventually, treatment in the twenty-first century will rely upon determining the specific DNA damage and custom fitting the medicinal cocktail to each tumor.[58] Some of the genes associated with cancers are listed in Table 4.5. Routine testing for genetic mutations is still a ways off. However, the National Cancer Institute (NCI) has a long-range project called the Cancer Genome Characterization Initiative (CGCI) that is working to create a DNA library of tumor types. For instance, researchers have studied the genetic alterations in medulloblastoma (MB), a brain cancer that occurs primarily in children and B-cell NHL, a white blood cell cancer.[66] Technology has allowed some of this more detailed research. Second-generation genome sequencing is allowing scientists to sequence the entire exome, that is, the known protein-coding regions of the genome. This can be useful to find variations in genes, by comparing the exome from different people. This type of second-generation sequencing is also being done with mRNA.

Individual screening for cancer risk has been associated with technological problems as well as the ethical questions discussed earlier. Still, the science of epidemiology and genetics (molecular epidemiology) is producing more knowledge on the common factors that influence genetic susceptibility to cancer.

Advances in Gene Editing

New technologies in gene editing may provide greater understanding and precision in treating cancer, especially around single point mutations. One of these technologies is called clustered regularly interspaced short palindromic repeats and CRISPR-associated protein 9 (CRISPR-cas9). It allows scientists to identify and cut out a specific portion of DNA, for instance, a DNA segment with a disease-causing mutation. Then, using the cell's own repair system, scientists incorporate a corrected DNA sequence, one without the mutation. So far, this has only been done in the laboratory.[67]

The CRISPR system was derived from a bacterial version of an immune system, in which bacteria "remember" invading viruses by copying the viral DNA. This DNA is added to the bacterial genome, creating a memory bank of viral attacks. If that same or a similar virus attacks the bacteria again, the bacteria reviews its virus memory files. If it finds a match, the bacterium transcribes the saved viral DNA into RNA then adds a defensive protein. This complex then attacks the virus, slicing the invading viral DNA into inactive bits. Scientists have developed a few different gene-editing systems, but so far CRISPR has gained the most attention for its simplicity and effectiveness.[68]

Major Cancer Risks

"Cancer" is a word no one wants to hear. Thankfully, cancer risk may be reduced by modifying certain behaviors (Table 4.6).

Smoking

Smoking cigarettes is recognized as one of the leading causes of cancer in the United States, causing more than 30 percent of all cancers and one third of cancer deaths. Lung cancer used to be a disease primarily of men, but the popularity of smoking among women has driven the cancer cases upward so that lung cancer surged past breast cancer as the leading cause of cancer deaths among women beginning in the 1980s.[36] This is true almost 40 years later, with lung cancer the number one cause of cancer death in both men and women.[29] Since the 1960s there has been a decline in smoking of more than 15 percent among American adults that is being reflected in a decline in lung cancer.[69] Also, the EPA has designated environmental tobacco smoke as a class A carcinogen. Unfortunately, there is a disturbing trend of increased cigarette smoking globally, with dramatic increases in countries of the former Soviet Union, China, South Korea, and Spain.[70] This increase is despite broad communication programs designed to educate the public about the risks of smoking and to monitor tobacco usage.[71]

Diet

Dietary factors also play an important role in cancer risk. Diets that are limited to 20 percent fat or less of total calories have shown dramatic decreases in prostate-specific antigen (PSA) circulating in the blood from the high 20s (indicating dangerous cancer activity) to levels of 0.4–2.5 (suggesting the tumors were no longer growing).[60] High fat intake has been linked with increased risks to colon and prostate cancer, while being excessively overweight encourages the growth of endometrial cancer.[72] The precise mechanism for how low-fat diets may lower risk to cancer is unknown, but studies on mice suggest that calorie restriction boosts apoptosis and gene repair.[60] Apoptosis may also be triggered by butyric acid

---

TABLE 4.6

Modify Behavior to Decrease Cancer Risk

1. Get regular cancer-screening tests
2. Get to and stay at a healthy weight
3. Exercise regularly
4. Eat a healthy diet
5. Avoid tobacco
6. Limit alcohol

Source: https://www.cancer.org/latest-news /6-steps-to-help-lower-your-cancer -risk.html, 2018.

produced when bacteria ferment fiber in the gut. Consumption of high-fiber foods instead of refined flour and processed foods also appears to reduce the risk of cancer.[72] Cruciferous vegetables including broccoli, cabbage, cauliflower, kale, and Brussel sprouts all produce a powerful isothiocyanate anti-carcinogen known as sulforaphane. Sulforaphane appears to increase the production of detoxifying enzymes that quickly jettison cancer-promoting substances from the body.[73] Fruits and vegetables rich in antioxidants such as vitamins C and E protect against a variety of cancers by reducing free radical damage to DNA or to membranes that would otherwise increase the access of environmental toxins. Garlic and onions have a long history of medicinal use throughout the world. Both garlic and onions have strong antimicrobial activity and also increase production of detoxifying enzymes through the allylic sulfides (allium) in the plant tissues. Many fruits, vegetables, and nuts contain ellagic acid, which binds to DNA preventing the powerful carcinogen aflatoxin from forming a DNA adduct. Research in Sweden has shown that human breast milk contains a protein with anti-cancer activity. This protein seems to be a natural protection in infants from cancer and other diseases such as bacterial infections. In laboratory tests this protein seems effective against many different types of cancer cells.[74]

Another cancer preventative is aspirin. Although not technically a food, aspirin was originally derived from a chemical found in plants, such as the willow tree. This chemical is salicylate and it has long been known for its fever-reducing properties. The form we use today, aspirin, or acetylsalicylic acid, was first synthesized in 1853 by a chemist who treated the chemical acetylchloride with sodium salicylate.[75] Persons who use aspirin regularly experience half the rate of colon cancer and lower rates of esophageal cancers than do persons who do not use aspirin.[60] Its mechanism is simple but effective: An enzyme called cox-2 prevents DNA-damaged cells from dying and aspirin inactivates this enzyme allowing colon-cancer cells in vitro (in a test tube) to die. Aspirin has been associated with reduced risk of colorectal cancer in humans, and it has been recommended by doctors that people at risk of colorectal cancer should take a single aspirin tablet (325 mg) every other day providing there are no contraindications.

Trends in Cancer

Cancer is one of the leading causes of death globally. A recent global WHO study revealed that in 2012, the most recent year of the WHO study, there were 14 million new cancer cases and 8.2 million cancer-related deaths in 2012.

The most recent global study indicated that cancer incidence was 454.8 per 100,000 men and women per year.[76] Cancer incidence is the number of new cases of cancer per 100,000 people. Similarly, cancer mortality is the number of deaths from cancer per 100,000 people. Between 2008 and 2012, the global cancer mortality rate was 171.2 per 100,000 men and women per year. Overall cancer mortality is higher among men than women, at 207.9 per 100,000 men versus 145.4 per 100,000 women. The highest cancer mortality rate, 261.5 per 100,000 men, was seen in African American men and the lowest rate was in Asian/Pacific Islander women, at 91.2 per 100,000 women

Cancer Trends in the United States

The rate of Americans diagnosed with cancer and dying from it continue to decline. The results published in 2017 represent a collaboration of several agencies including the ACS, The NCI, The Centers for Disease Control and Prevention (CDC), including the National Center for Health Statistics (NCHS). The data on cancer incidence was generated from the NCI's Surveillance Epidemiology and End Results (SEER) program representing 39 states' cancer registries and 89 percent of the U.S. population, for cancer diagnoses between 1999 and 2013.[77] The data on cancer deaths comes from the NCHS's National Vital Statistics System, which collects death certificates from every state, including information of age, sex, and race/ethnicity. This data covered the years 2000–2014.

The data show that for the period 2004–2013, in men, rates of cancer death decreased by 1.8 percent per year. For the same time period, the rate of cancer deaths of women decreased by 1.4 percent per year. For children ages 0–19 years, cancer mortality rates decreased 1.4 percent per year.[77]

Data is also available for cancer mortality rates of specific cancers. In men, for the period 2004–2013, eleven common cancers showed a decrease in mortality: Lung and bronchus cancer mortality decreased by 3.5 percent, prostate cancer by 3.4 percent, larynx by 2.5 percent, colon and rectal cancer by 2.5 percent, NHL by 2.0 percent, stomach, melanoma, esophagus, and leukemia mortality decreased between one and two percent mortality, myeloma and kidney cancer mortality between 0.9 percent and 0.8 percent respectively. Several cancers showed an increase in mortality rates in men. The greatest increase was for liver cancer, with a 2.6 percent increase in mortality among men. Oral cavity cancer had a 1.1 percent mortality increase. Brain/nervous system and pancreatic cancers had mortality increases of 0.5 percent and 0.3 percent, respectively.[78]

For the period 2004–2013, many of these cancer mortality rates decreased in women as well. The largest decrease was seen in colorectal cancer in women, where the cancer mortality rate decreased by 2.8 percent. Other mortality rate decreases include ovarian cancer by 2.3 percent, NHL by 2.2 percent, and lung and bronchus cancers by 2.0 percent. Cancers of the stomach, esophageal, breast, oral cavity, kidney, and leukemia mortality rates all decreased between 1.0 and 2.0 percent. Cervical cancer decreased by 0.8 percent. Bladder cancer and melanoma skin cancer mortality rates decreased by 0.6 percent each. Pancreatic cancer decreased slightly, by 0.2 percent. In women, the rate of liver cancer mortality had the largest increase, at 3.0 percent, with uterine cancer slightly below that at a 2.0 percent increase in mortality. Myeloma and brain/nervous system cancer had increased mortality rates of 0.5 and 0.4 respectively.[78]

Surviving cancer for 5 years is a benchmark measure in the successful treatment and cure of cancer. In the U.S., this survival rate has improved steadily since 1975, when the average five-year survival rate was only 50 percent.[79] In 2012, the most recent period studied, the average five-year survival rate for cancer was 66 percent. For specific diseases, some of these improved survival rates are marked. For breast cancer, the five-year rate in 1975 was 75 percent; in 2012, it was 91 percent. In prostate cancer, in 1975 the five-year survival rate was 68 percent; in 2012, it was 99 percent. Some other cancers show smaller improvements in the five-year survival rate, but the trend is overall good for survival. Note that survival rates can vary based on individual factors such as race, ethnicity, and access to health care.

Survival rates for children with cancer have improved dramatically over time. In 1975, the five-year survival rate for the most common childhood cancers combined was only 58 percent.[79] By 2012, it improved to 85 percent. Marked increases in five-year survival rates for specific cancers include Myeloid leukemia from 18 percent to 66 percent survival rate, NHL from 43 percent to 91 percent, acute lymphoid leukemia from 57 percent to 92 percent, and bone and joint cancer from 49 percent to 80 percent. Improved chemotherapies are responsible for most of these improved survival rates.

Managing the Risk of Environmental Disease

Recent and spectacular advances in the treatment of these childhood diseases has produced a decrease in the percent of deaths in spite of the increased number of children diagnosed with cancer. The causes of the cancers are uncertain, but some experts point to toxins in air, food, dust, and drinking water as probable suspects. Carol Browning, the former director of EPA, would like to see increased federal support to investigate the effects of environmental pollutants on children, and to begin looking at children's total cumulative risk from all exposures to toxic chemicals.[80] Children tend to be more susceptible to environmental toxins because (1) they take in more food, water, and air and accompanying carcinogens relative to body weight compared to adults; (2) nursing infants get higher exposures of substances such as dioxin in breast milk than adults exposed to background levels; (3) children have a higher internal dose of toxins and greater genetic damage than adults who have similar exposures to tobacco smoke and polycyclic aromatic hydrocarbons (PAH); (4) children have reduced detoxification and repair systems; (5) they have a higher rate of cell proliferation during early developmental stages; and (6) children have increased absorption and retention of toxins.[14] Just as health and regulatory policies are likely to target more specific at-risk groups because of enhanced genetic knowledge, children may well become another more specific target for such policies and regulations. It may take a number of years to confirm that children require lower doses of carcinogens to promote disease, but if this is confirmed, it is expected that more rigid controls on pesticides, toxic wastes, and other chemicals will occur. Congress has already started to move in this direction with the approval of new laws that incorporate potential exposures to children when standards are established for pesticide residues in food and contaminants in drinking water.[80]

Far-Reaching Effects

Even in the most remote parts of the world, there is evidence of environmental impacts from the factors that contribute to disease in humans. Scientists report that chlorinated organic compounds and heavy metals are ubiquitous and that even the most remote life forms in the Arctic Ocean demonstrate measurable levels of these toxins in their tissues. For instance, polar bears have been found to have 1,000 times the "safe" human amount of a chemical in their systems.[81] These chemicals include both older chemical likes polychlorinated biphenyls (PCBs), which are no longer used in the United States and Europe, but which persist in the environment, and newer chemicals such as PFOS, the same chemical implicated in the Australian cancer cluster mentioned at this chapter's introduction. The risks are greatest for the young. Just as children face greater health consequences from environmental factors such as pollution, young animals face similar health risks from environmental pollutants. In the following chapters we look closely at several of the environmental factors that contribute to disease, including specific toxins, water and air pollution, and ionizing radiation.

References

1. Carcinogenic chemicals found in cancer-riddled town's water, *9News.com.au*, August 15, 2017 (accessed February 15, 2018).
2. Chang, C., Williamtown water contamination highlights dangers of PFOS and PFOA, news .com.au, April 1, 2016.
3. Prüss-Ustün, A., Wolf, J., Corvalán, C., Bos, R., and Neira, M., *WHO Report,Preventing Disease through Healthy Environments: A Global Assessment of the Burden of Disease from Environmental Risks*, World Health Organization, Geneva, Switzerland, 2016.

4. Changing disease burdens and risks of pandemics, European Environment Agency, February 18, 2015. https://www.eea.europa.eu/soer-2015/global/health (accessed February 12, 2018).

5. Nelson, A., EU on brink of historic decision on pervasive glyphosate weedkiller, *The Guardian*, October 24, 2017. https://www.theguardian.com/environment/2017/oct/24 /eu-brink-historic-decision-pervasive-glyphosate-weedkiller.

6. Nelson, A., Controversial glyphosate weedkiller wins new five-year lease in Europe, *The Guardian*, November 11, 2017.

7. Sessions, S.K., Franssen, R.A., and Horner, V.L., Morphological clues from multi-legged frogs: Are retinoids to blame? *Science*, 284, 800–802, 1999.

8. Tanner, C., Parkinson's disease in twins: An etiologic study, *Journal of the American Medical Association*, 281, 341–346, 1999.

9. Whelan, E.M., *Toxic Terror: The Truth Behind the Cancer Scares*, 2nd ed., Prometheus Books, Buffalo, New York, 1993, Chap. 1.

10. Nader, R., Brownstein, R., and Richards, J. (Eds.), Who's *Poisoning America: Corporate Polluters and Their Victims in the Chemical Age*, Sierra Club Books, San Francisco, CA, 1981, Chap. 1.

11. Grossman, K., *The Poison Conspiracy*, The Permanent Press, Sag Harbor, NY, 1983, Chap. 1.

12. Naar, J., *Design for a Liveable Planet*, Harper and Row, New York, 1990.

13. Downs, H., *The Poisoning of America*, ABC-TV, New York, September 8, 1988.

14. *Tianjin, China, Explosions: 50 Dead, More Than 700 Injured*, The Associated Press, New York, August 13, 2015.

15. Rather, J., Contaminant from gas is found in water, *New York Times*, August 29, 1999.

16. Koren, H., *Illustrated Dictionary of Environmental Health and Occupational Safety*, CRC/Lewis Publishers, Boca Raton, FL, 1996, p. 407.

17. Thomas, C.L. (Ed.), *Taber's Cyclopedic Medical Dictionary*, 13th ed., F.A. Davis Co., Philadelphia, PA, 1977.

18. Parkinson's disease, National Institute of Health. https://www.niehs.nih.gov/health/topics/ conditions/parkinson/index.cfm (accessed February 17, 2018).

19. India tops the world in pollution-related deaths, *www.economictimes.indiatimes.com*, October 21, 2017, 01.02 AM IST.

20. Landrigan, P.J., Fuller, R., Acosta, N.J.R. et al., The lancet commission on pollution and health, *Lancet*, 391, 462–512, 2018.

21. Interview with John Higginson, *Science*, 1363, September 28, 1979.

22. Whelan, E.M., Cancer scares and our inverted health priorities, *IMPRIMIS*, 20(6), 1–4, 1991.

23. Preston, R., What new things are going to kill me? *Time*, November 8, 1999. www.time.com (accessed July 15, 2018).

24. Gadsby, P., Fear of flu, *Discover*, January 1999, pp. 82–89.

25. Lyme borreliosis in Europe, European Centre for Disease Prevention and Control/WHO fact-sheet. https://ecdc.europa.eu/en/borreliosis (accessed February 15, 2018).

26. Gorman, C., Beyond lyme, *Time*, July 26, 1999.

27. Bickley, H.C., *Practical Concepts in Human Disease*, The Williams and Wilkins, Baltimore, MD, 1975, Chap. 7.

28. Morgan, T.M., Gordon, L., Walker, B. et al., *Environmental Health*, 2nd ed., Morton Publishing, Englewood, CO, 1997, Chap. 15.

29. Kochanek, K.D., Murphy, S.L., Xu, J., and Tejada-Vera, B., Division of vital statistics, *National Vital Statistics Reports*, Vol. 65(4), 2014.

30. Grinshteyn, E. and Hemenway, D., Violent death rates: The US compared with other high-income OECD countries, *American Journal of Medicine*, 129(3), 266–273, 2016.

31. Accidents at Nuclear Power Plants and Cancer Risk. https://www.cancer.gov/about-cancer / causes-prevention/risk/radiation/nuclear-accidents-fact-sheet (accessed February 15, 2018).

32. Allen, M., Why do we build medical scanners from sunken battleships? http://www.science -madesimple.co.uk/curriculum-blogs/engineering (accessed February 8, 2017).

33. Cummings, C.H., Entertainment foods, *The Ecologist*, 29, January/February 1999, pp. 16–19.

34. Associated Press (AP), Mutation may raise cancer risk from secondhand smoke, *Cable News Network*, December 1, 1999. http://www.cnn.com/health/aids.
35. Sheperd, T.H., Detection of human teratogenic agents, *Journal of Pediatrics*, 101, 810, 1982.
36. Nadakavukaren, A., *Our Global Environment*, 4th ed., Waveland Press, Long Grove, IL, 1995, Chap. 6.
37. Underhill, W., Shaped by life in the womb, *Newsweek*, September 27, 1999.
38. Shreeve, J., Secrets of the gene, *National Geographic*, October 1999, pp. 43–75.
39. Tortora, G., Funke, B.R., and Case, C., *Microbiology, An Introduction*, 5th ed., The Benjamin Cummings Publishing Company, Redwood City, CA, 1995, Chap. 9.
40. Smith, C.S., China rushes to adopt genetically modified crops, *New York Times*, October 7, 2000.
41. U.S. Department of Agriculture, *Adoption of Genetically Engineered Crops in the U.S.* (accessed February 19, 2018).
42. Brower, L.P., Will biotechnology doom the monarch? *Defenders of the Wildlife*, Fall 1999.
43. Kaiser, J., Environmental institute lays plans for gene hunt, *Science*, 278, 569–570, 1997.
44. All about the human genome project, May 11, 2016. https://www.genome.gov/10001772/all-about-the-human-genome-project-hgp/ (accessed February 17, 2018).
45. National Institute of Environmental Health Science, Environmental Genome Project. egp.gs.washington.edu/welcome.html (accessed February 15, 2018).
46. Faherty, S., Reddit talk sparks questions on the ethical, legal and social implications of genomics research. Genome.gov (accessed February 15, 2018).
47. Jaroff, L., Keys to the kingdom, *Time, Special Issue: Frontiers of Medicine*, 124, Fall 1996.
48. The Washington Post, Belated gene therapy records reported deaths, *Daily Hampshire Gazette*, Vol. 214, No. 126, January 31, 2000.
49. *What Is Gene Therapy? How Does It Work?* http://www.fda.gov (accessed February 19, 2018).
50. FDA approves novel gene therapy to treat patients with a rare form of inherited vision loss. http://www.fda.gov (accessed February 19, 2018).
51. BBC Online, Gene therapy advance for kids, *British Broadcasting Company*, February 20, 2000.
52. Couzin, J., Quandaries in the genes, *U.S. News and World Report*, November 1, 1999, pp. 64–66.
53. Hyman, D. and Huang J., The courage to heal cancer survivor doesn't regret her difficult decision, *Springfield Sunday Republican*, October 24, 1999.
54. Berg, K., Ethical aspects of early diagnosis of genetic diseases, *World Health*, 5, September/October 1996, p. 20.
55. Glasser, R.L., *The Greatest Battle*, Random House, New York, 1976, Chap. 1.
56. https://www.cancer.gov/about-cancer/causes-prevention/genetics (accessed February 17, 2018).
57. Center for Devices and Radiological Health, Consumer Update on Mobile Phones, October 20, 1999. http://www.fda.gov/cdrh/ocd/mobilephone.html.
58. Gorman, C., The hope and the hype, *Time, Cancer: A Special Report*, May 18, 1998, p. 38.
59. Cavenee, W.K. and White, R.I., The genetic basis of cancer, *Scientific American*, March 1995, p. 72.
60. Kohitak, R., Cancer's Nemesis: The suicide gene, *Chicago Tribune*, 257, September 14, 1997, p. 1.
61. Harris, C.C., p53: At the crossroads of molecular carcinogens and risk assessment, *Science*, 262, 1980–1981, 1993.
62. Yu, Y., Wang, Y., Wang, Y. et al., Antiangiogenic therapy using endostatin increases the number of ALDH+ lung cancer stem cells by generating intratumor hypoxia, *Scientific Reports*, 6, 34239, 2016. http://www.nature.com/articles/srep34239.
63. Recer, P., Drug said to aid cancer therapy toleration, *The Boston Globe*, September 10, 1999.
64. Nash, J.M., The enemy within, *Time: Special Issue: The Frontiers of Medicine*, Fall 1996, p. 14.
65. Saltus, R., Cancer cells' gene signature may aid in diagnosis, *The Boston Globe*, October 15, 1999.
66. Cancer Genome Characterization Initiative, Office of Cancer Genomics. ocg.cancer.gov/programs/cgci/overview.
67. What are genome editing and CRISPR-Cas9? ghr.nlm.nih.gov/primer/genomicresearch/genomeediting (accessed February 16, 2018).
68. Molteni, M., Everything you need to know about CRISPR gene editing. http://www.wired.com/story/what-is-crispr-gene-editing/ (accessed December 5, 2017).

69. Wingo, P.A., Reis, A.G., Rosenberg, H.M., Miller, M.D., and Edwards, B.K., Cancer incidence and mortality, 1973–1995: A report card for the U.S., *Cancer*, 82, 1197–1207, 1998.

70. Brown, L.R., Kane, H., and Ayres, E., *Vital Signs 1993*, W.W. Norton, Worldwatch Institute, Washington, DC, 1993.

71. World Health Organization, *WHO Global Report on Trends in Prevalence of Tobacco Smoking*, World Health Organization, Geneva, Switzerland, 2015.

72. Willett, W.C., Diet and health: What we should eat, *Science*, 264, April 22, 1994.

73. Davis, D.L., Natural anticarcinogens: Can diet protect against cancer? *Health and Environment Digest*, 4, no. 1, February 1990.

74. Radetsky, P., Got cancer killers? *Discover*, June 1999.

75. Diarmuid, J., *Aspirin: The Remarkable Story of a Wonder Drug*, Chemical Heritage Foundation, Philadelphia, PA, 2008.

76. 'Cancer Statistics'. https://www.cancer.gov/about-cancer/understanding/statistics (accessed February 16, 2018).

77. Ahmedin, J., Ward, E.M., Johnson, C.J. et al., Annual report to the nation on the status of cancer, 1975–2014, featuring survival, *JNCI: Journal of the National Cancer Institute*, 109(9), 2017.

78. National Trends in Cancer Death Rates, Annual Report to the Nation, Part 1. https://seer.cancer .gov/report_to_nation/infographics/mortality2.html (accessed February 16, 2018).

79. Noone, A.M., Howlader, N., Krapcho M., Miller, D. et al. SEER cancer statistics review, 1975–2015, National Cancer Institute. Bethesda, MD. https://seer.cancer.gov/csr/1975_2015/ (accessed February 16, 2018).

80. Cushman, J.H., U.S. reshaping cancer strategy as incidence in children rises, *The New York Times*, p. 147.

81. Lockwood, D., Polar bear cubs at high risk from toxic industrial chemicals, *Scientific American*, January 23, 2017.

82. *Morbidity and Mortality Weekly Report*, June 19, 1992, p. 47.

83. Elmer-Dewitt, P., The genetic revolution, *Time*, January 17, 1994, p. 46.

84. Lehninger, A.L., *Bioenergetics: The Molecular Basis of Biological Energy Transformations*, W.A. Benjamin, New York, 1965, Chap. 11.

85. Perera, F.P., Environment and cancer: Who is susceptible? *Science*, 278, 1068–1073, 1997.

86. Chung, W., *Teratogens and their Effects*, Columbia University, New York, 2005.

87. The most famous genes in cancer research. https://www.rapamycinpress.com/blog/science -talk/most-famous-genes-cancer-research/ (accessed January 18, 2018)

88. https://www.cancer.org/latest-news/6-steps-to-help-lower-your-cancer-risk.html (accessed February 20, 2018).

5

Toxicity and Toxins

Introduction

As you pull away from the restaurant's drive-in window, you give the bag of fast food to your friend in the passenger seat. He hands out the burgers and fries to the guys in the back seat and to himself. Then he hands you a burger, which you partly unwrap to take a bite as you drive. It's delicious! Little do you know that your burger may be contaminated with a dangerous cancer-causing chemical found in the paper wrapped around your lunch. The culprit is a group of fluorinated compounds known as polyfluoroalkyl and perfluoroalkyl substances (PFASs) that make food wrappers resistant to grease. These compounds have been linked not only with cancer, but also with immune suppression and endocrine disruption.[1]

In the United States, there are currently more than 70,000 synthetic chemicals in commercial use, and for most of them, their toxicity is not widely understood. More than 60,000 of them are thought to be potentially harmful and their effects could be devastating, by causing cancer, neurological and reproductive disorders, or other injuries.[2] Some of these chemicals are so widely distributed and of such potential harm to human health and/or ecology that they serve as sentinel models for understanding the potential harm for the indiscriminate use and disposal of toxic agents. The compounds to be discussed in this chapter include dioxin, polychlorinated biphenyls (PCBs), and other endocrine disruptors; the heavy metals, lead and mercury; asbestos; and organic solvents. These compounds represent a broad array of chemical compositions, in mechanisms of toxicity, distribution, and in patterns of human exposure. Some substances, such as lead, have long been recognized as a hazard, since early civilization when lead was used to store wine, to pipe water, and even as vessels in which to cook.[3] Dioxin, PCBs, and other organochlorines, whether manufactured as products or formed as unwanted byproducts, have been implicated as agents in reproductive disorders or other diseases. Exposure to many of these chemicals at critical points in human development may produce lifelong adverse effects, or even effects that are passed on to subsequent generations.

All these potentially harmful chemicals in the environment and the exposure levels of toxicity are not understood? The reason for this is scarce resources, specifically time and money. Toxicity tests are done via animal studies and these animal studies require significant time and money. Further, each study deals with only one chemical at a time. Animal studies do not provide information on any interactive effects, which may occur during or after exposure to multiple chemical exposures. Additionally, animal studies often fail to detect future or delayed conditions. Human exposure studies are seldom performed because of the ethical problems associated with exposure of humans to various chemicals and metals. In addition to these challenges, the federal government is currently reducing support for further research, while private corporations are filling the gap, which may potentially lead to bias in both study designs and interpretations.[2]

If you are wondering how all these chemicals get into commercial use, to make the products we buy, it is because a chemical is often considered safe until proven otherwise by the appropriate federal agency, similar to the legal premise that a person is innocent until proven guilty. Conferring such a privilege on poorly understood chemical substances is not the best approach to protecting the public health. One to two thousand new chemicals are introduced each year into our society, allowing significant opportunity for untested materials to enter our environment and to expose humans, wildlife, and plants to toxic effects. Sometimes, decisions on what substances are allowed to enter the environment are political and economic, with little regard for the effects on human health. In other instances, the health effects are simply unstudied or not recognized.[2]

Organism

Absorption   Exposure

Dose
risk
hazard

Poison   Environment
Concentration
Toxic triangle

FIGURE 5.1
The toxic triangle. (Adapted from Malachowski, M.J., *Health Effects of Toxic Substances*, 2nd ed., Government Institute, Rockville, MD, 1999, Chapter 1.)

A potentially toxic substance produces its adverse effect by interacting with humans (or organisms) and the environment in a relationship referred to as the toxic triangle (Figure 5.1). Consequently, a poison or toxic substance doesn't constitute a hazard unless contact is made with the organism in a form and quantity that can cause harm. Furthermore, the environmental conditions of heat, humidity, or even noise may aggravate the response to the poison.[3] Hazardous substances or wastes may be explicitly defined under federal law, and many of them are listed by name and regulated under the Clean Air Act (CAA), the Toxic Substances Control Act (TSCA), and the Clean Water Act (CWA). A hazardous substance is defined in the Comprehensive Environmental Response, Compensation, and Liability Act (CERCLA) as any chemical regulated under the previously mentioned acts. Hazardous materials are defined and regulated under the U.S. Department of Transportation's (DOT) regulations (49 CFR). The definition is quite broad and pertains to "a substance or material, including a hazardous substance, which has been determined by the Secretary of Transportation to be capable of posing an unreasonable risk to health, safety and property when transported in commerce..." Hazardous wastes are solid wastes defined under 40 CFR.261 as "any discarded material garbage, refuse, sludge, or other discarded material, that is solid, liquid, or confined gas which is abandoned, inherently waste-like, or recycled." Listed wastes appear under 40 CFR 261.30. Wastes not listed may still constitute a hazard if it possesses defined characteristics of toxicity, corrosiveness, reactivity, and flammability. Toxic substances are those that (1) can produce reversible or irreversible bodily injury; (2) have the capacity to cause tumors, neoplastic effects, or cancer; (3) can cause reproductive errors including mutations and teratogenic effects; (4) produce irritation or sensitization of mucous membranes; (5) cause a reduction in motivation, mental alertness, or capability; (6) alter behavior; or (7) cause death of the organism.[4]

People continue to be exposed to a wide spectrum of toxic and hazardous substances that potentially threaten their health. In most cases, the consequences of such exposures are unknown. The following sections in this chapter outline the principle mechanisms of toxicity and discuss several of the more widely distributed toxic and/or hazardous chemicals in our society.

Exposure and Entry Routes

Exposure

In order for a toxic substance to produce its harmful effects on the human body, a person must first be exposed to the chemical. The harmful effect may occur immediately upon direct contact or the substance may be absorbed, after which the adverse effect is produced. Some toxic materials such as strong bases or acids can express their harmful effects by causing chemical lesions upon contact with the skin. Although no absorption of toxic substances takes place, the lesions produced are local and often result in killing areas of tissue in a process known as necrosis. Absorption refers to the passage of substances across the body's surfaces, either the skin or mucosal membranes, into body fluids and tissues via one of several processes, such as diffusion, facilitated diffusion, active transport, or special processes.[4,5] *Diffusion* is a passive process that occurs when molecules move from an area of high concentration of a molecule to one of low concentration. When this process involves water molecules across a semipermeable membrane, it is referred to as *osmosis*.

Not all molecules can passively diffuse from a high to low concentration. Generally, the chemical nature (basic or acidic), electrical charge, and the size and shape of the molecules determine their ability to passively diffuse across the membrane. Some molecules such as amino acids and sugars require specialized carrier proteins to be transported across a membrane. No high-energy phosphate such as adenosine triphosphate (ATP) is required in this process and so it is referred to as *facilitated diffusion*. When ATP is required in conjunction with special carrier proteins to move molecules through a membrane against a concentration gradient (i.e., from low concentration to high concentration), this process is called active transport. Typically, about 40 percent of the ATP in a human body cell is consumed in this process of actively transporting molecules across the cell membrane. This process is important in nerve impulse conduction in which sodium and potassium are actively transported across the nerve cell membrane by the sodium-potassium ATP*ase* pump. The energy from ATP is required because three sodium ions are actively transported out for every two potassium ions actively transported in.[4,5]

Endocytosis is a third method of molecule transport. Particles and large molecules that might otherwise be restricted from crossing a plasma membrane can be brought in to a cell or removed by the process of endocytosis. There are three major types of endocytosis: phagocytosis, pinocytosis, and receptor-mediated endocytosis. Phagocytosis (phag = eat; cyto = cell) occurs when specialized cells such as phagocytes extend out streams of cytoplasm much like an amoeba and engulf solid particles to form a sac around the particle(s). The sac detaches from the plasma membrane inside the cell and enzymes attack the particle(s). When the materials are liquid, a small drop adheres to the cell membrane surface where an indentation forms to surround the drop and form a vesicle, which then detaches from the membrane in a process called pinocytosis.[4]

Waste products, poisons, toxicants, nutrients, and hormones may be located in interstitial fluids (between cells) at lower concentrations than in the cells. These substances bind to receptor sites on the cell surface causing a vesicle to form around the substance in a process called receptor-mediated endocytosis. In a series of steps, the receptor is separated and the substance is enzymatically digested.[4,5]

Routes of Entry

Toxic substances may exist in any form, whether gas, liquid, solid. Consequently, there are several routes by which toxic substances may enter the body. They may enter through the lungs by inhalation; through the skin, mucous membranes, or eyes by absorption; or through the gastrointestinal tract by ingestion. The following section presents these routes of entry.

Respiratory System

The respiratory system is composed of the nose, pharynx, larynx, trachea, bronchi, and lungs (Figure 5.2). Its function is to supply oxygen to the body's cells and to expel carbon dioxide from the body in a process called respiration. The act of breathing or ventilation brings air into and out of the lungs. The exchange of oxygen and carbon dioxide

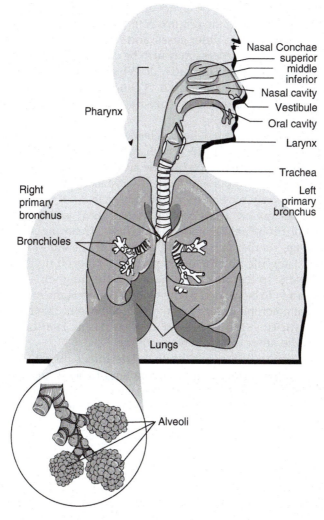

FIGURE 5.2
The respiratory system.

between the atmosphere and blood is known as external respiration, while the exchange of gases between blood and individual cells is called internal respiration. Air is drawn into the nasopharyngeal area through nasal hairs at the vestibule and then over mucous-membrane-covered bony plates called the nasal conchae. These structures combine to filter, warm, and moisten inhaled air. Additionally, nerve endings in this nasal area may be stimulated to cause a sneeze reflex helping to eliminate the mucous and the trapped particles. The inhaled air then enters the trachea-bronchial area where the bronchi branch into numerous bronchioles with even smaller diameters until they terminate into thin-walled, delicate alveoli where gas exchange takes place. The trachea, bronchi, and bronchioles are lined by a velvety layer of cilia with mucous cells throughout. Particles such as dust and pollen of 10 mm or larger are removed by the constant streaming of mucous propelled from the bronchial and tracheal passages by the cilia beating at over 1300 times per minute in a process known as mucociliary streaming. The tube-like structures of the bronchi and trachea are surrounded by layers of smooth muscle that contract in response to irritating substances and allergens (Figure 5.3). This bronchoconstriction narrows the lumen and restricts the flow of air, other gases, and particles from reaching more delicate tissues deeper in the lung. This process may also be combined with excess mucous secretions or other fluids that make breathing difficult. When prolonged and severe, this process can be life threatening as in asthmatic attacks. A cough reflex may be initiated at the level of the bronchi to eliminate accumulated mucous in which foreign substances have become lodged.[3,4,6]

Inhaled air ultimately reaches the delicate, elastic, and very thin alveolar sac at the terminus of the bronchioles (Figure 5.3). The alveoli are surrounded by tiny capillaries, where the red blood cells (RBCs) release carbon dioxide and take up oxygen in exchange. Within the lumen of the alveoli are macrophages, which serve as the final line of defense against microbes and a variety of organic and inorganic materials.[3]

Although particulates greater than 5 micrometers (μm) are mostly removed within the nasopharyngeal or bronchial areas by mucociliary streaming and either swallowed, sneezed, or coughed up, smaller particles in the range of one μm or less reach the alveoli. There, they may be (1) engulfed by the macrophages, (2) moved up the respiratory tree in a mucous sheet, (3) lodge in the interstitial tissues of the lung, or (4) enter the lymphatic spaces and lymph glands to be carried to the bloodstream.[4] Some inorganic particles cannot be digested by the macrophages and may prove to be lethal to them, resulting in scar tissue and loss of elasticity in the associated alveolar tissue. The size of particles largely determines where in the lungs they will be deposited.

For toxic gases and vapors, it is solubility that determines the fate in the lungs. Highly soluble gases such as sulfur oxides and ammonia tend to act in the upper moist airways of the respiratory tree, whereas less soluble nitrogen dioxide tends to reach deep into lung tissue where it can cause tissue destruction and be more easily absorbed into the bloodstream.

The Skin

The skin is normally thought of as a relatively impervious barrier to chemicals and biologic agents, and for the most part this is true. The skin is the body's largest organ, consisting of many interconnected tissues covering an area of nearly two M^2 in the average size adult. The skin is a critically important organ. It helps to (1) regulate body temperature through sweat glands; (2) provide a physical barrier to dehydration, microbial invasion, and some chemical insults; (3) excrete salts, water, and organic compounds; (4) serve as

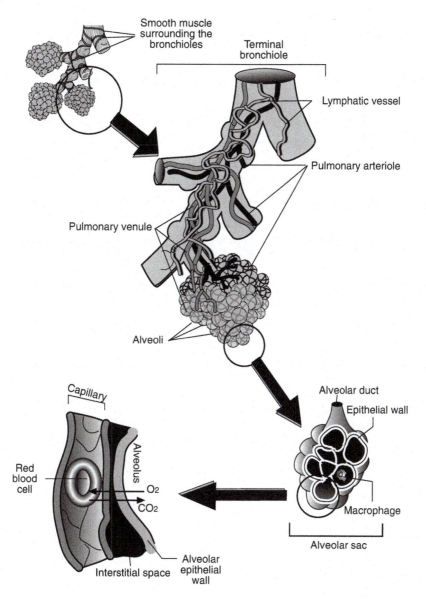

FIGURE 5.3
Terminal bronchiole and alveoli.

a sensory organ for touch, temperature, pressure, and pain; and (5) provide some important components of immunity.[7,8] The skin consists of an outer, thin layer of epithelial cells called the epidermis that is attached to an inner and much thicker layer called the dermis (Figure 5.4). Beneath this dermis layer is a subcutaneous layer of glandular and adipose tissues. The skin works to keep out water and water-soluble (hydrophilic) materials, as well as pathogens and toxins. Small amounts of substances may be absorbed through hair follicles or sweat glands. Materials may also enter breaks in the skin from cuts, injection sites, and insect or snake bites, and via high pressure steam or liquids.[8] Absorption of hydrophilic materials may increase as sweating increases or in the presence of moisturizers or

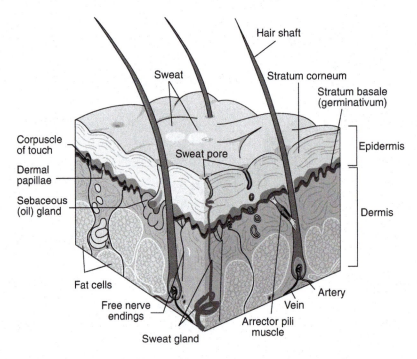

FIGURE 5.4
Cross-section of human skin.

oily materials. Substances that are fat soluble (lipophilic) such as organic solvents, petroleum compounds, and organochlorine compounds can be absorbed by passive diffusion through the skin. A variety of substances such as hexane, trichloroethylene, many insecticides, military nerve gases, and numerous other substances fall into this category.

Mucous membranes lining the nose and mouth are also made of epithelial cells, which are kept moist by glandular secretions. These membranes too may serve as a route of absorption.

Finally, the eye is also a possible route of absorption, especially via the moist epithelial tissue surrounding the eye.

The Gastrointestinal Tract

The gastrointestinal tract is a continuous tube of nearly 9 meters in length extending from the mouth to the anus (Figure 5.5). The components of the digestive tract include the mouth, pharynx, esophagus, stomach, small intestine, large intestine, and anus. The purpose of the digestive system is to ingest and move food for digestion and nutrient absorption. The system also functions to eliminate indigestible substances by defecation.[8,9] The gastrointestinal tract is a major route of absorption for many toxic agents including mercury, lead, and cadmium that appear in food and water. Most of this absorption occurs between the stomach and the upper portion of the intestine. Toxic substances may also be transferred from hand to mouth, then ingested.

The inner layer of the small intestine or mucosa consists of a velvety layer of finger-like projections of 0.5–1.0 micrometer length called villi. There are up to 40 of these villi per mm^2 giving a tremendous absorptive capacity to this section of the intestinal tract. Within

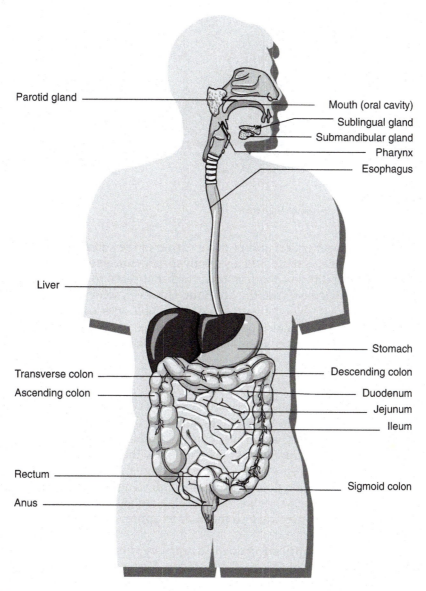

FIGURE 5.5
The gastrointestinal tract.

each villus there is a venule, a capillary network, an arteriole, and a lymphatic vessel. Nutrients as well as toxic agents can penetrate through the epithelial cells of the villus, enter the blood and lymph vessels, and be carried to various parts of the body (Figure 5.6). Nutrients, including carbohydrates, proteins, electrolytes, sodium, and calcium are absorbed with a combination of facilitated and active transport. Toxic materials that are similar in configuration, electrical charge, and molecular size may also be transported in this way. Such substances include lead, mercury, cadmium, and fluorides. The absorptive process can be influenced by physical health such as the presence of parasites, viruses, infectious bacteria, ulcers, cancer, or other digestive disorders. Absorption is also affected

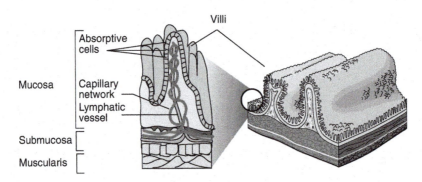

FIGURE 5.6
A cross-section of villi in the gastrointestinal tract.

by nutritional status, the chemical and physical nature of the substances in the intestinal tract, and level of physical activity and age. Both spastic movements and mucous secretion are defensive capabilities of the intestinal tract that serve to remove many noxious agents either through diarrhea, vomiting, or both. However, soluble toxic chemicals may be quickly absorbed into the bloodstream while less soluble ones are often carried into the lower bowels for excretion. It is also possible for particles up to 50 mm to penetrate the mucosa and enter the lymphatic channels and blood vessels to reach the internal organs.

Toxic Chemicals: Mechanisms of Action

Nearly 2500 years ago, Hippocrates described toxic effects of numerous poisons including lead, mercury, and arsenic. In the Middle Ages, the art of poisoning and the study of toxic effects became a formidable science in the hands of Catherine de Medici, who reported on the use of poisons, their clinical signs and symptoms, their potency, site of action, and how rapidly they produced their effects.[4] Today, we know that the toxic actions of pollutants vary dramatically based on dose, concentration, the length of exposure, and an individual's response. In fact, the symptoms that develop among a group of people exposed to the same toxic substance may vary so much as to appear to be caused by completely different substances.

A toxin can produce a harmful effect upon an organ by either stimulating or depressing the normal metabolic actions of that particular organ. Very small amounts of toxic agents have the ability to stimulate function(s) in a variety of organs, whereas larger doses may impede or even destroy the activity of the organ involved. The body is able to detoxify foreign substances principally through the liver and other organs such as the kidneys, intestines, and lungs. Sometimes adverse effects occur when the quantity of the toxic substance in the body overwhelms the body's detoxification mechanisms or when injury or illness has compromised the body's ability to detoxify substances. The harmful effects of environmental toxins are dominated by three principal mechanisms: (1) the toxins influence on enzymes, (2) direct chemical combination of the toxin with a cell constituent, and (3) secondary action as a result of the toxin's presence in the system.[13]

Effects of Toxic Agents on Enzymes

Enzymes are complex, three-dimensional proteins that have the unique ability of catalyzing (promoting or speeding up) metabolic activities in the body without undergoing any

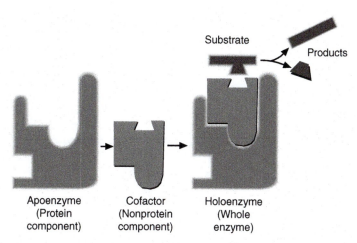

Substrate

Products

Apoenzyme
(Protein
component)

Cofactor
(Nonprotein
component)

Holoenzyme
(Whole
enzyme)

FIGURE 5.7
Components of an enzyme and mechanism of action.

changes themselves. Most enzymes are highly specific and selective. The enzyme "prototype" may be envisioned in the following manner: an enzyme has the capacity to hold molecules together by binding to them, keeping them stabilized in order for a reaction to occur between them, and then moving on to repeat this process with other molecules when the reaction is complete. Many enzymes require a protein component (apoenzyme), and a nonprotein component (cofactor) to become active. The apoenzyme and its cofactor combine to form a complete and active enzyme called a holoenzyme. Cofactors are usually metal ions such as manganese, magnesium, copper, or iron. They may also be an organic molecule called a coenzyme including vitamins such as nicotinamide adenine dinucleotide (NAD+) and nicotinamide adenine dinucleotide phosphate (NADP+). Enzymes act on substrates to add or remove molecules of water, oxygen, or hydrogen, or amino- or other functional groups. Enzymes may also rearrange atoms within a molecule or join molecules (Figure 5.7). Enzymes are important in nearly all life functions including most aspects of energy production and biosynthesis. Many toxic substances have the ability to (1) interfere with or block the active sites of the enzyme, (2) inactivate or remove the co-factor, (3) compete with the co-factor for a site on the enzyme, or (4) alter enzyme structure directly thereby changing the specific three-dimensional nature of the active site (Figure 5.8). Hydrogen cyanide and hydrogen sulfide inactivate enzyme cofactors by binding to the iron present in the cytochrome oxidase system and inhibiting enzymes critical to the final steps in glycolysis and ATP production.[3,10] Similarly, dithiocarbamate will bind to copper and prevent the activity of copper-dependent enzymes important in the breakdown of ethanol. Consequently, alcohol consumption results in severe nausea and headaches. Substances similar to dithiocarbamate are used in alcohol anti-abuse drugs.[9] Cadmium and Beryllium are believed to inactivate enzymes by blocking the sites on the enzyme where such co-factors as magnesium or zinc normally attach.[11]

The inhibition of enzymes may be reversible or irreversible. Usually, reversible enzyme inhibition occurs when the enzyme acts upon a substance similar to its normal substrate and so is unavailable to carry out its normal function. When a toxic chemical such as lead, arsenic, parathion, ozone, or fluorine covalently bonds to an enzyme, the bonding tends to be permanent for that particular enzyme molecule and is therefore considered an irreversible reaction. In most of these cases, the active group of the enzyme is blocked or destroyed.[10]

| Blocking the active site of the holoenzyme | Inactivate or remove cofactor | Compete with cofactor for the site on the enzyme | Altering active site of enzyme |

FIGURE 5.8
Actions of enzyme inhibitors.

The Direct Action of Toxicants on Cell Components

Many substances affect cells directly. Strong acids, bases, and phenols will combine directly with cellular materials and kill tissue, producing areas of necrosis. Nitrous and sulfuric oxides and ozone are powerful oxidizers that irritate mucous membranes. Inhalation of phosgene or chlorine gases can blister tissue and produce edema. Inhaled carbon monoxide reacts directly with hemoglobin to produce carboxyhemoglobin, thereby blocking the attachment of oxygen, which has a lower affinity to hemoglobin than carbon monoxide. The greater the blood concentration of carboxyhemoglobin, the more severe the symptoms will be. This is a reversible reaction. In an irreversible reaction, many sulfur-containing compounds including sulfonamides can combine with hemoglobin to produce sulfhemoglobin, a compound that is unable to carry oxygen in the bloodstream. Further, a number of aromatic amines, nitrites, and nitro compounds can oxidize hemoglobin to methemoglobin, which is unable to carry oxygen, resulting in oxygen-poor blood and a syndrome in infants called "blue baby."

Toxicants That Cause Secondary Actions

Many substances may themselves be harmless but can cause the formation of other chemicals in the body that are harmful or potentially lethal. As an example, fluoroacetate (rodenticide 1080) is metabolized as acetic acid in the respiratory metabolic chain causing the formation of fluorocitric acid. This substance prevents the formation of NADP and ATP through the inhibition of aconitase. The exposure is lethal in relatively small doses.[4,10] Another example is methanol, which, once degraded to formaldehyde, attacks the optic nerve and causes blindness.

Many people suffer severe health effects from seemingly innocuous substances such as animal dander, pollen, milk products, shellfish, peanuts, and even sunlight. In these susceptible individuals, exposure may trigger an exaggerated immune response, ranging from mild allergies associated with hay fever to rapidly fatal anaphylactic shock. The intermediary substances produced in these reactions include histamine, serotonin, and similar products. Not surprisingly, many toxic agents also have an allergic component. Toluene diisocyanate (TDI) is a component of polyurethane foam and is reported to produce allergic reactions.[6] The air pollutants nitrogen dioxide and sulfur dioxide are also reported to have an allergic component. In fact, there is an allergic or immune component to many environmental diseases. Parasites, bacteria, viruses, and toxic chemicals may also have an allergic component. How do otherwise harmless substances cause the release/production and final expression of intermediate substances in the body, which can result in significant discomfort and even death? What are allergies? To understand this mechanism, a discussion of specific immunity and allergies is in order.

Immunity and Allergies

The immune system is critical to the defense of the human organism against invading microbes and abnormal or cancerous cells. An immune system debilitated by disease, chemotherapy, or inherited defects will permit the rapid onset of overwhelming infections or possibly cancer. The movie titled *The Boy in the Plastic Bubble* was based on the true story about a boy who had inherited Severe Combined Immunodeficiency Syndrome (SCIDS). The absence of immune responses in this boy required him to live his life within an antiseptic, plastic-lined room in his home. Attempts to build an immune system in the boy through bone marrow transplants failed, and he died soon after. Just as threatening is an exaggerated response of the immune system as in anaphylactic shock from insect bites or food allergies, which can lead to death within minutes. Chemical pollutants such as ozone can depress the immune response by inactivating alveolar macrophages. Others such as nitrogen dioxide or sulfur dioxide have an allergic component that makes the reactions to exposure more severe. Immunity is a double-edged sword.

Origins of the Immune System

Immunity is based on the premise that certain immune cells in the body can recognize microbes, tissues and other substances that are "non-self" or foreign, and in an effort to protect the body, will destroy, encapsulate, or remove the foreign material. There are two separate but cooperating components of the immune system known as the humoral (antibody-mediated) immunity and the cellular (cell-mediated) immunity. Cell-mediated immunity (T cells) responds primarily to viruses, fungi, foreign tissue transplants, and cancer cells. In humoral immunity, antibodies produced by B cells are directed against bacterial and viral infections, and soluble antigens, while also being involved in anaphylactic or "Type I" allergic responses. The responses of cellular and humoral immunity are quite different. A rabbit injected with bovine serum albumin (BSA) and a purified protein derivative (PPD) from a modified tuberculosis bacterium will develop an amplified immune response to these materials within two weeks. When these substances are injected separately just under the skin (subcutaneously) in two separate areas, the immune response is quite different (Figure 5.9). The raised, hardened warm area is caused by the cellular immune system to the tuberculin PPD. This area is filled with macrophages and lymphocytes and takes 2–3 days to develop. The subcutaneous bleeding is caused by humoral (antibody) response to subcutaneously injected BSA, which causes localized vasodilation and leaking of blood components into the area. The mechanisms of these two components of the immune system are discussed in the following sections. Each component of the immune system is formed in the embryonic stages from lymphocytic stem cells that appear in bone marrow. During this development, some stem cells migrate to the thymus gland where they are influenced to develop into T cells and then populate specific areas in lymph glands, spleen, bone marrow, and other lymphoid tissue in the gastrointestinal tract. The remaining lymphocytic stem cells come under the influence of tissues (possibly gut-associated lymphoid tissue,

This reaction developed within 72 hours of subcutaneous injection of PPD to which the rabbit had been previously exposed. The reaction is characterized by influx of macrophages and lymphocytes producing a raised, hard area.

This reaction developed within minutes of subcutaneous injection of BSA to which the rabbit had been previously exposed. The reaction is characterized by subcutaneous bleeding.

Cellular immune response to purified protein derivative (PPD) from Mycobacterium tuberculosis.

Humoral immune response to bovine serum albumin (BSA).

FIGURE 5.9
Demonstration in the rabbit of humoral and cellular components of the immune system.

bone marrow) to become antibody-producing cells or B cells that also occupy lymphoid tissue, but in different compartments other than T cells (Figure 5.10).

The Initial Immune Response

The immune system responds to agents, cells, or substances that are foreign or non-self, and which are collectively called antigens. An antigen is a chemical, cell, or toxic agent that causes the immune system to respond with the production of antibodies (humoral immunity) or T cells (cellular immunity) or both, and then react with the antibodies or T cells. Most antigens are high molecular weight proteins often associated with lipids,

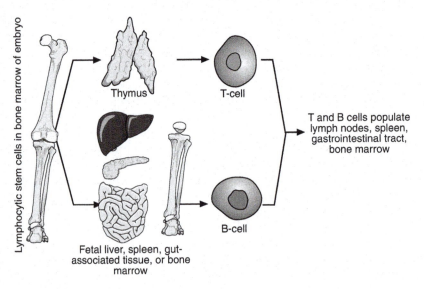

Lymphocytic stem cells in bone marrow of embryo

Thymus

T-cell

Fetal liver, spleen, gut-associated tissue, or bone marrow

B-cell

T and B cells populate lymph nodes, spleen, gastrointestinal tract, bone marrow

FIGURE 5.10
The origin of T cells and B cells.

carbohydrates, or nucleic acids. Some large polysaccharides may also be antigenic. Many molecules are too small to promote antibody formation. However, they may link to a larger carrier molecule such as a serum protein and function together as an antigen. The attached molecule is then called a hapten. Under most circumstances, an antigen such as a bacterial cell reaches one of the numerous lymph nodes in the body where a large white blood cell known as a macrophage (large eating cell) ingests and processes the antigen by partially digesting it and displaying it on the surface of the macrophage along with human leukocyte associated (HLA) antigens to B cells and T cells. The macrophage also produces lymphokines and interleukins that promote the growth and activity T cells and B cells.

Cellular Immunity

T cells respond to a particular antigen then enlarge, divide, and give rise to clones of several subpopulations of T cells including cytotoxic (killer T cells), helper T cells, memory T cells, suppressor T cells, delayed hypersensitivity T cells, and amplifier T cells (Figure 5.11).[12,13]

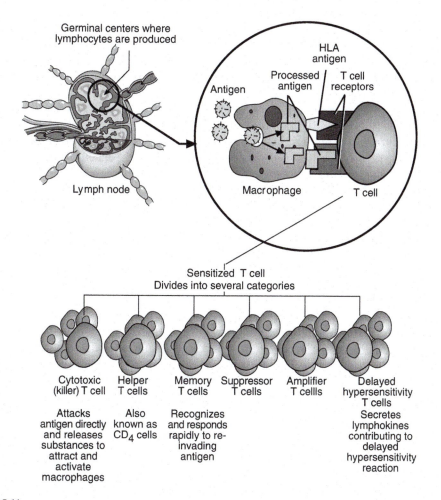

FIGURE 5.11

Action of killer T cells. (Adapted from Tortora G.J. and Anagnostakos, N.P., *Principles of Anatomy and Physiology*, 6th ed., Harper and Row Publishers, New York, 1990, Chapter 22; and Tortora, G.J. et al., *Microbiology: An Introduction*, The Benjamin Cummings Publishing Company, Redwood City, CA, 5th ed., 1995, Chapter 17.)

Killer T cells are sensitized to recognize particular antigenic groups on the surfaces of cancers cells, foreign transplanted tissues, viruses, and fungi. As the cells migrate through the body and recognize such chemical groups, they will respond by attaching to the cell and (1) secreting lymphotoxins that can kill the cell directly; (2) releasing a series of chemicals that attract, hold, and activate macrophages to assist in the destruction of the foreign cell; (3) secreting transfer factor that sensitizes nearby lymphocytes and recruits them to the battle; and (4) producing interferon which inhibits multiplication of viruses. Helper T cells stimulate other immune cells to activity. These helper cells are sometimes termed CD4 cells because of certain cell receptors on their surface. The CD4 or helper T cells are among the primary targets of the AIDS virus, thereby depressing the immune response. There also appear to be memory cells that are long-lived and will recognize the re-entry of a particular foreign cell or material and cause a rapid immune response.

Humoral Immunity

B cells do not normally leave the protection of their locations in the lymph nodes, spleen, gastrointestinal tract, or bone marrow. Their response to foreign invaders is to produce liquid proteins (humoral) known as antibodies and secrete them into the blood stream where they can travel to the affected site and carry out their destructive action. A foreign substance is processed by a B cell with antigenic receptors that recognize specific chemical groups on the foreign material. Each B cell is specific to one particular antigenic group. An antigen selects a specific B cell and attaches to its antibody receptor on the B cell. The antigen is processed and presented with its HLA antigens on the B cell surface to helper T cells (Figure 5.12). The helper T cells produce substances that promote B cell proliferation into a clone of identical plasma cells. These plasma cells produce antibodies to this specific antigen. Other B cells are recruited as memory B cells that are long-lived and will direct a rapid and aggressive response if the identical foreign material should reappear at a later time. Antibodies to a particular antigen are then released from the B cells into the blood stream where they are carried to the affected site and released to the area. The antibodies will attach to the antigenic groups on the foreign substance causing a number of possible actions including (1) neutralization; (2) precipitation or agglutination; or (3) activation of the complement system. Prior to explaining these activities, it is necessary to describe a typical antibody molecule and its mode of action.

The Antibody Molecule

Antibodies (also called immunoglobulins) are proteins with molecular weight of 150,000–900,000Kd. The simplest form of an antibody appears as a bivalent or "Y"-shaped structure consisting of two identical light peptide chains and two heavy peptide chains joined together by disulfide linkages (Figure 5.13). Each light and heavy chain has a constant region composed of specific sequences of amino acids. These regions remain largely unchanged and differ only among the five major classes of antibodies known as immunoglobulins IgG, IgA, IgM, IgD, and IgE, plus four subtypes of IgG(IgG1-4) and two subtypes of IgA(IgA1 & IgA2). Light chains exist in two classes: lambda and kappa. Each antibody molecule has either lambda or kappa light chains, not both. Each of these classes has a separate function. The stem of the constant region of the heavy chain is also known as the Fc region and is capable of attaching to a number of host cells including mast cells. The opposite end of the antibody molecule, the Fab portion, consists of variable regions in the heavy chains and the light chains. The combination of the variable regions on a single heavy chain and its attached light chain form a single antigen-binding site that will attach tightly to one specific antigenic group.

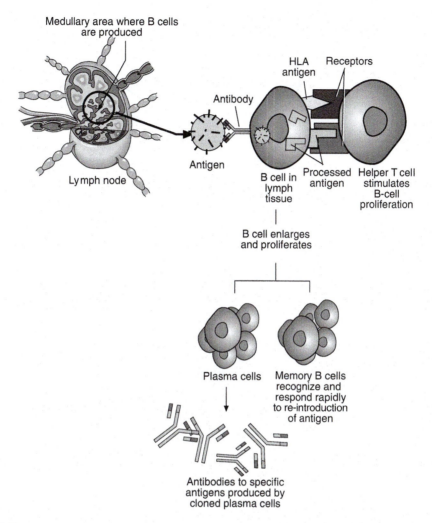

FIGURE 5.12

How B cells respond to antigens in humoral immunity. (Adapted from Tortora G.J. and Anagnostakos, N.P., *Principles of Anatomy and Physiology*, 6th ed., Harper and Row Publishers, New York, 1990, Chapter 22; and Tortora, G.J. et al., *Microbiology: An Introduction*, The Benjamin Cummings Publishing Company, Redwood City, CA, 5th ed., 1995, Chapter 17.)

Since there are two such sites on the simplest "Y" form antibody, it is called bivalent. Further, the peptide arms of the "Y" are flexible and may form a "T" shape as well. Each possible antigenic group on a foreign substance must have its own unique antibody. Consequently, the variable regions of the antibody are created in a specific three-dimensional form that is pre-configured in the B cell clone to only one antigenic group (Figure 5.14). The largest percent of antibodies belong to the IgG class. These bivalent antibodies have a molecular weight of 150,000, can cross the placenta to provide passive immunity in newborns, enhance phagocytosis, neutralize toxins and viruses, and activate the complement system. The role of IgE in allergic reactions will be discussed later.

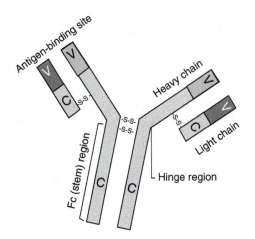

FIGURE 5.13
Typical bivalent antibody molecule. (Adapted from Tortora, G.J. and Anagnostakos, N.P., *Principles of Anatomy and Physiology*, 6th ed., Harper and Row Publishers, New York, 1990, Chapter 22; and Tortora, G.J. et al., *Microbiology: An Introduction*, The Benjamin Cummings Publishing Company, Redwood City, CA, 5th ed., 1995, Chapter 17.)

FIGURE 5.14
Specificity of antibody and antigenic determinants.

Antibody Activities

Antibodies can neutralize viruses and organic toxins (as from snake bites or bacterial toxins) by attaching to receptors and preventing their attachment to host cells or neutralizing toxins by blocking active sites. Antibodies will combine with cells or soluble proteins or haptens to promote agglutination or precipitation. This effectively enlarges the particle sizes and enhances the process of phagocytosis by macrophages. Yet another function of antibodies is opsonization in which the bacteria is coated by the Fab region, thereby facilitating subsequent phagocytosis by cells possessing an Fc receptor. The binding of an antibody with its specific antigen can activate the complement system.

The complement system is composed of several (at least 20) serum proteins that follow a sequence or cascade of actions beginning with the Ag-Ab binding. This sequence of action enhances phagocytosis, inflammation, and cell lysis (Figure 5.15). When exposed to an antigen, there are two types of responses: (1) primary response, which is slow in onset, low in magnitude, and short-lived, and (2) secondary response, which is rapid in onset, high in magnitude and long lived. There is a lag of several days before specific antibody becomes detectable. This antibody is IgM. After a short time, the antibody level declines. This is the

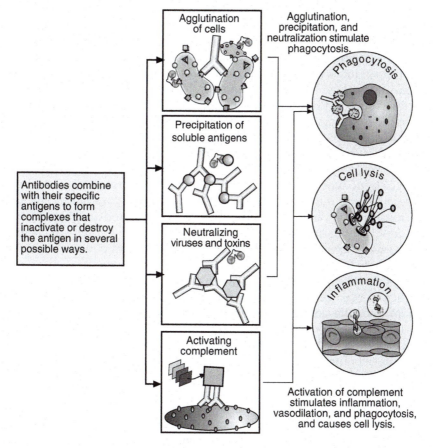

FIGURE 5.15

Antibody activities stimulated by Ag-Ab binding. (Adapted from Tortora, G.J. and Anagnostakos, N.P., *Principles of Anatomy and Physiology*, 6th ed., Harper and Row Publishers, New York, 1990, Chapter 22; and Tortora, G.J. et al., *Microbiology: An Introduction*, The Benjamin Cummings Publishing Company, Redwood City, CA, 5th ed., 1995, Chapter 17.)

primary response. If at a later date we are exposed to the same antigen, there is a far more rapid appearance of antibody and in greater amount. It is of IgG class and remains detectable for months or years. These are the features of the secondary response.

Hypersensitivity

An exaggerated immune response to the presence of an antigen is termed hypersensitivity or allergy. Allergies are the other side of the two-edged sword of immunity. These immune reactions are usually unnecessary, harmful, and sometimes fatal. People have developed allergies to most everything including insect bites, pollen and other plant products, environmental toxins, a variety of foods, sunlight, human semen and much more. Some people have become so sensitized to chemical substances they are unable to physically cope in routine surroundings. Such people are said to have multiple chemical sensitivities (MCS). There has also been a significant increase in the numbers and severity of asthma cases globally. The prevailing scientific thought is that environmental pollutants such as ozone and fine particulates contribute to this increase.[14]

There are four major types of hypersensitivity reactions: cytotoxic reactions, cell-mediated reactions, immune complex reactions, and the most serious, anaphylactic reactions. Cytotoxic hypersensitivity involves activation of complement. This is most often seen in blood transfusions when incompatible ABO and Rh antigens on blood cells are transfused to a person with existing antibodies for those antigens, resulting in the donor cells being lysed. Cell-mediated reactions involve T cells that respond to antigens on transplanted tissues and organs and to haptens associated with poison ivy or poison oak, cosmetics, latex gloves and condoms, nickel in coins and zippers, and bacterial extracts such as PPD from Mycobacterium tuberculosis. Immune-complex reactions involve the coupling of antibodies with soluble antigens circulating in the serum in specific ratios that permit the complex to become trapped in the basement membrane of blood vessel cells where they can activate complement and cause inflammatory responses.

The anaphylactic or Type I hypersensitivity is potentially the most serious type of hypersensitivity reaction and may result in death if an individual does not receive immediate treatment. This reaction is associated with hay fever, asthma, and anaphylactic shock from insect bites, certain foods such as shellfish and tree nuts, and a variety of chemicals. The reactions are usually immediate, occurring within moments in persons previously sensitized by exposure to the offending antigen, and sometimes severe.

The Type I hypersensitivity occurs when the antigen binds in specific ways to IgE antibodies attached to mast cells and basophils. Mast cells occur in the connective tissue of the respiratory tract, the skin, and the blood vessels. Basophils are a form of white blood cells circulating in the bloodstream. The IgE antibodies attach by their Fc fragment to the mast or basophil cells. Persons who have developed an excess of IgE antibodies run a higher risk of developing an anaphylactic type of allergic response because most sites on the mast cell or basophil cells will be covered by IgE antibodies. When an antigen such as animal dander or pollen attaches across two closely spaced IgE molecules specific to that antigen, this causes the mast or basophil cell to degranulate and release histamines, which stimulate vasodilation and bronchoconstriction (Figure 5.16). Vasodilation increases the permeability of blood vessels allowing blood components to seep into the affected area (Figure 5.7). Swelling and redness of mucous membranes in the eyes, nose, and throat are also common symptoms of this type of allergy when the allergens are confined to the

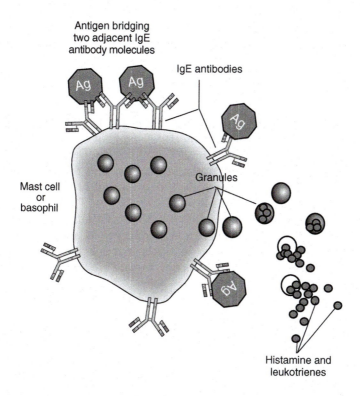

FIGURE 5.16

Mechanism of anaphylactic allergic responses involving IgE. (Adapted from Tortora, G.J. and Anagnostakos, N.P., *Principles of Anatomy and Physiology*, 6th ed., Harper and Row Publishers, New York, 1990, Chapter 22; and Tortora, G.J. et al., *Microbiology: An Introduction*, The Benjamin Cummings Publishing Company, Redwood City, CA, 5th ed., 1995, Chapter 17.)

respiratory tract. The histamines also cause smooth muscle contraction, which occurs in the bronchi and bronchioles of the lung. This bronchoconstriction decreases the lumens of the respiratory tract and makes breathing difficult. The mast or basophil cells, once triggered by an antigen, cause the synthesis of leukotrienes and prostoglandins that cause prolonged contractions of smooth muscle and increased mucous production, which further aggravates respiratory distress. Localized reaction may lead to hay fever, hives, and asthma. These allergic reactions can also be systemic (involving the whole body) producing breathing difficulties, a precipitous decline in blood pressure, cyanosis, and possibly death.[12,13]

Factors Governing Toxicity

A chemical's toxicity—the degree of harm it may inflict—has multiple influencing factors. These factors include (1) the chemical properties of the toxic agent; (2) the dose of the substance; (3) the length of time to which the person is exposed; (4) the route of exposure; (5) whether or not it is bio-transformed in the body; (6) any interaction with other pollutants; and (7) a number of factors about the exposed individual(s), including a person's age, level of activity and/or stress, and health status. Depending on these factors, adverse effects from exposure to a toxic agent may range from a slight irritation, such as itchy, red eyes, to rapid death.

Chemical Properties

The chemical properties of a toxic agent will define its level of toxicity. A chemical's properties are determined by its composition, by the elements it contains, by its structure, by its reactivity, and by its charge. These inherent properties may determine if a chemical is highly corrosive or highly flammable. For instance, chemicals that oxidize rapidly will be severely caustic to the skin while chemicals with an affinity for binding to certain molecules in the body may cause systemic harm.

A chemical's properties will also determine if it exists as a solid, liquid, or gas and if it is hydrophilic or hydrophobic. These properties will influence its ability to move through the environment and to persist. An example we discuss later in this chapter, PCBs, are highly persistent in the environment—they do not break down—and are extremely harmful to health.

Concentration/Dose

The concentration of a chemical is the amount of that chemical in a given volume. This term generally refers to liquids and gases. A chemical can be highly concentrated or dilute. The more concentrated a chemical, the more harmful it may be. For instance, in a very polluted drinking water supply, the concentration of the polluting chemical might be described as 10,000 parts per billion (10,000 ppb), meaning 10,000 molecules of the chemical per billion water molecules. Or if it is less concentrated, and therefore less polluted, it might be measured as 10 ppb. But depending on the chemical's overall toxicity, a concentration of 10 ppb might still be harmful to health.

Related to concentration, dose is the amount of a substance ingested, usually described as that amount ingested in relation to the weight of the organism (i.e., grams of a toxic agent per kilogram weight of an organism). If a toxin is inhaled, the dose may be described by the concentration of the substance. The dose is important, because different doses, from low to high, will yield different physiological responses. Mapping those health responses against varying dose levels (i.e., the amount of a substance ingested) helps scientists build a dose-response curve.

Each toxin has its own unique dose-response relationship. Some toxic agents require only very small doses to achieve a toxic effect while others become toxic only at very large doses. For example, an oral dose of sodium chloride would require 3,000 milligram/kilogram to achieve an LD_{50} in rats whereas the LD_{50} for strychnine is less than 2.5 milligram/kilogram. The LD_{50} is defined as the dose of a chemical that will cause death in 50 percent of a defined animal population (Table 5.1). Further, large acute doses will generally cause more harm than exposure to the same toxin in small amounts over a long period of time. In acute exposure, effects are often witnessed immediately, with characteristic symptoms occurring over a limited time. Chronic effects are usually the result of long-term exposure to smaller amounts of toxins, where health effects develop slowly and insidiously, without obvious identifiable symptoms.[3] For example, a sudden exposure to large amounts of carbon monoxide will lead to dangerous concentrations of carboxyhemoglobin in the blood with resulting unconsciousness and death. Persons exposed to the same toxin in smaller doses over longer periods may slowly develop headaches, dizziness, and chest pains; the toxin may also possibly produce damage to heart and blood vessels.[15]

TABLE 5.1

Comparing Levels of Toxicity: Approximate Oral
LD_{50} Values in Rats

Agent or Chemical	LD_{50} (mg/kg)
Sucrose (Table sugar)	29,700
Ethyl alcohol (Grain alcohol)	14,000
Sodium chloride (Table salt)	3,000
Vitamin A	2,000
Vanillin	1,580
Aspirin	1,000
Chloroform	800
2,4-D	375
Ammonia	350
Carbaryl	250
Caffeine	192
Phenobarbital (Barbiturate)	162
DDT	113
Sodium nitrite	85
Arsenic	48
Sodium cyanide	6.4
Strychnine	2.5
Nicotine	1.0
Dioxin (TCDD)	0.001
Botulinum toxin	0.00001

Source: Malachowski, M.J., *Health Effects of Toxic Substances*, 2nd ed., Government Institute, Rockville, MD, 1999, Chapter 1.

Usually, only small amounts of an absorbed chemical reach the target tissue, which becomes the site of action for the chemical when it attaches to molecular receptors. The effective dose is the amount that reaches the general circulation and ultimately the location where a particular negative effect is seen or felt. This dose, along with the duration of exposure, largely determines the physiological effect. Drinking five beers in an hour has a noticeably different effect than drinking five beers in four hours. Once in the general circulation, the toxic chemical may be absorbed, stored, biotransformed by other tissues, and/or excreted.

The route of entry is also critical to determining the toxic consequences of the agent. Some substances such as lead and mercury are more efficiently absorbed through the lungs than the intestines, making airborne forms of these substances of more concern.[16] When materials containing lead or mercury are incinerated or otherwise combusted, the potential consequences of even small amounts should be of greater concern.

The mechanisms of tissue absorption have been previously discussed and include passive and active processes. Some substances may be localized or stored in various body compartments characterized by having similar structures and functions. Lead tends to be stored in long bones, while organochlorines tend to be stored in fatty tissue. This is called bioaccumulation.

Not all toxic chemicals will bioaccumulate. An organism can metabolize a chemical, altering its structure and characteristics, making it easier to excrete out of the body. This is called biotransformation, which occurs when the toxic agent is metabolized. A wide number of enzymatic pathways are capable of biotransforming toxic substances into non-toxic substances, and these pathways exist in the liver, kidneys, lungs, intestines, and other organs. Pesticides such as malathion are detoxified in humans and other mammals by the enzyme carboxyesterase. However, such biotransformations are not always beneficial as discussed previously, as when sodium fluoroacetate is biotransformed into the more toxic substance fluorocitrate by mammalian cells, resulting in chemical asphyxia.[3,10] Absorbed toxic substances may also be excreted. The major excretory organs are the kidney, liver, and lungs. Chemicals that are not readily soluble in blood and that may have high vapor pressures will likely be exhaled via the lungs. Urine and bile are often the recipients of water-soluble toxic materials. Substances not eliminated in this way are generally bio-transformed or stored in the body.

Interactions

Information known about a chemical's toxicity is derived from animal studies that strive to control all variables but the effects of the single chemical under study. However, most human exposures involve multiple chemicals and these additional chemicals influence the outcome of the exposure. The effects of some substances are additive, in that the two substances taken together are the sum of the two substances taken separately. If you are exposed to chemical A and it produces a measurable response of a 5 percent decrease in RBCs, and chemical B produces a reduction of 3 percent in RBCs, the two together (A+B) would be expected to produce an 8 percent reduction if their effect was additive. Generally, substances behaving in this way employ a similar set of toxic mechanisms. In our example, if the two substances produce a 20 percent reduction in RBCs the effect would be much greater than the sum of the two substances, producing an effect that is synergistic. This has great significance in environmental health since many toxic agents act in this way. Sulfur dioxide is far more toxic in the presence of small particulates or sulfuric acid mists, which carry the pollutant deeper into lung tissue.[17] Again referring to our example, if substance A and substance B are combined and the effect is to produce a reduction of only 2 percent in RBCs, the effect is lesser than the sum of the two acting independently, and this is referred to as antagonistic. Although particulates enhance the toxicity of sulfur dioxide, they have quite the opposite effect when inhaled simultaneously with nitrogen dioxide by decreasing the toxicity.[18] Stimulants such as caffeine in some teas or coffee will tend to counteract the effects of barbituates.[10]

Personal Factors

Age

There are a number of personal factors that influence the effects of toxic substances. The age of the person is a major factor. The very young and the elderly tend to be at increased risk to toxic substances. The young are more susceptible for a number of reasons including (1) they inhale or ingest more toxic substance per kilogram of body weight than adults, (2) they are more likely to put dangerous substances into their mouths, (3) children often absorb materials into the tissues at a higher rate than adults, (4) they will tend to retain larger proportions of toxic substances than adults, and (5) children's nervous systems and detoxifying systems are not fully developed causing disproportionate effects from toxic materials.[19] The adverse effects of ozone, lead, mercury, and sulfur dioxide are more pronounced in children than in adults.[19,20] The elderly are also at greater risk to the effects of toxic substances since they tend to have pre-existing respiratory and cardiovascular diseases, depressed immune systems, and higher levels of stored toxic agents in their bodies than do younger people.

Exercise and Physical Stress

Most athletes are aware that you should not run in heavily polluted air. The consequences are reduced performance, pain in the area of the lungs, and potential long-term harm from inhaling pollutants such as ground-level ozone and exhaust fumes. There is a tendency when exercising heavily to take in air through the mouth in large amounts. Consequently, higher levels of pollutants bypass the nasopharyngeal area, which can filter out pollutants, and enter the sensitive tissues of the lungs more directly. In a study of the effects of ozone on exercise, rats exposed to 1.0 ppm of ozone while exercising periodically died while non-exercising rats experienced little discomfort from the same exposure.[20]

Health Status

The health status of a person also affects the outcome from toxic chemical exposure. Pre-existing cardiovascular disease will render a person more susceptible to a variety of air pollutants. Persons who have asthma or who smoke are also increasingly sensitive to the many pollutants. Alpha1-antitrypsin deficiency is a genetic defect in humans that markedly increases the risk of emphysema from a number of chemical exposures. Genetic polymorphisms, or variations of the same gene can produce subtle or dramatic differences in response to toxic chemicals. The P450 genes produce enzymes that detoxify some agents in tobacco smoke. Variations in P450 genes result in enzymes of varying activity that may increase the risk of lung cancer for many people.[21] Unfortunately, many human diseases are associated with genetic defects. Debilitating diseases, poor nutrition, and lack of exercise all contribute to increasing the risk of adverse effects from toxic substances.[6]

Examples of Toxic Agents

In this section we look at several toxic agents that cause a wide array of damaging health effects in humans and to the environment. The focus is on a group of chemicals that disrupt endocrine function, specifically two: Dioxin and PCBs. We will also look at the toxicity of heavy metals.

It may be useful to consider how toxic chemicals are grouped and classified. Just as the periodic table organizes the elements by similar properties, chemicals may be classified together by their similarities. For examples, chemicals may be grouped together by similarity of chemical structures, physical properties, mechanisms of action, or even levels of toxicity. In this section, we study several chemicals grouped together by the primary toxic effect they have on living organisms.

Endocrine Disrupters and Reproductive Health

Hormone Function

Hormones are a type of chemical messenger. They are made in the body's organs, glands, or tissues and conveyed through the blood to a target organ or tissue, where the hormone stimulates its target to increase or decrease functional activity or secretions. Hormones are active in very small quantities and minor disturbances of their concentrations in the body can disrupt normal functioning. They are critical in the regulation of many life processes, including sexual development, metabolic functions, brain development, human growth, and the fight-or-flight stress response. As examples, androgens regulate the development and maintenance of male sexual characteristics, while estrogens stimulate the development of female sexual characteristics. The predominant and most active androgen is testosterone, which is produced by the male testes. The other androgens that support the functions of testosterone are produced mainly by the adrenal cortex—the outer substance of the adrenal glands. Estrogen is produced primarily in the ovaries, with some minor production in the liver and breasts.

Hormones function by attaching to specific receptors on the inside or outside of cells. These receptors have a three-dimensional complementary fit to the active site of the hormone. Once the hormone links with its specific receptor site, a series of events is initiated that may involve the production of other chemical products or specific biochemical reactions.[2,22] When hormonal activity is disrupted, a number of unwelcome outcomes may occur.

Adverse Effects of Endocrine Disruption

In a report issued in 1997, the EPA concluded that endocrine disrupters may produce cancer, developmental problems, and sterility in animals.[23] There are many examples of adverse effects on reproduction in wildlife and fish in many geographic areas including Florida, the Great Lakes, and California.[24,25] In those areas where fish and wildlife have been exposed to endocrine-disrupting hormones, the effects on reproduction and development have been marked and include the mixing of male and female characteristics, or indeterminate sex; the feminization of male species and the masculinization of female species; unsuccessful reproduction; thyroid dysfunction in birds and fish; decreased fertility in birds, shellfish, and mammals; gross birth deformities in turtles, fish, and birds (e.g., twisted beaks in birds); and behavioral abnormalities.[26,27]

The effects of endocrine disruptors on humans are more problematic. Significant differences in opinion have erupted concerning the degree of this potential threat. Some authorities emphasize that there is no strong evidence that environmental disruptors cause adverse health effects in humans at the current ambient exposure levels, and that there is no causal link between such chemicals and perceived human illnesses.[22] Others believe the evidence supports such a linkage. Some evidence has implicated endocrine disruptors as the cause of a number of emerging human ailments, including (1) reduced sperm counts, (2) early puberty, (3) increase in non-Hodgkin's lymphoma, (4) marked increase in males having undescended testicles (cryptochordism), and (5) testicular cancer.[28,29] Studies have shown a reduction in sperm counts within the general population of many different countries to levels that suggest increased infertility.[30] Other studies suggest there may be regional problems of sperm decline, but challenge the assertion that sperm counts are declining worldwide.[31] The frequency of undescended testicles has more than doubled in England and Wales, and has increased in other countries as well.[32,29] More recent studies have not demonstrated an association between breast cancer and the endocrine disruptors DDE (a metabolite of DDT) and PCBs. In one study blood levels of PCBs and DDE were measured in 600 pregnant women. When their babies were born, the researchers measured levels in the mothers' breast milk. Finally, the team monitored the children as they grew and entered puberty. The most prominent effect reported was that boys exposed to DDE and girls exposed to PCBs were heavier than their unexposed peers at age 14. The study also noted that girls with high prenatal PCB exposure tended to hit the first stages of puberty earlier than others.[33] Numerous endocrine disruptors exist in the environment and the level of concern among public health professionals is great. The combination of increasing testicular cancer, undescended testicles, and falling sperm counts is thought to be caused by fetal exposures to endocrine-disrupting chemicals.[22] Studies of rats exposed in utero have shown that alkylphenols disrupt Sertoli cell development in the fetuses caus-ing decreased testes size and a decrease in daily sperm production of the rats growing to adults.[34] Because humans are exposed from many pathways at levels similar to exposure levels in rats, there is a significant concern that these and other endocrine disruptors may be the reason for the increasing reproductive disorders seen in males.[22] In the late 1990s, due to concern about the potential for endocrine disruption by several types of widely used chemicals, the EPA created a program to screen certain chemicals for endocrine dis-ruption effects. This program, the Endocrine Disruptor Screening Program (EDSP), has identified over 200 chemicals for screening. In 2013, the EPA released its second final list of chemicals to undergo Tier One screening for endocrine disruptive ability. Tier One screen-ing involves a multitude of screens for endocrine disruption of estrogen, androgen, and thyroid hormone systems. These include tests on cell systems and live animals. Chemicals that have disruption move on to Tier Two testing.[35]

What Are Endocrine Disruptors and How Do They Work?

Endocrine disruptors have been referred to as xeno-estrogens, estrogenic, hormone-mimicking, and endocrine-disrupting chemicals. The synthetic chemicals presently iden-tified include a number of pesticides such as DDT, lindane, and atrazine plasticizers such as phthalates and alkylphenols; bisphenol A; PCBs; dioxin; the common food antioxidant butylated hydroxyanisol (BHA); and several fungicides. A variety of naturally occur-ring plant compounds or phytoestrogens are also known to have estrogen compounds (Table 5.2). There are at least four different mechanisms by which endocrine disruptors can exert their adverse effects (Figure 5.17).

TABLE 5.2

Endocrine-Disrupting Chemicals

Chemical	Uses
Phthalates	Used as plasticizers; found in medical and household products, clothing, toys, and packaging including plastic wraps, and liners of metal food and beverage cans; phthalates leach from food wraps and lining of containers into food and beverages where they are ingested; phthalates may also be absorbed through the skin and have been associated with testicular toxicity
Dioxin	An unwanted byproduct in the production of some chlorinated hydrocarbons, paper production, and incineration of plastics
PCBs	Used in electrical transformers and capacitors, hydraulic fluids, plasticizers, and adhesives; production was banned in the U.S. in 1977 because of bioaccumulation, persistence, and toxicity
Alkylphenols	Present in detergents, paints, pesticides, plastics, food wraps, and plastic laboratory ware; they are known to be estrogens
Pesticides	Methoxychlor, lindane, dicofol, atrazine, and DDT are pesticides that exhibit weak to moderate estrogenic activity, but mixtures of two or more pesticides have been shown to produce potent activities, sometimes 1000 times that of individual pesticides
Natural sources	Grains, some vegetables, nuts, berries, grasses, and soybeans may contain phytoestrogens such as flavinoids and liganans with hormone-like properties

FIGURE 5.17
Mechanisms of endocrine disruption.

1. Some endocrine disrupters are similar enough to the native hormone that they can occupy its site on the receptor tissue causing a cascade of events as if the native hormone had bound to the site. Such a chemical is called a hormone mimic. Some chemicals may bind to the site but are not similar enough to activate a response. They may simply prevent normal hormonal action through a hormone block.

2. Endocrine disruptors may cause artificial and unwelcome changes in the numbers of hormone receptor sites on tissues and organs.

3. An endocrine disruptor may cause a change in the production of a hormone, either increasing its production or decreasing, thereby changing the concentration of the

hormone in the body. Small amounts of a change in a hormone can have large effects on the target organ.

4. Endocrine disruptors may replace hormones on the carrier proteins that transport hormones via the blood stream to various parts of the body. This makes some proportion of the population native functional hormone unavailable and produces adverse health consequences that may be barely noticeable to very severe.[21]

Endocrine disruptors were originally labeled as estrogens or androgens based on their ability to bind to estrogen receptors or androgen receptors, respectively. Little thought was given to the possibility of a disruptor being able to bind to both types of receptors. However, recent studies in vitro reveal that some environmental pollutants can bind to estrogenic receptors, androgenic receptors, and binding proteins. If this phenomenon occurs in living organisms, then the effects of environmental disruptors may occur across several different receptor sites and magnify the effects, making even very low concentrations quite potent.[36]

Reducing Exposure

Although no definitive evidence exists on the role of endocrine disruptors in adversely effecting human health, there is sufficient concern that federal agencies, industry, private organizations are attempting to respond to this complex dilemma with more answers through research and recommendations for risk reduction. Many of the endocrine-disrupting chemicals are chlorine-based and persist in the environment. Furthermore, they are ubiquitous and find their way into the food supply through (1) ingestion of contaminated grains and grasses by livestock, which then store the lipophilic chemicals in their fatty tissues; (2) contamination of fruits and vegetables by spraying with pesticides; and (3) leaching of endocrine disrupting chemicals from plastic wrappers, plastic liners of cans, and polystyrene containers. The majority of exposures may come from contaminated food. Individuals who are concerned about the exposure to endocrine disruptors may reduce that exposure by:

1. Reducing or limiting ingestion of dairy products and meat high in fat, where organochlorines tend to accumulate
2. Avoid synthetic pesticides by purchasing foods low in pesticide residues and switching to herbal or scent-based repellents
3. Keep children from vinyl toys or teething rings
4. Use detergents and shampoos that do not contain alkylphenols such as nonoxynol and octoxynol[23]

Dioxin

The dioxins are a family of related chemicals with a highly toxic profile. There are 75 separate chemical compounds that comprise the group of chemicals called dioxins. Dioxins are not manufactured for direct use in any product. Rather, these chemicals are unwanted by-products of certain manufacturing processes and of waste incineration. The bulk of dioxins are generated by heating mixtures of chlorine and organic compounds in industrial processes, such as the bleaching of paper pulp, the incineration of chlorine-containing materials or in the production of pesticides, herbicides and chlorophenol wood preservatives. They are formed by the incomplete combustion of wood products and both industrial and municipal wastes. Unfortunately, dioxin does not easily break down and

FIGURE 5.18
Chemical structure of dioxin.

will bioaccumulate (the accumulation of toxic chemicals in living things through the consumption of food or water) in the fatty tissue of humans and animals. This substantially increases the potential exposure of nursing infants whose mothers may have concentrated dioxin in the milk and fatty tissues of the breast.[37]

The most acutely toxic chemical within the dioxin family is known as 2,3,7,8-tetrachlorodibenzo-p-dioxin, or TCDD. The International Agency for Research on Cancer (IARC), part of the WHO, announced on February 14, 1997 that TCCD is now considered a class 1 carcinogen, meaning a "known human carcinogen."[38] Dioxin slowly breaks down in the environment when it is exposed to the ultraviolet rays of the sun. Otherwise, dioxin is a stable compound. Since it is highly insoluble in water, it is not readily able to migrate through the soil. Once in the soil, it persists.[39]

Dioxin is considered to be one of the most toxic manufactured chemicals (Figure 5.18). It is not an estrogen, however, but it can block the action of estrogens.[40] Dioxin is capable of lowering androgens and affecting the levels of thyroid hormones. It affects both the endocrine system and the immune system.[30] Dioxin does not attach to the estrogen receptor; rather, it attaches to a receptor called the Ah-receptor, whose function is unknown. Even though it does not attach to the estrogen receptor, it still exerts both estrogenic and anti-estrogenic activity, indirectly influencing estrogen activity. Its anti-estrogenic effects may result from (1) causing the DNA, which is attached via the receptor, to produce an enzyme that is capable of breaking down the body's normal estrogen or (2) simply by decreasing the number of estrogen receptors available for naturally occurring estrogen.[41]

Although dioxin is considered to be among the most potent synthetic toxic chemicals known, its effects are highly variable among species and even differ among age groups within the same species. Animal studies show wide variations in reproductive and developmental effects from dioxin in different species at different levels of exposures. Adult hamsters are several thousand times more resistant to the toxic effects of dioxin than adult guinea pigs, while there is much less difference among the fetuses.[42]

Most of the effects of dioxin are based on animal studies, but there have been unintended or accidental human exposures exceeding normal ambient levels. The largest known dioxin contamination occurred between 1963 and 1971, when 45 million liters of Agent Orange, a defoliant mixture consisting of 2,4,5,-T and 2,4-D contaminated with dioxin, was sprayed over parts of Vietnam. Anecdotal reports followed by media attention thrust the specter of Agent Orange into the public eye as a possible cause for veterans' health complaints. Most studies to date show little or no adverse health effects from such exposures. A study by the Centers for Disease Control showed Vietnam veterans have slight losses in sperm concentrations and somewhat more hearing loss than other veterans.[43] The health of the Air Force unit responsible for aerial spraying of the herbicides (Ranch Hands) was compared with units handling other cargo. The two groups were similar in health except Ranch Hands had slightly more basal cell carcinomas.[44] There were also no differences seen in major, minor, or suspected defects among children between Vietnam and non-Vietnam veterans.[45]

The danger of dioxin exposure was seen in an accidental release in Italy in July 1976. At that time, pressure built up in the reaction chamber of the ICMESA chemical company, which was producing trichlorophenol, the parent compound of hexachlorophene and 2,4,5-T. At noon, the pressure became too great and up to 1.8 kilograms of dioxin was released into the air, descending onto the town of Seveso and the patios where meals were being served. Many of the animals in the town, including birds, rabbits, and chickens died within a few hours of exposure. People developed chloracne (a disfiguring acne involving the sebaceous glands), nausea, and temporary nerve damage. Since that time, scientists have been monitoring the health of about 2000 families in Seveso and have documented increases in cardiovascular disease and questionable increases in certain types of cancers. Research carried out by the same group of scientists has strengthened the evidence that dioxin may be a carcinogen in humans. They have found that people living in the second most contaminated area, referred to as zone B, were nearly three times more likely to develop liver cancer than the general population. In the same zone, the occurrence of a form of myeloma was seen 5.3 times more often among women, while among men, cancers of the blood were 5.7 times more likely. The scientists did not find a greater number of cancers in the most heavily polluted area due to the small group of people who were most affected and who moved immediately, thereby shortening the duration of their exposure. People living in zone B had a lower albeit more prolonged exposure.[46]

Although these findings were not the first to associate dioxin with cancer in humans, the Seveso study has significance. This population was well monitored because new techniques had made blood levels of dioxin easy to measure, which enables researchers to more accurately determine exposure levels. The investigators also based their findings on extrapolations from soil data, which corresponded to the analyses of the blood samples.[46]

In the 1970s, a waste hauler sprayed dioxin-contaminated oil on several towns in Missouri and at the Shenandoah Stables near Moscow Hills in Missouri. More than 50 quarter horses died. Chickens, dogs, cats, and birds died. The owner's young daughter was hospitalized with aching joints, inflammation, headaches, and diarrhea.[47] In 1983, the Environmental Protection Agency announced that the riverside community of Times Beach, Missouri, should abandon their homes and property and evacuate. During the early 1970s, the roads were oiled to control the dust. Soil analyses revealed high levels of dioxin contamination; the oil had been scavenged from a trichlorophenol factory by a waste hauler.[39] Subsequent studies revealed no health problems associated with the former citizens of Times Beach.[39]

The EPA released a draft dioxin reassessment document in 1994 based on an extensive review of dioxin's toxicity. The document asserts that levels of dioxin present in the environment may be very close to those that can produce health consequences to humans. The report further states that dioxin poses increased risk for cancers, adverse reproductive and developmental effects, neurological damage including both cognitive and behavioral effects from in utero exposure, endocrine disruption, and reproductive and development effects.[48]

In a continuing effort to provide a balanced view, it should be reported that some authors report that (1) there are no formal studies to date that any person has died from environmental exposure to dioxin in the United States; (2) dioxin has been reported to cause chloracne and short-term reversible nerve dysfunction, and may be toxic to both the liver and the kidneys; (3) dioxin is highly toxic to some species of animals and minimally toxic to others; (4) the evidence that dioxin causes birth defects is inconclusive; and (5) occupational exposures to dioxin may increase the risk of soft-tissue sarcomas.[39]

FIGURE 5.19
Chemical structure of polychlorinated biphenyls (PCBs).

Polychlorinated Biphenyls

PCBs are relatively chemically inert, nonflammable fluids with high plasticizing ability and a high dielectric constant (Figure 5.19). They were commercially synthesized in the United States from 1929 to 1977 and were used in transformers, capacitors, hydraulic and heat transfer fluids, and solvents in adhesives and sealants. They were also used to coat electrical wires and as a protective coating on lumber, concrete, and metal surfaces. PCBs were used as an anti-oxidant in paints, varnishes, and rubber compounds.[48] The PCBs were found not only around us in air, food, soil, water, and in human and wildlife fat tissue, but also in plastics, food wraps, cosmetics, detergents, pesticides, and even baby bottles.[1] PCBs are a family of approximately 209 chemical compounds whose characteristics vary considerably. Some of the PCBs are light and oily fluids, while others are heavy, greasy, or wax-like. The properties that made PCBs desirable in electric equipment, such as their resistance to thermal degradation and their stability, are the very same properties that have led to their bioaccumulation in the environment.[29] PCBs were banned from most uses in the U.S. in the 1970s due to their persistence in the environment and their ability to bio-accumulate in animal tissue. There are various chemical alternatives to PCBs like silicone fluids, fluorocarbons, high-molecular-weight hydrocarbons, low-molecular-weight chlorinated hydrocarbons, and high-boiling oils and esters, which work well in electrical equipment. PCBs[49,50] remain widely spread in the environment including the fat tissue of humans and animals. More than 94 percent of fish collected in the U.S. show PCB residues at an average concentration of 0.53 ppm.[51,52]

Sources of PCB Exposure[52]

1. Using old fluorescent lighting fixtures and old appliances such as television sets and refrigerators; these may leak small amounts of PCBs into the air when they get hot during operation.
2. Eating food, including fish, meat, and dairy products containing PCBs.
3. Breathing air near hazardous waste sites that contain PCBs.
4. Drinking PCB-contaminated well water.
5. Repairing or maintaining PCB transformers.

While the U.S. stopped producing many of these PCBs, they are still being made in other parts of the world. These PCBs are binding to dust particles and are transported through the air, depositing on plants, which enter the food chain, or they land on water where they enter aquatic life and bio-accumulate.[29]

A few of the PCBs are dioxin-like, exerting similar biological activities. Some PCBs, or their metabolites, are estrogenic, while others can alter the levels of thyroid hormones in the bloodstream.[53] Some PCBs have the ability of binding to proteins in the blood, which would normally bind to thyroid hormone, interfering with the transport of the thyroid hormone and altering normal growth and metabolism.[41] PCBs may also be altered during metabolism, so one portion of the PCB molecule resembles a portion of an estrogen molecule, allowing for the PCB molecule to bind to the estrogen receptor. This activity may cause either an estrogen or anti-estrogen response.

Researchers are consistently finding delayed psychomotor development of children exposed to PCBs in the uterus. There is evidence of a connection between paternal dioxin exposure to birth defects in offspring. There are worldwide increases in diseases and conditions of the reproductive system in infants, children, and adults, which may be linked to environmental chemical exposure.[1]

There have been two major PCB poisoning incidents in which the adverse effects have been clearly observed in children born to women exposed to high levels of PCBs. The PCBs were also contaminated with dioxin analogues, known as polychlorinated dibenzofurans (PCDFs). In Japan, in 1968, a large outbreak of what was thought to be PCB poisoning occurred, when 1,300 Japanese developed chloracne, eye discharge, swelling in the joints, weakness, and other symptoms after consuming rice oil. Apparently, the rice oil had been contaminated when a PCB-containing heat exchanger leaked fluid from a pipe during processing. This incident was referred to as "Yusho (rice oil) disease."[54] A similar incident occurred in Taiwan in 1979, dubbed the "Yu-Cheng disease."[55]

Children were studied over time after the Yu-Cheng incident. Children at birth were observed to have abnormalities in teeth, nails, and pigmentation. Many of the children exhibited developmental delays in addition to being smaller than average size. The symptoms of a decreased body size and a 5-point IQ deficit was observed even after the children continued to mature.[55] Behavioral problems and cognitive deficiencies have also been evidenced in children born to mothers who have eaten PCB-contaminated fish.[55] Studies have also shown that in utero exposure to PCBs and related contaminants is associated with poorer short-term memory functioning in early childhood.[56]

Although the environmental levels of PCBs have declined over recent years, the risk of exposure from industrial toxic waste continues due to the current use of PCBs in older electric equipment or deposited in landfills. The PCBs in such landfills exceed the quantity that has escaped into the environment.[56] PCBs have been banned in this country since 1977, but the persistence and ubiquity of these compounds in the environment means that children continue to be exposed, potentially leading to deficiencies in cognitive, behavioral, and motor functions of the brain.[2]

Lead

Lead has been in documented use since the days of the Egyptian pharaohs. The Romans used it for piping water in parts of the viaduct, in cooking vessels, and as a sweetener and preservative in wines. Its dangers in affecting the human mind have been known for more than 2,000 years.[57] Despite its known dangers to human health, it continues in wide use throughout the world and especially in North America where it has had a history of use in agriculture in the form of lead arsenate for pesticide use, as solder in pipes, as a solder in food containers, and as an anti-knock compound in gasoline. It has also been used in paints, inks, and glazes for ceramics. Lead has been used so extensively, for such a long time that it has been widely dispersed throughout our environment in substantial quantities and appears

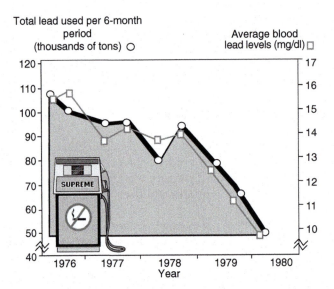

FIGURE 5.20
Comparison of lead in gasoline and average blood lead levels. (From Schwartz, H. et al., *Costs and Benefits of Reducing Lead in Gasoline: Final Regulatory Impact Analysis*, EPA-230-05-006, U.S. Government Printing Office, Washington, DC, 1985.)

in the flesh of every human, in nearly all living creatures, and virtually everywhere in the environment including the most remote locations of Antarctica.[58] Lead has been placed on the forefront of environmental health concerns due to compelling evidence of a wide range of adverse effects. In the 1970s, both federal regulatory and legislative efforts were begun to reduce lead hazards, including the limitation of lead in paint and gasoline (Figure 5.20).[59] Lead was also reduced in evaporated milk, juices, and by 1989, less than 4 percent of food cans manufactured in the United States contained lead solder.[57] Since 1980, there have been intensive actions to further reduce lead exposure from gasoline, paint, solder, and other sources. The EPA regulations requiring the reduction of leaded gasoline were highly successful in reducing the blood lead levels in people.[60] Secondary prevention programs, such as screening for early detection and lead education programs, have also been implemented.

The most significant sources of exposure to inorganic lead include food, water, soil, and lead-based paint.[57] The most current environmental exposures to lead in the U.S. are from lead paint in older homes, although people are exposed to lead in their drinking water supplies, due to the leaching of lead from pipes. Additional sources of exposure may include lead-glazed pottery, medicinal and cosmetic preparations, and food grown in lead-contaminated soil. The fact that solder containing lead is no longer used in commercial food canning and leaded gasoline has been phased out since 1983 has reduced lead exposures in the U.S. However, leaded gasoline is currently being used in other parts of the world and will continue to expose pregnant women and children for many more years to come.[58]

Lead may enter the body through ingestion and inhalation, while organic lead may also be absorbed through the skin. Lead does not have any normal function in living organisms, but lead does produce toxic effects. After lead enters the body, it is able to distribute to various organs, cross the blood-brain-barrier, and can cross the placenta. Blood lead levels in the fetus are up to 90 percent as high as blood lead levels in the mother.[61] Lead is also excreted into breast milk during lactation.

Whether lead enters the body via ingestion or inhalation, it then enters the bloodstream where its half-life is estimated at 36 days. Therefore, measurements of blood lead levels are calculated as micrograms of lead per deciliter of blood (microgram per deciliter), which gives the most accurate indication of short-term lead exposure. While some lead will make its way to soft tissues such as the brain and kidneys, 50–60 percent of the lead that enters the body is excreted relatively quickly. Over time, the lead will be distributed, deposited, and stored in bones, especially the long bones of the arms and legs.[57] Lead can remain in the long bones for years and may be released back into the bloodstream secondary to conditions such as high fever, osteoporosis, or pregnancy, which may result in acute lead poisoning.

Lead may cause numerous toxic effects on the body. One of the most significant is its effect on the formation of RBCs. It does this via two distinct mechanisms: (1) by slowing the normal maturation of RBCs in the bone marrow, decreasing the number of RBCs and possibly causing anemia; and (2) inhibiting the synthesis of hemoglobin by interfering with two important substances that are necessary for the formation of hemoglobin: delta-aminolevulinic acid and coproporphyrin III. In lead poisoning, both of these substances are subsequently excreted in the urine in excessive amounts. The inhibition of the activity of delta-aminolevulinic dehydrogenase in RBCs (an enzyme involved with the synthesis of hemoglobin) serves as a diagnostic marker in the early stages of lead poisoning before the symptoms are pronounced.[61–63]

Lead impairs fertility in both men and women when blood lead levels approach 50 microgram per deciliter. Lead may act directly on the testes, lowering the sperm count. At high blood lead levels, lead may cause spontaneous abortions and stillbirths. It is interesting to note that in the past, lead was actually used to induce abortions.[64]

In recent years, there has been more and more evidence that lead may have serious health effects even at exposure levels that were previously not considered harmful. The EPA and the CDC now suggest 5 microgram per deciliter as the maximum acceptable blood lead level for fetuses and young children. In 2012, the CDC recommended that the new blood lead *level of concern* for children be changed from 10 to 5 microgram per deciliter.[58] This *level of concern* continues to decrease as scientists learn more about the dangers to children of even the smallest exposure to lead. In 1960, the CDC level of concern was 60 microgram per milliliter; in 1975, it was halved to 30 microgram per deciliter; in 1986, it was lowered again to 25 microgram per deciliter. Now at its lowest level of 5 microgram per deciliter, this level of concern reflects decades of research on lead's health effects. Thankfully, lead levels continue to decline in children (Figure 5.21).

There is a socioeconomic factor to lead exposure. In the U.S., nearly 22 percent of black children ages 1–2 years old currently have blood lead levels over 10 microgram per deciliter.[63] In the workplace, employers are not required to remove workers from the exposure until their blood lead levels reach 50 microgram per deciliter. This occupational health standard will only protect against acute lead poisoning.[58] In the U.S., women of childbearing age have an average blood lead level of about 2.0 microgram per deciliter, while the average blood lead level for men of the same age is 4.2 microgram per deciliter.[63]

The results of numerous studies carried out over the last several decades continue to demonstrate that at levels previously considered safe, chronic low-level lead exposure can inhibit the normal development of a child's intellectual ability. A 1979 study in the Boston suburbs of Chelsea and Somerville, performed on second grade students, concluded that those whose baby teeth had a higher lead level also exhibited a greater incidence of unruly classroom behavior, lower IQ scores, and a diminished capacity to follow instructions than

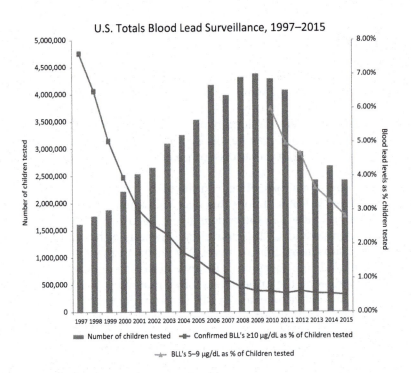

FIGURE 5.21
Blood lead levels in children, Centers for Disease Control, 2015.

did classmates with lower lead levels in their teeth (Figure 5.22).[65] A follow-up study was carried out in 1990.[66] All students in the 1979 study were now 18 years of age. In comparison to classmates whose baby teeth had minimal lead levels, the students with higher lead levels had a greater school drop-out rate, higher incidence of reading difficulties and a lower class rank, all in part due to childhood lead exposure. Several studies conclude that lead exposure at levels even below 10 microgram per deciliter is associated with aggressive and delinquent behavior, impaired neuro-behavioral development, and low birth weight.[66,67] Studies on rats confirmed that lead exposure may produce long-term cognitive deficits. Additionally, rats exposed to lead post-weaning did not show impairment in performance despite high body burdens of lead, whereas exposures during gestation and lactation (maternal exposure) resulted in significant impairment. The authors concluded that a window of vulnerability to lead levels exists that results in long-term cognitive impairment.[68]

Human exposure to lead has been reduced in recent years by the near phase-out of lead in gasoline, paint, food containers, and home plumbing. Nevertheless, preschoolers are continually exhibiting elevated lead levels, possibly secondary to lead paint and dust still present in their homes. Overall, young children are at a greater risk for elevated lead levels due to (1) their increased oral activity, (2) increased ability to absorb lead, (3) higher retention of absorbed lead, and (4) the incompletely developed nervous system.[57] Other children are exposed by family members involved in furniture, refinishing, pottery, or stained glass businesses who unintentionally carry lead dust home with them on their clothes.

Blood lead levels as low as 5 micrograms per deciliter have been associated with developmental delays, deficits in intellectual performance and neurobehavioral functioning, and decreased stature and diminished hearing activity.[59] There is a continuing need to identify and eliminate the source of lead exposure for every child.

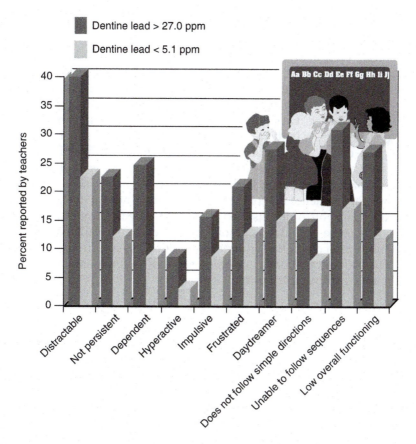

FIGURE 5.22

The negative effect of classroom behaviors associated with dentine lead levels. (Adapted from Needleman, H.L. et al., *J. Am. Med. Assoc.*, 251, 2956, 1984.)

Organic Solvents

Organic solvents are a group of simple organic liquids that have the capacity to evaporate at atmospheric pressure. Organic solvents are widely used in the U.S., in both the home and industry. They are components of many products, including paints, varnishes, paint removers, adhesives, glues, degreasing and cleaning agents, pharmaceuticals, plastics, and pesticides. Nearly 50 million tons of industrial organic solvents are manufactured yearly in the U.S.[57] Organic solvents have been known to contaminate drinking water supplies. Not only can solvents enter the body by ingestion from contaminated drinking water supplies, they also may enter the body via skin absorption and inhalation in the shower. In fact, a major route of organic solvent exposure is inhalation since these chemicals are so volatile. The total exposure from taking a 10-minute shower in contaminated water is actually greater than the exposure after drinking two quarts of the same water.[69]

Individuals may be exposed to organic solvents if they work with electronics or in the health care field, in auto repair, dry cleaning, laboratory work, or painting. Solvents may enter the body through a variety of ways such as inhalation, skin contact, or ingestion. The amount of solvent entering the body strictly depends on factors such as the concentration of the solvent in the air and the route of exposure.[57] Organs such as the brain, due to its blood supply and high fat content, tend to achieve high levels of solvents more quickly than organs with lower fat contents. All solvents are fat-soluble and will produce effects on the central nervous system.[70]

Researchers have examined the reproductive effects of solvents, in which many of the human studies have found harmful reproductive effects. The evidence remains consistent that maternal exposure to solvents during pregnancy increases the risk of spontaneous abortions by two- to fourfold.[71] Birth defects, including cleft palate and lip, cardiovascular malformations, and CNS defects, have also been reported in association with organic solvents.[72]

Studies of glycol ethers, in both animal and human studies, conclude that they damage testicular function, cause testicular atrophy, lower sperm counts, and may lead to infertility.[73] Solvents may also lead to acute leukemias in children of exposed parents as well as an increased risk of neurological and urinary cancers.

There is disturbing evidence that organic solvents, most notably the glycol ethers, cause spontaneous abortions, birth defects, and childhood cancers. It is estimated that nearly 10 million workers come in contact with organic solvents on a daily basis.[57] Workers and their families need protection through intensive education, informing workers about the health risks involved and mandatory training sessions to teach employees how to minimize exposure to hazardous substances.

Asbestos

Asbestos is a collective term for a group of six fibrous silicate materials—asmolite, chrysotile, tremolite, actinolite, anthophyllite, and crocidolite—that are found throughout most of the world. Nearly 95 percent of all asbestos in the United States comes from just one of the six types of commercially used asbestos: chrysotile. The "white" asbestos is used in wall insulation, ceilings, around pipes and vents, and for heat resistance in roofing shingles. The chrysotile-based asbestos is also widely used to reduce friction in automotive brake-shoe linings. Asbestos has been used as an essential component in various products and processes, including building materials, brake linings, textiles, and insulation, as well as floor tiles, cement, and potholders (Figure 5.23). There are approximately

FIGURE 5.23
Asbestos used in insulating pipes.

3,000 uses of asbestos in various industries.[74] Unfortunately, during 1950–1972, asbestos was used in spray insulation in buildings, which pose a significant environmental health hazard as they are being demolished. Asbestos can enter our water supplies by airborne settling, dumping of asbestos waste, and the dumping of effluent from mining operations. Asbestos may also leach from asbestos-cement pipes in municipal water systems. In cities where asbestos-cement pipes are leached heavily by the water, long fibers of asbestos have been known to clog household appliances.[75] Asbestos is a human carcinogen, which can be found in air, water, and food throughout the U.S. Asbestos is naturally released into the environment, but there are also manufactured emissions, from mining asbestos ores to the manufacture of asbestos products. Exposure to these fibers may occur throughout the environment, especially in our drinking water supply.

Asbestos, although extremely useful, represents an enormous health risk. It is estimated that of the 8–11 million current and retired workers who were exposed to vast amounts of asbestos, between 30 and 50 percent can be expected to die from asbestos-related cancers.[70] Their families are also at risk, since asbestos workers presumably carried dust on their clothes and their hair into their homes. Roughly 5–10 percent of all workers who were employed in asbestos manufacturing or mining will die from mesothelioma (an asbestos-caused cancer of the lining of the lung or abdomen). Asbestos workers who smoke have more than a 75 percent increase in the incidence of lung cancer over that in non-smokers.[76]

Exposure to asbestos is primarily achieved through the inhalation of tiny fibers that are suspended in the air. Once inhaled, these fibers often get trapped deep within the lungs. Most fibers are carried through the gastrointestinal tract and excreted within a few days, while the few that remain imbedded deep in the lungs lead to irreversible disease processes including asbestosis, lung cancer, and mesothelioma. Certain fibers such as cro-cidolite are more likely to produce disease than other forms of asbestos, but it is generally believed that the size of the fiber may be the more important characteristic in disease production than the fiber type.

Two types of disease commonly seen in those exposed to asbestos are asbestosis and mesothelioma. Asbestosis is a slowly developing disease normally occurring over 20–30 years, resulting in increasing breathlessness on exertion and the development of a productive cough. Decreased respiration leads to a barrel chest, bluish discoloration of the skin, and a restricted chest expansion.[74] The lungs are typically scarred, with thickening of the lining of the lungs, which is replaced by calcium deposits. Mesothelioma is a rare malignant tumor involving the pleura or the abdominal wall and abdominal organs, caused only by exposure to asbestos. The disease is a diffuse cancerous tumor that spreads rapidly, is usually fatal, and results in death within 1–2 years after diagnosis.[74] Cancer can also develop in the digestive tract as a result of ingesting asbestos in food, in beverages, or swallowing the fibers in contaminated saliva.

Asbestos is durable, flexible, strong, inflammable, and an excellent insulator. These useful qualities led to it being widely used in construction. Buildings constructed between 1945 and 1978 are likely to contain asbestos material, while pipe insulation manufactured from 1920 to 1972 contained asbestos to prevent heat loss from pipes. In schools, the EPA estimated that 15 million students and 1.4 million teachers and other employees are in buildings that contain asbestos.[77]

In 1971, the EPA classified asbestos as a hazardous air pollutant as defined by the Clean Air Act of 1970. Although this regulation has helped to control asbestos emissions from manufacturing, roadway surfacing, and demolition projects, the main concern and target from asbestos regulations have been school systems. Congress passed the Asbestos School Hazard Abatement Act (ASHAA) in 1984 to provide financial assistance to schools having significant asbestos problems. The Act was reauthorized in 1990 and required additional training requirements for abatement workers along with revisions to accreditation plans. President Reagan authorized the Asbestos Hazard Emergency Response Act (AHERA) on October 22, 1986, which required schools to inspect for asbestos, develop asbestos management plans, and implement actions involving repair, encapsulation, or removal of the asbestos. This was all to take place prior to 1988 and be conducted by accredited asbestos removal personnel. Nearly $300 million has been distributed to schools under these various Acts for abatement, training, and to conduct other activities related to asbestos.[78] Despite this enormous cost and effort, the average level of asbestos in schools was less than 1/3000 the level (0.0007 fibers per milliliter) than the federal limit permitted by OSHA in the workplace environment. Work on the removal of asbestos from schools continues today at very significant costs. The EPA estimates that the cost of removing all asbestos from public and commercial buildings throughout the United States may exceed $51 billion.[79]

Mercury

There are three different forms in which mercury is found in the environment: elemental mercury vapor, inorganic mercury compounds, and organic (usually methyl) mercury. The three different forms are significantly different from one another. They are each absorbed by the body differently, they are used and manufactured for different purposes and they have different effects on reproduction and development.[69]

Mercury is used in the production of chlorine, and also in thermometers, batteries, and fluorescent light bulbs. For centuries mercury was an essential part of many different medicines such as diuretics, antibacterial agents, antiseptics, and laxatives. More recently these drugs have been substituted and drug-induced signs of mercury toxicity are rare.[80] Mercury is also a constituent of dental amalgams. Concerns regarding mercury stem from its volatile nature, resulting in global dispersal throughout the air and water. To make matters even worse, inorganic mercury may be methylated by microorganisms that are present naturally in the environment to an even more toxic form, methyl mercury, which tends to bio-accumulate in marine life.[81]

In the 1940s and 1950s, mercury became known as the product that caused acrodynia also known as Pink disease. Manifestations of acrodynia included pain and erythema of the palms and soles, irritability, insomnia, anorexia, diaphoresis, photophobia, and skin rashes.[80] Debate continues about the lowest levels at which mercury causes human health effects. Studies have found that the health effects of organic mercury are similar in both animal and human populations. Studies have shown in mice that the cells that are in the developing brain stop in the middle of cell division when exposed to organic mercury.[69] Mercury poisoning usually is misdiagnosed due to the insidious onset, nonspecific signs and symptoms, and lack of knowledge within the medical profession.

Organic mercury or methylmercury is the most toxic form of mercury, due to its ease of absorption, and the ability to cross the blood-brain-barrier, where it can destroy the cells that control coordination. Symptoms of poisoning appear 1–2 months after exposure, with initial symptoms including numbness of the tongue, lips, and fingertips. Speech gradually becomes slurred, and difficulty swallowing soon follows. It is a developmental toxin. In cases involving pregnant women, often a pregnant mother will remain relatively free of symptoms while children born to her will demonstrate gross physical abnormalities and severe mental retardation.[81–82,84]

The most infamous episode of organic mercury poisoning occurred during the years 1953–1961 in the town of Minamata on the Japanese island of Kyushu. A local vinyl chloride manufacturing facility had been dumping inorganic mercury waste into the Minamata Bay where bacteria in the sediment converted it into organic mercury. It then moved through the aquatic food chain, contaminating the fish. The population of Minamata relied heavily on a fish diet and were ingesting organic mercury when eating the local fish. When tested for levels of organic mercury, these fish showed concentrations 100,000 times that of the water levels. By 1956, people started to exhibit symptoms of central nervous system damage including tremors, convulsions, numbness of hands and feet, and losses in speech, hearing, and vision.[82] More than 1,000 Minamata Bay sufferers have since died. Minamata's population has shrunk since the time of the poisoning from 50,000 to less than 33,000. The chemical company (Chisso Corporation) implicated in this disaster did not stop dumping until 1968.[82] More than 50 years later, the prefectural government has declared that the fish in Minamata Bay now meet safety standards and the net that confined the fish to the bay has been removed. However, the social stigma and discrimination for victims of Minamata will likely continue.[82] By 1991, more than 2,000 cases of Minamata disease had been confirmed.[81] A similar epidemic occurred again, in 1965, but this time in riverside villages along the Agano River, Niigata, Japan.[81]

Elemental mercury is the type of mercury found in dental amalgam fillings, thermometers, and batteries. It is hazardous when inhaled. Elemental mercury can easily pass the blood-brain barrier and cross the placenta to the fetus once it is in the body. Individuals can be exposed to mercury in the air from waste incinerators that burn medical waste or batteries, and also from oil and coal burning, since mercury is a contaminant of these fuels.[83] It is far more dangerous to inhale mercury fumes than to swallow inorganic mercury. If swallowed, the mercury passes through the gastrointestinal tract and is excreted in a few short days.

Inorganic mercury vaporizes at room temperature and its volatility increases markedly with each degree increase in temperature. This makes it especially hazardous when spilled on floors or surfaces where dirt or grease may exist and combine with the mercury to form tiny drops that expose larger surface area to air. Vaporized inorganic mercury can enter the body either by inhalation or absorption through the skin following prolonged contact. Symptoms may include damage to kidney tubules and the liver, coordination problems, generalized tremors, and possible hallucinations. These symptoms are described among hat makers in seventeenth-century France who soaked animal hides in a solution of mercuric nitrate in order to soften the hides. This practice exposed the hat makers to mercury vapors for long periods of time. Due to this exposure, hat makers developed the typical symptoms of inorganic mercury poisoning referred to as "Mad Hatter's Disease" (featured in the children's story *Alice in Wonderland*).

Chronic and acute exposures cause cutaneous and neurological symptoms. The classic triad found in chronic toxicity is tremors, gingivitis, and erethism (i.e., a constellation of neuropsychiatric findings that includes insomnia, shyness, memory loss, emotional instability, depression, anorexia, vasomotor disturbance, uncontrolled perspiration, and blushing).

Arsenic

Since the Middle Ages, arsenic has been a popular way of poisoning people. Arsenic itself is a gray semi-metal that is not very poisonous. Arsenic and related compounds occur naturally in granite and mountains as arsenopyrites or arsenic sulfide ("Arsenikon" is the

Latin word for yellow pigment). Other sources include microbiological activity, pesticides, herbicides, fertilizers, human waste, volcanic eruptions, and industrial sources.[85]

A dangerous form of arsenic is white arsenic or arsenic oxide. This is extremely poisonous, although the symptoms of arsenic poisoning can be confused with those of other illnesses. Arsenic poisoning causes a variety of systemic problems like diaphoresis, muscle spasms, nausea, vomiting, abdominal pain, garlic odor to the breath, diarrhea, anuria, dehydration, hypotension, cardiovascular collapse, aplastic anemia, and death.[87] Perhaps unsurprisingly, given all these symptoms, historically, arsenic was used successfully as a poison, because it was effective and very difficult to detect after death. Politicians, popes, and other powerful people have all been victims at different times. Indeed, white arsenic became known as "Inheritance Powder."

Arsenic poisoning occurs when arsenic binds very strongly to sulfur groups. Many proteins contain one or more sulfur atoms, and enzymes (also proteins) are responsible for regulating the body's internal chemistry. An enzyme with an arsenic atom bound to sulfur may not be able to function.

Bangladesh and parts of the Indian state of West Bengal are facing a disaster of astronomic proportions. Beginning in the 1970s, tube wells were sunk in Bangladesh and West Bengal to avoid cholera and other bacterial infections present in aboveground water supplies. However, arsenic has slowly crept in on unsuspecting villagers. Over 6 million people in West Bengal and 30 million Bangladeshis are drinking water containing arsenic over five times the WHO limit of 10 ppb.[86] A decrease of the water table resulted in exposure to air in the zone of aeration. This exposure led to the oxidation of arsenic minerals previously below the water table in the Bengal sediments. The arsenic oxides then migrated to the groundwater and were reduced to the poisonous forms in the reducing environments below the water table.

Do you want to learn more about toxicity and toxins? Point your browser to <https://people.umass.edu/~envhl565/> and click on any of the Chapter 5 Web links associated with toxicity and toxins including:

- e.hormone: Gateway to environmental signaling
- EPA Toxic Release Inventory
- State of Endocrine Disruption Science—WHO

References

1. Schaider, L.A., Balan, S.A., Blum, A., Andrews, D.Q., Strynar, M.J., Dickinson, M.E., Lunderberg, D.M., Lang, J.R., and Peaslee, G.F., Fluorinated compounds in U.S. fast food packaging, *Environment Science and Technology Letters*, 2017, 4(3), 105–111.
2. Schettler, T., Solomom, G., Burns, P., and Valenti, *Generations at Risk: How Environmental Toxins May Affect Reproductive Health in Massachusetts*, Greater Boston Physicians for Social Responsibility, and Massachusetts Public Interest Research Group (MASSPIRG), Education Fund, Boston, MA, July 1996, Chap. 1.
3. Waldbott, G.L., *Health Effects of Environmental Pollutants*, 2nd ed., The CX Mosby, St. Louis, MO, 1978, Chap. 5.
4. Malachowski, M.J., *Health Effects of Toxic Substances*, Government Institute, Rockville, MD, 1995, Chap. 1.

5. Tortora, G.J. and Anagnostakos, N.P., *Principles of Anatomy and Physiology*, 6th ed., Harper & Row Publishers, New York, 1990, Chap. 3.

6. Tortora, G.J. and Anagnostakos, N.P., *Principles of Anatomy and Physiology*, 6th ed., Harper & Row Publishers, New York, 1990, Chap. 23.

7. Tortora, G.J. and Anagnostakos, N.P., *Principles of Anatomy and Physiology*, 6th ed., Harper & Row Publishers, New York, 1990, Chap. 5.

8. Malachowski, M.J., *Health Effects of Toxic Substances*, Government Institute, Rockville, MD, 1995, Chap. 2.

9. Tortora, G.J. and Anagnostakos, N.P., *Principles of Anatomy and Physiology*, 6th ed., Harper & Row Publishers, New York, 1990, Chap. 24.

10. Malachowski, M.J., *Health Effects of Toxic Substances*, Government Institute, Rockville, MD, 1995, Chap. 6.

11. Simon, F.P., Potts, A.M., and Gerard, R.W., Action of cadmium and thiols on tissue enzymes, *Archives of Biochemistry and Biophysics*, 12, 283, 1947.

12. Tortora, G.J. and Anagnostakos, N.P., *Principles of Anatomy and Physiology*, 6th ed., Harper & Row Publishers, New York, 1990, Chap. 22.

13. Tortora, G.J., Funke, B.R., and Case, C.L., *Microbiology: An Introduction*, The Benjamin/Cummings Publishing Company, Redwood City, CA, 5th ed., 1995, Chap. 17.

14. Friebele, E., The attack of asthma, *Environment Health Perspectives*, 104, 1, 22, January 1996.

15. Astrop, P., Kjeldsen, K., and Wanstrup, J., Enhancing influence of carbon monoxide on the development of atheromatosis in cholesterol-fed rabbits, *Journal of Atherosclerosis Research*, 7, 343–354, 1967.

16. Schroeder, H.A., Metals in the air, *Environment*, 13, 18, 1971.

17. Amdur, M.O., The physiologic response of guinea pigs to atmospheric pollutants, *International Journal of Air Pollution*, 1, 170, 1969.

18. Labelle, C.W., Long, J.E., and Christifano, E.E., Synergistic effects of aerosols, *Archives of Industrial Health*, 11, 297, 1955.

19. U.S. Department of Health and Human Services, *The Nature and Extent of Lead Poisoning in Children in the United States: A Report to Congress*, Public Health Service, Agency for Toxic Substances and Disease Registry, Washington, DC, 1988.

20. Stokinger, H.E., Ozone toxicology: A review of research and industrial experience, 1954–1964, *Archives of Environmental Health*, 10, 719, 1965.

21. Perera, F.P., Environment and cancer: Who are susceptible? *Science*, 278, 1068, November 7, 1997.

22. Crapo, L, *Hormones: Messengers of Life*, WH Freeman, New York, 1985, Chap 1.

23. Biondo, B., Are common chemicals scrambling your hormones? *USA Weekend*, 15, 18, February, 13, 1998.

24. Fry, D.M., Speich, S.M., and Peard, R.J., Sex ratio skew and breeding patterns of gulls: Demographic and toxicologic considerations, *Studies in Avian Biology*, 10, 26, 1987.

25. Colburn, T., von Saal, F.S., and Soto, P.M., Developmental effects of endocrine-disrupting chemicals in wildlife and humans, *Environmental Health Perspectives*, 101, 378, 1993.

26. Guillette, L.J., Gross, T.S. Masson, G.R. et al., Developmental abnormalities of the gonad and abnormal sex hormone concentrations in juvenile alligators from contaminated and control lakes in Fla., *Environmental Health Perspectives*, 102, 680, 1994.

27. Colburn, T. and Clement, C., Advances in modern environmental toxicology, Vol. XXI *Chemically Induced Alterations in Serial and Functional Development: The Wildlife/Human Connection*, Princeton Scientific Publishing, Princeton, NJ, 1992, Chap. 1.

28. McLachlan, J.A., Functional toxicology: A new approach to detect biologically active xenobiotics, *Environmental Health Perspectives*, 101, 386, 1993.

29. Birnbaum, L., Endocrine effects of prenatal exposure to PCBs, dioxins, and other xenobiotics: Implications for policy and future research, *Environmental Health Perspectives*, 102(8), 676, 1994.

30. Kimmee, CA., Approaches to evaluating reproductive hazards and risks, *Environmental Health Perspectives*, 101(supp 2), 137, 1993.
31. Forum, New evidence in sperm counts, *Environmental Health Perspectives*, 104(9), 919, 1996.
32. Jackson, M.B., Chilvers, C., Pike, M.C. et al., Cryptochordism: An apparent substantial increase since 1960, *British Medical Journal*, 293, 1401, 1986.
33. Teens before their time, *Time Magazine, Health*, 156(18), October 30, 2000.
34. Sharpe, R.M., Fisher, J.S., Miller, M.M. et al., Gestational and lactational exposure of rats to xenoestrogens results in decreased testicular size and sperm production, *Environmental Health Perspectives*, 103(12), 1136, 1995.
35. *Endocrine Disruptor Screening Program Tier 1 Assays: Considerations for Use in Human Health and Ecological Risk Assessments, Office of Chemical Safety and Pollution*, Office of Pesticide Programs (OPP) 2013. https://www.epa.gov/endocrine-disruption/endocrine-disruptor-screening -program-edsp-tier-1-assessments#test-conducted (accessed March 3, 2018).
36. Raloff, J., Hormone mimics get harder to pigeonhole, *Science News*, 151, 254, April 26, 1997.
37. Smith, A.H., Infant exposure assessment for breast milk dioxins and fumas derived from waste incineration emissions, *Risk Analysis*, 7, 3, 347, 1987.
38. World Health Organization, The international agency for research on cancer, *Polychlorinated Dibenzo-para-Dioxins and Polychlorinated Dibenzofurans, Summary of Data Reported and Evaluation*, Vol. 69, August 12, 1997. http://www.iarc.fr.
39. Whelan, E.M., Toxic terror, *Prometheus Books*, New York, 1992, Chap. 9.
40. Safe, S., Astroff, B., Harris, M. et al., 2,3,7,8-Tetrachlorodibenzo-p-dioxin (TCDD) and related compounds as antiestrogens: Characterization and mechanism of action, *Pharmacol Toxicol*, 69, 400, 1991.
41. McKinney, J.D., and Waller, C.L., PCBs as hormonally active structural analogues, *Environmental Health Perspectives*, 102(3), 290, 1994.
42. Olson, J.R., Holscher, M.A., and Neal, P.L.A., Toxicity of 2,3,7,8-tetrachlorodibenzo-p-dioxin in the golden Syrian hamster, *Toxicology and Applied Pharmacology*, 55, 67, 1980.
43. Centers for disease control Vietnam experience study, health status of Vietnam veterans, Part 11, Physical Health, *The Journal of the American Medical Association*, 259, 2708, 1988.
44. Wolfe, W.H., Michalek, J.E., Miner, J.C. et al., Health status of air force veterans occupationally exposed to herbicides in Vietnam, *The Journal of the American Medical Association*, 264, 14, 1824, 1990.
45. Centers for disease control Vietnam experience study, health status of Vietnam veterans, Part 111, reproductive outcomes and child health, *The Journal of the American Medical Association*, 259, 2715, 1988.
46. Holloway, M., Dioxin Indictment: A growing body of research links the compound to cancer, *Scientific American*, 270(1), 25, January 25, 1994.
47. Monks, V., The truth about dioxin, *National Wildlife*, September 4, 1994.
48. Schecter, A., Dai, L.C., Bich, L.T. et al., Agent orange and the Vietnamese: The persistence of elevated dioxin levels in human tissues, *American Journal of Public Health*, 85, 4, April, 516, 1995.
49. World Health Organization, The international agency for research on cancer, *Polychlorinated Dibenzo-para-Dioxins and Polychlorinated Dibenzofurans, Summary of Data Reported and Evaluation*, Vol. 69, August 12, 1997. http://www.iarc.fr.
50. Jensen, S., The PCB story, *Ambio*, 1, 123, 1972.
51. Schmitt, C.J., Zajicek, J.L., Ribick, M.A., National pesticide monitoring program. Residues of organochlorine chemicals in freshwater fish, 1980–1981, *Archives of Environmental Contamination and Toxicology*, 14, 225, 1985.
52. U.S. Department of Health and Human Resources, Agency for toxic substances and disease registry, (ATSDR). *Toxicological Profile for Selected PCBs (update)*. ATSDR/TP-92/16, Atlanta, GA, 1993. https://www.atsdr.cdc.gov/csem/csem.asp?csem=30&po=6.
53. Safe, S.H., Polychlorinated biphenyls (PCBs): Environmental impact, biochemical and toxic responses, and implications for risk assessment, *Critical Reviews in Toxicology*, 24, 1, 1994.

54. Kuratsume, M., Yusho, A poisoning caused by rice oil contaminated with polychlorinated biphenyls, *H.S.H.M.A.S. Health Reports*, 86, 1083, 1971.

55. Gun, Y.L., Ju, S.H., Chen, Y.C. and Elsu, C.C., Sexual developments and biological findings in Yucheng children, *Chemosphere*, 14, 235, 1993.

56. Jacobson, J.L., Jacobson, S.W., and Humphrey, H.E.B., Effects of in utero exposure to polychlorinated biphenyls and related contaminants on cognitive functioning in young children, *Journal of Pediatrics*, 116, January 1990.

57. U.S. Congress, Office of Technology Assessment, *Neurotoxicity: Identifying and Controlling Poisons of the Nervous System*, OTA-BA-436, U.S. Government Printing Office, Washington, DC, April 1990, Chap. 10.

58. *Blood Levels in Children*, CDC Fact Sheet, Center for Disease Control, Washington DC, 2018.

59. Pirkle, J.L., Brody D.J., Gunter, E.W. et al., The decline in blood lead levels in the United States: The national health and nutrition surveys (NHANES), *The Journal of the American Medical Association*, 272, July 1994.

60. Schwartz, H., Pitcher, H., Levin, P.T. et al., *Costs and Benefits of Reducing Lead in Gasoline: Final Regulatory Impact Analysis*, EPA-230-05-006, U.S. Government Printing Office, Washington, DC, 1985.

61. Clarkson, T.W., Nordberg, G.E, and Safar, P.R., Reproductive and developmental toxicity of metals, *Scandinavian Journal of Work, Environment & Health*, 11, 145, 1985.

62. Waldbott, G.L., *Health Effects of Environmental Pollutants*, 2nd ed., The CX Mosby, St. Louis, MO., 1978, Chap. 12.

63. Brody, D.J., Pirkle, J.L., Kramer, R.A. et al., Blood lead levels in the US population: Phase I of the third national health and nutrition examination survey (NHANES III), 1988–1991, *The Journal of the American Medical Association*, 272, 277, 1994.

64. Agency for Toxic Substances and Disease Registry (ATSDR), *Toxicological Profile for Lead, U.S.* Department of Health and Human Services, Atlanta, GA, April 1993.

65. Needleman, H.L., Rabinowitz, M., Leviton, A. et al., The relationship between prenatal exposure to lead and congenital anomalies, *The Journal of the American Medical Association*, 251, 2956, 1984.

66. Needleman, H.L., Schell, A., Bellinger, D. et al., The long-term effects of exposure in low doses of lead in childhood: An 11-year follow-up, *The New England Journal of Medicine*, 322, 83, 1990.

67. Needleman, H.L., Reiss, J.A., Tobin, M.J. et al., Bone lead levels and delinquent behavior, *The Journal of the American Medical Association*, 275, 363, 1996.

68. Kuhnnann, A., MeGlothan, L., and Guilarte, T.R., Developmental lead exposure causes spatial learning deficits in adult rats, *Neuroscience Letters*, 223, 101–104, 1997.

69. Schettler, T., Solomom, G., Bums, P., and Valenti, M., Generations at risk how environmental toxins may affect reproductive health in Massachusetts, *Greater Boston Physicians for Social Responsibility, and Massachusetts Public Interest Research Group (MASSPIRG) Education Fund*, Boston, MA, July 1996, Chap 5.

70. Dick, R.B., Short duration exposures to organic solvents: The relationship between neurobehavioral test results and other indicators, *Neurotoxicology and Teratology*, 10, 39, 1988.

71. Taskinen, H., Kyyronen, P. Hemminki, K. et al., Laboratory work and pregnancy outcome, *Journal of Occupational Medicine: Official Publication of the Industrial Medical Association*, 36, 311, 1994.

72. Holmberg, P.C., Hernberg, S., Kurppa, K. et al., Oral clefts and organic solvent exposure during pregnancy. *International Archives of Occupational and Environmental Health*, 50, 371, 1982.

73. Wess, J.A., Reproductive toxicity of ethylene glycol monomethyl ether, ethylene glycol monoethyl ether and their acetates, *Scandinavian Journal of Work, Environment and Health*, 2, 43, 1992.

74. Waldbott, G.L., *Health Effects of Environmental Pollutants*, 2nd ed., The CX Mosby, St. Louis, MO, 1978, Chap. 16.

75. Archer, S.R. and Blackwood, T.R., *Status Assessment of Toxic Chemicals: Asbestos*, U.S. Environmental Protection Agency, Monsanto Research Corporation, Dayton, OH, December 1979.

76. Selikoff, Q., Churg, J., and Hammond, E.C., Asbestos exposure and neoplasia, *The Journal of the American Medical Association*, 188, 22, 1964.
77. Koren, H. and Bisese, M., *Handbook of Environmental Health and Safety: Principles and Practices*, Vol. 1, 3rd ed., CRC Lewis Publication, Boca Raton, FL, 1995, Chap. 7.
78. Whelan, E.M., Toxic terror, *Prometheus Books*, New York, 1992, Chap. 8.
79. EPA, *Study of Asbestos Containing Materials in Public Buildings: A Report to Congress*, United States Environmental Protection Agency, Washington, DC, February 1988.
80. U.S. Department of Health and Human Resources, Agency for toxic substances and disease registry, (ATSDR), *ToxFAQs:Chemical Agent Briefing Sheets (CABS), Arsenic*, January, 2006. http://www.atsdr.cdc.gov/cabs/arsenic/arseniccabs_info.html.
81. Ratcliffe, H.E., Swanson, G.M., and Fischer, L.J., Human exposure to mercury: A critical assessment to the evidence of adverse health effects, *Journal of Toxicology and Environmental Health*, 49, 221, 1996.
82. Pollack, A., Mercury, mostly gone from bay in Japan, still poisons Town's life, *New York Times*, 1, August 23, 1997. www.nytimes.com (accessed July 11, 2018).
83. Agency for Toxic Substances and Disease Registry (ATSDR), *Draft Toxicological Profile for Mercury. US Department of Health and Human Services;* October 1992.
84. Cox, C., Clarkson, T.W., Marsh, D.O. et al., Dose-response analysis of infants prenatally exposed to methyl mercury: An application of a single compartment model to single strand hair analysis. *Environmental Research*, 49, 318, 1989.
85. Dhar, R.K., Biswas, B.K., Samanta, G. et al., Groundwater arsenic calamity in Bangladesh. *Current Science*, 73(1), 48, 1997.
86. Mandal, B.K., Chowdhury, T.R., Samanta, G. et al., Chronic arsenic toxicity in West Bengal. *Current Science*, 72(2), 114–117, 1997.
87. Mazumder, N.G., Gupta, J.D., Santra, A. et al., Non-cancer effects of chronic arsenicosis with special reference to liver damage, *Book: Arsenic: Exposure and Health Effects*, C.O. Abernathy, R.L. Calderon and W.R. Chappell (Eds.), Chap. 10, pp. 112–123, 1997, Chapman & Hall, New York.

6

The Trouble with Pests

Introduction

According to a report prepared by Bechtel, Parsons and Brinckerfoff, co-managers of the massive Central Artery/Third Harbor Tunnel project in Boston, Massachusetts, rats were sighted at the rate of 500 per hour shortly after 9:00 p.m. on November 12, 1993, in the open-air market known as Haymarket in downtown Boston.[1] Fears were so great that construction would displace the rats into nearby neighborhoods that the construction managers were required to develop plans and countermeasures for exterminating the rats. They weren't the first. The blunt-nosed, muscular rat (*Rattus norvegicus*) arrived from Norway as stowaways on boats in the 1600s and numbered in the tens of thousands by the mid-1700s. Nearly 10,000 rats were killed in the last 4 months of 1742 as the town of Boston paid 4 pence for each rat killed. An all-out effort in 1917 to rid Boston of the more than 700,000 rats failed dismally, and all efforts throughout the following years have failed to eradicate them.[1] Boston is not alone in these efforts. In one 18-meter-deep pit in Bombay, India, there are 442,186 dead rats of the variety *Rattus rattus*. After more than half a century of relief from the plague, the return of the rat in huge numbers has promoted the return of the plague to many of India's cities.[2] In most areas of the world, rats are feared and detested, and they periodically become the targets of military-style campaigns to destroy them as pests. This is not so everywhere. In Mexico's impoverished northern desert region, rats are hunted down as a favorite food where they boil them up in soup or fry them golden brown, although the government views the selling of rats for meat (about $1 each) as illegal.[3] The prospect of hunger and starvation is a powerful motivation in adjusting attitudes towards what is food and what is a pest.

Also on the comeback trail after a pause in years of effort to destroy them are the mosquitoes. *Aedes aegypti* has returned in enormous numbers to Latin America after funds for mosquito spraying were diverted elsewhere, human populations exploded, sewage and waste were improperly handled, and sanitation suffered. The result has been a dramatic increase in dengue fever.[4] For similar reasons, increasingly larger epidemics of dengue fever and other vector-borne diseases have returned to India, Sri Lanka, Taiwan, China, and Africa.[4] Despite heroic attacks against the female *Anopheline* mosquito to control malaria, these programs have also proved ineffective, and once more malaria affects 500 million people worldwide and causes 2 million deaths per year.[5] Deer mice have proliferated in New Mexico and spread hanta virus to the human population in that area, while on the east coast of the United States the deer tick, *Ixodes scapularis*, carries the disease agent for Lyme disease in a complicated life cycle involving the tick, deer, and mice. Rats, mice, mosquitoes, and ticks all have something in common. Most of humanity perceive them as undesirable because of their ability to transmit diseases and have labeled them as pests to be targeted for destruction.

What Are Pests?

Pests are unwanted plants and animals.[6] Pests are not limited to organisms that transmit disease but may be any living thing that negatively affects human interests. The interests usually include (1) a loss of resources such as agricultural crops, food and property damage, and damage to lawns and gardens; (2) agents of disease; and (3) sources of annoyance and discomfort. Pests may be weeds that compete for agricultural space; freshwater mussels that clog intake pipes; rabbits that decimate field crops; moles that burrow through lawns; starlings that threaten collisions with aircraft; wasps that sting; moths that eat through clothes; cockroaches and flies that carry disease organisms onto food; or insects, fungi and viruses that attack crops or people. Pests cover the entire spectrum of the plant and animal kingdoms ranging from small viruses to large mammals. Although well represented across the animal and plant kingdoms, the actual number of pests is only a small percent of the total members. The vast majority of plants and animals are viewed by most as beneficial, worthwhile, or at least neutral. The most unwelcome pests from a public health perspective tend to include arthropods and rodents. The arthropods (phylum Arthropoda) are invertebrate animals with jointed and paired appendages, a chitinous exoskeleton, and segmented bodies. It is the largest animal phylum with over 700,000 species, including insects, arachnids, myriapods, and crustaceans (Figure 6.1).[7] Within the arthropods are included the insects and arachnids that are medically important as vectors of disease, disease agents themselves, or noxious pests. Certain arthropods and rodents have been associated with numerous disease outbreaks, crop damage, annoyance from bites, and contamination of food sources. Insects, ticks, and mites are involved in the majority of important human vector-borne diseases, and most of these diseases cannot be prevented by vaccines or chemotherapy. Consequently, control of vector-borne diseases often relies upon reducing the number of vectors.[8] Early success in the control of insect vectors has been met with reversals, such as (1) insects have developed resistance to insecticides; (2) insect control programs have been halted or under-funded; (3) the use of less-expensive pesticides such as dichlorodiphenyltrichloroethane (DDT) have been reduced because of environmental concerns and political pressures; (4) the combination of poverty and overpopulation has led to poor sanitation with greater opportunity for insect proliferation; (5) destruction of forested areas has eliminated natural insect predators; and (6) climate changes, including warming trends, have promoted increases in some insect populations.[4] For these reasons, many insects are rebounding to large numbers and there have been significant increases in vector-borne diseases such as malaria, yellow fever, dengue fever, chagas disease, typhus, relapsing fever, the encephalitides, schistosomiasis, and plague. These diseases affect hundreds of millions of people worldwide and suppress the vitality and economies of many nations. Understanding the role of insects in this process is important to the development of public health policy and programs. In order to develop workable biological, physical, or chemical controls, it is critical to understand the biology and mechanisms for the spread of disease by these organisms.

Do you want to learn more about pests? Point your browser to <https://people.umass .edu/~envhl565/> and click on any of the Chapter 6 Web links associated with pests, including:

- Insect Fact Sheets
- Insects on the WWW
- CDVIK
- Crab Louse or Pubic Louse
- Rats

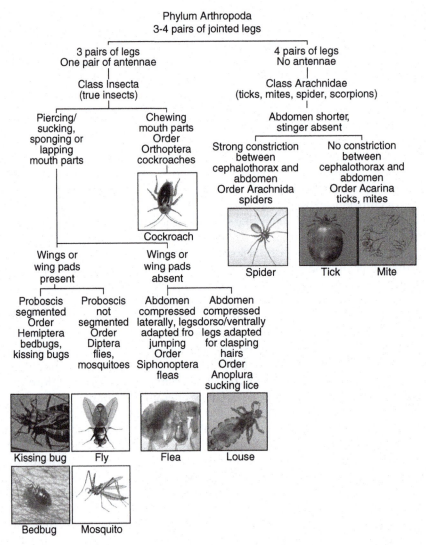

FIGURE 6.1
Pictorial key to selected orders of arthropods of public health significance. (Adapted from U.S. Department of Health, Education and Welfare (HEW), *Pictorial Keys to Some Arthropods of Public Health Importance*, Washington, DC, U.S. Department of HEW Public Health Services, 3, 1964.)

Insects and Other Arthropods

General Structure and Development

Most insects start their life cycle with the fertilization of an egg, and they pass through either a complete or incomplete metamorphosis. Incomplete metamorphosis refers to insects such as roaches, body lice, and grasshoppers, which go through three developmental stages including egg, nymph, and adult stages. Mosquitoes, flies, and butterflies go

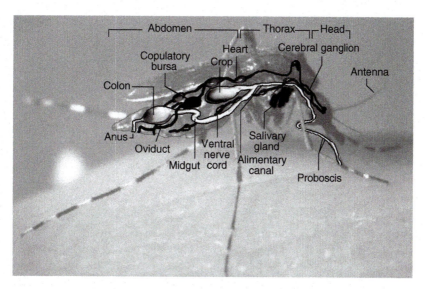

FIGURE 6.2
Internal and external structures of typical insect. (Adapted from Koren, H., and Bisesi, M., *Handbook of Environmental Health and Safety Principle and Practices*, 3rd ed., CRC Press, Lewis Publishers, Boca Raton, FL, Chapter 4.)

through complete metamorphosis and pass through four developmental stages including egg, larvae, pupa, and adult stages. As insects grow and develop, they shed their protective exoskeleton while a new one develops underneath in a process called molting. The exoskeleton is a hard outer skin that protects internal organs and serves as an attachment site for muscles to provide movement. The hardened outer body wall of the exoskeleton may be covered with hairs, scales, or spines, and is normally divided into segments joined by flexible intersegmental structures. In insects, these segments consist of a head, thorax, and abdomen as seen in Figure 6.2. Members of the order Arachnida have only one or two main segments as in ticks, mites, or spiders. The head usually contains mouth parts, antennae, large compound eyes, and/or simple eyes. The numbers of eyes and antennae help to distinguish among species. Mouth parts vary and are adapted to forms of feeding including piercing/sucking (mosquitoes and fleas), chewing and grinding (roaches), and lapping/sponging (houseflies, blowflies). Houseflies and similar sponging insects regurgitate fluids onto solid foods, liquefy them, and suck them up. They can carry disease organisms on body parts or in the digestive tract to food or open wounds and so serve as mechanical vectors of disease. Mosquitoes and fleas can pierce the skin of animals and humans to reach blood. In many instances, the microbial disease agents go through a developmental stage in the vector involving sexual reproduction (i.e., malaria). In these instances, the vector is known as a biological vector. In the case of malarial transmission, the infective stage or sporozoites migrate to the salivary glands where they are inoculated into the bloodstream of the next human victim as the mosquito takes a blood meal (Chapter 7, Figure 7.10). The thorax of insects bears three pairs of jointed legs and may also contain two sets of wings that are often reinforced with structures called veins. The abdomen, or third body part, contains external openings to the respiratory system known as spiracles, and also contain external reproductive organs. Internally, the abdomen is composed mostly of reproductive and digestive organs (Figure 6.2).

kissing bug

bedbug

FIGURE 6.3
Example of typical kissing bug and bedbug.

Bedbugs and Kissing Bugs

The insects belonging to this category have three pairs of jointed legs, sucking mouth parts, wings or wing pads, a segmented proboscis, and belong to the order Hemiptera. The kissing bug has well-defined wings while the wings are absent in the dorso-ventrally flattened bed bug (Figure 6.3). True to their name, bedbugs hide in bedsprings, mattresses, and cracks in the wall coming out at night to feed on humans and warm-blooded animals. People sensitive to the bites will develop large and very itchy welts. No disease transmission has been associated with bedbugs, although here has been a significant increase in the numbers of reported infestations and irritations.

The kissing bug is the arthropod vector for American trypanosomiasis or Chagas disease caused by the flagellated protozoan parasite *Trypanosoma cruzi*. The disease occurs predominantly in Mexico and South/Central America where reservoirs include rodents, armadillos, and opossums. The kissing bug lives in the crevices and cracks of stone or mud huts with thatched roofs. This reduviid bug appears at night while people are sleeping and bites them near the lips (hence "kissing bug"). The Trypanosome is deposited to the skin surface in the bugs' feces. The irritated skin is scratched by the person who inadvertently rubs the organism and feces into the open bite wound. The disease is affecting more than 50 percent of the population in some rural South American communities. Once inoculated, the Trypanosomes are carried by the blood to tissues throughout the body where they can multiply and produce a severe inflammation. Fatalities occur in about 10 percent of those infected. These are usually children, and the heart appears to be the primary target organ disrupted in these fatalities. Damage to the central nervous system (CNS) with paralysis of certain body functions is also common.[9]

Flies

Flies belong to the same order of insects (Diptera) as mosquitoes. Adult flies have one pair of functional wings and three distinct body parts including the head, thorax, and abdomen (Figure. 6.4). The middle thoracic section is larger than the others because it supports the highly developed wing muscles. Flies may have sponging mouth parts as does the house fly, while other fly species may have rasping or sucking mouth parts. All have compound eyes and a single pair of antennae. There are thousands of species of flies, and they are significant vectors of human disease and discomfort. Black flies, deer flies, sand flies, horse flies, and stable flies can all inflict painful bites or bleeding punctures. Humans have died from anaphylactic shock produced by the bulk of foreign protein injected into the skin from the simultaneous bites of hundreds of flies. Some species may lay eggs or larvae in wounds resulting in invasion of the living tissue by fly larvae that produce a condition known as myiasis where maggots grow in the tissue or in open cavities. The common house fly, *Musca*

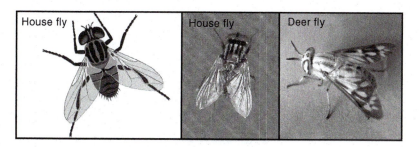

FIGURE 6.4

Typical domestic fly (*Musca domestica*) and the deerfly (*Chrysops discalis*), which is a biting fly that transmits deerfly fever, a form of tularemia.

domestica, is an important vector of infectious disease organisms and many foodborne illnesses. House flies can travel several kilometers in a single day, feeding upon human waste(s) and transporting disease organisms on their bodies and in their feces to human food and beverages. Diseases that may be mechanically transmitted by flies include cholera, bacterial and amoebic dysentery, typhoid and paratyphoid fevers, salmonellosis, hookworm, pinworm, and whipworm. Biting flies have been responsible for the transmission of a number of diseases including sandfly fever, onchocerciasis (blinding filariasis), African sleeping sickness, deerfly fever (a form of tularemia), and loiasis (African eyeworm disease).

Flies breed rapidly with the female laying several batches of 75–150 eggs onto garbage, manure, or any moist organic material within 1–2 weeks after reaching the adult stage. Flies undergo complete metamorphosis, with the eggs hatching in 24 hours to a larval stage that lasts 4–7 days. The larvae crawl into the soil or under debris where they enter the pupa stage for 3 days to several weeks depending on the temperature. The adult stage emerges from the pupa stage ready to start breeding. Several generations can be produced during the warm seasons.

The control of flies begins with the elimination or covering of breeding materials such as garbage, feces, or dead animals. These materials need to be buried, tightly covered, or incinerated to prevent access by adult flies. Tightly fitting screens of a 16-mesh size or smaller are effective physical barriers. In heavily trafficked areas, air shields that blow down and away from the entrance have been successfully used to limit access by flies to a room or building. Biological controls have involved the radiation of male screw worm flies in Florida and on the island of Curacao. The screw worms mate once, and if their mating takes place with a sterile male, no progeny develops. This process requires sterilization of large numbers of males by radiation or chemicals and their release into the environment. This is most effective when the area is somewhat isolated making re-invasion from other areas less likely. Chemical attractants can be used to draw flies onto sticky "fly-strips," or DDVP resin strips, that are useful in killing flies in areas of 28.3 meters³ per strip. As with any of the insecticides, precaution must be used around foods and sensitive persons.[8]

Mosquitoes

Mosquitoes have one pair of wings with scales, belong to the order Diptera (two wings), and the females are characterized by having piercing/sucking mouth parts with an elongated proboscis. Mosquitoes undergo complete metamorphosis. Some of the female species may lay several batches of 50–200 eggs, while others lay single eggs. Depending on the species, the eggs may be deposited in still water such as swamps, wet depressions, or in the

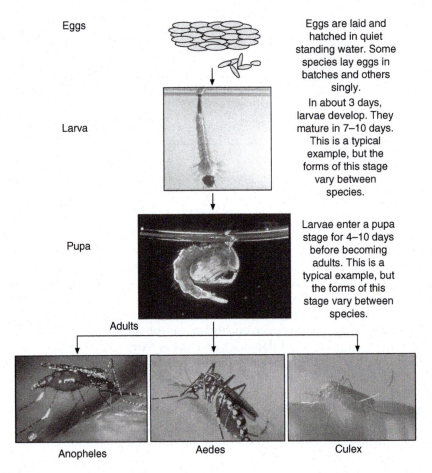

Eggs

Eggs are laid and hatched in quiet standing water. Some species lay eggs in batches and others singly.

Larva

In about 3 days, larvae develop. They mature in 7–10 days. This is a typical example, but the forms of this stage vary between species.

Pupa

Larvae enter a pupa stage for 4–10 days before becoming adults. This is a typical example, but the forms of this stage vary between species.

Adults

Anopheles   Aedes   Culex

FIGURE 6.5
Development stages of a typical mosquito. (Adapted from Koren, H. and Bisesi, M., *Handbook of Environmental Health and Safety Principle and Practices*, 3rd ed., CRC Press, Lewis Publishers, Boca Raton, FL, Chapter 4.)

collected water of old tires or containers. The eggs hatch in about 3 days to a larval stage that may molt three to four times over a 7–10-day period. The larvae enter a pupa stage in several more days becoming adult mosquitoes (Figure 6.5). The adult mosquitoes may travel from 100 yards to several kilometers depending on the species in order to feed on preferred hosts that may include humans, cattle, horses, or other domestic animals, or a variety of wildlife. Once a host is found, the female mosquito takes a blood meal during which time it injects saliva into the skin. The saliva prevents clotting and also causes the itching that is experienced by most people. The time required for digestion of the blood meal, laying more eggs, and then seeking another blood meal may be as little as 2 days. Some disease agents require 10–14 days of development in the mosquito before they become infectious to humans or animals. A mosquito may therefore breed multiple times before passing on the disease. There are more than 2600 known species of mosquitoes. Many of the species of public health interest belong to the genera Anopheles, Culex, and Aedes. *Aedes aegypti* is an important vector for yellow fever. The human virus circulates in the blood only 3–4 days during the first stages of illness and when the Aedes mosquito takes a blood meal during this time, the virus enters the mosquito. The disease can then be transmitted in 10–14 days. *Aedes albopictus* is also known as the Asian Tiger mosquito and is thought to have been inadvertently transported from Japan before it was found breeding in the pooled water of tires being transported over Texas highways in 1985.[8,10] This mosquito bites severely, survives cold winters, and is capable of transmitting the agents of dengue fever and dengue hemorrhagic fever, the encephalitides group of viruses, and yellow fever. Although it has not yet become an important agent of disease in the United States, *Aedes albopictus* has the potential to become a significant public health problem just as it is in Asia.

Species of the female Anopheline mosquito are important vectors of malaria. Malaria is biologically transmitted from human to human in a process that begins when male and female gametocytes of the malarial parasite Plasmodium sp. are taken in with a blood meal by the female Anopheline mosquito. The parasite goes through a sexual stage in the developmental cycle within the mosquito that results in the formation of slender sporozoites in a few weeks (see Chapter 7, Figure 7.10). These sporozoites migrate to the salivary glands of the mosquito and enter the host blood during the mosquito's next blood meal. Here, the parasites pass through an asexual cycle characterized in some forms of the disease by the synchronous growth of the parasites in the red blood cell. The periodic destruction of the red blood cells with release of parasites into the circulating blood often causes characteristically periodic episodes including headaches in the prodromal stage followed by chills and then fever. The chills and fever often occur in regularly timed waves of 36–72 hours, depending on the malarial species involved and result from inflammatory responses of the body to the explosive release of foreign bodies into the bloodstream.

Culex sp. or house mosquitoes tend to bite fiercely and continuously on humans, birds, horses, and other animals. They can breed in most any standing water, especially when contaminated with organic matter such as waste food or sewage. Culex sp. is a night feeder and rests by day. They have the capacity to travel several kilometers though they usually stay within 1.6 kilometers of the breeding area. They have been implicated primarily in the spread of viral encephalitis.

Control of mosquitoes can occur at the larval and adult stages. Larval stages may be controlled by chemicals, biological controls, or mechanical means. The most effective and environmentally friendly control involves the elimination of standing or stagnant water. Old tires, tin cans, plastic bottles, and land depressions where water can stagnate need to be buried or filled. Irrigation ditches should be smoothed and straightened to permit water to flow easily and drain quickly. In bodies of water too large to fill in, top-feeding minnows have been used to control the larvae. The addition of chemicals such as oil or pesticides to surface water has been effective in the past, but environmental concerns and regulations may restrict most applications. The adult mosquitoes can be temporarily or seasonally controlled by area spraying with compounds such as carbaryl or malathion. Most community spraying programs now occur in conjunction with a public health threat such as the potential spread of mosquito-borne viral encephalitis. Within the home, screens are an effective control when in good repair and having at least a 16–20 mesh. In many lesser-developed countries (LDCs) and in areas where screens are unavailable, mosquito netting is a useful deterrent. In LDCs a contact poison such as DDT has been used on the netting to increase its effectiveness. Repellents such as diethyl toluamide (DEET) and Indalone offer some short-term protection against mosquito bites but will vary in effectiveness, depending on the mosquito density, the species, and many other factors.

Fleas

Fleas are true insects with three pairs of legs and no wings. They are small and in the range of 1–8 millimeters with a siphon or tube as a mouth part that is used to feed on the blood of warm-blooded animals (Figure 6.6). There are several species of fleas that have the capacity

FIGURE 6.6
Oriental rat flea (*Xenopsylla cheopis*). The dark area in the abdominal area is where the plague bacillus *Yersinia pestis* has blocked the intestinal tract.

to bite humans; these include the dog flea and cat flea (Ctenocephalides sp.), the human flea (*Pulex irritans*), and the rat flea (*Xenopsylla cheopis*). The female flea takes a blood meal from its host, mates, and then lays eggs. The eggs drop to the nearest surface such as carpet, concrete floors, or furniture where they undergo complete metamorphosis to an adult stage within 2–3 weeks. At the adult stage, the fleas look for a blood meal and will feed on animals or humans, multiplying in large numbers and spreading rapidly through living quarters, livestock pens, and yards. Their ability to take blood meals and move swiftly among and humans makes them important vectors of disease including plague (*Yersinia pestis*), murine typhus, tularemia, and even salmonellosis. The rat flea carries the plague bacillus in its intestinal tract, spreading it from rat to rat and from rat to human. The dark blue-black hemorrhages under the skin seen in bubonic plague caused the disease to be called the "black death," when it ravaged Europe in the fourteenth century and killed more than 25 percent of the population. India continues to see periodic episodes of plague with nearly 10 million people dying from it in the early 1900s, and many have died in more recent outbreaks in 1994 in the Indian city of Surat.[9] The disease is endemic in the wild rodent population in the far west and southwestern United States where it is found in prairie dogs, chipmunks, and squirrels. The disease has continued to spread eastward since its introduction into the wildlife rodent population early in this century, and now has been identified in 13 western states with more than 360 cases of plague identified from these areas over the last 50 years (Figure 6.7).[9,10]

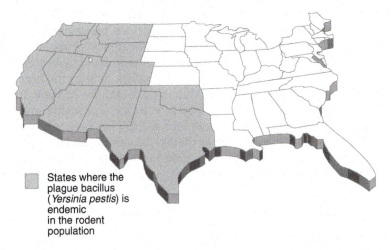

States where the plague bacillus (*Yersinia pestis*) is endemic in the rodent population

FIGURE 6.7
States in which plague is endemic in the wildlife rodent population.

Flea control usually involves treating infested domestic animals and the domiciles in which they live with a pesticidal dust containing either carbaryl or methoxychlor. The mildly toxic pesticides have a prolonged residual time. Powder should be rubbed thoroughly into the animal fur and especially around the head and neck areas.

Lice

Lice that feed on humans or animals are called sucking lice. They have the capacity to bite severely and inject saliva into the bite, resulting in irritation and severe itching to the skin. Lice have no wings, a dorso-ventrally depressed abdomen (flat), and have legs adapted for clasping hairs (Figure 6.8). Lice pass through an incomplete metamorphosis to adults. The egg or nit attaches to the head, pubic hair, or skin depending on the species. There are three kinds of human lice including the head louse, *Pediculis humanis capitis*; the body louse, *Pediculis humanis corporis*; and the crab louse, *Pthirius pubis*. The eggs hatch in the presence of body heat and the resulting nymph molts three times before becoming an adult within 1–4 weeks depending on how long the nymphs are in contact with the body. The lice are easily and rapidly transferred through person-to-person contact, bedding, brushes, and combs.

Louse-borne typhus is caused by *Rickettsia prowazekii* and carried by the human body louse. The organisms grow in the louse intestinal tract and are excreted by it. When the person scratches the itching bites, the feces and bacteria are rubbed into the wound, causing an infection. Inflammation from the infection causes local blockage of small blood vessels and subcutaneous hemorrhaging. Untreated diseases can result in high mortality. The disease is normally seen in very crowded and unsanitary conditions. In addition to spreading epidemic typhus, lice also cause trench fever (*Rickettsia quintana*) and relapsing fever (spirochetes belonging to the genus *Borrelia*).

Roaches

Cockroaches are true insects having three pairs of legs, one pair of antennae, and chewing mouth parts (Figure 6.9). Roaches are oval and flattened dorso-ventrally with the head usually not apparent or hidden. There are several common species of roaches including the

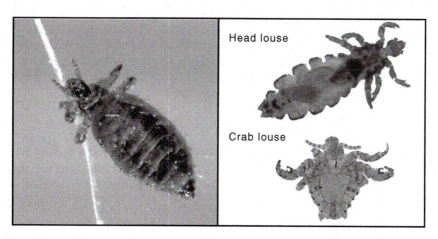

FIGURE 6.8
Typical louse shown clinging to hair shaft, a head louse (*Pediculus humanis capitis*), and crab louse (*Pthirius pubis*).

FIGURE 6.9
Typical American cockroach (*Periplaneta americana*).

American roach, *Periplaneta americana*; the German roach, *Blattella germanica*; and the oriental roach, *Blatta orientales*. The largest of these widely distributed roaches is the American variety, which reaches lengths of 5 centimeters. The adults are dark brown to reddish brown in color. Roaches go through incomplete metamorphosis involving egg, nymph, and adult stages. The eggs are usually encased in a capsule, carried for a few days to a month depending on the species, and then attached to some object out of sight. The nymphs will hatch in 2–3 months and pass through many molts over several months before reaching the adult stage. Roaches prefer the dark and are usually concealed during daylight or when bright lights are shining on an area. The roaches will conceal themselves behind cabinets, in drawers, behind moldings or furniture, and any place where there is a crack or crevice. The appearance of roaches in lighted areas usually means the area is heavily populated. Roaches feed on most organic material including fecal material, decayed food remnants, glue, and any accessible food. During the process of feeding, they regurgitate. Disease producing microbes may be carried on the external body parts or in the intestines of the roach and mechanically transferred to food. No specific disease outbreaks have been attributed to roaches but they may serve as vectors and create a level of discomfort for most people.

Roaches are best controlled by making certain no potential food sources exist. Waste foods should be immediately cleaned and leftovers tightly covered and/or refrigerated. All areas, including the dark reaches of cabinets and closets, should be periodically cleaned to remove potential food stuffs from surfaces. Items from food stores including bags or cartons should be discarded as they may harbor roaches. Borax powder has proven effective in controlling roaches by dusting in areas where roaches are likely to hide. Organophosphate poisons such as malathion have also proven effective for roach control.

Ticks and Mites

Ticks and mites have four pairs of legs, no antennae or wings, and belong to the class Arachnida. A shorter abdomen and the absence of a constriction between the cephalothorax and abdomen place organisms such as ticks and mites in the order Acarina.

Mites

Mites do not have a clearly differentiated head, thorax, and abdomen (Figure 6.10). They are extremely small and may not be easily seen without the aid of magnification.

FIGURE 6.10
Typical scabies mite (*Sarcoptes scabiei*).

Mites pass through several stages of development that include eggs, larvae, nymphs, and adult within 2–3 weeks, which is about their average length of life. Because mites may proliferate in areas where human body cells are shed and where body oils accumulate (mattresses, couches, beds), they can infest and irritate humans as well as transmit diseases. The mite *Sarcoptes scabiei* burrows under human skin to lay its eggs. This burrowing causes slightly elevated, serpentine lines, or a variety of inflammatory skin lesions resulting from secondary infections as a result of scratching. Intense itching, rashes, and redness often develop. The mites are transmitted through close personal contact with an infected person. Dogs and cats are often infested and develop a condition called mange. The mites frequently infest birds and rodents, which become a major source for infestations of homes and domestic animals. A type of mite called a chigger or redbug is the six-legged larvae of mites. Chiggers attach to the skin and inject a salivary secretion that dissolves the surrounding tissue and is then sucked up the tubular stylostome of the chigger. Scrub typhus is a rickettsial disease passed to rodents and humans by the chigger causing death in up to one half of those infected. The mite infesting mice has been associated with the transmission of rickettsial pox caused by *Rickettsia akari*. Certain forms of hemorrhagic fever and encephalitis have also been associated with transmission by mites.

Preventing mite infestations begins with eliminating rodent and bird populations in the house or building, keeping everything clean and vacuumed while allowing plenty of sunlight into the living spaces. When mite infestations have occurred, these steps should be preceded by using a pyrethrum bomb, the area cleaned and vacuumed, and malathion applied around windows, doors, and baseboards for residual action.

Ticks

Just as mites, ticks belong to the class Arachnida. Consequently, they have four pairs of legs, no antennae, and a fused head, thorax, and abdomen (Figure 6.11). Ticks are particularly adapted to feed on blood by (1) the use of barbed feeding organs that pierce the skin to reach the blood and also anchor the tick, (2) a flexible leathery body that easily distends when filled with blood, and (3) uniquely adapted pharyngeal muscles for sucking blood. There are two categories including soft (family Argasidae) and hard (family Ixodidae) ticks. Ticks undergo a complete metamorphosis often involving a fairly complex cycle such as the deer tick, *Ixodes scapularis*, which is the vector of Lyme disease caused by the spirochete *Borrelia burgdorferi* (Figure 6.12). The adult female Ixodes tick feeds once on a deer and then mates. She lays several thousand eggs in a batch. The eggs hatch to larvae in a few weeks to several months. The larvae of Ixodes fall to the ground where they come into contact with rodents and other small animals. The larvae (now possibly infected with

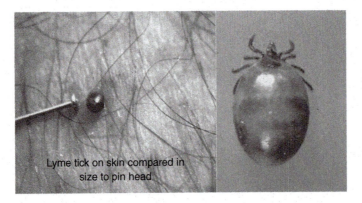

FIGURE 6.11
An engorged tick associated with Lyme disease (*Ixodes scapularis*).

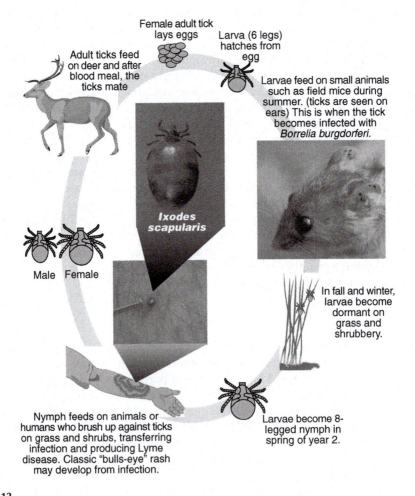

FIGURE 6.12
The life cycle of *Ixodes scapularis*, the tick vector of Lyme disease. (Adapted from Tortora, G.J. et al., *Microbiology: An Introduction*, 5th ed., The Benjamin/Cummings Publishing Company, Redwood City, CA, 1995, Chap. 23.)

Borrelia sp.) would be deposited on grass and leaves where they become dormant over the winter. During spring of the second year, the larvae develop into eight-legged nymphs that appear on the tips of grass and brushes at the forest edge, lawns, and gardens where they can cling to humans or animals that brush by them. The ticks will spread the disease to humans on which they feed.

Ticks have the capacity to transmit diseases mechanically since disease organisms may be carried on tick mouth parts or other body parts and easily infect other persons who are bitten by the tick. Many other organisms pass through developmental stages in the tick by a process known as transstadial transmission. In this process organisms are passed from the tick to the egg and back through the adult tick developing from that egg. Thus, the disease agent may be carried over several generations. Among some of the more important diseases transmitted by ticks are (1) the rickettsial diseases including tularemia, Q fever, and Rocky Mountain spotted fever; (2) the disease caused by *Borrelia sp.*, including Lyme disease and relapsing fever; (3) viral diseases such as Colorado tick fever; and (4) bacillus diseases such as *Pasteurella tularensis*. Ticks also produce a disease known as tick paralysis.

Keeping ticks off of exposed skin is the best control method. Before hiking, biking, or working in high grass or brush it is good to protect yourself by tucking in pants to the socks, tightly buttoning or taping long-sleeved shirts at the wrist, and wearing clothing that fits tightly to the neck. It's also good practice to wear light, uniformly colored clothing so that ticks can be easily spotted and removed. Spraying clothing and exposed skin with DEET insect repellent offers some protection. Some people have treated clothing with benzyl benzoate or indalone pesticides. Family dogs and cats should be frequently inspected for ticks using a fine-toothed tick comb especially around the neck area. Tick and flea collars are helpful, but don't guarantee against some infestation. When ticks are found on humans or animals they can be removed by pulling directly out from the body while trying to keep the mouth parts intact to the tick. This process can sometimes be aided by applying nail polish or Vaseline to the tick. Malathion dust or DDVP have proven effective in controlling ticks when applied over several months.

Rodents and Pests

Rodent Characteristics

Rats and mice are members of the order Rodentia, which include mammals with teeth and jaws adapted to gnawing. Rodent teeth can grow 12–15 centimeters a year and must be constantly used to keep them short. The constant gnawing of rodents is the source of much destruction. Other members of Rodentia include beaver, squirrels, porcupines, chinchillas, and lemmings. The two major rat species in the United States and many parts of the world include the roof rat, *Rattus*, and the Norway rat, *Rattus norvegicus*. The house mouse, *Mus musculus*, and deer mouse, *Peromyscus maniculatis*, are also pests in the United States. The two rats mentioned, and the house mouse, are commensals in that they live in or near humans, eat our food, and share diseases without providing any benefits. Because of their close association with humans, they are classified as murine rodents belonging to the subclass Murinae. The roof rat prefers to live in barns, attics, warehouses, and other human structures. *Rattus* is highly adapted to balancing along narrow beams and squeezing through cracks and crevices or tiny holes that it can create by gnawing. The roof rat

Thin agile body

Ears prominent and stick out above fur

Tail longer than body

Pointed snout

FIGURE 6.13
The roof rat (*Rattus rattus*).

is slender, has a pointed snout, large prominent ears, and a tail longer than its body. It weighs between 226 and 340 grams and reaches lengths of 20–43 centimeters (including tail) (Figure 6.13). The Norway rat tends to dig burrows under foundations, along banks of rivers, in fields, under solid waste, and under lumber piles. They live in sewers and between floors and walls. These are muscular rats with thick bodies, a blunt snout, with a tail shorter than its body. They reach lengths of 30–45 centimeters (including tail) and weigh up to 453 grams (Figure 6.14). These features are adapted to burrowing. Rats have several abilities that help them to survive and adapt around humans. Rats can (1) pass through a quarter-size hole, which is about the size of their heads; (2) climb straight up a brick wall providing there are tiny holds for their toes; (3) survive falls up to five stories high; (4) jump over obstacles 1 meter high; (5) cross over and balance on wires; (6) and swim more than 0.8 kilometers in open water while periodically diving to locate possible escape openings.[1,8]

The house mouse thrives most everywhere worldwide that people live. It weighs from 14 to 21 grams with a length of 15–17 centimeters (including a tail longer than its body). Its eyes and ears are large and prominent, and the nose is pointed (Figure 6.15a). They have a short gestation period of about 29 days producing up to 8 liters of 5–6 offspring per liter. They tend to build small nests of shredded paper, fabric, or grass anywhere including in stored goods, cabinets, furniture, and drawers, while outside they live in weeds, rubbish, and tall grass.

Ears not prominent, tend to lie close to body

Thick, muscular body

Tail shorter than body

Blunt snout

FIGURE 6.14
The Norway rat (*Rattus norvegicus*).

FIGURE 6.15
(a) House mouse, *Mus musculus,* and (b) deer mouse, *Peromyscus maniculatis.*

The deer mouse, *Peromyscus maniculatis* is a vector of the hanta virus (See Chapter 7). It is a thin agile creature weighing up to 35.4 grams and reaching lengths of up to 22 centimeters with a tail longer than its body. Its eyes and ears are large and prominent, and the nose is pointed (Figure 6.15b). It has white feet and a white underside with a brownish back. It tends to live apart from humans in the forest and at the forest edge or in meadows. They produce 2–4 liters per year with 3–5 offspring per liter. The deer mouse lives mostly on nuts, acorns, fruits, and insects.

Importance as Pests and Vectors of Disease

Rats and mice are considered to be pests by most societies because (1) they harbor a variety of organisms that can produce diseases in humans, (2) they produce crop damage throughout the world, (3) they eat and contaminate food produce and cause structural damage, (4) they afflict harm directly by biting humans, and (5) they cause extreme discomfort in some people by their visible presence.

Carriers of Disease

Rodents harbor and transmit a variety of disease organisms including bubonic plague, *Yersinia pestis*; rickettsial pox, *Rickettsia akari*; murine typhus, *Rickettsia typhi*; rat bite fever, *Streptobacillus moniliformis*; Weil's disease, *Leptospira incetrohaemorrhagiae*; salmonellosis, *Salmonella enteriditis* and other species; and trichinosis, *Trichinella spiralis*. Hantavirus is an emerging disease (Chapter 7) of severe consequences with a mortality in humans exceeding 50 percent. It is normally transmitted by inhaling virus particles in the dried urine and feces of the deer mouse.

Crop Damage

Rodents produce crop damage throughout the world. In some areas of the world, rats destroy as much as one third of the entire harvest.[11] In the Philippines there have been a number of rat population explosions such as those occurring in 1952 and 1984 with rat populations reaching 200–2000 rats per hectare and destroying 90 percent of rice crops and 20–50 percent of corn.

Contaminate Food

Rodents eat and contaminate food and produce structural damage. Rats contaminate ten times more food and materials then they use, damaging from one to ten dollars' worth of food for each of the estimated tens of millions of rats in this country alone. Rats and mice also gnaw their way through walls and doors and eat the coating off of wires, causing electrical shorts and fires. Rodents are thought to produce one billion dollars in damage every year in the United States.[8]

Rat Bites

Rats bite an estimated 45,000 people each year in the United States, producing disfiguring injuries, infections, and some fatalities. In some reported cases, the rat had to be beaten unconscious to release its human victim.[8] Since there are no requirements for reporting of rat bites, the actual number of bites may be much higher. The number of rat bites may likely increase since rats thrive in areas of poverty and urban decay where garbage and trash, poor sanitation, and inadequate rat control measures abound along with increasing population density.

Discomfort

Rats and mice cause extreme discomfort in some people. Persons experiencing rat bites or who have a direct confrontation with rats may be traumatized psychologically by the encounter and some have lasting emotional consequences.[8] Others are simply terrified by the visible presence of rats or mice. I still recall my otherwise normally adjusted mother standing on a kitchen chair screaming as a mouse scurried along the kitchen baseboard.

Rodent Control

Establishing Presence

The murine rodents tend to live close to humans and have a very limited range of 30–45 meters for rats and 3–9 meters for the house mouse. Deer mice range over 0.4–1.6 hectares. Rodents tend to be most active at night when there is plentiful food. If rats and mice are seen during the day, that is usually evidence of many more in the vicinity. Evidence of the rodents is often established by (1) seeing the rodent or its droppings; (2) seeing nesting materials in drawers, cabinets, near foundations, or other places around or in the structure; (3) seeing holes gnawed into walls, doors, or other structures; or (4) seeing grease trails or runways, tracts, swing, or rub marks along the wall edges (Figure 6.16). Rodents prefer to travel familiar and established routes along room edges and if travelled frequently, will leave their body oils along those edges.

Rodent Proofing

Rodents live in proximity to humans because they have easy access to food. Any effective control program must begin with the removal or securing of food and denial of harborage. Whether a private home, commercial establishment, or an entire community, every effort must be made to store food and food waste in areas or containers inaccessible to rodents. Refuse should be stored in 113-liter metal containers with tight-fitting lids that are placed on platforms about 30 centimeters off the ground. The areas should be frequently

FIGURE 6.16

Rub marks or grease trails left by rats frequently traveling along the edge of a structure. (From Training Materials from HEW, *Control of Domestic Rates and Mice*, Public Health Service, Communicable Disease Center, 22, 1964.)

cleaned and inspected for signs of rodent activity. Dried or canned foods should be stored in clean, dry areas on shelves above the floor or on palettes placed 60 centimeters from the wall so that inspection of all sides is made easy and cleaning is facilitated. Brush, weeds, and grass should be cut outward from the building to 30 meters if possible. Any rubbish, debris, brush piles, or old lumber should be removed. The structure should be examined for any holes greater than .6 centimeters in diameter. The building can be made resistant to rodent entry by (1) installing metal doors or metal edges around doors, (2) installing metal screens of 19 gauge or smaller, (3) installing metal cones around pipes or wires entering the building, (4) placing sheet metal plates where pipes or wires enter the building, and (5) installing a curtain wall around the foundation made of sheet metal or concrete (Figure 6.17). An L-shaped curtain wall of concrete at least .6 meters deep may be poured along the foundation walls to prevent burrowing. Alternatively, sheet metal can also be used that is 30 centimeters above the ground and 15 centimeters below the ground.[12]

Poisoning or Trapping of Rodents

The poisoning or trapping of rodents is often necessary to complete the task of control. There are numerous varieties of traps available for the homeowner including some that are disposable plastic shelters where the bait may be inserted at the inside rear, the trap set by pulling a tab at the outside rear, and pushed into place where the rodents frequent. When sprung, the mouse and its container may be discarded without contacting the dead rodent. Traps for rats are less successful because the rats drag them or chew through them or themselves to escape (Figure 6.18). No rodents should be handled with unprotected skin.

Rat and mouse poisons available to consumers usually are composed of powerful anticoagulants such as warfarin, fumarin, chlorophacinone, PNP, and diphacinone. The anticoagulants cause internal hemorrhaging resulting in the death of the rodent over a period of several days. Consequently, the rats or mice seldom associate the baited food with their illness and the colony continues to feed. The baits may be purchased as ready-to-distribute. Further discussion of rodent poisons is given in this chapter under the Pesticides section.

Once the areas of concern have undergone rodent proofing, the removal of food and denial of harborage, and poisoning or trapping has occurred, then continued vigilance is necessary. Cleanliness must be maintained, and food made unavailable to the rodents. There should be periodic checks for signs of rodent activity.

Conical metal guard to prevent rodents from reaching structure by wires.

Screening should be galvanized hardwire of 19 gauge (10.16–11.43 cm.)

Keep garbage in tightly sealed metal container off of ground on palette.

Openings around pipes covered with sheet metal or concrete mortar.

Metal flashing around wooden, earthen, or adobe dwelling to a height of 30 cm above ground and 15 cm below ground. Concrete curtain can also be used.

Channeling and cuffing of door edges with metal guards.

Opening below door must be less than 1.27 cm.

Holes greater than 0.63 cm in diameter should be covered with metal sheeting, or plugged with concrete.

FIGURE 6.17
Methods of rodent proofing. (From Training Materials from HEW, *Control of Domestic Rates and Mice*, Public Health Service, Communicable Disease Center, 22, 1964.)

FIGURE 6.18
Evidence of rat gnawing itself free from trap. Rodents have teeth and jaws adapted to gnawing. (From Training Materials from HEW, *Control of Domestic Rates and Mice*, Public Health Service, Communicable Disease Center, 22, 1964.)

Pesticides

History

Pests compete with humans for food, serve as vectors of disease, destroy crops or depress their market quality, cause structural damage to buildings and homes, and attack people directly causing annoyance, injury, or even death. Prior to the 1940s, there was very little most people could do but suffer the consequences of pest invasions and co-habitation. In the middle-to-late 1800s copper arsenate compounds (Paris Green) and lead arsenate (Bordeaux mixture) were introduced as fungicides and pesticides. These substances remain indefinitely in the environment and are toxic to insects and humans while creating potential harm to the environment through contamination of soils and water.[10] Compounds made from copper, chromium, and zinc were used to control insects while hydrogen cyanide was used to control red scale on plants. All of these compounds are highly toxic, most are long-lived in the environment, and resistant species developed against many of them. All arsenicals were banned following the introduction of better alternatives.[8] The advent of the chemical revolution in pesticides probably began in 1939 with the discovery of the usefulness of DDT by Paul Muller of the Geigy Corporation which was found to be highly toxic to most insects but not to humans. The use of DDT became critical to the effort of the allied troops in WWII who were combating malaria, yellow fever, and a host of mosquito-borne, louse-borne, and other vector-borne diseases. Within the next 15 years, more than 25 new pesticides were synthesized including many chlorinated organics (chlordane, heptachlor, dieldrin, endrin, and iodrin) and the toxic organophosphate, parathion. These compounds were credited with saving millions of lives by suppressing insect vectors of malaria, filariasis, dengue fever, and typhus.[13]

These pesticides were widely used in the United States since their introduction after WWII until the 1970s when it became evident that many of the chlorinated organics produce significant effects on ecology and human health. Consequently, the use of most organochlorine insecticides were canceled for use in the United States although DDT and a few other of the organochlorines introduced about the same time remain in use in other parts of the world such as LDCs, where vector-borne diseases remain a major threat and the need for low-cost insecticides is great.[11] Nevertheless, the promise of a world free of vector-borne disease and unmarred crops free of insects and fungi has vanished in the avalanche of unforeseen events over the last 40 years. These problems include (1) the resistance of vectors to pesticides; (2) the adverse health and ecological effects of pesticides; and (3) the proliferation of pesticides globally.

Insect Resistance

The previous four decades witnessed a dramatic shift between people and pests and raised our expectations that the battle over insects, rats, and fungi could be won. After all, there was a revolution in chemical warfare against these organisms that promised low toxicity to humans and high toxicity to insects. These chemicals were more effective and less toxic than arsenicals and heavy metals, while being effective against a broad array of insects. They were revolutionizing agriculture and freeing humanity of vector-borne diseases. However, almost immediately after their introduction, DDT and other organochlorine pesticides surfaced as potential villains to ecology and human health. Just as important was the realization that many important pests were developing resistance. United

States production of chemical pesticides rose from 210,000 kilograms in 1951 to levels approaching 680,000 kilograms today without any decline in crops lost to pests.[15,16] From 1970 to 1980, the number of arthropods resistant to insecticides nearly doubled from 224 to 428 while the numbers of resistant species of rodents, bacteria, fungi, and weeds are increasing as well.[15] There are presently more than 500 species of resistant arthropods. Farmers use ten times the amount of pesticides, while crop losses are almost double that when DDT was first introduced.[10] Furthermore, many weed species are becoming resistant to herbicides and it is becoming an international problem since many of the genetically engineered plants associated with the "Green Revolution" may be easily overgrown by weed grasses. Such overgrowth could reduce crop yields by one third or more.[17]

The types of resistance most often encountered in arthropods are physiological or behavioral. Physiological resistance may occur from (1) enzymatic detoxification of pesticides into a less harmful form, (2) reduced permeability of exoskeleton to pesticides, and (3) storage or excretion of pesticides without harm. Some insects survive pesticide exposure through such behavior as (1) landing less frequently or changing landing areas or (2) avoidance of baits such as fly-paper.[8] Some insects such as mosquitoes and flies have developed cross-resistance and are now resistant to both organochlorines and organophosphates.

A related issue with resistant species is the destruction of competitor organisms. Predators and competitors restrain pest populations. The application of pesticides upsets this ecological relationship by killing off predatory and competitive species, which often permits the original target pests to resurge dramatically to even greater numbers when the pesticides are diminished. This process is known as target pest resurgence. Very often the target pest is controlled by the pesticide but once harmless species grow numerous and become pests themselves to produce a secondary pest outbreak. Examples of this include eruption of the tobacco budworm on crops in Texas and Mexico, the cottony cushion scale in citrus trees in California.[10,14]

The Health and Ecological Effect of Pesticides

Exposures

Humans are exposed to pesticides through (1) contaminated drinking water; (2) eating foods contaminated with pesticides; (3) pesticide use in the home, garden or lawn: (4) exposure on transcontinental flights; (5) use in agriculture; (6) in the production of pesticides; and (7) in other occupations.[18] Nearly 75 percent of pesticides are used in agriculture, accounting for 66 percent of the $7 billion spent annually on pesticides.[19] There are 4–5 million Americans working in agriculture with more than half of these being migrant or seasonal workers, and many of these are children under the age of sixteen.[18] Poisonings from pesticides have been estimated to affect between 500,000 and 2.9 million people worldwide annually,[20] causing fatalities for one percent of the cases or more than 10,000 deaths per year worldwide.[21] There may be 300,000 cases of pesticide-related illness in the United States although less than 2 percent are reported, and most of these are neurotoxic pesticides.[18] There are reports that only 25 deaths occur in the U.S. each year from exposure to agricultural chemicals.[22] The major neurotoxic pesticides are organophosphate insecticides, which make up nearly 40 percent of all pesticide use in the United States, and which are the most commonly reported source of worker illness. Many organochlorine pesticides used prior to the 1970s are now banned in the United States.[18] Most acute or one-time exposures include such symptoms as headaches, dizziness, muscular weakness, and fatigue. Poisoning from organophosphates may lead to diarrhea, vomiting, severe abdominal pain, muscular twitching, breathing difficulty, sweating, and in severe cases, convulsions, coma, and death.[18] Four to nine percent of acutely poisoned persons experience delayed or persistent neurological or psychiatric effects including agitation, insomnia, confusion, and depression.[18] The consequences of chronic or low-level pesticide exposure are uncertain and unresolved. The concerns for the present tend to be directed at the possibilities of cancer and/or reproductive problems associated with organochlorine pesticides. There is concern that pesticides may be depressing the immune response in the general population especially in countries continuing to use pesticides banned in the United States. Studies in the former Soviet Union found dose-dependent effects of pesticide exposure on the immune function and the development of chronic and infectious diseases, while a Canadian study demonstrated effects of pesticide-laden breast milk on babies' immunologic deficiencies.[23] The major groups of people at risk from the adverse risks of pesticides are children and persons in LDCs.

Children at Risk

Children are exposed to pesticides through (1) prenatal maternal exposure; (2) food and water sources; (3) presence in agricultural, fields with parents, or working the fields themselves; (4) contact with residues on parents' clothing; or (5) exposure to common household pesticides while in the home. Based on data from the American Association of Poison Control Centers, 74,000 children a year are involved in poisonings in the home from household pesticides.[24] Nearly half of all households with children under 5 had at least one pesticide stored in an unlocked cabinet less than 1.2 meters off the ground, with bathrooms and kitchens the most likely areas for improper storage. Common pesticides included roach sprays, bath disinfectants, insect sprays, flea and tick shampoos and dips, chlorine bleach, insect repellents, weed killers, and pool chemicals. Children may be protected from exposures to household pesticides according to the steps outlined in Table 6.1. Children are at greater risk to the toxic effects of these chemicals because they (1) absorb more pesticides per kilogram of body weight, (2) tend to receive higher doses than adults, (3) have immature detoxification systems compared to adults, and (4) may be more susceptible to neurotoxic effects because the CNS is still developing in children.[18] Food may

TABLE 6.1

Recommendations for Preventing Accidental Poisoning

Preventing Accidental Poisonings in Children

1. Store out of reach in a locked cabinet or shed, use child-proof safety latches
2. Read label and follow instructions for use including precautions and restrictions
3. Before applying pesticides, remove children, toys, and pets until pesticide dries, or according to recommendations on label
4. If use is interrupted (as by phone call), reclose the package and place out of reach of children
5. Never transfer pesticides to other containers
6. Teach children that pesticides are poisonous—something not to touch
7. Keep the number of the Poison Control Center near telephone
8. Alert caregivers to potential hazards of pesticides

Source: U.S. EPA, Office of Pesticide Protection (OP), *Pesticides and Child Safety: Tips to Protect Children from Poisonings Around the Home, Prevention, Pesticides, and Toxic Substances*, EPA#735-F-3 -050R, update, November 19, 1997, http://www.epa.gov/oppfead/vhildsaf.htm.

be another potential source of pesticide exposure to children. The Natural Resources Defense Council (NRDC) released a report in February 1989 that attempted to analyze the extent of children's exposure to pesticides and the hazards based on information from the EPA, the Food and Drug Administration (FDA), and the Department of Agriculture (DOA). The report concluded that 17 percent of preschoolers, or 3 million children in that age group, are being exposed to hazardous levels of pesticides in fruits and vegetables exceeding federal safe level standards.[25] The concern about cancer-causing pesticides in our food supply and the risk to children continues to be of public health interest,[26] despite the fact that the FDA found that 99 percent of 19,962 food samples contained no illegal residues.[27] Prompted by concerns of pesticides in foods, President Clinton signed the Food Quality Protection Act in August, 1996. The new law requires (1) special consideration of infants and children when setting standards, (2) consideration of endocrine disruptors, (3) consideration of pesticide's benefits, and (4) expanded consumer right-to-know efforts.[28]

Exposures in LDCs

Pesticides are widely misused in LDCs and it has become a significant problem. It is estimated that less than 25 percent of the world's production of pesticides is used in these countries, but more than 50 percent of acute poisonings of pesticide applicators occur in these countries while 73–99 percent of worldwide deaths due to acute pesticide poisoning occur in LDCs.[18,29] The World Health Organization has estimated that every minute some person in an LDC is poisoned by pesticides.[30] The reasons for this are numerous, but may include (1) failure to use protective clothing (sometimes workers are clad only in bathing suits) when applying pesticides; (2) workers may be in the field during pesticide spraying; (3) pesticides are being used in LDCs that are banned or severely restricted in the developed countries; (4) pesticide containers are often recycled to store food, milk, or cooking oil; and (5) pesticide products are labeled in languages not readable to the native populations, many of whom are also illiterate.[18] In an effort to control part of this problem, the United Nations Food and Agriculture Organization (FAO) adopted in 1985, and amended in 1987, a code that included a provision for prior informed consent (PIC). Under PIC, any pesticide restricted or banned in one country cannot be exported to another without its consent after being informed of the dangers. In May 1995, countries agreed to establish this as a legally binding document and targeted 1997 as the year to make PIC a legally binding treaty.[31]

While DDT has not been used in the U.S. since it was banned in 1972, it is still in wide use in LDCs. In 1999, it was being used in at least 27 countries. Its main purpose in these countries, however, is not to protect crops, but to prevent the spread of illness and death from malaria. Malaria kills 2.7 million people each year, most of whom are children living in LDCs. Normally, environmental protection and efforts to protect the health of populations go hand in hand. In the case of DDT use in LDCs, however, environmentalists and many international public health advocates are at odds.

The United Nations has been attempting to draft a treaty banning the use of DDT (as well as a number of other pesticides) worldwide for several years. A number of prominent medical researchers, however, are fighting to ensure that the treaty will allow for DDT to be sprayed in small quantities on the interior walls of homes, where it acts as a repellent to potentially malaria-carrying mosquitoes. Advocates of the ban (including the World Wildlife Fund and Physicians for Social Responsibility) argue that DDT has been found in the breast milk of mothers living in homes sprayed in this fashion. Unfortunately, many of the malaria victims living in LDCs cannot afford alternatives such as drugs available to treat malaria—and drug resistance is increasing in any case. Other pesticides that are less harmful to the environment are available, such as pyrethroids, but there are two significant problems with their use. First, spraying a house with pyrethroids is three to four times more expensive than DDT, and few of the LDCs can afford the cost. Second, a reliance on only pyrethroids may ensure an increase in resistance to the pesticide among mosquitoes. Environmental groups such as the World Wildlife Fund argue, however, that more creative solutions must be pursued, as the short-term benefits of using DDT are not worth the long-term consequences on the environment.[32]

The Counter Arguments

The arguments have been advanced that pesticides are designed to be toxic, they are made specifically to kill insects, fungi, and other pests. Pesticide in our foods and environment are claimed to cause a variety of toxic effects including reproductive damage, birth defects, interference with the immune and endocrine systems, nervous system damage, and cancer.[23,33] The counterview is that:

1. Trace exposure to pesticides show no evidence of increasing the risk of cancer, birth defects, or other human ailments.[22]

2. No documented cases of deaths or illness from pesticide use in the U.S. have occurred when used in standard and acceptable ways.[22]

3. Many of the banned organochlorine pesticides have shown carcinogenicity only in mice.[22]

4. Pesticides help prevent grains and seeds from being contaminated with fungi that produce carcinogenic aflatoxins, while also providing an abundance of low-cost attractive fruits and vegetables, which are humanity's best defense against cancer and heart disease.[33,34]

5. Naturally occurring carcinogens may cause an estimated 38,000 of the nation's 500,000 annual cancer deaths, while pesticidal residues are thought to contribute to less than 40 cancer deaths per year.[33]

Ecological Concerns

Much more than 499,000 kilograms of pesticides are used in the United States each year with more than five times that amount used globally. The greatest potential for the unwelcome effects on health and ecology come from the contamination of surface and groundwater. Water is one of the major pathways by which pesticides are transported long distances to other parts of the environment moving from streams, to rivers, to lakes and oceans.[16] Nearly 50 percent of the nation's drinking water is provided by surface supplies, while surface water also provides vital aquatic ecosystems for fish, birds, plants, and other wildlife. Pesticides enter the waterways from streams through agricultural runoff, from wastewater discharges, groundwater inflow, atmospheric deposition, and spills. Pesticide residues, especially of the persistent organochlorines, are now detected in the tissues of animals in virtually every location on Earth from the Antarctic to the Everglades, small New England streams, to the deepest ocean trenches. Such is the concern, that a program was designed known as the U.S. Geological Survey's National Water Quality Assessment (NAWQA) Program to provide information on the extent and consequences of such pollution. The program is designed to provide consistent and reliable information on water

resources in 60 important river basins and aquifers across the nation and was fully implemented in 1997. Early findings on pesticides in the atmosphere show "the pesticides were
ubiquitous and generally detected wherever they were sought."[16] Further, a review of
studies indicated that most water treatment plants are ineffective in removing the pesticides, and that drinking water derived from some surface water sources contain pesticides
above federal guidelines for at least part of the year.[16] Additionally, both the EPA and the
National Academy of Sciences and Engineering (NAS/NAE) have developed or recommended water quality criteria for a total of 43 pesticides in order to protect aquatic life
from short- and long-term exposure. One or more of all of these pesticides have exceeded
aquatic-life criteria in the river basins tested. There are few such criteria for herbicides, and
none for fungicides.[16] The significance of pesticides in surface waters has not been assessed
because little is known about most pesticide toxicity at low-level exposures, nor the possible cumulative effects of multiple pesticide exposures.[16]

The consequences of surface water contamination are usually evident in their effects
on wildlife. At low concentrations, the persistent, bioaccumulative, organohalogen pesticides have disrupted the endocrine systems of birds, fish, turtles, and mammals causing
decreased fertility, metabolic and behavioral abnormalities, feminization of male species,
and masculinization of female species. The effects have been manifested in the offspring
because of exposures at critical times during embryonic development.[35]

Acute exposures to larger concentrations of pesticides can produce immediate and
dramatic effects including massive fish kills in streams where pesticide runoff or spills
have occurred, deaths of geese browsing on pesticide-treated golf courses, or a variety
of more than 2 million birds dying from carbamate pesticides contaminating water and
plants.[36] Further, the destruction of certain insects or plants may have negative effects
when they serve as a food source or habitat for other organisms. This is especially evident
in birds that feed on insects that have been killed by pesticides or use particular plants for
nesting that have been eliminated by herbicides.

Groundwater can also be contaminated. Extracting water by pumping it from depths
within the ground is no guarantee of chemical or biological purity. Thousands of private
wells, and those serving entire communities have been contaminated with pesticides and/or
organic solvents. The EPA conducted the National Pesticide Survey in all 50 states between
1988 and 1990. The preliminary results revealed 10 percent of the nation's public well supplies had detectable levels of at least one pesticide. Federal guidelines were exceeded in
one percent of the community wells and 0.8 percent of the private wells.[8] In 1992, the EPA
reported that more than 30 percent of private (rural) wells exhibited unacceptable levels of
pesticide contamination resulting mostly from agricultural use. The complex and usually
slow movement of groundwater plumes, combined with the highly variable application
of pesticide amounts and types makes it difficult to assess the full extent of the problem.
The increasing use of pesticides in agriculture combined with garden and lawn use may
combine with the impending already-contaminated plumes to cause more extensive contamination in the future. The Food Quality Protection Act of 1996 established a new safety
standard for pesticide residue limits in food, and the EPA is required to periodically reas-
sess these 9,721 limits (tolerances). By August 3, 1999, the EPA, as required, had completed
3,290 of these reassessments, two thirds of which fell into the category of pesticides most
threatening to humans. These reassessments included lower tolerance levels and restricted
use for some of the pesticides that represent the greatest risk to humans, including two
of the most toxic organophosphates: methyl parathion and azinphos methyl. Still, certain
environmental advocacy groups have felt that the EPA's new regulations fall short of the
EPA's mandate to protect the public safety.[37]

The Proliferation of Pesticides Globally

More than 272 megagrams of DDT were shipped to Peru in 1992, 20 years after it was banned in the United States. The U.S. shipped nearly 10,886 megagrams of other restricted or banned pesticides in the same year. Since then, the export of pesticides banned for use in the U.S. have escalated by nearly 50 percent. According to studies performed by the Foundation for Advancement in Science and Education (FASE), about 226,796 megagrams of pesticides are exported from the U.S. annually, only a quarter of which are reported.[38] Nearly 4,535 megagrams of the exported pesticides were never approved for use in this country, and more than 11,000 tons were severely restricted for use. Many of the LDCs still continue to import harmful pesticides because they are inexpensive and effective. In Kenya, small farms have increased output significantly in the last few years, making it the third largest source of income from foreign sources. The consequence has been increasing reliance on pesticides banned or restricted in the United States.[31] The continued use of DDT in Mexico is evidenced in increased levels of related compounds (DDE) in breast milk and tissues indicating to the authors of the research article that DDT use in Mexico is a public health problem.[39]

The banned insecticides being exported to LDCs are thought to be evaporating or volatilized to the atmosphere in the warmer climates and "hopping" in stages about the globe to cooler climates such as the Arctic through the global distillation effect.[40] Many of the chemicals banned for decades in the U.S. are being emitted from the hotter climates in LDCs, carried to colder climates in the atmosphere and condensing where it shows up in the bark of shrub trees near the Arctic Circle. The more volatile the chemical, the faster it travels these distances to eventually condense in colder climates. It is thought to be the best explanation for how large quantities of volatile pesticides and other chemicals find their way into the flesh of polar bears and the breast milk of Inuit natives in the Arctic.[6]

Types of Pesticides

Insecticides

Insecticides generally produce their toxic effect on insects by poisoning the nervous system. Well-designed insecticides are harmless to humans, highly toxic to insects, easily dispensed and inexpensive, non-hazardous, and rapidly degradable to non-toxic forms. They often (1) act as contact poisons by penetrating the foot pads or body wall; (2) enter the insect breathing pores as a fumigant; (3) act as a stomach poison after ingestion; or (4) desiccate (dry out) the body wall, causing it to crack or break.[6] Insecticides are also capable of exerting their toxic effects on humans, and the major difference between its use as an insecticide or human poison is the dose. There are several types of insecticides that include organochlorines, organophosphates, carbamates, and a variety of botanical and biologic insecticides.

Organochlorines

The chlorinated hydrocarbons attack the CNS, sending insects into hyperactivity, tremors, convulsions, and death. The organochlorines may be categorized into three subgroups including DDT, hexachlorocyclohexanes, and chlorinated cyclodienes.

FIGURE 6.19
Chemical structure of p,p'-DDT.

Dichlorophenylethanes

This subgroup of pesticides consists of DDT (Figure 6.19), methoxychlor, and dicofol. DDT is more acutely toxic and more slowly cleared from the body then methoxychlor or dicofol and is also more environmentally persistent. Consequently, methoxychlor, being less toxic and persistent, has been used in many household insect sprays. DDT appears to disrupt sodium and potassium permeability in the nerve axon.[14] The storage of DDT in adipose (fat) tissue; biomagnification in the food chain; and disruption of calcium metabolism in fish-eating birds, causing thinning egg shells and declines in bird populations, led to the banning of DDT for use in the United States in the 1970s. A closely related pesticide, DDD, is also banned for use in the United States.

Hexachlorocyclohexanes

There are many isomers for benzene hexachloride (HCB) with the gamma-isomer known as lindane being the most toxic. Lindane (Figure 6.20) is used primarily for treating seeds in the United States and is effective in the treatment of ectoparasites such as fleas, lice, and ticks. As such it is available in a powder, shampoo, lotion, or cream and is often is used in the treatment of scabies, as well as the treatment of fleas on household pets. Part of the toxicity of lindane is based on its ability to inhibit g-aminobutyric acid (GABA)-dependent chloride flux into the neuron.[14]

Chlorinated Cyclodienes

Chlordane (Figure 6.21), aldrin, dieldrin, endosulfan, and heptachlor belong to this class of compounds that are characteristically more toxic than DDT and its related

FIGURE 6.20
Chemical structure of lindane.

FIGURE 6.21
Chemical structure of chlordane.

FIGURE 6.22
Chemical structure of kepone.

compounds. Because of their toxicity, rapid absorption through the skin, and animal carcinogenicity, the EPA suspended the use of most chlorinated cyclodienes in the United States in the mid-1970s. These substances are effective as stomach poisons, fumigants, and contact poisons and have been used effectively against roaches, silverfish, ants, and termites.[14,36]

Chlorinated Cyclodiene

Kepone (Figure 6.22) and mirex cause liver tumors in mice, and both pesticides were banned for use in the United States although they were never widely used in the nation. Kepone has caused severe neurological damage among production workers in Virginia and contamination of the nearby waters including the Chesapeake Bay.[41]

Organophosphates

Organophosphates are chemical derivatives of phosphoric acid developed in Germany during WWII as a poison for the control of aphids. Many of the organophosphates are short-lived in the environment because they are easily biodegraded and are effective against pests resistant to the organochlorine pesticides.[36] These compounds vary dramatically in their toxicity. Parathion is the pesticide most likely to be involved in human fatalities, although used extensively in agriculture (Figure 6.23). A single drop to the eye may be sufficient to kill a person, and so only licensed pesticide operators should use the more toxic organophosphate pesticides. In an ongoing search to find organophosphates less toxic to humans, more than 100 such compounds have been developed.[42] These compounds vary in their toxicity with the most toxic being phosdrin, parathion, dichlorvos, and TEPP. Moderately toxic organophosphates include DDVP, dimethoate, baytex, and diazinon. The least toxic of the organophosphates include ronnel, gardona, abate, and malathion.[36] Toxicity of these compounds is based on cholinesterase inhibition at the neuromuscular junction of nerve fibers. Acetylcholine is a neurotransmitter compound that accomplishes the transmission of a nerve impulse from the nerve ending to smooth

FIGURE 6.23
Chemical structure of parathion.

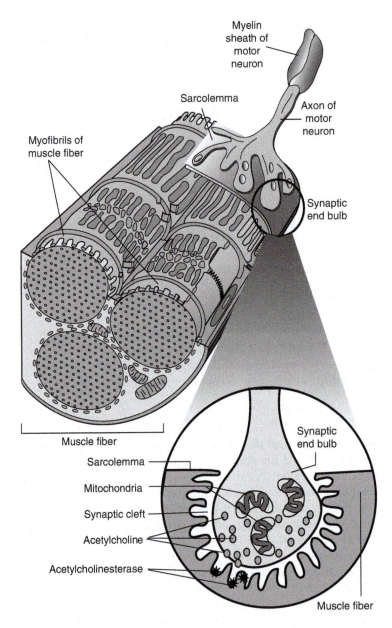

FIGURE 6.24
Acetylcholine, cholinesterase, and the neuromuscular junction.

muscle, the heart, and endocrine glands (Figure 6.24). The motor neuron stimulates muscle tissue by releasing the neurotransmitter acetylcholine from synaptic vesicles in the synaptic end bulbs at neuromuscular junctions occurring between the motor neuron and muscle fibers. A nerve impulse causes acetylcholine to be released that diffuses across the synaptic cleft and combines with receptor sites on the muscle fiber. This causes changes in the flow of sodium and potassium ions across the sarcolemma, resulting in a muscle action potential that travels down the sarcolemma and results in muscle contraction. Under normal conditions, the acetylcholine is immediately hydrolyzed by the enzyme acetylcholinesterase.

Organophosphates inhibit acetylcholinesterase, causing acetylcholine to accumulate at the neuromuscular junction, producing a variety of symptoms that include involuntary twitching, muscle weakness, paralysis of respiratory muscles, constriction of pupils, involuntary urination, vomiting, abdominal cramps, increased salivation and lacrimation, bronchoconstriction, increased sweating, and increased peristalsis. Symptoms that result from accumulation of acetylcholine in the CNS may include anxiety, restlessness, headache, insomnia, nightmares, slurred speech, tremors, and coma. Fortunately, therapy is available for diagnosed cases of organophosphate poisoning that includes the administration of atropine plus 2-PAM. Parathion weakly inhibits cholinesterase but is quickly converted to paraoxon in tissues, which is several thousand times more toxic. Malathion is much less toxic than parathion because humans produce a carboxyesterase enzyme that renders malathion nontoxic.[14] The highly volatile organophosphate DDVP (also known as vapona or dichlorvos) is often used as a spray or in pest strips. Diazinon is an effective insecticide for use in vegetables and fruits and is useful against roaches. Some of the more toxic organophosphates have been banned by EPA while others such as parathion are rarely used because of its toxicity. However, organophosphates continue to cause more deaths and poisonings of humans than any other category of pesticides.[14,36]

Carbamates

Carbamates are derivatives of carbamic acid and are contact poisons that inhibit cholinesterase in a manner similar to organophosphates. They vary widely in toxicity depending on the specific carbamate pesticides, as do the organophosphates, but the treatment is not similar. Examples of carbamates include (1) carbaryl, which is mildly toxic to mammals and which is used to control fleas on pets and is also effective in killing the rat flea; and (2) propoxur or baygon, which is both a contact and stomach poison that is effective against ticks, flies, sandflies, roaches, and mosquitoes.

Botanical and Biological Insecticides and Other Alternatives

The chrysanthemum plant produces natural, highly insecticidal alkaloids known as pyrethrins. This is nontoxic to mammals. Because it is difficult to get enough chrysanthemums to supply the materials needed, a number of pyrethrum compounds have been synthesized including permethrin, deltamethrin, and cypermethrin, pyrethroids exhibited a toxicity to insects similar to the mechanism of DDT.

Another alkaloid naturally derived from the roots of plants grown in South America or South East Asia is known as rotenone. This pesticide paralyzes insects by closing down chemical respiration in the nerve cell but is fairly harmless to humans. For this reason, rotenone is widely used in home-use insect sprays and other insecticide formulations for the household.

Methoprene is a growth regulator for insects that causes the mosquitoes to die in the pupa stage by halting normal metamorphosis. It was first developed in the 1970s to control mosquito larvae but several versions have been subsequently produced to control some beetles, flies, and fleas.[8,14]

Bacillus thurengensis produces a protein ("Bt" toxin) that is selectively toxic to insects by disrupting ion transport in the gut of the insects after they ingest it on the plants they are eating. It has been used with great success against a wide variety of leaf-eating insects including gypsy moth larvae, although some insects appear to be developing resistance to the Bt toxin. It is also a bioengineered pesticide in that the gene for its production has been inserted into a plant genome that then manufactures the protein Bt toxin that selectively kills insects feeding on the plants. The Bt gene has been inserted into corn, tomatoes, and potatoes while targeting insects that feed on these plants.

A variation on bioengineered crops is the development of soybeans that are resistant to a powerful herbicide manufactured by Monsanto called "Roundup." Consequently, the crops can be sprayed with herbicide to kill weeds without effecting soybeans. Critics are concerned about the herbicidal resistance being spread to unwanted weeds. They are also concerned that pests will develop resistance to the Bt toxin engineered into the plants.[43] The critics may be justified since similarly bioengineered cotton crops were severely attacked by cotton bollworms that appeared unperturbed by the Bt toxin bioengineered into the cotton. This is especially distressing since several pests commonly migrate between the cotton and corn crops.[44]

The introduction of corn that has been genetically modified to include the Bt gene, however, reduced the need to spray crops with the Bt organism, but also raised concern about potential environmental effects. Use of Bt crops reduces use of pesticides and insecticides applied to corn and cotton crops to prevent destruction of the crops by the European corn borer. As mentioned previously, the use of this technology could lead to an increase in the population of Bt-resistant corn-borers. The *"high dose/refuge" strategy* is designed to delay the development of resistant pests. By surrounding Bt crops with non-Bt hybrids, the very small number of pests who are resistant to the toxin are likely to mate with non-Bt resistant pests that thrive in the adjacent, non-Bt treated corn. The offspring they produce should have only low or moderate resistance to the Bt toxin. This strategy, however, depends on certain assumptions about the mating patterns of Bt-resistant pests.[45]

Some concern has arisen over this seemingly harmless pesticide because of its close genetic relation to the anthrax bacterium, *B. anthracis*. Some scientists argue that the genetic similarity points to the potential of the Bt bacteria to mutate into strains that are more dangerous to humans.[46]

Naturally occurring disease agents such as fungal spores (*Metarhizium flavoviride*), pathogenic viruses, and protozoan parasites have been formulated as bait or sprayed on crops. In the case of protozoans and fungi, they grow inside the target pest and dissolve or digest internal organs. A fungal pathogen, *Lagenidium giganteum*, produces a motile biflagellate zoospore that infects mosquito larvae in their native water state, and has been recently shown to decrease populations of selected mosquitoes by 77 percent in trial assessments.[47]

A novel approach to control of vector-borne diseases is being developed by a team of researchers at the Colorado State University in Fort Collins. Instead of developing methods to kill the mosquito vector for dengue fever, they are using bioengineering techniques to prevent the dengue virus from replicating in the mosquito. The researchers attached a small section of the dengue virus RNA to a harmless virus called Sindbis and inserted it into the mosquito where it is replicated. The dengue RNA component halted any replication attempts by complete dengue viruses in the mosquito, thereby establishing a form of intracellular immunity. If the researchers can figure out how to get the mosquitoes to pass the disease resistance onto offspring, this may represent a worthwhile approach to the control of the disease.[48]

A significant alternative to the indiscriminate and runaway use of pesticides is the use of integrated pest management (IPM). IPM is a process that combines the best current knowledge about host-plant resistance, farming practices, and biological control, with some limited chemical controls as required to effectively control weeds and insect pests in an environmentally sound way. It's a process requiring flexibility, strategy, and knowledge. The agriculturist needs to know which pests to treat, when to treat, and how long to treat in order to keep crop losses from becoming economically unacceptable. Organic farming, which employs several different strategies to control pests, is increasing in popularity.[49] While conventional farmers have and are using genetically modified crops to control pests, organic farmers are using naturally occurring pesticides, predators, crop rotation, cover crops, and other non-petrochemical means of pest control.[50] Conventional farmers are also incorporating some of these methods, using IPM. In part, the conversion to natural methods is due to numerous pests that have evolved resistance to pesticides. Increases in consumer awareness of the adverse effects of pesticides is also increasing market demand for agricultural products grown with reduced pesticide use.

Herbicides

Most herbicides are thought to have low toxicity to humans although their role as possible carcinogens has been an ongoing controversy.[51] There have been some reports of Hodgkin's lymphomas and other soft tissue sarcomas associated with occupational exposure to phenoxyherbicides.[14] There are sufficient differences in plant species such that herbicides can be selective by killing weeds that are broad-leafed dicotyledonous plants while leaving members of the desirable grass family untouched. Atrazine, alachlor, 2,4-D and 2,4,5-T are examples of such selective herbicides (Figure 6.25). These are chlorophenoxy agents that destroy the plants by stimulating abnormal growth and interfering with the transport of nutrients. Here again, there is extraordinary controversy over the potential teratologic effects of these herbicides spurred on by the use of Agent Orange to defoliate areas of jungle in Vietnam.[52] Agent Orange contained 2,4-D, 2,4,5-T and an unwanted but always present contaminant 2,3,7,8-tetrachlorodibenzo-p-dioxin (TCDD) also known as dioxin. Dioxin is an environmentally persistent substance that is highly toxic to humans and other mammals and is discussed in more detail in Chapter 5. Other selective herbicides include paraquat, diquat, and atrazine, used on such crops as sugar cane, pineapple, maize, cotton, and potatoes.

Rodenticides

Although rodent control has been discussed earlier in the chapter, a more detailed discussion of the rodenticide warfarin is offered here. Warfarin is among the most widely used rodenticides because it is effective in small quantities, reasonably non-toxic to humans, and does not promote bait shyness because it is a slow-acting poison. Warfarin is a coumadin compound and therefore a vitamin K antagonist that inhibits prothrombin synthesis. Repeated dosing over several days causes an overall reduction in the blood level prothrombin, which is necessary for coagulation. The consequence of such dosing is internal hemorrhaging, blood in the sputum and stools, and externally visible hematomas. The animals ultimately bleed to death. Some populations of rats have developed a resistance to warfarin. Consequently, more powerful versions known as super-warfarin, with a potency up to 200 times that of warfarin, have been developed that are more effective against these resistant strains.[53-55] Such compounds include brodifacoum,

FIGURE 6.25
Chemical structure of 2,4,5-T herbicide.

FIGURE 6.26
Chemical structure of warfarin.

difethialone, and flupropadine. As might be expected, these compounds have much higher toxicity to humans than warfarin (Figure 6.26).[54] Other multiple dose poisons effective in controlling rats include fumarin, PMP, and diphacinone.

References

1. Vandermark, P., Oh, Rats, *The Boston Globe Magazine*, 24, November 5, 1995.
2. Blank, J., Bombay's rat patrol takes on the plague, U.S. *News and World Report*, 14, October 14, 1994.
3. Mrotta, A., Rat meat, poor man's delicacy in Mexico, *Reuter's News Service*, July 14, 1998. http://www.infoseek.com/Content?am=a2873LBY209reulbl 9980713&qt--rats&sv=IS&Ik=noframes&col=NX&kt=A&ak=newsl486.
4. Dispatch, Dengue hemorrhagic fever: The emergence of a global health problem, *Eid al-Fitr*, 1, 2, April–June, 1995.
5. World Health Organization, *Control of Tropical Diseases (CTD): Malarial Control*, WHO Office of Information, Geneva, Switzerland, 1995.
6. Koren, H., *Illustrated Dictionary of Environmental Health and Occupational Safety*, CRC Press/Lewis Publishers, Boca Raton, FL, 407 p.
7. U.S, Dept of Health, Education and Welfare (HEW), *Pictorial Keys to Some Arthropods of Public Health Importance*, U.S. Department HEW Public Health Services, Washington, DC, Vol. 3, 1964.
8. Koren, H. and Bisesi, M., *Handbook of Environmental Health and Safety, Principle and Practices*, Vol. I, 3rd, ed., CRC Press/Lewis Publishers, Boca Raton, FL, Chap. 4.
9. Tortora, G.J., Funke, B.R., and Case, C.L., *Microbiology: An Introduction*, 5th ed., The Benjamin/Cummings Publishing Company, Redwood City, CA, 1995, Chap. 23.
10. Nadakavukaren, A., *Our Global Environment: A Health Perspective*, 4th ed., Waveland Press, Prospect Heights, IL, 1995, Chap. 8.
11. Morgan, M., Gordon, L., Walker, B. et al., *Environmental Health*, 2nd ed. Morton Publishing, Englewood, CO, 1997, Chap. 8.
12. U.S. Department of Health, Education and Welfare (HEW), *Control of Domestic Rats and Mice, Public Health Service*, Communicable Disease Center, Washington, DC, 22, 1964.
13. Joy, R.M., Chlorinated hydrocarbon insecticides. In: *Pesticides and Neurological Diseases*, Ecobichon, D.J., and Joy, R.M. (Eds.), CRC Press, Boca Raton, FL, 1982, pp. 91–150.
14. Saunders, D.S., and Harper, C., Pesticides, In: *Principles and Methods of Toxicology*, 3rd ed., Hayes, A.W. (Eds.), Raven Press, New York, 1996, Chap. 11.
15. Dover, M., Getting off the pesticide treadmill, *Technology Review*, November/December, 53, 1985.
16. United States Geologic Survey (USGS), Pesticides in Surface Waters, U.S. *Geological Survey Fact Sheet*, FS-039-97, May 30, 1997. http://wwwrvares.er.usgs.gov/nawqa/NAWQA.ofr94.70.html.

17. Ganesse, L.P. and Puffer, L.A., Herbicide-resistant weeds may threaten wheat production in India, *Resources, No. I I I (Spring)* Resources for the Future, Washington, DC, 1993.
18. U.S. Congress, Office of Technology Assessment, *Neurotoxicity: Identifying and Controlling Poisons of the Nervous System*, OTA-BA-426, U.S. Government Printing Office, Washington, DC, April, 1990, Chap. 10.
19. U.S. Environmental Protection Agency (EPA), Office of pesticide programs, *Pesticide Industry Sales and Usage:1987 Market Estimates*, U.S. EPA, Washington, DC, 1988.
20. Jeyaratnam, J., Health problems of pesticide usage in the third world, *British Journal of Industrial Medicine*, 42, 505, 1985.
21. World Health Organization (WHO), Pesticide deaths, what's the toll?, *Ecoforum*, 6, 10, 1981.
22. Whelan, E.M., *Toxic Terror: The Truth Behind the Cancer Scares*, 2nd ed., Prometheus Books, Buffalo, New York, 1993, Chap. 4.
23. Repetto, R. and Baligo, S.S., Pesticides and the immune system: The public health risks, *Central European Journal of Public Health*, 4, 263, 1996.
24. U.S.EPA, Office of Pesticide Protection (OPP), *Pesticides and Child Safety: Tips to Protect Children from Poisonings Around the Home, Prevention, Pesticides, and Toxic Substances*, EPA#735-F-93-050R, update November, 19,1997. http://www.epa.gov/oppfead/childsaf.htrn.
25. Siontral Resource Defense Council (NRDC), *Intolerable Risk: Pesticides in Our Children's Food*, Natural Resources Defense Council, Washington, DC, 1989.
26. Vegej, J., Cancer causing pesticides abound in supermarkets'bounty—Washing, peeling, buyag organic can help, *Health News Review,* June 12, 14, 1996.
27. Yess, N.J., Food and drug administration pesticde program, residues in food, 1990, *Journal of Association of Official Analytical Chemists*, 74, 745, 1991.
28. Hanson, D., Regulation after delaney—New health-based standard replaces controversial clause dealing with cancer-causing pesticides in foods, *Chemical and Engineering News*, 74(39), 38–39, September 23, 1996.
29. Conservation Foundation, *Opportunities to Assist Developing Countries in the Proper Use of Agricultural and Industrial Chemicals*, Washington, DC, Vol. 1, 1988.
30. U.S. Department of State, *Proceedings of the US. Strategy Conference in Pesticide Wanagement*, U.S. Government printing Office, Washington, DC, June, 1979.
31. Ngunjiti, P., Environment: Stemming the flow of dangerous chemicals. *Interpress Service English Newswire*, September. 25, 1996. dnV://www2.elibrary.com/id/2525/getdoc.cgi?id=52564954xO y445&oids=OqOOI 0012&forW=Tl>
32. Stolberg, Sheryl DDT, target of global ban, finds defenders in experts on malaria, *The New York Times*, August 28, 1999.
33. Ames, B. and Wiles, R., Pesticide pros and cons: Are residues really bad for your health?, *Vegetarian Times*, 68, November, 1995.
34. Avery, D., How pesticides help prevent cancer, *Consumer Research*, 10, June, 1995.
35. Colburn, T. and Clement, C. (Eds.), *Advances in Modern Environmental Toxicology, Vol. XXI— Chemically Induced Alterations in Sexual and Functional Development The Wildlifefluman Connection*, Princeton Scientific Publishing, Princeton, NJ, 1992.
36. Koren, H. and Bisesi, M., *Handbook of Environmental Health and Safety, Principle and Practices*, Vol. II, 3rd, ed.,CRC Press/Lewis Publishers, Boca Raton, FL, Chap. 6.
37. Wald, M, Citing children, *The New York Times*, EPA is Banning Common Pesticide, New York, 1999.
38. Raloff, J., The pesticide shuffle, *Science News*, 149, 174, March, 16, 1996.
39. Lopez-Carillo, L., Tores-Arreola, L., Torres-Sanches, L. et al., Is DDT use a public health problem in Mexico? *Environmental Health Perspectives*, 104(6), 584–588, June, 1996.
40. Greentumble editorial team, What is the global distillation effect? Greentumble, July 5, 2016. Retrieved July 6, 2018, https://greentumble.com/what-is-the-global-distillation-effect/.
41. Reich, M.R., and Spong, J.K., Kepone: A chemical disaster in Hopewell, Virginia, *International Journal of Health Services*, 13, 227, 1983.

42. Meerdink, G.L., Organophosphorous and carbarnate insecticide poisoning in large animals, *Veterinary Clinics of North America: Food Animal Practice*, 5, 374, 1989.

43. Tortora, G.J. and Anagnostakos, N.P., *Principles of Anatomy and Physiology*, 6th ed., Harper and Row Publishers, New York, 1990, Chap. 10.

44. Greene, R., Hope, concern, sprout over biotech crops, *Los Angeles Times, Science, Medicine and the Environment*, October 20, 1996.

45. Beardsley, T., Picking on cotton, *Scientific American, News & Analysis*, 43, October, 1996.

46. Mackensie, D., Friend or foe?, *New Scientist*, October 9, 1999, 22–23.

47. Renner, R., Will Bt-based pest resistance management plans work?, *Environmental Science and Technology*, 33, 410, October 1, 1999.

48. Cuda, P., Hornbey, J.A., Cotterill, B., and Cattell, M., Evaluation of *Lagenidium giganteum* for biocontrol of *Mansonia* mosquitoes in Florida (Diptera: Culicidae). *Biological Control*, 8, 124, 1997.

49. Adler, T., Keeping mosquitoes healthy for human's sake, *Science News*, 149, 295, May 11, 1996.

50. Bourne, J., The organic revolution, *Audobon*, 101, 64–70, March 1999.

51. Bourne, J., Bugging out, *Audobon*, 101, 71–73, March 1999.

52. Morrison, H.I., Wilkins, K., Semenciw, R. et al., Herbicides and cancer, *Journal of the National Cancer Institute*, 84, 1866, 1992.

53. Timoc, R., Parsonnet, J., and Halperin, D., Paraquat poisoning in southern Mexico: A report of 25 cases, *Archives of Environmental Health*, 48, 78, 1993.

54. Haug, B., Schjadtlversen, K., and Rygh, J., Poisoning with long-acting anticoagulant, *Tiasskr Nor Laegeforen*, 112, 195(8), 1992.

55. Routh, C.R., Triplett, D.A., Murphy, M.J. et al., Superwarfarin ingestion and detection, *American Journal of Hematology*, 36, 50, 1991.

7

Emerging Diseases

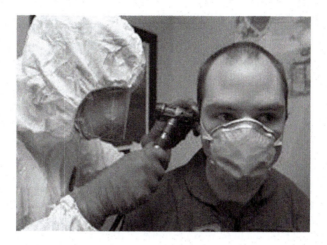

OBJECTIVES

A student reading this chapter will be able to:

1. Differentiate the emerging infectious diseases in the United States and those occurring worldwide.
2. Recognize the six major reasons associated with the emergence of infectious diseases.
3. Explain the likely reasons for the emergence of specific infectious diseases.
4. Identify the etiological agents, the epidemiology, and the disease characteristics of the major emerging infectious diseases including influenza, hanta virus, dengue fever, Ebola, AIDS, cryptosporidiosis, malaria, Lyme disease, tuberculosis (TB), streptococcal infections, *E. coli* infections, MERS, SARS, chikungunya, avian flu, swine flu, and more recently Zika.
5. Recognize the practical approaches to limiting the emergence of infectious diseases.

Introduction

When infections have recently appeared in a human population or those infections have suddenly escalated in incidence or geographic area it is referred to as an emerging infectious disease. This can apply to a disease that also threatens to increase in the near future. These emerging infections can happen when the disease originates from an infectious agent not previously detected or known. It can also happen when the agent was previously known but its role as a disease was not known until recently. Some diseases known to have declined may suddenly emerge again. That is considered a re-emerging disease. When disease agents spread to a new area and/or a new population it is considered an emerging disease for that area.[1]

Ten years ago, the World Health Organization (WHO) reported that infectious diseases are emerging at a rate that has never before been seen. There have been about 40 infectious diseases newly described in the last four decades, and more recently we have seen MERS, SARS, Ebola, chikungunya, avian flu, swine flu and, most recently, Zika. These emerging diseases are difficult to contain because of the millions of people that travel around the world almost daily. Because populations are denser, and poor sanitation is an increasing problem, along with overlapping of human populations with wild animal habitats, emerging diseases are born more frequently and spread more easily.[1]

Infectious diseases continue to be the foremost cause of death worldwide. The Centers for Disease Control and Prevention (CDC) reported a 58 percent rise in deaths from infectious diseases since 1980. Infectious diseases account for about 1 in 4 deaths worldwide, including approximately two-thirds of all deaths among children younger than age 5.[2]

Infections and parasitic diseases have caused more than 17 million deaths in a year, more than a quarter of all human deaths.[3] Deaths from all other infectious diseases but AIDS rose a dramatic 22 percent.[4] These figures are even more extraordinary because the International Code of Diseases lists many infectious diseases in categories other than infectious. For example, meningitis and cirrhosis are classified as diseases of the nervous system and liver, respectively, and only 17% of deaths attributable to infections are actually included in the codes for parasitic and infectious diseases.[5] About a quarter of the visits to physicians in the United States each year are attributable to infectious diseases, and the costs associated with these infections are estimated at more than $120 billion.[6]

Despite more than 100 years of scientific progress and the development of sanitary methods, infectious diseases still cause enormous human suffering, deplete scarce resources, impede social and economic development, and contribute to global instability. The potential for even greater human misery looms as a growing cloud. Recent outbreaks underscore the potential for the sudden appearance of infectious diseases in currently unaffected populations in the United States and the world.

Emerging Diseases in the United States

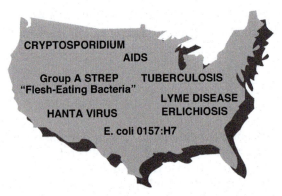

We're being continuously challenged in the public news media with the terrifying prospect of new and emerging infections. The British tabloids coined the term "flesh-eating bacteria" to describe an invasive necrotizing infection caused by Group A streptococci (GAS) and suggested that we are about to experience epidemics from this organism producing high mortality, amputations, and grotesque disfigurement.[7] New parasitic threats to water supplies have erupted into the nation's view. One of those parasites, cryptosporidium, has caused more than five major outbreaks in the U.S. with the most severe being in Milwaukee in 1993. This outbreak affected more than 400,000 people.[8] Avian influenza has captured worldwide attention. Avian influenza presents a potential threat because the H5N1 virus has spread across eastern Asia and other countries as an evolving highly virulent virus easily transmitted among migrating bird populations with some capacity to afflict mammals including humans in close circumstances. Most of the human cases appear related to exposure to infected poultry, and no sustained human-to-human transmission has yet occurred.[9]

The first cases of what would later become known as AIDS were reported in the United States in June of 1981. Today, there are more than 1.1 million people living with HIV and more than 700,000 people with AIDS have died since the beginning of the epidemic.[10]

AIDS is the ninth leading cause of death among persons 25 to 44 years of age in the U.S., accounting for 19 percent of deaths from all causes in this age group.[11,12] Infection with *Escherichia coli* serotype O157:H7 was first described in 1982 and rapidly emerged as the major cause of bloody diarrhea and acute renal failure. The American public became aware of this disease when 700 persons became infected and four died after eating undercooked hamburger from a regional fast food chain in a 1993 outbreak.[13] There have been more than 30 outbreaks in the United States since, with many additional cases worldwide.[14] In 1993 a six-year drought followed by heavy rains and flooding contributed to the increase of deer mice in the American Southwest, leading to an outbreak of hantavirus pulmonary syndrome. First appearing in the Navajo reservations, the disease has since spread to 20 additional states and taken the lives of 45 people, nearly half of those infected.[15] Lyme disease is the second most prevalent emerging infectious disease in the United States; more than 65,000 cases have been reported to the CDC since the disease was first described by Steere and colleagues in three Connecticut communities in 1977.[16] Eastern equine encephalitis (EEE) is a viral disease transmitted from birds to humans by the bite of a mosquito such as *Culex* or *Aedes*. Reporting of the diseases began in 1964. There is only an average of five cases per year in the United States, but there is 30 percent mortality and half of those that survive will experience mild to severe permanent neurologic damage.[17] West Nile virus (WNV) is a flavivirus occurring commonly in the Middle East, West Asia, and Africa. It is thought to have been first recognized in New York State in over-wintering *Culex* mosquitoes during 1999 before the widespread transmission in 2000.[18] The virus can infect humans, birds, mosquitoes, horses, and some other mammals. It is transmitted to humans primarily by mosquitoes and tends to be seasonal. The disease may be mild or very severe resulting in death. West Nile encephalitis and West Nile meningitis are forms of severe disease that affect a person's nervous system. Encephalitis refers to an inflammation of the brain, while meningitis is an inflammation of the membrane around the brain and the spinal cord. There were 33 cases reported to ArboNET by state and local health departments from January 1, 2006 to July 29, 2006. There were 16 (48%) encephalitis (neuroinvasive disease), and 17 (52%) West Nile fever (milder disease).[18] ArboNET is the national, electronic surveillance system established by CDC to assist states in tracking WNV and other mosquito-borne viruses.

Emerging Diseases Worldwide

These and other infectious diseases have potential worldwide significance. As an example, the possibility of dengue fever reaching the United States is very real. According to the WHO, there are almost 30 "new" recognized infectious diseases including HIV, Ebola, hantavirus, hepatitis-C, the Legionella bacterium and the cyclospora parasite."[3] Until 1995, dengue fever had been isolated in Costa Rica's coastal mountain regions along the Pacific shore. But rising temperatures permitted the mosquitoes responsible for dengue fever to break the coastal mountain barriers and invade the rest of the country. By September, the epidemic had infected 140,000 people and claimed the lives of 4,000. Dengue fever, a debilitating tropical disease transmitted by mosquitoes, is reemerging with such force that it has now reached epidemic proportions in Central America and is bearing down on the United States.[19] Each year 35–60 million people worldwide contract dengue fever.[20]

The Ebola virus grabbed global attention with a flurry of news articles and films as one the deadliest viruses known, killing nearly 90 percent of those infected. There is no treatment or vaccine. Ebola and other hemorrhagic fever viruses typically start suddenly with malaise, fever, and flu-like symptoms, which are often followed with rashes and bleeding along with kidney and liver failure as these organs and other body parts dissolve into mush.[21] According to the Pan American Health Organization, Chagas is the most important parasitic illness in Latin America causing over 40,000 deaths each year. The protozoan disease is transmitted by a beetle known as the kissing bug (*Rhodnius prolixus*), which thrives in dark houses with cracked floors and walls and comes out at night to feed on blood sucked from the face of the human host. The inoculated Trypanosome parasite invades the stomach, heart, and muscles of the human host to produce disabling fatigue, energy loss, and often death.[22] In the 1990s, epidemic cholera reappeared in the Americas after being absent for nearly a century; from 1991 through June of 1994 more than one million cases and nearly 10,000 deaths were reported.[23]

In 1994 an extended period of monsoon season in northern India followed by 3 months of extremely hot weather (100°F) drove rats into the cities, causing pneumonic plague from *Yersinia pestis* in Surat. The subsequent panic claimed the lives of many people and cost India $2 billion.[15]

A cluster of Creutzfeldt-Jakob disease (CJD) cases with a unique neuropathologic picture among unusually young patients in the United Kingdom was reported during a widespread epizootic of bovine spongiform encephalopathy (BSE).[24,25] This became notoriously referred to by the press as "mad cow disease" and prompted the uncommon action of destroying thousands of cows. This appearance of the disease alerted many countries, including the United States, to update their surveillance for CJD and look for similar cases.[26] This increased attention, and the fact that CJD is rapidly and invariably fatal, raised public attention about this emerging disease to a heightened level.[27]

Malaria is the world's largest widespread mosquito-borne illness and is potentially the worst threat to humans in the future. Global warming provides an environment for mosquitoes to multiply, migrate, and increase their metabolism, forcing them to feed (bite) more often. Despite the fact that mosquitoes die when temperatures rise above 40°C, so do people and crops. Programs to eradicate malaria, and socioeconomic development, have reduced endemic areas, but malaria has seen an increase in many parts of the world despite continued efforts to eradicate it. More than 100 million clinical cases occur annually and more than 1 million people, mostly children, die each year in tropical Africa.[28] Mainly due to drug-resistant strains, malaria has reemerged as a major international infectious disease with nearly half a billion people worldwide developing the disease each year.[29] Malaria has also begun to reappear in the United States and is now reemerging at a geometric rate in South America.[30]

The WHO reported that efforts to control TB have failed and that drug resistant strains of mycobacterium TB are rising. TB is considered one the world's leading infectious diseases and kills over 3 million worldwide each year with many more becoming ill.[31]

What Is an Emerging Infectious Disease?

New infections are emerging and old ones are expanding their geographical range faster than we can deploy resources to stop the organisms responsible. International travel and commerce are increasing and technology is rapidly changing the conditions that increase the risk of exposure to infectious agents. The spectrum of infectious disease is changing quickly in conjunction with dramatic societal and environmental changes. Worldwide, explosive population growth in developing nations, with expanding poverty and urban migration, are contributing to proliferation of both new and old infectious diseases. At the turn of the century only 5% of the population lived in cities but by the year 2025 it is estimated that this will increase to 65%. Therefore, more and more microbes will migrate from rural areas to urban areas.[32] Once expected to be eliminated as a public health problem, infectious diseases remain the leading cause of death and disability-adjusted life years (DALYs). Some of these diseases may be new to the human species. Others may simply have never been seen in North America before. The term "emerging infectious diseases" is now being applied to this phenomenon, and awareness of these emerging diseases has reached international levels. The term "emerging infectious diseases" refers to diseases of infectious origin whose incidence in humans has either increased within the past two decades or threatens to increase in the near future.[32]

Emerging infectious diseases, often with unknown long-term public health impact, continue to be identified. Table 7.1 lists major diseases or etiologic agents identified just within the last 20 years. New agents are regularly added to the list, particularly with the availability of nucleic acid amplification techniques for detecting and identifying otherwise non-cultivable microorganisms.[33] In many cases, diseases became better recognized or defined (e.g., Legionnaires' disease, Lyme disease, human ehrlichiosis). Others are relatively new. A previously unknown and deadly disease, acquired immunodeficiency syndrome (AIDS), originated in one part of the world and spread globally at a speed that would have been unthinkable in earlier times.[33] Now, decades after the first AIDS cases emerged, scientists have confirmed that the HIV virus plaguing humans originated in wild chimpanzees, in a corner of Cameroon.[34]

TABLE 7.1

Major New and Remerging Etiologic Agents of Infectious Diseases and the Reasons for Their Emergence

VIRAL

Viral diseases that have been identified since 1973[a]

1977 Ebola, Marburg
Origin undetermined (importation of monkeys associated with outbreaks in these primates in Europe and the U.S.)

1980 HTLV Influenza (pandemic)
Pig duck agriculture thought to contribute to reassortment of avian and mammalian influenza viruses

1983 HIV
Transmission by intimate contact as in sexual transmission, contaminated hypodermic needles, transfusions, organ transplants. Contributing conditions that spread the disease include war or civil conflict, urban decay, migration to cities, and travel

1989 Hepatitis C
Transmission in infected blood such as by transfusions, contaminated hypodermic needles, and sexual transmission

1993 Hantaviruses
Increased contact with rodent hosts because of ecological or environmental changes

Viral diseases that have reemerged
Argentine, Bolivian hemorrhagic fever
Agricultural changes that promote growth of rodents
Bovine spongiform encephalopathy (cattle)
Alterations in the rendering of meat products
Dengue, dengue hemorrhagic fever
Travel, transportation, urbanization, and migration
Lassa fever
Conditions such as urbanization that favor rodent host, increasing exposure (usually in homes)
Rift Valley fever
Irrigation dam building, agriculture: possibly change in virulence
Yellow fever
Conditions favoring mosquito vector (in "new" areas)

BACTERIAL

Bacterial diseases that have been identified since 1973[a]

1977 Legionella disease
Cooling and plumbing systems that allow the organism to grow in biofilms that form on water

1982 Hemolytic uremic syndrome (*Escherichia coli* – O157:H7)
Modern food processing on a large scale permitting contamination through meat storage tanks and unsterile plumbing

1982 Lyme borreliosis (*Borrelia burgdorferi*)
Close contact between homeowners encroaching on forested areas and the mice and deer (a secondary reservoir host) that maintain the tick vector for Borrelia

1983 *Helicobacter pylori*
Newly recognized as agent involved with gastric ulcers, probably widespread before recognition

(Continued)

TABLE 7.1 (*Continued*)

Major New and Remerging Etiologic Agents of Infectious Diseases and the Reasons for Their Emergence

1987 Toxic shock syndrome (*Staphylococcus aureus*)
Ultra-absorbency tampons

1992 Cholera (type 0139)
Likely introduced from Asia to South America by ship, with spread made possible by reduced water chlorination; Strain (type O139) from Asia newly spread by travel (similarly to past introductions of classic cholera)

Bacterial diseases that have reemerged

Tuberculosis
Breakdown in public health measures such as reduction in prevention programs, inadequate sanitation, homelessness, AIDS
Streptococcus, group A (invasive necrotizing)
Unknown, may be increased use of aspirin and related substances

PARASITES

Parasitic diseases that have been identified since 1973[a]

1976 Cryptosporidium, other waterborne pathogens
Contaminated surface water, lack of proper filtration methods

Parasitic diseases that have reemerged

Malaria (in "new" areas)
Spread of mosquito vectors, worldwide travel or migration, "Airport" malaria

Schistosomiasis
Agriculture, dam building, deforestation, flood/drought, famine, climate changes

Source: Adapted from Morse, S., *EID*, 1, 1995.
[a] Compiled by CDC staff. Dates of discovery are assigned on the basis of the year the isolation or identification of etiologic agents was reported.

Do you want to learn more about emerging diseases? Point your browser to https://people.umass.edu/~envhl565/ and click on any of the Chapter 7 Web links associated with emerging diseases:

- Emerging Diseases (Michigan)
- Emerging Infectious Diseases Home Page
- CDC Office of Infectious Diseases

Reasons for the Emergence of Infectious Diseases

The emergence of so many new diseases and the re-emergence of those thought to be controlled appear difficult to understand or explain. However, rarely do emerging infections surface without reason. There are a number of specific explanations responsible for disease emergence that can be identified in most all cases.[35] Most emerging infections, and even antibiotic-resistant strains of common bacterial pathogens, usually originate in one geographic location and then disseminate to new places. Morse suggested that infectious disease emergence can be viewed operationally as a two-step process: (1) introduction of the agent into a new host population (whether the pathogen originated in the environment, possibly in another species, or as a variant of an existing human infection) followed by (2) establishment and further dissemination within the new host population.[23,33] The majority of emerging infections are thought to be caused by pathogens formerly existing in the environment, and released from obscurity or presented with a selective advantage by altered conditions and given an opportunity to infect new host populations. Occasionally, a variant may evolve and cause a new disease.[32,34] The term "microbial traffic" is given to the process by which infectious agents transfer from animals to humans or spread from isolated groups into new populations.[36,37] There are identifiable activities that increase microbial traffic and promote emergence and epidemics. In a number of instances, including many of the more unusual infections, the agents are zoonotic, crossing from their natural animal hosts into the human population. In separate instances, pathogens existing in geographically isolated populations are able disseminate further. Very often, the emergence of a disease results from human actions, usually unintended. Changes in climate and other natural causes can also at times be responsible.[38]

Factors responsible for emerging infectious diseases include (1) ecological changes, such as those due to agricultural or economic development or to anomalies in climate; (2) human demographic changes (urbanization) and behavior; (3) travel and commerce; (4) technology and industry (globalization); (5) microbial adaptation and change (resistance); and (6) breakdown of public health measures.[23]

Ecological Changes (Agriculture, Climate)

Agriculture

Changes in ecology, such as from agricultural or economic development, are among the most commonly identified reasons in the emergence of infectious diseases. Ecological changes are often present in outbreaks of formerly unrecognized diseases with high case-fatality rates. These diseases are often introduced from animal hosts because people are placed in close contact with a natural reservoir or host where the disease organism is present (often a zoonotic or arthropod-borne infection). People may expand into an area where the animal host thrives, the animal host may expand into human living areas, and/or the infectious agent may have increased in the animal host population.[23,36,37]

Deforestation is reported to be a major reason for the emergence of Lyme disease in the United States and Europe because it increased the population of deer and the deer tick, the vector of Lyme disease. The movement of people into these areas placed a larger population in close proximity to the vector.[39] In Asia, the field mouse, *Apodemus agrarius*, is abundant in rice fields where people contract Korean hemorrhagic fever usually during the rice harvest from contact with infected rodents. Argentine hemorrhagic fever has increased almost in direct relation to the expansion of maize agriculture in Argentina and favoring the growth of a rodent that carries the disease organism.[40] Current researchers report that pandemic influenza arises when two influenza strains recombine resulting in a new virus that is infectious for humans. Ducks are a major source of avian influenza. The avian flu virus may enter pigs that are in close proximity where the gene segments of the viruses may then recombine (or re-assort).[40,41] The agricultural practice of pig-duck farming as practiced in China places these animals in contact and creates an ideal setting for the creation of influenza recombinants capable of infecting humans.[42]

Water is also frequently associated with disease emergence. Infections transmitted by mosquitoes or other arthropods include some of the most serious and widespread diseases.[43] Mosquitoes are often stimulated by expansion of standing water, simply because many of the mosquito vectors breed in water. There are many cases of diseases transmitted by water-breeding vectors, most involving dams, water for irrigation, or stored drinking water in cities.

Climate

Virtually all climate scientists believe that increase in potentially fatal diseases is likely to be the single most important threat that climate changes poses to human health.[43] Extreme weather patterns, as a result of natural fluctuations in the atmosphere or manufactured changes (i.e., global warming) have routinely been followed by outbreaks of disease. "Dry spells in wet areas or torrential rains in normally dry spots tends to favor so-called opportunistic pests—rodents, insects, bacteria, protozoa, viruses—while making life more difficult for the predators that usually control them."[13] "The same synergies that empower microbes also weaken our defenses against them. Heat, increased ultraviolet radiation resulting from ozone depletion, and pollutants like chlorinated hydrocarbons all suppress the disease-battling immune systems—both for humans and for other animals."[15]

The earth's climate is becoming a critical factor in enhancing the spread of infectious diseases. The WHO predicts the temperature of the planet will rise 2.7°F–8°F by the year 2030, and profoundly affect human health. One of the predictable results will be the expansion of mosquito-borne disease from the tropics into temperate zones. The expansion may move northward as far as New York and could include malaria, dengue, and yellow fever (this latter has a 50 percent mortality). In 1990, WHO warned that global warming would enlarge the territories of disease-carrying insects and increase the at-risk population by 620 million people by the year 2050.[4]

The Pacific Ocean warming events influence weather patterns and jet streams across North and South America, Western Europe, Asia, and Africa. El Niño events normally begin in the latter part of December and determine weather patterns for a year. The El Niño warming events have historically occurred about twice every 10 years.[44] These sea warming events cause evaporation of water, producing extensive rainfall in some areas and drought in others. The power of the El Niño events have increased dramatically in recent years, producing +2°C to 3°C temperature increases in the ocean well into the summer months, and with the possibility of extending over much longer periods of time. It is predicted that El Niño events will be more frequent with greater intensity, onset, and duration consistent with a continued increase in greenhouse gases.[45] A warmer climate initiated by increased amounts of carbon in the atmosphere has caused El Niño events to last longer and to be more dramatic.[46] The sudden eruption of hantavirus infection in the southwestern United States in 1993–1994, killing 48 percent of the 94 persons infected, has been attributed to the sudden population explosion of deer mice, which harbor the organism. The probable events are that heavy rains fell following 6 years of drought, causing pine nuts and grasshoppers to flourish, which fed the population eruption of deer mice. This enlarged population was driven from the flooded burrows, increasing the possibility of contact with people.[47]

It has been suggested that certain organisms in marine environments are natural reservoirs for cholera vibrios and that large-scale effects on ocean currents may cause local increases in the reservoir organism with consequent flare-ups of cholera. Pollution and global climate changes induced by humans adversely affect the health of marine populations and have the potential to harm the humans because of the ocean's ability to act as incubator for diseases like cholera.[48] Changes in coastal ecology are generating "hot systems" in which random mutations of *Vibrio cholera* may be selected and amplified and then transferred to human hosts through the consumption of contaminated fish and shellfish.[48] Aquatic plants, seaweeds, and free-floating phyto- and zooplankton are known to harbor vibrios.[49] The phytoplankton or algae are growing in epidemic proportions spurred on by climate-related warmer sea surfaces and the increase of nitrogen-rich wastewaters from coastal cities.[50] Additionally, the species of algae that are proliferating are less palatable to fish, and the numbers of fish have shown a decline in 14 of the world's 17 major fishing grounds.[48]

Human Demographic Changes (Urbanization) and Behavior

Changes in human demographics and behavior, or human population movements and upheavals caused by migration or war, are often significant factors in disease emergence. Increased population density in urban areas—migration in hopes of a better, more comfortable lifestyle—has surpassed basic services, including clean water supplies, sanitary conditions such as sewage disposal, and adequate housing.

Some of the most profound effects were among the poor of the inner-city slums. It was a frustrating paradox for many in public health that those most susceptible to infection had the worst facilities. Their socio-economic status led to a heightened risk of disease because of little or no access to clean water, running sewers, and pest control. In one example, until the mid-1950s, large areas around and within Washington DC had no running water and were served by outhouses.[51] These areas not only infected the people who lived there but they also acted as a reservoir of disease that would periodically spread across entire regions of the country. The diseases were incubated by unhygienic practices as well as brought in by immigrants who could often not afford to live anywhere else.

Depressed economic conditions and the disappearance of viable agricultural land in rural areas in many parts of the world are prompting the mass movement of workers from rural areas to cities. It has been estimated by the United Nations that 65 percent of the world's population will be living in cities by the year 2025 as this migration continues.[30] Diseases such as HIV are thought to have originated in isolated rural areas and to have been brought to the densely populated cities where the disease could spread locally and travel by infected humans along highways and air routes to cross-state and national boundaries. Public health measures in overcrowded cities are often strained or unavailable to large groups of the urban impoverished living in inner city slums or in shanty towns on the periphery, thereby increasing the opportunity for emerging infectious diseases.[32,51,52]

Emerging infections such as HIV, cholera, and dengue have followed this pattern. An example of this is HIV-2, which was isolated from a man living in rural Liberia and thought to have originated from contact with animals where a virus similar to HIV-2 was isolated from the Sooty Mangabey monkey.[53,54] This monkey is often used for food. Such zoonotic infections may often occur and remain isolated, but the swelling migration of people from rural to urban environments increases the probability that the disease will spread to densely populated areas and be transported regionally along highways and airways to reach distant locations permitting a pandemic as was the case for HIV-1. Simply transporting the virus with naturally low transmissibility would not be sufficient to produce a worldwide disease.

There were social behaviors that permitted intimate contact with the HIV organism through sexual transmission, intravenous drug use, or the use of technology in the form of blood transfusions and blood products. These factors associated with the very long duration of infectivity have combined to increase the opportunity for transmission resulting in a pandemic that has affected much of the world and is now spreading through Asia.[23] Human activity influences a number of factors associated with the emergence of disease; the control of such diseases will require the motivation of appropriate individual behavior and constructive action regionally and even internationally. Unfortunately, the continued increase of infectious diseases on many fronts points to the need for additional scientific knowledge in human behavior.

Travel and Commerce

The main factor effecting the emergence and spread of new diseases is believed to be transportation.[3] Increased economic growth into national and international boundaries has led to increased travel, contributing to the notion of "diseases without boundaries." The incubation period for infectious diseases can be a period of hours to weeks. As a result, diseases have the potential to spread from one continent to another in a matter of hours before symptoms occur.

Travel

The spread of HIV through travel has been suggested.[53] Historically, infections introduced into a population in a remote area might, occasionally, be transported to a new place through commerce, travel, or war.[55] The bubonic plague, carried by the rat and transmitted to humans by the rat flea, is thought to have been introduced into Europe with the silk trade routes from Asia and continuing with the Crusades. Smallpox was introduced by the Spanish conquistadors into the Aztec nation in the fifteenth century and decimated more than 80 percent of the highly susceptible population. The sixteenth and seventeenth centuries saw ships transporting slaves from West Africa to the New World infected with yellow fever and carrying the mosquito vector, *Aedes aegypti*, to the new territories. Two centuries later, cholera traveled with its human hosts from its likely origin in the Ganges plain to the Middle East. From there cholera traveled to Europe and a substantial part of the remaining world.[23] The Asian tiger mosquito, *Aedes albopictus* entered the United States, Brazil, and parts of Africa in shipments in the collected water within used tires shipped from Asia.[56] The mosquito's introduction is problematic because it bites aggressively, survives in both forest and suburban habitats, and is a vector for many human pathogens. It is a vector for epidemic dengue fever transmission in Asia and is a competent laboratory vector of La Crosse, yellow fever, and other viruses.[57] Fourteen strains of EEE virus have been isolated from *A. albopictus* in Florida.[58] The mosquito has become established in at least 21 of the contiguous states in United States and in Hawaii.[59] Malaria is another disease transmitted by a mosquito vector and is one of the most repeatedly imported diseases in non-endemic-disease areas. Cases of malaria contracted in airports are sometimes identified. Rats have carried hantaviruses virtually worldwide.[60] These infections were originally localized but were carried to exotic and foreign parts of the world. Similar histories are being repeated today with the added problem of greatly increased volumes of traffic moving at enormous speed. Travel may involve short distances or the crossing of international borders. In 1998 an emigrant from Ukraine in a plane from Paris to New York infected 13 other passengers with drug-resistant TB.[32] The volume is huge. In the early 1990s, in excess of 500 million persons crossed international borders yearly on commercial airplane flights.[59] As many as 70 million persons are thought to work either legally or illegally in other countries, and the majority are from developing countries. Many are migrant workers following crops and so moving seasonally.[61] Large populations are moved among countries as part of military maneuvers. The consequence of armed conflict and political unrest also displaces millions. In the early 1990s, there were an estimated 20 million refugees and 30 million internally displaced persons (IDPs) worldwide.[59,62] That number has not changed significantly over the years. The difference between refugees and IDPs is that refugees cross international borders while IDPs stay within their own country but may be at greater risk since they are likely to be persecuted by their own government.

Commerce

Marine organisms may be transported on the hulls of ships and in their ballast water. There were 367 different species identified in the ballast water of ships traveling between Japan and Coos Bay, Oregon.[63] There is evidence that disease organisms are also transmitted by shipping such as the introduction of *Vibrio cholerae* into South America.[64] *Vibrio cholerae* was isolated from the ballast, sewage, and bilge of 21 percent of cargo ships docked at ports in the Gulf of Mexico. The ships were previously in Brazil, Colombia, and Chile.[65] *Vibrio cholerae* O139 has traveled along waterways in Asia carried by people and boats generally engaged in commerce.[66] A strain of *Vibrio cholerae* identical to an epidemic strain found in Latin America was detected in Mobile Bay, Alabama, within oysters and the fish that eat oysters.[67]

Researchers have regularly demonstrated that mosquitoes are present on and can survive international flights. When random searches were made of airplanes in London, mosquitoes were located on 12 of 67 airplanes from tropical countries.[59] Arthropods can survive even more extreme environments. In another study, mosquitoes, house flies, and beetles placed in wheel bays of Boeing 747B aircraft survived flights of 6–9 hours with external temperatures of −42°C.[65] Airplanes have also carried infective mosquitoes that caused human infection outside malaria-endemic areas (in Europe, e.g.).

More Than Transportation Required

Travel will permit a potentially pathogenic organism to be introduced into a new geographic area. But, if it is to be established and cause disease, the organism must survive, reproduce, and enter a susceptible host. Fortunately, the majority of such introductions do not result in disease. Microbes that survive exclusively in the human host and are contagious through intimate contact, and which droplet nuclei, can be readily transmitted worldwide. Many diseases such as TB, AIDS, pertussis, measles, diphtheria, and hepatitis are routinely transported around the globe. In those populations where a majority of people have been vaccinated, the diseases find little opportunity to progress. Organisms are much more difficult to transplant when they have complicated life cycles that may involve animal hosts, vectors, or specific environmental requirements. Yellow fever, malaria, and dengue fever require specific mosquito vectors or they cannot proliferate in a particular geographic area. A snail intermediate host is required for the development of certain life stages in schistosomiasis, and so the disease organism cannot spread in the absence of this host. Even when all the conditions are present to permit the successful introduction of a parasite in a geographic area, the development of human disease is not an absolute consequence. The hog tapeworm, *Taenia solium*, is often not successfully transmitted in developed nations because there are adequate methods for confining and treating human sewage that prevent the eggs from infecting other persons. Briefly, the probability of transmission involves many biological, social, and environmental variables.[59]

Technology and Industry (Globalization)

High-volume rapid movement characterizes not only travel but also other industries in modern society. In operations that include food production, the use of products of biological origin, or the use of modern production methods, increase the chances of accidental contamination and amplify the effects of such contamination.[68] The problem is complicated by globalization, because agents are introduced from far away. Globalization has been defined as the process of denationalization of markets, laws, and politics in the sense of interlacing peoples and individuals for the sake of the common good.[69] Countries are finding it increasingly difficult to solving the problem of emerging infectious diseases and public health policy is becoming denationalized to attack this problem. Globalization is influencing public health in three ways.[68] First, the diseases are moving rapidly around the globe because of technology and economic interdependence, which has increased international travel and the international nature of food processing and handling.[59] As an example of technology increasing the risk of emerging diseases, the concentrating effects that occur with blood and tissue products have inadvertently disseminated infections unrecognized at the time, such as HIV and hepatitis B and C. Medical settings are also at the leading edge of exposure to new diseases, and many emerging infections have spread nosocomially in health care settings. Additionally, the common use and misuse of antibiotic treatments has permitted the development of drug-resistant pathogens.[70,71] Food processing is also a contributory factor to emerging diseases. As an example, BSE, which emerged in Britain within the last few years, was probably an interspecies transfer of scrapie from sheep to cattle.[71] This probably occurred when changes in rendering processes resulted in incomplete inactivation of scrapie agent in sheep byproducts fed to cattle.[72] Second, the funding of public health programs has been reduced because of increased competition in the global market and increased pressures to cut expenditures, thereby negatively affecting public health programs.[73] Third, public health programs have become international through WHO and health-related nongovernmental organizations. Worldwide health has improved through these international efforts with the elimination or reduction of many diseases. Unfortunately, these successes have contributed to a population crisis and produce overcrowding, inadequate sanitation, and overstretched public health infrastructures leading to the expansion of disease.[74]

Microbial Adaptation and Change (Resistance)

Antibiotic Resistance

Different mechanisms of genetic change for microbes are being routinely discovered. Spontaneous mutation is just the beginning. Microbes most often survive in enormous populations and are influenced by a myriad of physical and chemical mutagens. Haploid organisms are able to instantly express their genetic variations. They have numerous repair mechanisms, which themselves are subject to genetic control. Some strains that do not repair their DNA are therefore highly mutable; others are relatively more stable. Some organisms are highly flexible in responding to environmental stresses. There are abundant processes that allow bacteria and viruses to very readily exchange genetic material.

Plasmids now spread throughout the microbial world. They can intersect the boundaries of yeast and bacteria. Lateral transfer is very substantial in the development of microorganisms influencing the expressions of pathogenicity, toxicity, and antibiotic resistance. Microbes have a wide variety of mechanisms that allow them to exchange genetic information.[75] Microbes are, therefore, constantly evolving.

Antibiotic-resistant bacteria are emerging from the environment in response to the wide distribution of antimicrobials. This is an evolutionary lesson on microbial adaptation and a display of the power of natural selection. Selection for antibiotic-resistant bacteria [76] and drug-resistant parasites has become common, generated by the wide and often unsuitable use of antimicrobial drugs.[77,78] Pathogens can also attain antibiotic resistance genes from other, often nonpathogenic, species in the environment, selected or perhaps even driven by the ongoing pressure of antibiotics.[76] Many viruses demonstrate a high mutation rate and can rapidly evolve to yield new variants.[79] A classic example is influenza.[80] Yearly epidemics result from antigenic drift in a previously circulating influenza strain. An alteration in an antigenic site of a surface protein, usually the hemagglutinin (H) protein, permits the current variant to re-infect formerly infected persons because the immune system no longer recognizes the altered antigen. Rarely, the evolution of a new variant may result in a new expression of disease.[36] It is possible that some recently reported characteristics of disease by Group A streptococcus, such as rapidly invasive infection or necrotizing fasciitis, may also fall into this category.[23]

Only during the last 50 of 3,000 years of recorded history have humans been able to cure an individual of a communicable disease. To underline the significance of this apparently mild statement consider that in 1925 the number one killer in America was diarrhea and enteritis caused by bacterial infection; 109 of every 100,000 men, women and children were killed. This is roughly three times the current death rate from all causes combined. In the same year, the top ten causes of death included scarlet fever, pneumococcal pneumonia, whooping cough, diphtheria, and TB. Sixty years later diarrhea and enteritis in this country were responsible for one in 1,000,000 deaths.[81]

There is growing concern that bacterial pathogens are developing a resistance to antibiotics as a result of patients not completing the prescribed course of treatment or the inappropriate and over-prescribing of common antibiotics by physicians. Most of the bacteria are destroyed in the early course of treatment, generally the weakest bacteria. However, the strongest survive and mutate (build resistance) so that the drug becomes ineffective. Thus, bacteria that develop a resistance to a specific antibiotic will not respond to the drug if it is prescribed later.

The use of unsupervised prophylactic tetracycline administration to 100,000 pilgrims enroute to Mecca from Indonesia is thought to have been significantly responsible for the fact that 50% of cholera strains in that country are now tetracycline resistant. For a country as poor as Indonesia with an average of $4.52 annual medical expenditure per capita, this is a devastating blow.[82] Tetracycline is one of the easiest and cheapest drugs to produce, which makes it safe and widely available in the Third World. Other drugs often need to be imported and are so expensive as to be out of reach for the general populace.

A different example of resistance developed in shigella. This disease is the cause of dysentery throughout the developing world. After the development of multi-drug resistant shigella in Africa, it appeared almost immediately in Bulgaria (resistant to four drugs) and on the Navajo and Hopi reservations in the United States (resistant to two drugs) and in Vietnam where 90% of the cases were resistant to three or more antibiotics.[83] Shigella was being carried to the four corners of the earth and breaking out in areas with insufficient water treatment facilities. This demonstrates the disturbing properties of drug-resistant bacteria to travel vast distances.

While it took 15 years from the introduction of penicillin for widespread penicillin resistance to *N. gonorrhoea* to be reported in 1962, physicians who began using the flouroquinolone group of antibiotics (ciproflaxacin, ceftriaxone) in the late 1980s began reporting large number of cases resistant to the drug by 1992.[23] This acceleration in conversion times was not limited to gonorrhea but is universally seen across most common bacterial infective organisms. It suggests something greater than a cause and effect of antibiotic misuse by physician "A" leads to a mutant strain in patient "B"; these drug-resistant strains are displaying an evolutionary trait. Over the last 50 years we have not only bred resistant organisms but have selected for organisms that have the ability to develop resistance quickly.

By the last quarter of the twentieth century the threat of bacterial infections had begun to wane from the consciousness of the American mind. People who died from diseases such as pneumonia, like Jim Henson, creator of the Muppets, caused a stir because they were the exception to the rule that modern medicine could tackle bacterial infections. Part of the popular faith in medicine came from the perception of research as a constant fountain of new compounds. An array of new drugs seemed to appear every few years to be aggressively marketed by the pharmaceutical industry. In the early 1950s there were three classes of antibiotics to choose from, including the beta-lactams (penicillin), tetracyclines and aminoglycocides. In the 1990s there are seven classes—although several are close cousins to each other.

Drug-resistant infections could be treated with other more powerful drugs like vancomycin and Rocephin, which many hospitals consider their "big guns" in the disease war. The appearance of vancomycin-resistant strains in hospitals during the late 1980s has encouraged a view among many epidemiologist and bacteriologists that hospitals themselves are the source of many forms of drug resistance. A recent report by the CDC found that vancomycin resistance measured at 0.3 percent in 1986, rose to 7.9 percent across several facilities in 1994.[83] Intravenous vancomycin is the only treatment for methicillin-resistant *Staphylococcus aureus*. MRSA are staphylococci that are resistant to the antibiotic, methicillin, and other commonly used antibiotics such as penicillin and cephalosporins. There is a unique gene in the staphylococcus that makes them resistant to most currently used antibiotics. Therefore, alternate antibiotics must be used to treat persons infected with MRSA. Vancomycin has been the most effective and reliable drug in these cases but is used intravenously and is not effective for treatment of MRSA when taken by mouth.[84] The first infection with methicillin-resistant *Staphylococcus aureus* (MRSA) in the United States was reported in 1968.[85] Since then, the proportion of *S. aureus* isolates resistant to methicillin causing infections in hospitalized patients has risen significantly from 2 percent in 1974 to about 40 percent in 1997. The MRSA infections involved patients in hospitals or long-term care facilities, but recent reports of "community-acquired" MRSA infections raise concern. These community MRSA infections are occurring in healthy, non-hospitalized persons who have no contact with healthcare personnel or other colonized patients.[86] Should MRSA become widespread in the community, treatment of these infections will be much more difficult.[84]

Antibiotics in Livestock Feed

Many studies from various sources including the CDC, the FDA, and the Department of Agriculture suggest a strong link between the use of antibiotics in livestock feed and the transmission of drug-resistant bacteria to humans. About 40 percent of antibiotics that are manufactured today are given to livestock with 22.6 million kilograms being made each year.[32] Antibiotics have been added to livestock feed for many years to promote growth. Farmers have found that animals are able to absorb more nutrients from the feed and therefore grow more if the bacteria in the gut are killed. Unfortunately, organisms such as salmonella and campylobacter, which are harmful diseases if contracted by humans, are beginning to appear increasingly more resistant to the antibiotics.[87] While people are not susceptible to all of these organisms, the bacteria are able to transfer their resistance to disease-causing bacteria through genetic material called plasmids.

In response to this discovery, the British Ministry of Health adopted a White Paper on Livestock Feed and Antibiotics that regulated the administration of antibiotics to animals intended for human consumption. While the CDC has considered similar guidelines, they have never been adopted by Congress due to pressure from livestock interests who fear a competitive burden if they are not able to use prophylactic levels of antibiotics in feed. While these practices do help to prevent some diseases in animals, over the long run they have been shown to select for resistant organisms that can harm animals as well as humans.

Viruses

Antibiotics have no effect on viruses, and vaccines are often ineffective against bacterial infections. We have traditionally been able to combat such viral pathogens as rabies, polio, measles, and influenza, and have even eradicated smallpox through preventive vaccines. However, there are few antiviral drugs that exist, mainly because drugs that kill viruses also harm or kill their host cells (i.e., healthy cells). In addition, new viruses are continuously emerging, making it difficult for vaccines to be developed in sufficient time.

Breakdown of Public Health Measures

During the eighteenth and nineteenth centuries advancements in public health vastly improved the overall health of the populace, particularly in urban settings. Improvements included reducing the exposure of drinking water and food to contaminated waste as well as improvements in the populations' natural mechanisms of resistance through better nutrition.[82] Vector control, chlorination of water, pasteurization of milk, immunization, and proper sewage disposal are classical public health and sanitation measures that have successfully minimized the spread of infectious diseases in humans. The pathogens, however, may exist in smaller numbers within reservoir hosts or the environment, and represent a continuing threat for re-emergence if these sanitary methods are not maintained. Well understood and recognized diseases such as cholera are rapidly increasing because once active public health measures have lapsed. This retrogression in sanitary methods unfortunately now appears frequently in both developing countries and the inner cities of the industrialized world. The reemergence of infectious diseases may, therefore, signal the disruption of public health measures and should demand vigilance against indifference in the war against infectious diseases.[23] For example, cholera has been fiercely erupting in South America and Africa for the first time in this century.[88] The swiftly expanding cholera outbreak in South America may have been aided by recent reductions in chlorine levels used to treat water supplies.[89] Cholera and other enteric diseases follow the fecal–oral route of transmission, and so are frequently promoted by the lack of a reliable water supply. Problems such as these are aggravated in developing countries but are not confined to these areas. The U.S. outbreak of waterborne *Cryptosporidium* infection in Milwaukee, Wisconsin, in the spring of 1993, with over 400,000 estimated cases, was partially the result of a nonfunctioning water filtration plant.[90] Water purification is a problem found in many other cities in the United States.[91]

Specific Emerging Diseases

Viruses

According to Stephen S. Morse, the viruses with the greatest potential for emergence in the near future include hantaan (hantavirus), dengue, influenza, and HIV.[33] Therefore, these will be discussed in more detail. I have included a discussion of Ebola because it has become globally feared. The numbers of persons who have acquired the disease are exceptionally small compared with all other new and emerging diseases, but the gruesome manifestations of the disease, the high mortality, and the lack of treatment methods have spiraled this organism into the public eye with books, televisions shows, and movies. Therefore, it is discussed here to provide realistic contrast to the real worldwide killers. There are viral diseases that have emerged since the last edition of this book and are discussed in this section. These include MER-CoV, SARS-CoV, chikungunya, and Zika viruses, with an update to the 2108 flu season.

Hantavirus

Background: The saga of hantavirus outbreak in the United States began in 1994 when a physician was called to care for a man who collapsed at a funeral in New Mexico and died of acute respiratory failure. Soon after this, the physician found that several other persons in the Four Corners region (New Mexico, Arizona, Colorado, and Utah) of the Southwest had also died from similarly aggressive respiratory illness dating back to 1993. Initially, the victims were Native Americans located on a huge Navajo reservation. The media called it the "Navajo Flu." However, new cases were reported in the following months from California, Rhode Island, and Florida. The CDC was contacted since the etiology of the disease couldn't be determined. CDC personnel descended on the area and began a systematic effort to identify the disease. They collected blood and tissue specimens, victim histories, performed database searches, and tested for many different groups of potential pathogens. They then suspected a hantavirus, which is carried by rodents. CDC personnel then trapped and tested rodents from the area including the deer mouse, *Peromyscus maniculatis*. They found this to be the primary reservoir in New Mexico. The outbreak was brought under control when people in the area were educated on how to rid their homes of the mice and their droppings.[36] Although there have been isolated outbreaks in New York and Louisiana associated with different species of rodents, they have been controlled by practicing acceptable rodent control techniques and avoiding enclosed areas where the rodents may thrive.[92]

The disease: The hantaviruses that emerged in the Four Corners region of the United States were determined to be the cause of an acute respiratory disease now termed hantavirus pulmonary syndrome (HPS). The identification was delayed because the symptoms varied remarkably from the characteristic hantaviral diseases reported in the Europe and Asia. Therefore, it was an unexpected finding when sera from patients in this area reacted with hantaviral antigens and genetic identification revealed a novel hantavirus in patients' tissues and in rodents trapped near patients' homes.[93,94] Initial symptoms of the American version of HPS were flu-like and manifested as fever, chills, headache, muscular aches, and pains. Within a short time, however, the disease turns ugly and the lungs fill with fluids causing severe respiratory distress for which there is no specific treatment. In HPS, death occurs from shock and cardiac complications, even with adequate tissue oxygenation.

Epidemiology: Contraction of HPS in the United States is uncommon and fewer than 100 cases have been reported, but the mortality is high and there were 52 deaths among those reported to have contracted the disease. In contrast, people in China have been afflicted with a strain for thousands of years that annually causes 100,000 illnesses and perhaps as many as 5,000 to 10,000 deaths (5% to 10% mortality).[95] The hantavirus outbreaks in the eastern hemisphere (Asia) are identified with kidney failure and have been referred to as "hemorrhagic fever with renal syndrome" (HFRS). In the 1950s, Hantaan (HTN) virus (isolated near the Hantaan river) caused Korean hemorrhagic fever with renal syndrome in thousands of United Nations troops.[96] Still others have been isolated in Scandinavia and called Puumala (PUU). About 5–10 deaths per year (0.1% mortality) are caused by this strain.[97] Since the discovery of the HPS-causing hantaviruses, intense investigation of the ecology and epidemiology of hantaviruses has led to the discovery of many other novel hantaviruses. None of the states where this virus is located wanted it to carry the name of their state or community, so there was some bitter controversy over the naming of the virus. Scientists at the CDC finally decided on the term Sin Nombre or "No Name" for the virus. Most of the 174 cases of HPS in the United States and Canada have been caused by Sin Nombre (SN) virus.[98] Cases of HPS in the southeastern United States, as well as many in South America, are caused by a newly recognized clade (a group that shares a common ancestor) of viruses that includes Bayou (BAY), Black Creek Canal (BCC), and Andes viruses. These viruses are widespread and present a significant public health issue.[99]

The virus is spread to humans from contact with rodents. The pathway was first revealed in the early 1940s, when a viral etiology for HFRS was suggested by Russian and Japanese investigators who injected persons with filtered urine or serum from patients with naturally acquired disease.[96] These studies yielded the first evidence that the natural reservoir of hanta viruses was a rodent. The Japanese investigators were able to produce disease in humans by injecting bacteria-free filtrates of tissues from *Apodemus agrarius* or mites that fed on the Apodemus mice. In 1978, investigators demonstrated that patient sera reacted with antigen in lung sections of wild-caught *Apodemus agrarius* and so confirming a rodent reservoir for HFRS-causing viruses. They also demonstrated that the virus could be passed from rodent to rodent.[97] The successful propagation of HTN virus in cell culture in 1981 provided the first opportunity to study this pathogen systematically.[100] It is now established that hantaviruses can be carried by at least 16 various rodent species including rats, mice, and voles. Additionally, chickens and cats may as well be infected.[101] The viruses tend to be identified with a specific rodent host. In the Four Corners region the host is *Peromyscus maniculatis*, while in Florida the cotton rat (*Sigmodon hispidus*) is thought to be the primary carrier. Hantavirus infection does not appear to incapacitate its rodent reservoir host, which tends to produce a significant antibody response against the virion envelope and core proteins while maintaining chronic, probably lifelong infection.[99]

The emergence of the HPS virus in the United States is reported to be associated with environmental factors that promote the multiplication of a natural reservoir (field mouse) and so increases opportunities for human infection.[96] Researchers now report that this outbreak of HPS was probably initiated by climate irregularities associated with El Niño. The warming of the seas off the West Coast resulted in extensive rainfall in many areas including the Four Corners of the Southwest. This dramatically increased the natural foods (pine nuts and grasshoppers) of field mice and their numbers expanded 10-fold.[102] The HPS outbreak in the population of field mice subsequently invaded buildings to produce greater opportunities for indoor contact. The most common route of transmission to humans is by aerosolized mouse droppings containing the virus particles, although there is evidence that bites may also transmit the disease. Investigations have linked virus exposure to such activities as heavy farm work, threshing, sleeping on the ground, and military exercises. Exposure indoors was associated with invasion of homes by field mice during cold weather or to the nesting of rodents in or near dwellings. Persons of lower socioeconomic status appear to be at higher risk because housing conditions and agricultural activities permit more immediate contact with the mice. However, suburbanization, wilderness camping, and other outdoor recreational activities have spread infection to persons of middle and upper incomes.[96,103]

Dengue Fever

Background: Dengue is considered as one of the most consequential mosquito-borne viral diseases to affect humans. This disease has a global distribution that rivals malaria and has the capacity to infect the more than 2.5 billion people on the planet at risk from epidemic transmission. There are in excess of tens of millions of cases of dengue fever annually, with several hundred thousand cases of the more severe dengue hemorrhagic fever (DHF). DHF accounts for the majority of the 5–15 percent mortality that occurs worldwide, mostly in children.[104]

The disease: There are four antigenically distinct viral serotypes that cause dengue and DHF. The serotypes are DEN-1 through DEN-4 and belong to the genus Flavivirus.[105] The expansion of dengue/DHF is aided by the fact that there is no cross-protective immunity with any of the viruses, so that it is possible for a person to acquire multiple dengue infections during their lifetime.[104] Dengue is primarily an urban tropical disease with severe flu-like symptoms that causes high fevers, frontal headache, severe body aches and pains, nausea, and vomiting. The "critical stage" for DHF occurs at the point at which people start to feel better. When the fever eases, patients start to develop "leaky capillary syndrome" in which the blood vessels leak and untreated patients will go into shock and die. Some cases show no outward symptoms, while others show bruising or pinpoint rashes—a signal of internal bleeding.[106]

Epidemiology: The greatest emerging health menace from dengue/DHF has been in Central and South America where previous campaigns in the 1950s and 1960s to eliminate the mosquito vector *Aedes aegypti* to combat yellow fever met with success. However, complacency in mosquito control programs has allowed *Aedes aegypti* to return with a vengeance and it is a competent vector for dengue viruses. In the late 1980s, memories of mosquito-borne diseases faded, and other public health programs competed for the monies normally given to mosquito spraying. A rapid population increase throughout Latin America combined with rapid urbanization, poverty, and insufficient sewage and waste treatment methods resulted in an explosive growth in the mosquito vectors and a 60-fold increase in dengue/DHF as seen in Figure 7.1.[104] Further, air travel has increased the probability of the

FIGURE 7.1
Distribution of *Aedes aegypti* (shaded areas) in the Americas in 1970, at the end of the mosquito eradication program, and in 1995. (Adapted from Dispatch, Dengue hemorrhagic fever, the emergence of a global health problem, *EID*, I, April–June 2, 1995.)

virus spreading. Based on viral envelope gene sequencing, the DEN-3 virus strain in Latin America appear different than the DEN-3 strains previously isolated in these areas, and identical to the DEN-3 virus causing the major dengue epidemics in India and Sri Lanka.[104] Travel, therefore, appears to have been an important component in the spread of this disease.

Dengue fever and DHF has returned or emerged in many areas of the world (Figure 7.2). A dengue pandemic that began in Southeast Asia after WWII has increased dramatically since the 1980s where it has emerged as the leading cause of hospitalization and mortality among children. Taiwan and China experienced a series of epidemics after being free of the scourge since the 1950s. When India and Sri Lanka experienced their first major epidemics, DHF once again re-emerged in Asia. In 1994, Pakistan reported its first major epidemic with dengue. All four serotypes have been involved in the epidemics in China. In Asia, the Pacific Islands, Africa, and the Americas, the last 15 years have seen increasingly larger epidemics.[104–108] In 1998, 770,000 people in the Americas became infected.[109] In the past 20 years, dengue transmission and the frequency of dengue epidemics has increased greatly in most tropical countries of the American region.[110]

The likelihood of major outbreaks occurring in the United States is considered doubtful, but the possibility remains because the competent mosquito vectors are present including *Aedes aegypti* and the Asian tiger mosquito, *Aedes albopictus*. These mosquitoes have the capacity to carry and transmit the dengue virus and such transmission has been detected in south Texas in 1980 and 1986 and in northern Mexico.[111] Dengue fever and DHF have been found within 16 kilometers of the United States border in Reynosa, Mexico, just across the line from McAllen, Texas.[106] Additionally, there have been more than 2,000 suspected cases of imported dengue since 1977 in the United States with more than 20 percent of these cases confirmed as dengue.[112] The last reported case of dengue fever in the United States was in Texas in 1999.[110] The Unites Stated needs to maintain vigilance in controlling mosquito populations and maintaining the public health infrastructure that prevents outbreaks: uncontrolled urbanization; substandard housing; and inadequate water, sewage, and waste management.

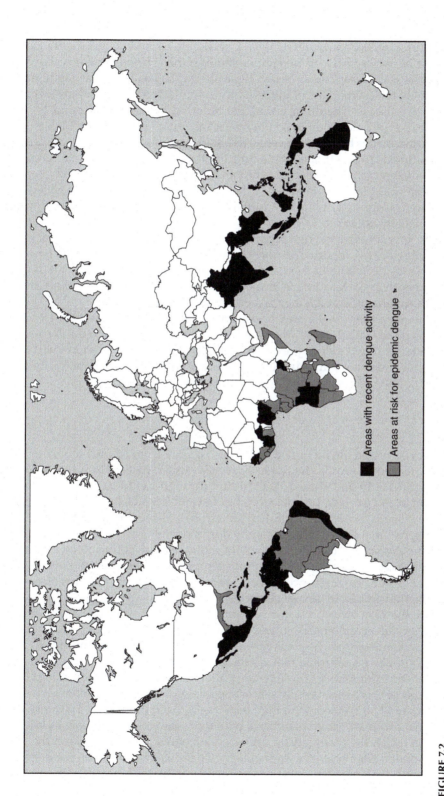

FIGURE 7.2
World distribution of dengue viruses and their mosquito vector, *Aedes aegypti*, in 1995. (Adapted from Dispatch, Dengue hemorrhagic fever, the emergence of a global health problem, *EID*, I, April–June 2, 1995.)

Chikungunya Virus

Chikungunya (pronounced chik-un-GUN-ya) is a disease caused by a virus transmitted among humans by two species of mosquito including the *Aedes aegypti* or yellow fever mosquito and the *Aedes albopictus* or Asian tiger mosquito. The virus has similarities to dengue virus in that the symptoms of the disease are similar and the same species of mosquito transmit the disease. The disease was first reported in Tanzania around 1952, and in the language of an ethnic group in southeastern Tanzania known as the Makonda, the name chikungunya means bent over or contorted. That's an apt description for the infected persons who suffered from extreme joint pain from the disease. They also suffered from fever, muscle pain, headaches, swelling in the joints, fatigue, nausea, and rash.[113]

This small spherical virus is a member of the *Togaviridae* family and is classified as an alphavirus. Once transmitted by the mosquito from an infected to an uninfected person, the symptoms usually appear in 3–7 days.

It was nearly 50 years after the first reported recognized outbreak in Tanzania in 1952 that a large outbreak was reported in the Democratic Republic of Congo, followed by outbreaks in islands of Southeast Asia and the Indian Ocean. The in 2013, chikungunya appeared in strength in St. Martin and throughout the Caribbean Islands. From there, it spread to South America, Central America, and North America.[114]

There are now more than 700 laboratory-confirmed cases of chikungunya infections in at least 44 states in the United States. Most are travel related with most in California, New York, and Florida. Of the locally acquired 200-plus cases, most have been in Puerto Rico. There is no licensed vaccine or antiviral medication that can treat or prevent the infection. Pain medicine, fluids, and rest are used to treat the symptoms.[114]

Zika Virus

The Zika virus was originally identified in 1947 in Uganda from Rhesus monkeys living in the Zika (overgrown) forest. It did not emerge as a human disease in full force until it erupted in Brazil in 2015 and spread like wildfire through Central and South America and the Caribbean. A year later it showed up in the United States during the summer of 2016. The virus belongs to the same family (Flaviviruses) as the dengue virus and WNV. It is then no surprise that the virus is transmitted by mosquitos and here again, the *Aedes aegypti* and *Aedes albopictus* mosquitos play an important role by taking blood meals from an infected person and transmitting it to uninfected persons. These are the same species that spread dengue and chikungunya viruses. There is also evidence that this disease can uncommonly be transmitted by sexual contact.[115]

Infection with the Zika virus produces symptoms in about 20 percent of those infected. Those who do develop symptoms usually exhibit a mild illness with joint pain, mild fever, a skin rash, and conjunctivitis. The virus can also produce a neurologic condition in adults known as Guillain-Barre syndrome that results in muscle weakness or even paralysis in the worst cases. The symptoms begin a few days after being bitten by a mosquito carrying the virus and can last two to seven days. Most people probably don't realize they have a Zika virus infection. The problem with the Zika virus is that it produces a grave risk to the fetus of a pregnant woman. Mothers infected with the virus can give birth to a child with microcephaly in which the brain fails to fully develop while in the womb and the baby's head is much smaller than that of a normal baby. These babies will often have developmental delays, and may experience more severe symptoms that threaten life, and have vision problems including blindness.[115]

The Zika virus outbreak in Brazil in 2015 is thought to have involved around 500,000 cases with more than 1,500 cases of microcephaly or other fetal abnormalities associated with the Zika virus. Since 2015, 60 counties have reported Zika virus transmission. Since September 2017, there have been more than 5400 cases of Zika virus disease with symptoms. Most of these were related travel to areas where mosquitos carry the virus. Most cases have been in New York, Florida, and California. Local transmission of the disease has been seen primarily in Puerto Rico, the U.S. Virgin Islands, and Samoa with over 37,000 infections of which 4,000 were in pregnant women. Zika virus has been spreading locally in the U.S. territory of Puerto Rico since the beginning of 2016 and began spreading locally in the United States in the summer of 2016. There is no vaccine or medicine for treatment. Avoidance of mosquito bites by staying clear of areas with the infected mosquitos, using sprays, netting, screens, and extended clothing, are recommended. As an example, the CDC took the unusual step of advising pregnant women in 2016 to avoid travel to areas of Miami-Dade County in Florida where Zika virus was transmitted by local mosquitoes.[116]

SARS and MERS

Coronaviruses are tiny globe-like virus particles with spikes all around. The SARS (severe acute respiratory syndrome) and MERS (Middle East respiratory syndrome) are two coronaviruses that have caused fatalities in humans. SARS emerged from China in 2002 and outbreak spread quickly to surrounding countries and as far as Canada. The good news is that with international initiatives it was quickly contained with no other reported outbreaks. SARS infections can produce severe symptoms similar to pneumonia leading in many cases to fatalities. It is spread by droplet infection from the sneezing and coughing of infected persons or though contact with surfaces contaminated by infectious droplets. About 800 people died out of the 8000 persons infected in the initial outbreak. First persons with the disease were reportedly infected through direct contact with palm civet cats (toddy cat), wildgame in the markets of Guandong Province in China. The small wild cats were infected with a very similar coronavirus that likely mutated to adapt to humans allowing for human to human transmission. The lethality of this SARS-CoV appears to be related to its ability to block an enzyme in human lungs critical to regulating the balance of body fluid. Blocking this enzyme allows fluid to leak into the alveoli (air sacs of the lung) causing severe respiratory distress.[117]

The MERS-CoV (Middle East Respiratory Syndrome Coronavirus) was first reported in the fall of 2012 in the Arabian Peninsula. The disease causes by MERS-CoV is very similar to that caused by the MERS virus. About 30–40 percent of those contracting the disease died. Persons with renal disease. The virus may have spread from bats and camels to humans. Those that are immunocompromised have chronic lung disease, have diabetes, or are elderly tend to be at increased risk for the most severe forms of the disease. Most cases of MERS have occurred in Saudi Arabia with some infections in surrounding countries such as Jordan, Kuwait, and Yemen. One infected person spread the disease to South Korea causing 36 deaths from the 185 persons infected in 2015. There have no other major outbreaks since then. Over 1000 persons in Saudi Arabia contracted the disease with 1370 cases documented worldwide resulting in 480 deaths.[117]

Influenza

Influenza (flu) is a serious respiratory illness that can quickly spread through a human population infecting the nose, throat, and even the lungs. In some people such as the elderly, children, and people compromised by other health conditions, and pregnancy, the illness can be severe and even cause death. If you have a fever, cough, chills, feel very tired, and have muscle or body aches during the flu season, you may very likely have the flu. The flu is mostly spread by tiny droplets spread through the air when a person with the flu coughs, sneezes, or talks. Less often it can be communicated by touching a surface or object including food previously contaminated by a person with the flu, and then putting your now contaminated hands to your eyes, nose, or mouth. Symptoms can begin 1–4 days after you are exposed to the flu virus, and you can infect others even before symptoms develop and then several days after symptoms develop. You can reduce your risk of exposure to flu by avoiding people who are sick, washing your hands often, wearing an antiviral face mask, and getting a vaccine. The effectiveness of flu vaccines varies from year to year depending the antigen shifts and drifts of the virus. The effectiveness of the vaccines in flu season of 2018 was estimated to be about 30 percent.[117,118]

During a CDC teleconference on January 17, 2018, CDC Director Brenda Fitzgerald said the entire country is experiencing a widespread and intense influenza activity. The flu virus during the 2018 flu season is predominantly influenza A (H3N2) and linked to severe illness that strikes children and persons over 65 very seriously causing many hospitalizations and multiple deaths. The current batch of flu vaccines against H3N2 has been in the low 30s percentage range. However, the recommendation is that people still receive vaccinations because it may reduce the length and severity of illness and reduce the mortality.[119]

Background: There are three types of influenza viruses designated as A, B and C, plus subtypes of A. Influenza types A and B are responsible for the epidemics of respiratory influenza that occur annually and are frequently connected with heightened rates of hospitalization and death.[110] Type C is quite different than A or B because it either produces very mild symptoms or none at all, and consequently has not been involved in any epidemics. Therefore, most public health efforts have been focused on controlling types A and B. Influenza is commonly referred to as "the flu" and is caused by these viruses that attack the respiratory system. There are presently three strains of influenza viruses that are circulating globally and these include two type A viruses and one type B (Figure 7.3). The viruses contain proteins that can be detected immunologically and this serves as a useful tool for identification. The two main proteins used for identification are hemagglutinin (H) and neuraminidase (N). Type A viruses are separated into subtypes based on these two proteins and are designated as A(H1N1) and A(H3N2).

The influenza type A viruses are also capable of undergoing gradual changes with a succession of mutations or undergoing a sudden and significant change in the hemagglutinin (H) or neuraminidase (N) proteins. A gradual shift in the viral proteins is referred to as antigenic "drift," whereas abrupt changes in H or N proteins is called an antigen "shift." Changes in the type B virus appear limited to antigenic drift. Changes in the viral protein through either of these mutation mechanisms permit the virus to evade immune reactions that may have been formed in response to contact with previous but slightly different influenza viruses. The altered viral proteins are no longer recognized by the immune cells and/or antibodies, resulting in a weakened response or no response at all. Influenza A and B viruses are continually expressing antigenic drift resulting in seasonal flu outbreaks. When the more abrupt antigenic shift occurs, entire populations are left immunologically unprotected and pandemics may occur.[120]

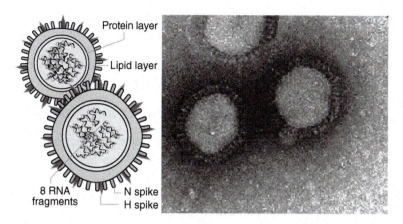

FIGURE 7.3

Photomicrograph of influenza virus. (From World Health Organization (WHO), Emerging and other communicable diseases, Influenza A(H5N1) Facts sheet 188, January 9, 1999.)

The disease: The symptoms of influenza can vary from none to fatal. Influenza is normally characterized by a fever (100°F–103°F); respiratory symptoms that include cough, sore throat, and stuffy nose; muscle aches and pain; and extreme fatigue. The course of the disease is usually self-limited and hospitalization is usually not required. Sometimes the disease produces vomiting and diarrhea, but this is normally in children and gastrointestinal symptoms are not the routine.[109] Most people recover within 1–2 weeks, but complications such as pneumonia may develop as a life-threatening complication. There are about 20,000 deaths annually in the United States with the majority of serious illness and death occurring in the aged, very young, and debilitated. This is probably strain dependent, however, and recent infections with influenza A(H5N1) indicate severe symptoms in other age groups as well. However, the small number of cases involved with this strain may not reflect unreported and/or milder cases.

Epidemiology: Influenza is an ancient disease, first described by Hippocrates in 412 BC. There have been, to this date, more than 30 pandemics with three occurring within the last 80 years. There were epidemics in 1918, 1957, and 1968. The viruses involved were all from animals (pigs or birds). The mortality associated with these pandemics was enormous, with the worst pandemic occurring in 1918–1919 involving influenza A(HINI) known as the "Spanish flu." This disease caused an estimated 500,000 deaths in the United States and 20 million deaths worldwide. The 1957–1958 "Asian flu" or A(H2N2) resulted in 70,000 deaths in the United States followed by the 1968–1969 "Hong Kong flu" or A(H3N2) that killed 34,000 people in the U.S. The emergence of the "Hong Kong flu" in 1968 was relatively mild in terms of mortality, but it continues to circle the globe and to demonstrate antigenic drifting. Since the virus emerged in 1968, there have been an estimated 400,000 deaths worldwide with most of the mortality in elderly people. A(H3N2) remains the most destructive of the currently circulating influenza viruses.[120] An influenza virus, A(H5N1) was isolated from a child in Hong Kong in May of 1997 after the child died from Reye's Syndrome associated with taking aspirin in the presence of the flu virus. This is an avian virus that had a history of killing ducks and chickens. The avian flu incubated in East Asia for 8 years where it killed tens of millions of birds and 68 people. Then suddenly in 2005 and for reasons unknown, it spread north into Mongolia and Siberia and west into the Ukraine and Turkey.[120] There have been more than 232 human cases worldwide since 2003, with nearly 134 deaths. Ninety-three of the cases occurred in Vietnam followed by Indonesia with 18 cases.[121] Health professionals around the globe are concerned that this continuing spread of H5N1 represents a major threat for humans because (1) it is highly virulent; (2) it has shown limited transmission to mammals and some humans, but could mutate to a highly infectious form for humans; (3) it is being spread around the globe by migratory birds; and (4) we have no immunologic defense for it nor a proven effective vaccine.[9]

Following increased surveillance of A(H5N1) in Hong Kong and Guandong Province in China, there appears to be no risk of this virus being transmitted from raw, chilled, or frozen poultry foods. Investigations are now concentrating on the characteristics of the virus isolated from human cases, whether there are other persons who are becoming infected, and other natural reservoirs of the virus.

Additionally, efforts are under way to develop high-growth re-assortment H5 viruses for vaccine production, but no vaccine is being developed specifically for A(H5N1). This is made difficult because the viruses that have been isolated kill the fertilized eggs that are normally used to develop the vaccines. In order to reduce the risk of pandemic in the event the virus should become transmissible human to human, WHO has established a collaborative team that is providing global surveillance, identification of potential pandemic viruses, promotion of high-growth seed virus for vaccine, and the facilitation of vaccine production and distribution.[122,123]

Ebola

Background: Ebola and Marburg viruses belong to a family of viruses called Filoviridae. These viruses have the appearance of long threads or filaments (Figure 7.4) and cause severe and often fatal hemorrhagic fevers in humans and nonhuman primates. Some strains have produced fatalities of nearly 90 percent in human outbreaks. Their extreme pathogenicity combined with the lack of effective vaccines or antiviral drugs classify them as biosafety level four agents. These viruses are even more mystifying because their natural reservoir remains unfamiliar and the pathogenicity of the disease is unknown.[124] The Ebola virus was discovered in 1976 and was named for a river in Zaire, Africa, where it was first detected. The widespread public fear of the virus is probably due to the gruesome symptoms and high mortality, which has led to a flurry of articles, books, and movies. This fear is greatly exaggerated. There have been only 1400 deaths from Ebola in the last 20 years with nearly all of them being in Africa. This is contrasted with the millions of cases of HIV that occur in Africa and worldwide.[125]

The disease: The Ebola virus is usually spread by close personal contact with a person who is symptomatic with the disease. This has most often occurred among hospital workers in Africa where sanitary conditions and infection control were limited and the reuse of hypodermic needles is common. Ebola fever typically starts suddenly 4–16 days after infection with malaise, fever, and flu-like symptoms, which can be followed by rashes, bleeding, and kidney and liver failure. Most patients appear overtly ill, dehydrated, apathetic, and disoriented. The fever

FIGURE 7.4
Ebola virus (simulated).

associated with Ebola appears to begin with chills, headaches, muscle pain and aches, abdominal pain, sore throat, and chest pain. Generalized bleeding occurs with massive internal hemorrhaging of the internal organs, with bleeding into the gastrointestinal tract, from the skin, and even from injection sites as the clotting ability of the blood is diminished. The Ebola virus produces lesions in the liver, kidney, and spleen with areas of necrosis on the organs and the lymph nodes. The death of the patient usually occurs from shock within 7–16 days and is accompanied by extreme blood loss. The virus is spread through bodily fluids and secretions though not through casual contact.[124]

Epidemiology: Infections from Ebola virus were first reported in 1976 when two outbreaks occurred at the same time but in different locations and with different subtypes of the Ebola virus. One occurred in northern Zaire and the other in southern Sudan.[126] The total number of cases in these two outbreaks was 550 with 340 deaths. The case fatality rate from the Zaire subtype Ebola virus was 90 percent and case fatality rate for the Sudan subtype was 50 percent. There was also an outbreak in Sudan with the Sudan subtype in 1979 resulting in 22 deaths among the 34 cases identified for a case fatality rate of 60 percent, similar to the 1976 outbreaks. The Ebola Zaire strain emerged again on April 10, 1995 in an outbreak in Kikwit, Zaire. A patient admitted for suspected malaria infected the attending medical personnel with the Ebola virus, which spread rapidly initially but was brought quickly under control by a coalition of international health organizations including CDC and WHO. The outbreak eventually resulted in 293 cases with 233 deaths. The United States escaped a major epidemic in 1989 when a new Ebola virus infected a colony of macaques at a Reston, Virginia, CDC facility that were imported from the Philippines.[127] This new virus was named Reston virus and was shown to be antigenically and genetically different than the African varieties. The Reston virus was highly pathogenic for the primates but produced no disease in the humans although some people developed antibodies to the virus.[124] There was a cluster of chimpanzee deaths in the Tai forest on the Ivory Coast (Cote d'Ivoire) of western Africa in November of 1995. A Swiss researcher contracted the disease and subsequently recovered. Researchers have been actively trapping small animals in a 51.7-square-kilometer section of the canopied Tai National Forest in order to reveal the natural reservoir for the Ebola virus. Although there are plenty of suspects including bats, red Colobus monkeys, and other small mammals, there has been no success in identifying the virus in any of the animals captured.[125] The reservoir for the Ebola virus is reported to extend throughout the rural and urban areas of central Africa and might therefore be a small terrestrial mammal or a flying animal (bat or bird). No good candidate species has yet been identified, despite extensive studies.[128]

The combination of impoverished populations, inadequate public health services, rapid transmission of the disease, and greater incursions into the African forests increase the possibility of more outbreaks. CDC and WHO have developed sensitive enzyme immunoassays and polymerase chain reaction (PCR) assays for the detection of filoviruses. The availability of rapid and reliable diagnostic tools will permit more effective disease surveillance programs that are necessary to minimize filovirus outbreaks.

AIDS/HIV

Background: AIDS is a virus that has been recently discovered to have originated from an endangered species of chimpanzee from Central Africa. It appears that the transmission of the virus from chimps to humans is related to rainforest destruction and the practice of human chimpanzee consumption.[129] The AIDS virus (Figure 7.5) belongs to a special group of viruses known as retroviruses and is referred to as human immunodeficiency virus (HIV). Its genetic information is not encoded as DNA, but instead as ribonucleic acid (RNA) and therefore has to be reverse transcribed into DNA. The tools for this are provided by the host cell itself, except for a little helper protein (reverse transcriptase) that the virus has brought with itself. When the virus enters the body, it prefers to attack certain types of white blood cells that are a component of the cellular immune system. These cells are called helper T cells and are an integral component of our immune system. The AIDS virus almost exclusively focuses on these white blood cells since these helper T cells have CD4 molecules on the surface to which the AIDS virus binds. The virus attaches itself to a special protein (CD4) on the surface of the helper T cell resulting in a fusion of the viral membrane to the host T cell's membrane. The viral genetic information is then able to enter the cell and is transferred to the nucleus. This process may be accomplished in less than

FIGURE 7.5
Human immunodeficiency virus (HIV).

12 hours after infection. The viral piece of DNA is then inserted randomly into the host DNA and ready for transcription. At the beginning of AIDS, the viral DNA is being transcribed to form many RNA molecules. The signal that causes this is yet unknown. The accumulating RNA is transported to the cytoplasm of the cell, where it can begin making proteins. The RNA, with the help of the host's resources, begins to make many copies of the different parts of the AIDS virus (the protective shell and the helper and anchor proteins). Once everything has been copied, thousands of partial virus particles migrate to the surface of the cell membrane and fuse with it. Finally, a copy of the RNA genetic information is added to the viral particle. Then this section of the cell membrane turns inside out and new viruses leave the cell.[130]

The disease: HIV is transmitted most commonly by sexual contact with an infected partner and can enter the body through the vaginal lining, vulva, penis, rectum, or mouth. Since the virus appears in the blood and many body fluids, it can be transmitted by infected blood as through contaminated needles. The risk of spreading HIV through transfusions has been significantly reduced since the introduction in 1985 of heat-treating techniques. HIV has been transmitted to fetuses during pregnancy and birth. Fortunately, the administration of the drug AZT during pregnancy has reduced the risk of transmitting the virus to the fetus dramatically from its non-treatment level of 20–30 percent infection rates. HIV has also been transmitted to babies through the breast milk of HIV-infected mothers. There is no evidence that casual kissing or contact will result in infection. Simply touching objects such as telephones, towels, and the like that have previously been used by HIV infected individuals will not spread the disease to uninfected individuals. There is also no known insect vector for HIV.[131]

Many people remain without symptoms for months or years after acquiring the infection. Symptoms do develop in some people within 1–2 months after infection. The symptoms may include enlarged lymph nodes, fever, headache, and malaise. The more severe and lasting manifestations of HIV infection may surface from months to years after infection, but during the asymptomatic period the virus is infecting and destroying T4 cells. Gradually, a number of symptoms may emerge that include prolonged enlargement of lymph nodes, energy and weight loss, recurrent sweating and fevers, skin rashes, or flaky skin, yeast infections, and pelvic inflammatory disease. A number of victims develop herpes infections with painful sores about the mouth and genitals, or a more painful disease known as shingles. During the early and intermediate periods of the disease, diagnosis depends on the detection of HIV antibodies in the blood. The antibodies may take as many as 6 months to appear in sufficient quantities in the blood to be detected by available tests such as ELISA and Western Blot.

The disease will often advance to a stage referred to as AIDS. The CDC have defined AIDS to include those people who have evidence of HIV and also fewer than 200 T4 cells. The definition also includes more than 25 opportunistic infections that have historically affected people with AIDS because of the immuno-compromised state. One of the most common and frequently fatal opportunistic infections is *Pneumocystis carinii* pneumonia. The opportunistic infections produce a myriad of debilitating symptoms from respiratory distress, severe headaches, extreme fatigue, nausea, vomiting, to wasting and coma. The continued decline in T4 cells and associated immunity increases the opportunity for the development of cancer. Many cancers may develop in AIDS patients, with some of the most common being Kaposi's sarcoma or lymphomas. Kaposi's sarcoma is often evidenced by pigmented spots that develop on the skin.[130]

Although treatments continue to enter the war on AIDS including reverse transcriptase inhibitors (AZT), and protease inhibitors, these drugs are likely to produce severe side effects, while HIV may become resistant to many of them. A number of drugs are also available to treat many of the opportunistic infections that are likely to arise when the immune system is compromised. There are currently 86 medications used to treat or reduce the symptoms of HIV.[132] Treatment with HIV medicines is called antiretroviral therapy (ART). ART is recommended for everyone with HIV. People on ART are usually treated with a combination of HIV medicines (called an HIV regimen) every day. Treatment normally begins with an HIV regimen that includes three HIV medicines from at least two different drug classes. These are treatments are not a cure but can help people live longer and reduce the risk of transmitting the disease to another person.[133] No vaccine has been successfully introduced to date, and the best recommendation is to avoid high risk behaviors that include unprotected sex or the sharing of needles.

Epidemiology: A joint surveillance effort by UNAIDS and WHO estimated in 1997 that over 30 million people were living with HIV infection.[131,134] UNAIDS estimated that in 2006 there were 40.3 million people living with HIV and 25 million people have died of AIDS.[123] This translates to 1 in every 100 sexually active adults in the ages 15–49 worldwide with about 90 percent of the HIV-infected people living in the developing nations. This figure also includes over 1 million children under the age of 15. HIV continues to spread rapidly at the rate of 16,000 new cases daily. AIDS is responsible for about 20 percent of all deaths since the epidemic began in the 1970s. More than two thirds of the total number of people in the world living with HIV are from sub-Saharan Africa (Figure 7.6). This is contrasted against the 6% decrease in new AIDS cases in the United States for the first time ever in 1996 and the continuing decline since then. Unfortunately, AIDS cases rose in some sections of American society including heterosexual African-Americans and Hispanics.[131,135] Today in the United States there are more than 1.1 million people living with HIV and more than 700,000 have died since the beginning of the epidemic in 1981.[10] The number of HIV infections has decreased by 18% from 2008 through 2014, and the HIV age-adjusted death rate has dropped by more than 80 percent mostly because of ART. Still, it remains the ninth leading cause of death in those aged 25–44 years.[11]

West Nile Virus

Background: The WNV was first isolated from a woman in the West Nile district of Uganda in 1937. In the 1950s, the virus became associated with severe cases of meningitis and encephalitis. During the early 1960s, WNV was also shown to afflict horses. The WNV virus has been found in Africa, Europe, the Middle East, west and central Asia, Oceania, and then in North America. WNV made its first appearance in the United States in 1999 and appeared as encephalitis in horses and humans. Since then, the virus has spread by birds and mosquitoes to every state except Hawaii, Alaska, and Oregon.[136]

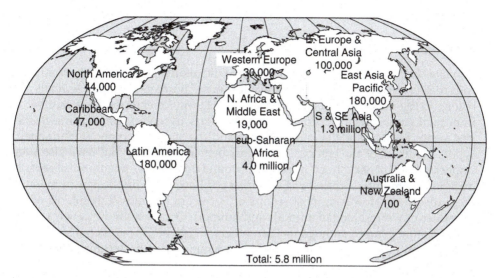

FIGURE 7.6
Distribution of HIV-infected persons globally for 1997. (UNAIDS and WHO, Report on the Global HIV/AIDS epidemic, December 1997. http://www.unaids.org/unaids/document/epideino/report97.html.)

The disease: People are usually exposed to WNV through the bite of a mosquito that has fed on an infected horse or bird. However, a very small number of cases have been linked to organ transplants, breastfeeding, and pregnancy from mother to child.[136] About 80 percent of those people will remain asymptomatic. About 20 percent of those exposed will be develop mild symptoms in 2–15 days after exposure that include fever, headache, tiredness, and body aches, occasionally with a skin rash (on the trunk of the body) and swollen lymph glands. This is self-limiting and does not usually require treatment. This is known as West Nile fever and can last days to weeks. In about 1 of every 150 people infected, the disease may present as encephalitis (inflammation of the brain) or meningitis (inflammation of the membrane around the brain and spinal cord) in humans and horses and may also cause mortality in domestic and wild birds. Many state public health agencies have programs to collect and test dead birds for WNV as part of an active surveillance program. Severe symptoms appear in about 1 in every 150 people infected with WNV. The symptoms can include fever, headache, stiffness of the neck, disorientation, coma, vision loss, numbness, paralysis, and even death. Those surviving may have lasting neurological effects.[137]

Epidemiology: There have been a number of outbreaks of WNV encephalitis in humans, horses, and birds reported since the early 1990s. Outbreaks among humans have been reported in Algeria, the Czech Republic, the Democratic Republic of the Congo, Romania, Russia, and more recently in the United States from 1999–2003.[136] Neuroinvasive WNV characterized by encephalitis and/or meningitis was reported in 9 states with a total of 16 cases for 2006 as reported to CDC by July 25, 2006.[138] The typically less severe form of WNV characterized as West Nile fever without a neuroinvasive component appeared in 9 states during the same period in 2006. It is very likely that far more cases occurred but were underreported because persons may not have sought medical attention, or their cases were not authenticated by laboratory tests, or the attending physician did not report mild cases.[138]

Bacteria

Escherichia coli

Background: *Escherichia coli* are gram negative, facultative anaerobic, short straight
rods that characteristically inhabit the intestines of humans and other animals and
belong to the family Enterobacteriaceae. This family of enterics has a number of
important pathogens including agents of bacillary dysentery (*Shigella dysenteriae*),
gastroenteritis (Salmonella and Shigella), and the agents of typhoid (Salmonella
typhi). Members of the enterics cause gastroenteritis, mostly, but have also been
implicated in urinary tract infections, wound infections, pneumonia, septicemia, and meningitis. Strains of *Escherichia coli* have been implicated in many of
these disorders. However, most strains of *E. coli* are harmless inhabitants of the
intestines of warm-blooded animals and humans. Occasionally, however, a strain
develops that is more pathogenic than others. *E. coli* 0157:57 is such a strain. The
combination of letters and numbers in the name of this bacterium refers to the
specific markers found on the surface of the organism and permits it to be distinguished by serological and other techniques from the hundreds of other strains.
Infections with *E. coli* O157:H7 were first described in 1982 and it has emerged
along with a few other strains of *E. coli* as the predominant cause of hemorrhagic
colitis in humans. The strains of *E. coli* capable of causing hemorrhagic colitis are
referred to as Enterohemorrhagic Escherichia coli (EHEC).[14]

The disease: *Escherichia coli* O157:H7 is pathogenic for humans and has characteristically produced bloody diarrhea with abdominal cramps; sometimes the infection causes non-bloody diarrhea with very few symptoms. There is usually little
or no fever and the illness resolves in 5–10 days. In some persons, particularly
children under 5 years of age and the elderly, the infection may progress into a
more severe and life-threatening form of the disease known as hemolytic uremic
syndrome (HUS). HUS is characterized by the destruction of red blood cells and
kidney failure. HUS is a life-threatening condition usually treated in an intensive
care unit with blood transfusions and kidney dialysis. Rapid and intensive care
such as this can reduce the mortality rate to lower than 5 percent. About one third
of the persons with HUS have abnormal kidney function many years later, and
a few may require long-term dialysis. Another 8 percent of persons with HUS
have other lifelong complications such as high blood pressure, seizures, blindness, paralysis, and sometimes removal of part of the bowel. In the United States,
HUS is the leading cause of acute kidney failure in children, and most cases are
caused by serotype 0157:H7. This serotype produces a group of cytotoxins known
collectively as Shiga-like toxins (SLTs) that are thought to be factors producing
some of its virulence.[14] The toxins are very similar to toxins produced by a pathogenic Shigella, another member of the family Enterobacteriaceae.[139]

Epidemiology: The American public was startled into awareness with this disease in 1993
following a foodborne outbreak traced to the undercooked hamburgers eaten at a
fast food chain restaurant.[13] The outbreak involved the infection of 700 persons from
four different states with 51 of these persons developing HUS and 4 people dying
from the syndrome. Serotype O157:H7 was implicated in this outbreak, but it was
first isolated 11 years earlier in 1982. There were more than 30 outbreaks to follow
in the 10 years before the 1993 tragedy.[140] Since the 1993 outbreak, there has been
increased reports of O157:H7 infections giving rise to more than 60 outbreaks. Part of

this increase may be due to elevated awareness among doctors, microbiologists, and consumers. As there is little evidence that this organism caused infections before the 1970s, it is an example of the emergence or evolution of a foodborne pathogen.

The majority of infections with serotype O157:H7 have come from eating under-cooked beef products, but many other sources of infection have been identified. Outbreaks have been associated with acidic foods such as fresh-pressed cider in which 27 persons were infected with serotype O157:H7.[141] The apples used to make the unpasteurized cider without preservatives were thought to have been con-taminated with cow manure.[142] In another outbreak involving several restaurants and 48 infected persons, the implicated food was mayonnaise, which is also acidic to a pH of 3.6.[143] The serotype O157:H7 organism was thought to have been trans-ferred to the mayonnaise by cross contamination with meat products, and this bacteria has been shown capable of surviving up to 55 days in this acidic environ-ment.[143] Serotype O157:H7 infections have also been traced to both drinking water and recreational water. An outbreak in Missouri involving more than 240 people with 4 deaths occurred in 1989 and was associated with contaminated drinking water.[144] Fecal contamination of lake water in Portland, Oregon was thought to be responsible for an outbreak of serotype O157:H7 among swimmers, where 21 chil-dren were infected. Based on these findings, an infection with serotype O157:H7 is most often associated with eating undercooked beef products. Contamination can also come from drinking contaminated raw milk or other unprocessed raw beverages that might come into contact with human or bovine feces. Swimming in or drinking water contaminated with sewage is also a risk. Finally, the organism has been found capable of surviving in highly acidic foods once thought to be low risk for the transmission of bacterial foodborne infection.

Lyme Disease

Background: Lyme disease is caused by the spirochete *Borrelia burgdorferi*, a gram nega-tive, slender, flexible bacterium that is helically coiled (Figure 7.7). The organism is anaerobic and fermentative in its energy metabolism and it is spread to humans by

FIGURE 7.7
The spirochete, *Borrelia burgdorferi* (Simulated).

FIGURE 7.8
Female *Ixodes scapularis*. This tick species is capable of transmitting Lyme disease to humans.

the bite of ticks of the genus Ixodes (Figure 7.8). The tick responsible for the spread of Lyme disease in the Northeastern and North Central United States is known as the deer-tick and characteristically feeds on the white-tailed deer, the white footed mouse, other mammals, birds, and even reptiles.[145,146] The black-legged tick is responsible for the transmission of the disease along the Pacific coast and in the southeastern United States. The Ixodes ticks are very small, and about the size of the head of a pin in their lymph and larval stages. The adult stages are just slightly larger.

The lymph stage is most likely to be the infective stage of the vector since it actively feeds, and its smaller size allows it to remain undetected for the 24–48 hours needed to complete a blood meal and transmit the infective spirochete. The tick is usually transmitted from the tips of grass and brushes to persons who brush against them. The ticks will search out any unprotected body parts but often attach to hidden and hairy areas such as the scalp, groin, or armpits. Here they will insert their mouth parts into the skin and begin a several-day process of feeding on the blood, which enlarges their bodies. The life cycle of these ticks is complex and requires 2 years to complete. Typically, the adult ticks will feed on large animals such as the deer in the fall and early spring. The females will eventually drop to the ground where they will lay their eggs, which hatch into larvae by the summer. The larvae will feed on mice and other small animals and birds throughout the summer and fall. They will become inactive through the winter and will molt to nymphs in the spring. This stage will feed on small rodents, other small mammals, birds, and the unprepared human that comes into contact with them on the tips of grass and brush, usually at the forest edge, but also lawns and gardens. During the normal cycle, the nymphs will develop into adults, completing the two-year cycle.

The disease and epidemiology: Lyme disease was first reported in 1975 near Lyme, Connecticut, following a mysterious outbreak of arthritis. Reports of Lyme disease escalated since then, placing Lyme disease in the category of an emerging disease. First reported in 1975, the Center for Disease Control and Prevention reported 226 cases in 1980 and more than 500 cases in 1983. The number of Lyme disease cases has escalated each year since then, climbing from a total of 12,801 in 1997 to 21,273 in 2003. The states experiencing the highest morbidity form Lyme disease and listed in order of magnitude include New York, Connecticut, Pennsylvania, New Jersey, and Massachusetts.[147]

The disease occurs primarily in the coastal regions of the northeastern United States including Connecticut, Rhode Island, New York, Maryland, and Massachusetts. It is also being reported in the western states of California, Nevada, Oregon, and in the many mid-western states. The early stages of Lyme disease are characterized by headache, fever, chills, swollen lymph glands, muscle and joint pain, and a characteristic skin rash (erythema migrans). This red rash is usually circular and appears 3–30 days after the tick bite. The rash often appears on the thigh, trunk, groin, or armpits. This rash (or rashes) will expand to include a large area and will often show clearing in the center to give a bulls-eye appearance. Diagnosis of Lyme disease is made difficult in the absence of the typical rash, and serologic tests may be required. Untreated, Lyme disease may progress to arthritis with pain and swelling in the large joints, paralysis of the facial muscles, and meningitis characterized by stiffness in the neck, fever, and severe headache. Lyme disease rarely results in death but chronic Lyme disease can lead to permanent damage to joints or the nervous system.

A newly discovered bacterium carried by the same tick that carries Lyme disease is also emerging as a significant agent of disease. The organism causes human granulocytic Ehrlichiosis (HGE). It produces high fever, headache, muscle aches, chills, and sometimes death. Most cases have been identified in Wisconsin and Minnesota, but the disease can be found everywhere there is Lyme disease, and about 10 percent of patients with Lyme disease may also have HGE. This is important, because HGE responds only to tetracycline and doxycycline.[148]

Prevention is best accomplished by avoiding tick-infested areas in the summer periods. Wearing light-colored clothing that covers arms and legs and is tucked in at the sock and belt line (with tape at the sock line) will help to prevent the ticks from crawling underneath clothing while making them easier to spot against a light background. Spraying with DEET on exposed skin surfaces other than the face will also be helpful. A vaccine has been approved by the FDA called LYMErix.[149] It must be administered in three injections spread out over the course of several months. It is limited to people 15–70 years of age and has only an 80 percent rate of effectiveness. Booster shots may also be necessary to keep the vaccine effective. The FDA recommends that all other preventative measures be taken whether or not one has received the vaccine. One out of five people who have had the vaccine will still be at risk for acquiring the disease.

Streptococcus

Background: Streptococci are gram positive cocci (spheres) arranged in chains or in pairs and include many species that are pathogenic. There are several groups of streptococci based on serological reactions to surface antigens according the original work of Lancefield in 1933. These groups include A–H, J, and K. The Group A streptococci (GAS) are commonly found in the throat and on the skin. They may cause no symptoms but may also cause infections that range from mild to severe and even life threatening. The major pathogens are included in groups A and B, and their pathogenicity is associated with certain enzymes and surface proteins including hemolysins, erythrogenic toxins, and M-protein.[150] Hemolysins are enzymes capable of breaking or lysing blood cells and when Streptococci are grown on sheep blood agar, the types of hemolysis produced place these organisms into three groups. There are gamma hemolytic strep that produce no hemolysis, alpha hemolytic strep that produce a zone of green discoloration around the colony, and beta hemolytic strep that produce a zone of clear hemolysis around the colony. Beta hemolytic, group A strep are most often associated with strep throat infections. The ability to hemolyse blood cells is associated with powerful hemolysins such as oxygen-sensitive streptolysin O and oxygen stable streptolysin S enzymes. Streptococci may also produce erythrogenic toxins (A, B, and C) that harm blood vessels beneath the skin and appear to be produced in streptococci that carry a provirus with the genetic materials capable of coding for the production of these toxins.[150] The streptococci may produce a broad of array of enzymes including neuraminidases, hyaluronidases, streptokinases, ATPases, DNAses, and many others that participate in the destruction and invasion of human tissue.

The disease: The GAS produce a variety of diseases that include strep throat, impetigo, and scarlet fever. Occasionally, the strep invade deeper tissues such as blood, lungs, or even fat and muscles. The more severe of these invasions results in necrotizing fasciitis and/or streptococcal toxic shock syndrome. The popular media characterized the strep that caused necrotizing fasciitis as "flesh-eating bacteria" because of their ability to produce a progressing destructive infection of the underlying tissues. Streptococcal toxic shock syndrome is caused by a highly invasive strep that produces shock and injury to internal organs such as the kidneys, liver, and lungs. Necrotizing fasciitis usually begins with a trivial or even unnoticed trauma that appears in 24 hours as a lesion with swelling and redness. This is a deep-seated infection of the subcutaneous tissue that progressively destroys the underlying connective tissue and fat while sometimes sparing the skin and muscle. The early signs and symptoms of necrotizing fasciitis include fever, severe pain and swelling, and erythema (redness) at the site of the wound. The wound deeply colors to blue with blisters that contain a clear yellow fluid. Within 4–5 days, the purple areas become gangrenous, and at 7–20 days, dead skin separates at the margins of infection revealing marked necrosis of the underlying subcutaneous tissue. In a few cases, this process quickly escalates until several large areas of gangrenous skin and the patient may become dull and unresponsive due to the toxins in his/her system. Many of the present GAS cases of necrotizing fasciitis inevitably lead to severe systemic illness with high morbidity even in otherwise healthy patients who receive extensive support in the form of antibiotics, dialysis, and surgical techniques.[150] These recent developments along with the increased mortality from GAS suggest the emergence of a more virulent streptococcus.[151]

Streptococcal toxic-shock syndrome (strep TSS) is defined as any Group A streptococcal infection associated with the early onset of shock and organ failure. The initial symptoms of strep TSS involve sudden and severe pain in an extremity or may mimic peritonitis, pneumonia, or heart attack. About one fifth of the patients develop flu-like symptoms with fever, chill, nausea, vomiting, and diarrhea. Fever and evidence of soft tissue infection are present in most patients, with a majority of these progressing to necrotizing fasciitis.[150] Patients are often admitted already in shock, and despite treatment, this shock continues combined with renal dysfunction and acute respiratory distress in about 55 percent of the patients. Mortality rates of 30–70 percent have been reported.[152]

Epidemiology: Beginning in the 1980s there has been a sudden elevation in the reporting of a highly invasive Group A streptococci infection with or without necrotizing fasciitis associated with shock and organ failure. The number of these cases appeared to increase from almost none to over 2,000 per year.[153] There are about

10,000–15,000 cases of invasive GAS in the United States annually resulting in 2,000 deaths. The mortality rate for streptococcal TSS is about 60 percent of the 2,000–3,000 cases reported per year. Annually, about 20 percent of the 500–1,500 patients who acquire streptococcal fasciitis have died. These figures should be contrasted to the millions of people who acquire strep infections annually. In fact, GAS is an organism with historic significance that has produced many outbreaks in the past, and the number of people involved were far greater than the several thousand cases of invasive GAS that are presently seen on a yearly basis. The argument has been made that GAS may have become more virulent and so produced these severe cases. This, however, may not be the only reason for the more intense invasions. Stevens argues that the emergence of a more virulent strain of GAS would have produced many more cases in the form of an epidemic.[151] It has been suggested that people are becoming more susceptible to GAS infections because of the use of nonsteroidal anti-inflammatory drugs (NSAIDs) such as ibuprofen. These observations have been anecdotal but reported from more than one medical group.[153] According to Stevens, NSAIDs block the production of prostoglandins, prevent fever, and encourage overproduction of certain chemicals (TNF or tumor necrosis factor) in macrophages that mask symptoms associated with strep TSS, which contributes to shock and organ failure.[154]

Tuberculosis

Background and disease: TB is a chronic infectious disease of the lower respiratory tract caused by *Mycobacterium tuberculosis*, a slender, acid-fast rod with cell walls containing high lipid levels. The slow-growing bacilli are transmitted by aerosols from persons with active disease and are obligate aerobes and intracellular parasites that survive inside alveolar macrophages in the lungs. After multiplying in the macrophages, they enter the lymph nodes through the lymph channels where they continue to multiply intracellularly. They can spread from here to other organs including the lungs. Symptoms normally begin to develop at this stage from a cell-mediated immunity that walls off the pathogen within multinucleated giant cells surrounded by lymphocytes and macrophages. These structures can be seen on X-ray and are known as tubercles (Figure 7.9).

FIGURE 7.9
X-radiograph of patient with tuberculosis.

Such tubercles may become scarred or calcified over time while the bacilli remain dormant but alive within. The disease is clinically manifested when there is an immune lapse or re-infection that allows the organisms to multiply. This stimulates the immune response causing the tubercles to break down and erode through the bronchi wall spilling mycobacterium-laden liquid caseum into the bronchi. The characteristic cough of active TB derives from this process and is also the reason why mycobacterium appears in the aerosols created from the cough of an infected person.[150]

Epidemiology: In 1999, there were 18,000 cases of TB in the U.S., with 1 percent having been resistant to antibiotics.[155] TB causes more death globally than any other single infectious disease with more than 8 million cases being reported each year.[155] The number of TB cases increased by 20 percent from 1985 through 1992 due in part to the onset of the HIV epidemic, an increase of TB cases among foreign-born people, and the appearance of TB strains that are resistant to multiple drugs.[155] Since 1992, the incidence of TB decreased by 40 percent with a decline every year to a level of 5.6 cases per 100,000 for the lowest recorded incidence in the country.[155] It is thought that this reduction may be due to more effective TB-control programs that lead to prompt identification and treatment of persons with efforts to make certain that therapy is completed.[155] After the introduction of antibiotics, TB was thought to have been under control, but the disease persisted in the poorer reaches of society (inner city dwellers, the homeless), and it is re-emerging dramatically among the foreign born. TB was declared a U.S. public health emergency in 1992, and the CDC is attempting to help control the spread of the world's top infectious disease killer. TB kills over 3 million people worldwide each year, and many more become ill from it.[156] Unfortunately, a WHO report released in March of 1995 reveals that efforts to control TB globally have been ineffective and that drug-resistant strains of *M. tuberculosis* are on the rise.[30] Even though chemotherapy and effective measures of prevention exist, TB is increasing in the developing world as well as in the industrialized world.[157] WHO estimates 8 million new cases and 3 million deaths per year from TB with 450,000 of those deaths in children under 15 years of age.[158] WHO estimated that we will see 90 million new cases and 30 million deaths with annual rates in infection in developing countries exceeding 2 percent.[159] South Africa currently has the highest rate of TB in the world and little is being done to curb the outbreak. There are 350 cases of TB per 100,000 population in South Africa with a rising incidence of "multi-drug-resistant" TB and a death rate of 54 percent of those who develop the disease. In Russia, there were 100,000 cases in a year with 30 percent having been resistant to antibiotics.[160] Certain environmental factors appear to accelerate the disease. These factors include crowded hospital rooms shared with TB patients, lack of sanitary conditions, unfiltered air, closed windows at night, and crowded living and working conditions.[161]

Patients often feel better after a few weeks of medicine and stop taking it. The consequence is a reinfection with TB that has a greater probability of being resistant to medication. World TB Day, which commemorates both the struggles and successes in the worldwide fight against TB, was March 24, 2006. TB claims the lives of more than 1.7 million people each year and hinders many of the 15.4 million individuals currently afflicted with this disease.[162] The reasons for this continued level of infection may be related to co-infection with HIV, the steady increase in the

number of multi-drug-resistant cases of TB, and the realization that available pharmaceuticals may be failing.[162] To combat the outbreaks and re-emergence of TB the
WHO report titled "Stop TB at the Source" recommends international adoption of
a control strategy called Directly Observed Treatment-Short-Course (DOTS), which
is intended to assure that all patients complete their entire course of treatment. This
program requires that patients take their medicine under the eyes of a guardian
every day for 6 months and are given regular sputum tests to see if the pills are
working. The program has been effective in New York where TB has been reduced
by over 20 percent to about 3000 new cases per year. Unfortunately, results from
studies conducted by the WHO have questioned the effectiveness of DOTS.[162] There
is increasingly more support for a new method of treatment called DOTSPlus, which
requires taking up to seven drugs daily for 18–24 months.[162] Despite these efforts,
there have been a small number of patients resistant to all anti-TB drugs. Starting
in 1993, the New York Health Department has placed patients who resist taking the
complete regimen of medicine under "detention until cured."[143] Multiple resistant
strains of TB will become increasingly less responsive to even the more vigorous
treatment programs such as DOTSPlus.[163] Seventeen million new cases and 60 million deaths due to multi-resistant TB are expected between 1998 and 2030.

Parasites

Cryptosporidium

Background: Cryptosporidium is a single-celled microscopic protozoan parasite
   that belongs to the Class Sporozoa. These organisms are not normally motile
   and asexually reproduce by multiple fission and sexually by gamete fusion. The
   resistant form of the parasite is called an oocyst, which is characterized by an
   outer protective shell that protects the organism against environmental extremes
   such as heat, cold, dryness, and chemical insult. There are several species of
   Cryptosporidium, but there is only one pathogenic species, *Cryptosporidium
   parvum*. This organism is widely distributed in the intestines of animals and
   humans, and normally causes infections in humans by the fecal oral route. It
   is estimated that as few as 30 or even one oocyst(s) may cause infection when
   swallowed.[164] Cryptosporidium is therefore a highly infectious enteric parasite,
   widely distributed, that is resistant to chlorine and difficult to filter thereby
   making it a serious threat to water supplies.

Disease: Persons who swallow Cryptosporidium cysts may develop symptoms
   within 2–10 days. The oocysts reach the upper small bowel where they excyst and
   produce four infectious sporozoites that attach to the surface epithelium of the
   digestive tract and reproduce, forming more oocysts and sporozoites. Many of
   the organisms are excreted while others maintain the infection. The symptoms
   include watery diarrhea, stomach cramps, nausea, and a slight fever. These symptoms are self-limiting and will normally last for 2 weeks or less. The organism
   may be shed from the intestines for up to 2 months after the initial illness and so
   spread this disease to others through fecal-oral transmission. Persons who have
   a weakened immune system such as those with HIV/AIDS, cancer and transplant
   patients, or those on immuno-suppressive drugs, are at increased risk from infection, and may develop serious and life-threatening illness from this organism.
   There is no pharmaceutical treatment for Cryptosporidiosis and the more serious
   illness is managed by fluid replacement.[165]

Epidemiology: Contamination of drinking water by Cryptosporidium is a growing concern especially after the outbreak in Milwaukee in 1993 that affected some 400,000 people. There have been five major outbreaks associated with public water supplies of Cryptosporidium gastroenteritis in the U.S. and seven in the UK since 1983.[8] There have also been numerous smaller outbreaks. Most of these outbreaks have been drinking and recreational waterborne in nature. Cryptosporidium has also been spread by unpasteurized cider, and by person-to-person in households, day care centers, and hospitals.[165,166] Cryptosporidium is found in animal droppings and human feces, soil, drinking water and recreational water, food, hands, and surfaces contaminated by such wastes. Many persons have become infected even swimming in chlorinated and filtered pools because the organism has the capacity to survive chlorination and to pass through most pool filters. Cryptosporidium was first reported as a human pathogen in 1976 and came to prominence in the United States in 1982 with the increased infections in the highly immunocompromised population of people living with AIDS.[167] Cryptosporidium is not a new organism but is thought to have emerged because it was newly spread among immunocompromised patients and became symptomatic.[168] In many areas of the world, the organisms has been shown to have infected a large portion of the population. In some areas of Brazil, as many as 90 percent of the children under five have been infected, and over 50 percent of the children in certain rural areas of China show serological evidence of infection.[169,170] Similar seroepidemiologic studies suggest that 17–32 percent of persons in Virginia, Texas, and Wisconsin with healthy immune systems showed evidence of Cryptosporidium infection by the time they reached young adulthood.[171]

Malaria

Background: Malarial diseases are caused by protozoan parasites belonging to the genus Plasmodium. There are four species known to infest humans: *P. falciparum, P. vivax, P. ovale*, and *P. malariae*. These parasites are transmitted from human to human by the bite of a female anopheline mosquito in which the parasite has gone through a complex development taking several days to three weeks depending on the ambient temperature. Since the sexual reproduction of the parasite occurs in the mosquito, it is the definitive host, and humans are the intermediate host and reservoir. The female anophelines become infected when they feed on blood from a person infected that contains male and female gametocytes of the parasite. A complex cycle takes place that involves the union of the gametocytes in the stomach of the mosquito and results in the development of slender, microscopic sporozoites that appear in the salivary glands, and are infective for humans (Figure 7.10). This maturation or sporogenic cycle must be completed within the mosquito before transmission to humans can take place. The night-feeding female anopheline takes a blood meal and injects the malarial sporozoites into the human bloodstream. The sporozoites enter the liver where they infect liver cells and begin the process of development. Depending on the species of malaria, the stages in the human show some variation. In general, once in the liver cells, the parasites begin the process of schizogeny (splitting stage) and they divide in the liver cells to produce merozoites. These merozoites are infective for red blood cells. The merozoites enter the bloodstream where they attach to red blood cells and invade them. Inside the cell, the

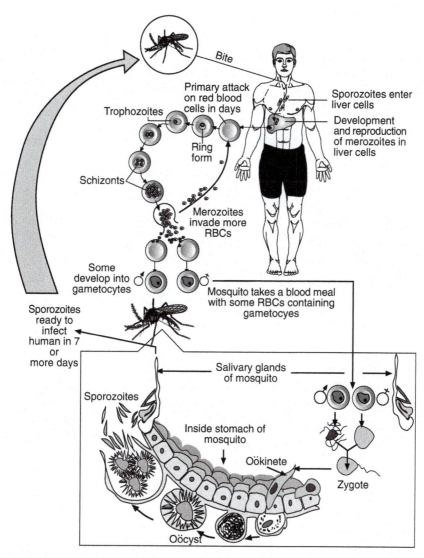

FIGURE 7.10

Life cycle *of Plasmodium* species in female *Anopheline* mosquito.

parasites form the classical signet ring stage (Figure 7.11) and feed on the cells contents as they grow through the stages of trophozoites and schizonts. Female and male gametocytes are also occasionally produced, and these are the sexual forms of the parasite that continue the cycle of the organism when taken into the anopheline mosquito through a blood meal. While developing within the red blood cell, the parasite converts hemoglobin to hemozoin, thereby depleting the hemoglobin content of the blood and giving rise to many of the symptoms of malaria. The parasites proliferate in the red blood cells until they can no longer be contained, and the red blood cells are destroyed as they break free into the blood stream where they can reinvade more red blood cells. This cycle often becomes synchronous causing regularly timed waves of parasitemia from 36 to 72 hours depending on the particular malarial parasite.[172]

FIGURE 7.11
Ring stage of malarial parasite in red blood cells.

Disease: The symptomology and pathogenesis of malaria infection is related to the parasite's stage of growth and the host's parasitemia. High parasitemias result in decreased hemoglobin and a lower oxygen carrying capacity. This quickly leads to fatigue and repository distress. The detritus of lysed red blood cells as well as the explosion of foreign bodies in the bloodstream from logarithmic growth of malaria results in massive immune responses, primarily driven by mononuclear phagocyte released cytokines. Since the explosions of parasitemia occur in regularly timed waves (from as little as 36 hours in *P. falciparum* to 72 hours in *P. malariae*), the characteristic periodic fever develops. The fever is a response to the massive interferon and interleukin release, which subsides as the body clears the infection and the merozoites invade new cells to begin a new cycle. The systemic damage of malaria is not only from septicemic reactions to cytokines, but also the proliferation of knobs on the red blood cell surface, sometimes called pigmotic cells. These projections can result in agglutination of parasites at capillary pathways and ischemia to many organ areas. Untreated infections lead to splenomegaly (enlarged spleen) and particularly in *P. falciparum* to cerebral malaria and death. Children are particularly susceptible to cerebral malaria and the bush doctor cure for urgent cases is an immediate blood transfusion from a close member of the family. This has resulted in a high rate of pediatric AIDS in Africa.

Epidemiology: The WHO estimates that there are 300–500 million people worldwide infected with malaria.[173] More than 805,000 deaths occur in sub-Saharan Africa alone on an annual basis. The impact of malaria on the human condition is even more startling when measured in terms of DALYs. This is an indicator developed to estimate the disease burden so the cost-effective prevention strategies can be developed.[174] On a worldwide basis, malaria has caused the loss of over 35 million DALYs, which is slightly behind TB and slightly less than half of all cancers globally. The contribution of malaria to global disease and disability is of major public health significance.

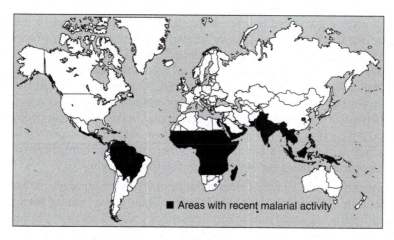

Areas with recent malarial activity

FIGURE 7.12
Global distribution of malaria. (Adapted from Zucker, J.R., Changing patterns of Autochthonous malaria transmission in the United States: A review of recent outbreaks, *EID*, 2, 1, January–March, 1996.)

The majority of malarial transmission occurs in tropical and subtropical countries (Figure 7.12). The re-emergence of malaria as a major threat in these areas is due to a number of reasons. Mosquito control programs have historically been the primary mechanism of attack. However, such programs alone have not been effective in many areas such as sub-Saharan Africa where the applications of spraying were neither complete nor periodic at effective intervals.

Decreased spraying of the homes with DDT is also blamed for the dramatic reemergence of malaria in South America where there has been a geometric growth in malaria incidence since the 1970s.[29] A major reason for this decreased use is the concern for ecological consequences of DDT spraying. Some authors reason that the use of DDT in homes or on nets used over the sleeping areas at night represents barely a fraction of the DDT used in agriculture. In fact, the mount used to treat 404 hectares of cotton could be used to treat the homes for the entire country of Guyana. However, most countries have looked to alternative treatments and found them too expensive. Furthermore, there has been a steady decline in international and U.S. support for malarial control in these countries, and similar reductions in the budgets of the endemic countries themselves.[175] The problem is aggravated by the development of pesticide-resistant mosquitoes in the endemic areas. There is little probability that more money will be allocated for more expensive insecticides.

There are other forces at work to power the reemergence of malaria in the world. These include the appearance of drug-resistant malaria and the predicted expansion of malaria associated with global warming. The malarial parasite has a family of genes called var genes for variability that permit it to change its signature proteins just enough to stay one step ahead of the host antibodies aimed against it in a process called antigenic variation. These var genes account for nearly 6 percent of the parasites genome and permit it to mix and match genes to create millions of antigenic proteins. The malarial parasite developed resistance to anti-malarial drugs as well. In particular, it developed resistance to the inexpensive and highly effective drug chloroquine, and resistant strains began appearing in the 1960s. Unfortunately, the alternative drugs are too expensive for most people in the malaria-endemic areas, and so the disease continues to kill nearly 2 million people a year.[28]

Global warming spurred by greenhouse gases is an internationally debated concern. One of the potential consequences of such a warming trend may be the increase in the distribution and incidence of malaria.[176] A team of four scientists from the Netherlands created a complex mathematical model that can potentially predict the impact of global climate changes on malarial risk. A simulation experiment using this model showed a widespread increase in the potential for malarial transmission.[177] A global increase of only a few degrees could expand malaria's domain from 42 percent to over 60 percent of the planet. Further, the increasing temperatures would increase the metabolism of the mosquito making it a more voracious biter while speeding up the development of the malarial parasite inside the mosquito.[15]

Americans should not be complacent about this re-emergence. Malaria has also begun to reappear in the United States. There have been at least three outbreaks of locally acquired malaria in densely populated areas of the country that point to the danger of a significant outbreak. Malaria was a once a major disease in the United States, with over 600,000 cases occurring in 1914 and more than 125,000 cases in 1934, when systematic reporting was in effect.[178] The processes of urbanization, increased drainage, DDT spraying, screening of houses, and introduction of quinine all contributed to the substantial control of malaria. Malaria was eradicated as a problem disease in the United States by the 1950s. Nearly all cases of malaria detected in the United States since then have been imported. There have been 76 cases of malaria reported from 1957 to 1994 including the three outbreaks occurring in the densely populated areas of New Jersey (1991), New York (1993), and Texas (1994). In these cases, the weather was more humid than usual, there were larger mosquito populations, and the patients lived in areas with large migrant populations from countries with endemic malaria.[172] The CDC received 1,278 malaria case reports occurring among persons in the United States and its territories, representing a 4.4 percent decrease from the 1,337 cases reported in the previous year. In 2003, United States civilians accounted for 767 cases and there were 306 cases among foreign civilians. In recent years, cases among U.S. civilians have increased and cases among foreign-born civilians have decreased.[177,179] The likelihood of widespread malarial outbreaks in the United States is limited by adequate measures of sanitation, health care, and targeted mosquito control. However, a continued warming trend predicted by global climactic changes could expand the range of malaria by permitting the parasite to complete its cycle in the mosquito. Greater pressures would be exerted on the public health support systems within the United States.[172]

Practical Approaches to Limiting the Emergence of Infectious Diseases

The emergence of more than 30 new infectious diseases and re-emergence of many others are creating national and international crises. The control of these diseases is made difficult because of the large number of disease-causing organisms, the enormous diversity of geographic areas from which they can emerge, the potential for rapid global dissemination, and the tremendous number of ecological and social factors that influence emergence.[33,37] Primary responsibility for addressing new and re-emerging infectious diseases rests squarely with the custodians of public health. Indeed, the fundamental maxim of public health must guide current prevention programs: the health of the individual is best ensured by maintaining or improving the health of the entire community.

D.A. Henderson, the driving force in the successful international smallpox eradication program, proposes the establishment of an international network consisting of disease centers strategically located in tropical areas, especially near metropolitan areas. Each center would have a clinical facility, diagnostic and research laboratory, an epidemiological unit to investigate disease and local response capabilities and a professional training unit. This would also require an academic and government laboratory necessary for coordinating data collection and analysis, conducting research, providing feedback and support and when necessary, activate an "international rapid response system."[23] The WHO has created a Division on Emerging Diseases in 1995 to try to deal with a global problem of new and emerging infectious diseases. WHO offices, including the Pan American Health Organization, are developing a regional plan and setting goals that include strengthening regional and national laboratory and surveillance networks; establishing national and regional infrastructures for early warning and rapid response to infectious diseases; promoting and developing applied research for rapid diagnosis, epidemiology and prevention; and strengthening regional capacity for effective implementation of prevention and control strategies.[180]

The CDC began four Emerging Infections Programs in Oregon, California, Connecticut, and Minnesota to deal with some of the nation's under-funded health threats. Each of these states has different projects that range from foodborne illnesses and meningitis to Valley Fever, Ehrlichia, and Cryptosporidium. The programs collaborate with federal agencies, local health departments, academic institutions, professional societies, and international organizations. A Prevention Strategy Plan for the United States was developed with four major goals (1) promptly investigate and monitor emerging pathogens, the diseases they cause, and factors of emergence; (2) integrate laboratory science and epidemiology to optimize public health practice; (3) enhance communication of public health information about emerging diseases and ensure prompt implementation prevention strategies; and (4) strengthen local, state, and federal public health infrastructures to support surveillance and implement prevention and control programs.[6]

Ewald reports that control could be made more manageable if it was possible to focus attention on the pathogens most likely to become dangerous. There is disagreement about the ability to predict and prevent the emergence of these new pathogens.[181] Ewald emphasizes that the identification of these pathogens rests upon two major characteristics that include (1) an ability to spread easily from human to human through vectors and (2) transmission features that select for high levels of virulence. The transmission features that promote virulence include one or more of the following: (a) being vector-borne with humans as part of the life cycle, (b) durability in the external environment if directly transmitted, (c) being needle-borne or borne by attendants in medical care, and (d) being sexually transmitted and being mutation-prone with an attraction to critical cell types.[162] Knowledge of these evolutionary characteristics should help to distinguish the long-term threat of the emerging disease organism and so adjust the amount of resources invested in keeping with the anticipated threat.

Unfortunately, we cannot accurately predict where, when, or how the next disease will emerge or re-emerge and what impact it will have on the individual, society, nation, or the world. Environmental factors such as unusual weather patterns, agricultural practices, deforestation, and other human behavioral factors such as increased population density, compliance and over prescribing of antibiotics, and mere chance will continue to play a significant role in the emergence or re-emergence of disease. Increased vigilance is necessary and the application of more resources is the ultimate answer—contradicting the current erosion of resources to public health being observed nationally and internationally.

References

1. Baylor College of Medicine, Emerging Infectious Diseases, acquired February 2018. https://www.bcm.edu/departments/molecular-virology-and-microbiology/emerging-infections-and-biodefense/emerging-infectious-diseases.
2. National Institutes of health, U.S. Department of Health and Human Service, Infectious Diseases, December 13, 2016. https://www.nih.gov/about-nih/what-we-do/nih-turning-discovery-into-health/infectious-diseases.
3. Mulvaney, K., The scourge of mankind: Infectious diseases, once near\eradication, show remarkable persistence, *The Environmental Magazine*, 42, 42–44, 1999.
4. Walsham, C., Our planet is environmentally sick, *PlanetENN*. http://www.enn.com/planetenn/I 11 196/feature3.htm (accessed November 11, 1996).
5. Bennett, I.V., Holmberg, S.D., Rogers M.F., and Solomon S.L., Infectious and parasitic diseases, *Closing the Gap: the Burden of Unnecessary Illness*, R.W. Antler, H.B. Dull (Eds.), Oxford University Press, New York, 1987.
6. Centers for Disease Control and Prevention, *Addressing Emerging Infectious Disease Threats: A Prevention Strategy for the United States*, U.S. Department of Health and Human Services, Public Health Service, Atlanta, GA, 1994.
7. Stevens, D.L., Streptococcal toxic-shock syndrome: Spectrum of disease, pathogenesis, and new concepts in treatment, *Emerging Infectious Diseases (EID)*, 1(3), 69, 1995.
8. Newman, A., EPA Considering quick action on *Cryptosporidium*, *Environmental Science & Technology*, 29(1), 17a, 1995.
9. U.S. Department of Health and Human Services, Pandemicflu.gov, August 2006. http://www.pandemicflu.gov/general.
10. CDC. *HIV Surveillance Report, Diagnoses of HIV Infection in the United States and Dependent Areas, 2016*, Vol. 28; November 2017. HIV diagnosis data are estimates from 50 states, the District of Columbia, and 6 U.S. dependent areas. Estimates for 2016 are preliminary and are not included in trend calculations.
11. CDC. NCHS. *Deaths: Leading Causes for 2015*; November 2017.
12. Centers for Disease Control and Prevention, Update: Trends in AIDS incidence, deaths and prevalence, United States, 1996, *Morbidity and Mortality Weekly Report*, 46, 165, 1997.
13. Centers for Disease Control and Prevention, Update: Multi-state outbreak of *Escherichia coli* 0157:117 infections from hamburgers-Western United States, 1992–1993, *Morbidity and Mortality Weekly Report*, 42, 258, 1993.
14. Feng, P., Synopsis: *Escherichia coli* Serotype 0157:H7: Novel Vehicles of infection and emergence of phenotypic variants, *Emerging Infectious Diseases*, 1(2), 47–52, 1995.
15. Linden, E., Global fever, *Time*, July 8, 56, 1996.
16. Steere, A.C., Maliwista, S.E., Snydman, D.R., Shope, R.E., Andirnan, W.A., Ross, M.R. and Steele, F.M., Lyme arthritis: An epidemic of oligoarticular arthritis in children and adults in three Connecticut communities. *Arthritis Rheum*, 20, 7, 1977.
17. CDC, Division of Vector-Borne Infectious Diseases, Arboroviral Encephalitides, April, 2006. http://www.cdc.gov/ncidod/dvbid/arbor/eeefact.htm.
18. CDC, Division of Vector-Borne Infectious Diseases, West Nile Virus, April, 2006. http://www.cdc.gov/ncidod/dvbid/westnile/qa/overview.htm.
19. Rohter, L., U.S. is now threatened by epidemic of dengue, *New York Times, International*, 5, September 23, 1995.
20. World Health Organization/WHO/ Vector-borne diseases, October 31, 2017. Retrieved July 7, 2018, https://www.who.int/news-room/fact-sheets/detail/vector-borne-diseases.
21. Altman, L.K., 56 die in a month in Zaire outbreak of Mystery disease, *New York Times International, Al*, May 10, 1995.
22. Mint, C., Insect-borne disease: Curing the carrier, *Science News*, 151, 223, April 12, 1997.

23. Satcher, D., Perspectives: Emerging infections: Getting ahead of the curve, *Emerging Infectious Diseases*, 1(1), 1, 1995.
24. Willq, R.G., Ironside, J.W., Zeidler, M. et al., A new variant of Creutzfeldt-Jakob disease in the UK, *Lancet*, 347, 921, 1996.
25. Centers for Disease Control and Prevention, World Health Organization consultation on public health issues related to bovine spongiform encephalopathy and the emergence of a new variant of Creutzfeldt-Jakob disease, *Morbidity and Mortality Weekly Report*, 45, 295–303, 1996.
26. Brown, P. and Cathala, F., Creutzfeldt-Jacob disease in France, *Slow Transmissible Diseases of the Nervous System*, W.J. Hadlow, S.B. Prusiner (Eds.), Academic Press, New York, 1979, vol. 1, 213.
27. Martens, W.J., Potential Impact of global climate change on malaria risk, *Environmental Health Perspectives*, 103(5), 458, 1995.
28. Angier, N., Malaria's genetic game of cloak and dagger, *New York Times*, C I, August 22, 1995.
29. Roberts, D.R., Laughlin, L.L., Hsheih, P., and Legters, L.J. Global strategies, and a malaria control crisis in South America, *Emerging Infectious Diseases*, 3(3), 1997.
30. Nowak, R., WHO calls for action against TB, *Science*, 267, 1763, March 1995.
31. Drexler, M. and Welch, M.B., Public health-in our battle to stay healthy, we are up against a daunting new array of…clever germs, *The Boston Sunday Globe*, November 21, 1999.
32. Morse, S., Perspectives: Factors in the emergence of infectious diseases, *Emerging Infectious Diseases*, 1, 7–15, 1995.
33. Morse, S.S., Examining the origins of emerging viruses, *Emerging Viruses*, S. S. Morse (Ed.), Oxford University Press, New York, 1993, p. 10.
34. Associated Press, Scientists trace AIDS origin to wild chimps: Gene tests match virus in primates in Cameroon to first known human case, *MSNBS*, May 25, 2006. http://msnbc.msn.com/id/12966623/.
35. Morse, S.S., Regulating viral traffic, *Issues in Science and Technology*, 7, 81, 1990.
36. Morse, S.S., Toward an evolutionary biology of viruses, *The Evolutionary Biology of Viruses*, Morse, S.S. (Ed.), Raven Press, New York, 1994, p. 1.
37. Morse, S.S., Emerging viruses: Defining the rules for viral traffic, *Perspectives in Biology and Medicine*, 34, 387, 1991.
38. Rogers, D.J. and Packer, M.J., Vector-borne diseases, models, and global change, *Lancet*, 342, 1282, 1993.
39. Barbour, A.G. and Fish, D., The biological and social phenomenon of Lyme disease, *Science*, 260, 1610, 1993.
40. Johnson, C.M., Emerging viruses in context: An overview of viral hemorrhagic fevers, *Emerging Viruses*, Morse, S.S. (Ed.), Oxford University Press, New York, 1993, p. 46.
41. Webster, R.G., Bean, W.J., Gorman, O.T., Chambers, T.M., and Kawaoka, Y., Evolution and ecology of influenza A viruses. *Microbiological Reviews*, 56, 152, 1992.
42. Scholtissek, C. and Naylor, E., Fish farming and influenza pandemics, *Nature*, 331, 215, 1988.
43. World Health Organization, Geographical distribution of arthropod-borne diseases and their principal vectors, World Health Organization (WHONBC/89.967), Geneva, Switzerland, 138, 1989.
44. Galntz, M.H., Katz, R.W., and Nicholls, N., (Eds.), *Teleconnections Linking WorldwideClimate Anomalies*, Cambridge University Press, New York, 1991.
45. Manabe, S. and Stouffer, R.J., Century-scale effects of increased atmospheric CO_2 on the ocean-atmosphere system. *Nature*, 364, 215, 1993.
46. Kluger, J. By any measure, Earth is at the tipping point, Time: Special Report on Global Warming, pp. 32–42, April 3, 2006.
47. Harvell, C.D. Kim, K., Burkholder, J.M., Colwell, R.R., Epstein, P.R., Grimes, D.J., Hofmann, E.E., Lipp, E.K., Osterhaus, A.D., Overstreet, R.M., and Porter, J.W. Emerging marine-diseases -climate links and anthropogenic factors, *Science*, 285, 1505, 1999.
48. Kingsnorth, P., Human health on the line, *The Ecologist*, 29, 92, 1999.
49. Epstein, P.R., Ford, T.E., and Colwell, R.R., Marine ecosystems, *Lancet*, 342, 1216, 1993.

50. Anderson, D.M., The fifth international conference on toxic marine phytoplankton: A personal perspective. *Harmful Algae News International Marine Science (UNESCO)*, Paris, France, Vol. 62(suppl:6–7), 1993.
51. The News Hour with Jim Lehrer, An essay in the latter half of the show on the troubled history of Washington D.C. in the 20th century, February 17, 1995.
52. United Nations, *World Urbanization Prospects*, 1990, United Nations, New York, 1991.
53. Han, J.S., Short, M., Taylor, M.E. et al., Species-specific diversity among simian immunodeficiency viruses from African green monkeys, *Journal of Virology*, 65, 2816, 1991.
54. Gan, F., Yue, L., White, A.T. et al., Human infection by genetically diverse SIVSM-related HIV-2 in West Africa. *Nature*, 358, 495, 1992.
55. McNeill, W.H., *Plagues and Peoples*, Anchor Press/Doubleday, New York, 1976.
56. Centers for Disease Control and Prevention, *Aedes albopictus* introduction into continental Africa, 1991. *Morbidity and Mortality Weekly Report*, 40, 836, 1991.
57. Moore, C.G., Francy, D.B., Eliason, D.A., and Monath, T., *Aedes Albopictus* in the United States: Rapid spread of a potential disease vector, *Journal of the American Mosquito Control Association*, 4, 356, 1991.
58. Mitchell, C.J., Niebylski, M.L., Smith, G.C. et al., Isolation of eastern equine encephalitis virus from *Aedes albopictus* in Florida, *Science*, 257, 526, 1991.
59. Wilson, M., Perspectives: Travel and the emergence of infectious disease, *Emerging Infectious Diseases*, 1, 2, 1995.
60. LeDuc, J.W., Childs, J.E., and Glass, G.E., The hantaviruses, etiologic agents of hemorrhagic fever with renal syndrome: A possible cause of hypertension and chronic renal disease in the United States, *Annual Review of Public Health*, 13, 79, 1992.
61. Siem, H. and Bollini, P., Migration and health in the 1990s, *International Migration*, 30, 1992.
62. Interaction: American Council for Voluntary International Action, *Assistance to Internally Displaced Persons: Issue Briefs/Facts and Recommendations*, Humanitarian Policy and Practice Committee, 2002. http://www.interaction.org/refugees/wpdisplaced.html.
63. Carlton, J.T. and Geller, J.B., Ecological roulette: The global transport of non-indigenous marine organisms, *Science*, 261, 78, 1992.
64. World Health Organization, Cholera in the Americas, *The Weekly Epidemiological Record*, 67, 33, 1992.
65. McCarthy, S.A., McPhearson, R.M., and Guarino, A.M., Toxigenic *Vibrio cholerae* and cargo ships entering Gulf of Mexico, *Lancet*, 339, 624, 1992.
66. Rarnamurthy, T., Garg, S., Sharma, R. et al., Emergence of novel strain of *Vibrio cholerae* with epidemic potential in southern and eastern India, *Lancet*, 341, 703, 1992.
67. DePaola, A., Capers, G.M., Moters, M.L. et al., Isolation of Latin American epidemic strain of *Vibrio cholerae* 01 from US Gulf Coast, *Lancet*, 339, 624, 1992.
68. Fidler, D., Perspectives; Globalization, international law, and emerging infectious diseases, *Emerging Infectious Diseases*, 2, 2, 1996.
69. Delbruck, J., Globalization of law, politics, and markets/implications for domestic law; a European perspective, *Indiana Journal of Global Legal Studies*, 1, 9, 1992.
70. World Health Organization, Communicable disease prevention and control: New, emerging, and re-emerging infectious diseases, WHO *Doc. A48115*, February 22, 1995.
71. Morse, S.S., Looking for a link, *Nature*, 344, 297, 1992.
72. Russel, R.C., Survival of insects in the wheel bays of a Boeing 747B aircraft on flights between tropical and temperate airports, *Bull WHO*, 65, 659, 1992.
73. Wilesmith, J.W., Ryan, J.B.M., and Atkinson, M.J., Bovine spongiform encephalopathy: Epidemiological studies on the origin, *Veterinary Record*, 128, 199, 1992.
74. Berkelman, R.L., Bryan, R.T., Osterholm, M.T., Leduc, J.W., and Hughes, J. M., Infectious disease surveillance: A crumbling foundation, *Science*, 264, 368, 1992.
75. Lederberg, J., Infectious disease as an evolutionary paradigm, *Electronic Frontier Foundation*, 3(4), December 1997.
76. Davies, J., Inactivation of antibiotics and the dissemination of resistance genes, *Science*, 264, 375, 1994.

77. Bloona, B.R. and Murray, C.J.L., Tuberculosis: Commentary on a re-emergent killer, *Science*, 257, 1055, 1994.
78. Neu, H.C., The crisis in antibiotic resistance, *Science*, 257, 1064, 1994.
79. Domingo, E. and Holland, J.J., Mutation rates and rapid evolution of RNA viruses, in *The Evolutionary Biology of Viruses*, Morse, S.S. (Ed.), Raven Press, New York, 1994, pp. 161–184.
80. Kilbourne, E.D., The molecular epidemiology of influenza, *The Journal of Infectious Diseases*, 127, 478, 1978.
81. Bureau of the Census, Mortality statistics, 1927, part 1, Washington, DC, 1927 U.S. Government Printing Office; National Center for Health Statistics: Vital statistics of the United States, 1986, col 11—Mortality, part A, US. *Government Printing Office*, Washington, DC, 1988.
82. Garret, L., *The Coming Plague*, Harper Collins Canada, New York, 1994, p. 445.
83. Jungkind, D.L., *Antimicrobial Resistance: A Crisis in Health Care. Advances in Experimental Medicine and Biology*, Plenum Press, New York, 1995, p. 33.
84. CDC, National Center for Infectious Diseases, Arctic Investigations Program, Methicillin-resistant *Staphylococcus aureus* (MRSA), July 2006. http://www.cdc.gov/ncidod/aip/research /mrsa.html#what_is_mrsa.
85. Barrett, F.F., McGhee, R.F., Finland, M., Methicillin-resistant *Staphylococcus aureus* at Boston City Hospital, *New England Journal of Medicine*, 279(9), 441–448, 1968.
86. Herold, B.C., Immergluck, L.C., Maranan, M.C. et al. Community-acquired methicillin-resistant *Staphylococcus aureus* in children with no identifiable predisposing risk. *The Journal of the American Medical Association*, 279, 593–598, 1998.
87. Torassa, U., Antibiotic overuse a growing problem, September 28, 1999. http://www. Sfgate .com/cgi-bin/article.cgi?file=/examiner/hotnews/stories/29/animal.dtl.
88. Glass, R.I., Libel, M., and Brandling-Bennett, A.D., Epidemic cholera in the Americas. *Science*, 265, 1524, 1992.
89. Moore, P.S., Meningococcal meningitis in sub-Saharan Africa: A model for the epidemic process, *Clinical Infectious Diseases*, 14, 515, 1992.
90. MacKenzie, W.R., Hoxie, N.J., Proctor, M.E. et al., A massive outbreak in Milwaukee of *Cryptosporidium* infection transmitted through the water supply. *The New England Journal of Medicine*, 33(1), 161, 1994.
91. Centers for Disease Control and Prevention, Assessment of inadequately filtered public drinking water Washington, DC, *Morbidity and Mortality Weekly Report*, 43, 661, 1994, December, 1993.
92. Ryan, M., They track the deadliest viruses, *Parade*, 10, April 23, 1995.
93. Nichol, S.T., Spiropoulou, C.F., Morzunov, S. et al., Genetic identification of a hantavirus associated with an outbreak of acute respiratory illness, *Science*, 262, 914, 1995.
94. Schmaljohn, A., Li, D., Negley, D.L. et al., Isolation and initial characterization of a newfound hantavirus from California, *Virology*, 206, 963, 1995.
95. Hurt, H., A deadly new virus, *Self Magazine*, 88, April 1993.
96. Gajdusek, D., Virus hemorrhagic fevers, *The Journal of Pediatrics*, 60, 841, 1962.
97. Lee, H.W., Epidemiology and pathogenesis of hemorrhagic fever with renal syndrome, *The Bunyaviridae*, R.M. Elliott, (Ed.), Plenum Press, New York, 1996, pp. 253–267.
98. World Health Organization, Haemorrhagic fever with renal syndrome memorandum from a WHO meeting, *Bull World Health Organ*, 61, 269, 1983.
99. Schmaljohn, C. and Hjelle, B., Synopses Hantaviruses: A global disease problem, *Emerging Infectious Diseases*, 3(2), 95, 1997.
100. French, G., Foulke, R., Brand, O., and Eddy, G., Korean hemorrhagic fever: Propagation of the etiologic agent in a cell line of human origin, *Science*, 21, 1046, 1981.
101. Epstein, P.R., Emerging diseases and ecosystem instability: New threats to public health, *American Journal of Public Health*, 80/2, 16, 1995.
102. Lee, H., Lee, P., and Johnson, K., Isolation of the etiologic agent of Korean hemorrhagic fever, *The Journal of Infectious Diseases*, 137, 298, 1978.
103. Wenzel, R.P., A new hantavirus infection in North America, *The New England Journal of Medicine*, 330, 1004, 1994.

104. Dispatch, Dengue hemorrhagic fever: The emergence of a global health problem. *Emerging Infectious Diseases*, 1, 2, 1995.

105. Gubler, D.J., Dengue, *Epidemiology of Arthropod-Borne Viral Disease*, T.P.M. Monath, (Ed.), CRC Press, Boca Raton, FL, 1988, pp. 223–260.

106. Manning, A., Painful, flu-like tropical illness can be deadly, USA *Today*, 1, September 18, 1995.

107. Henig, R.M., The new mosquito menace, *New York Time*, A21, September 13, 1995.

108. Gubler, D.J., Dengue haemorrhagic fever: A global update [Editorial], Virus, *Information Exchange News*, 8, 2, 1995.

109. Associated Press, TB, other disease reemerge to threaten millions in the Americas, September 27, 1999. http://my..../Go?template.html.

110. CDC, Division of Vector-Borne Infectious Diseases, *Dengue and Dengue Hemorrhagic Fever: Information for Health Care Practitioners*, June 30, 2005. http://www.cdc.gov/ncidod /dvbid/dengue/dengue-hcp.htm.

111. Gubler, D.J., Dengue and dengue haemorrhagic fever in the Americas, WHO, *Regional Office for SE Asia, New Delhi*, Monograph, SEARO, 22, 9, 1995.

112. Rigau-Plkrez, J.G., Gubler, D.J., Vomdarn, A.V., and Clark, G.G., Dengue surveillance—United States, 1986–1992, *Morbidity and Mortality Weekly Report*, 43, 7, 1994.

113. Baylor College of Medicine, Chikunguya, Retrieved February 2018. https://www.bcm.edu /departments/molecular-virology-and-microbiology/emerging-infections-and-biodefense /chikungunya.

114. Centers for Disease Control and Prevention (CDC), Chikungunya Virus, Page last updated: August 29, 2017. https://www.cdc.gov/chikungunya/.

115. Baylor College of Medicine, Zika Virus, Retrieved February 2018. https://www.bcm.edu /departments/molecular-virology-and-microbiology/emerging-infections-and-biodefense/ zika.

116. Centers for Disease Control, Zika Virus, Retrieved February 2018. https://www.cdc.gov/zika /about/index.html.

117. Baylor College of Medicine, SARS Virus, Retrieved February 2018. https://www.bcm .edu/departments/molecular-virology-and-microbiology/emerging-infections-and-biodefense /sars-virus.

118. CDC About the Flu, Page last reviewed: October 5, 2017, Centers for Disease Control and Prevention, National Center for Immunization and Respiratory Diseases (NCIRD). https://www.cdc.gov /flu/about/index.html.

119. Chris Crawford, AAFP, Flu Vaccination Still Recommended During Current Epidemic, January 17, 2018. https://www.aafp.org/news/health-of-the-public/20180117fluupdate.html.

120. Adler, J., The fight against the flu, *Newsweek*, 146(18), 39–46, 2005.

121. World Health Organization (WHO), *Cumulative Number of Confirmed Humans Cases of Avian Influenza A/(H5N1) Reported to WHO*, July 26, 2006. http://www.who.int/csr/disease /avian_influenza/country/cases_table_2006_07_26/en/index.html.

122. World Health Organization (WHO), Emerging and other communicable diseases, *Influenza A(H5NI) Fact Sheet 188*, January 9, 1999.

123. Centers for Disease Control (CDC), 1998. http://www.edc.gov/ncidod/diseases/flu/fluinfo.htm.

124. Commentary, Re-emergence of Ebola virus in Africa, *Emerging Infectious Diseases*, 1, 2, 1995.

125. Troy, K., The plague that wasn't, *Newsweek*, 46(6), 917–936, December 9, 1996.

126. Bowen, E.T.W., Platt, G.S., Lloyd, G. et al., Viral haemorrhagic fever in southern Sudan and northern Zaire: Preliminary studies on the aetiologic agent, *Lancet*, 1, 571, 1977.

127. Jahrling, P.B., Geisbert, T.W., Dalgard, D.W. et al., Preliminary report: Isolation of Ebola virus from monkeys imported to USA, *Lancet*, 335, 502, 1977.

128. Allela, L., Bourry, O., Pouillot, R. et al., Ebola virus antibody prevalence in dogs and human risk. *Emerg Infect Dis* [serial on the Internet]. 2005 Mar [*date cited*]. Retrieved http://www.cdc .gov/ncidod/EID/vol11no03/04-0981.htm.

129. Clover, C. and Highfield, R., AIDS started by humans eating chimps..., *The Daily Telegraph (London)*, February 1, 1999.

130. HU, W. and Hughe, S.H., HIV-1 reverse transcription, *Cold Spring Harbor Perspectives in Medicine*, 2(10), a006882, 2012.

131. National Institute of Allergy and Infectious Diseases (NIAID), HIV infection and AIDS, US Dept of Health & Human Services, Public Health Service, May 1997, http://www.aegis.com /topics/whataidsis.html.

132. WebMD, Drugs and Medication Search, retrieved 2018, https://www.webmd.com/drugs /condition-157-HIV+Infect.

133. U.S. Dept of Health and Human Services, AIDS info, HIV treatment, Retrieved February 2018. https://aidsinfo.nih.gov/understanding-hiv-aids/fact-sheets/21/58/fda-approved-hiv-medicines.

134. UNAIDS and WHO, Report on the Global HIV/AIDS epidemic, December 1997. http://www .unaids.org/unaids/document/epideinio/report97.html.

135. Gerlach, D., AIDS: 25 years later, *Newsweek*, p. 41, May 15, 2006.

136. CDC, Division of Vector-Borne Infectious diseases, *West Nile Virus: Background: Virus History and Distribution*, April 6, 2004. http://www.cdc.gov/ncidod/dvbid/westnile/background.htm.

137. CDC, Division of Vector-Borne Infectious diseases, *West Nile Virus: Questions and Answers*, August 16, 2004. http://www.cdc.gov/ncidod/dvbid/westnile/qa/symptoms.htm.

138. CDC, Division of Vector-Borne Infectious diseases, *West Nile Virus: Statistics, Surveillance and Control*. http://www.cdc.gov/ncidod/dvbid/westnile/surv&controlCaseCount06_detailed .htm.

139. Karmali, M.A., Infection by Verocytotoxin-producing *Escherichia coli*, *Clinical Microbiology Reviews*, 2, 15, 1989.

140. Griffin, P.M. and Tauxe, P.V., The epidemiology of infections caused by *Escherichia* coli 0 1 57:H7, other enterohemorrhagic *E. coli*, and the associated hemolytic uremic syndrome, *Epidemiologic Reviews*, 13, 60, 1991.

141. Besser, R.E., Lett, S.M., Weber, J.T. et al., An outbreak of diarrhea and hemolytic uremic syndrome from *Escherichia coli* O157:H7 in fresh-pressed apple cider. *The Journal of the American Medical Association*, 269, 2217, 1993.

142. Zhao, T., Doyle, M.P., and Besser, R.E., Fate of enterohemorrhagic *Escherichia coli* O157:H7 in apple cider with and without preservatives, *Applied and Environmental Microbiology*, 59, 2526, 1993.

143. Weagant, S.D., Bryant, J.L., and Bark, D.H., Survival of *Escherichia coli 015TH7* in mayonnaise and mayonnaise-based sauces at room and refrigerated temperatures, *Journal of Food Protection*, 57, 629, 1994.

144. Swerdlow, D.L., Woodruff, B.A., Brady, R.C. et al., A waterborne outbreak in Missouri of *Escherichia coli* O157: H7 associated with bloody diarrhea and death, *Annals of Internal Medicine*, 117, 812, 1992.

145. Peisman, J., Schreifer, M.E., and Burkot, T.R., Ability of experimentally infected chickens to infect ticks with the Lyme disease sprichete, *Borrelia burgdorferi*, *The American Journal of Tropical Medicine and Hygiene*, 54, 294, 1996.

146. Levin, M., Levine, J.F., Howard, P., and Apperson, C.S., Reservoir competence of the southeastern five-lined skink (*Eumeces inexpectatus*) and the green anole (*Anolis carolensis*) for *Borrelia burgdorteri*, *The American Journal of Tropical Medicine and Hygiene*, 54, 92, 1996.

147. American Lyme Disease Foundation (ALDF), *Reported Lyme Disease*, 2006.http://www.aldf .com/usmap.shtml.

148. Manning, A., Lyme-causing tick carries worse disease, *USA Today*, p. 1, September 18, 1995.

149. Colburn, D., The clock is TICK-ing, New Lyme vaccine effectively sterilizes the blood-sucking tick, *Daily Hampshire Gazette*, July 6, 1999.

150. Centers for Disease Control and Prevention. Group A streptococcal (GAS) disease, September 16, 2016. Retrieved July 7,2018, https://www.cdc.gov/groupastrep/diseases-public/index.html.

151. Stevens, D.L., Invasive group A streptococcus infections, *Clinical Infectious Diseases*, 14, 2, 1992.

152. Demers, B., Shno, A.E., Vellend, H. et al., Severe invasive group A streptococcal infections in Ontario, Canada: 1987–1991, *Clinical Infectious Diseases*, 16, 792, 1993.

153. Seachrist, L., The Once and future scourge, *Science News*, 148, 234, October 1996.

154. Stevens, D. L., Invasive group A streptococcal infections: The past, present and future, *The Pediatric Infectious Disease Journal*, 13, 561, 1994.

155. National prevention Information Network (NPIN), *TB Today*, 2006. http://www.cdcnpin.org/scripts/tb/tb.asp.

156. Meyer, T., U.S., Mexico must fight disease that knows no boundaries, *The Boston Globe*, 1, 15, November 30, 1996.

157. Starke, J.R. and Correa, A.G., Management of mycobacterial infection and disease in children, *The Pediatric Infectious Disease Journal*, 14, 455, 1995.

158. Kochi, A., The global tuberculosis situation and the new control strategy of the World Health Organization, *Tubercle and Lung Disease*, 72, 1, 1991.

159. Raviglione, M.C., Snider, D.E., and Kochi, A., Global epidemiology of tuberculosis. Morbidity and mortality of a worldwide epidemic. *The Journal of the American Medical Association*, 273, 220, 1995.

160. Hayden, T., Tuberculosis is making a comeback, *Newsweek*, November 8, 1999.

161. McNeil, D., South Africa slow to abate spread of tuberculosis, *The New York Times*, October 13, 1996.

162. Sizemore, C.F., Laughon, B., Fauci, A.S., *World TB Today*, March 24, 2006. National Institute of Allergy and Infectious Diseases (NIAID), March 24, 2006. http://www3.niaid.nih.gov/about/directors/news/tb_06.htm.

163. Miller, J., Study says new TB strains need an intensive strategy, *The New York Times*, A6, October 28, 1999.

164. DuPont, H.L., Chappell, C.L., Sterling, C.R. et al., The infectivity of *Cryptosporidium parvum* in healthy volunteers, *The New England Journal of Medicine*, 332, 855, 1995.

165. Centers for Disease Control and Prevention (CDC), *Cryptosporidium and Water: A Public Health Handbook*, NCID, Atlanta, GA, 30341–3724, 1997, 134 p.

166. Newman, R.D., Zu, S.-X., Wuhib, T. et al., Household epidemiology of *Cryptosporidium parvum* infection, *Annals of Internal Medicine*, 120, 500, 1994.

167. Nime, F.A., Burek, J.D., Page, D.L. et al., Acute enterocolitis in a human being infected with the protozoan *Cryptosporidium*, *Gastroenterology*, 70, 592, 1994.

168. Guerrant, R., Synopses Cryptosporidiosis: An emerging highly infectious threat, *Emerging Infectious Diseases*, 3(1), 1997.

169. Zu, S.-X., Li, J.-F., Barrett, L.J. et al., Seroepidemiologic study of Cryptosporidiurn infection in children from rural communities of Anhui, China, *The American Journal of Tropical Medicine and Hygiene*, 51(1), 1–10, 1994.

170. Ungar, B.L., Gilmn, R.H., Lanata, C.F., and Peres-Schael, L., Seroepidemiology of Cryptosporidium infection in two Latin American populations, *The Journal of Infectious Diseases*, 157, 551, 1988.

171. Ungar, B.L.P., Soave, R., Fayer, R., and Nash, T.E., Enzyme immunoassay detection of immunoglobulin M and G antibodies to Cryptosporidium in immunocompetent and immunocompromised persons, *The Journal of Infectious Diseases*, 153, 570, 1986.

172. Zucker, J.R., Changing patterns of autochthonous malaria transmission in the United States: A review of recent outbreaks, *Emerging Infectious Diseases*, 2, 1, 1996.

173. World Health Organization, *Control of tropical Diseases (CTD): Malarial Control*, WHO Office of Information, Geneva, Switzerland, 1995.

174. Murray, C.J.L., Quantifying the burden of disease; the technical basis for disabilityadjusted life years, Bull *World Health Organ*, 72, 429, 1994.

175. Pan American Health Organization, Status of malaria programs in the Americas. XLIV *Report*, *PAHO*, Washington DC, 1996.

176. Doll, R., Health and the environment in the 1990s, *American Journal of Public Health*, 82, 933, 1992.

177. Martins, W., Niessen, L., Rotmans, J., Jetten, T., and McMichael, A., Potential impact of global climate change on malaria risk, *Environmental Health Perspectives*, 103(5), 458, 1995.

178. Pan American Health Organization, Report for registration of malaria eradication from United States of America, Washington, DC, December, 1969.
179. Eliades, M.J., Snehal, S., and Nguyen-Dinh, P. et al., Malaria Surveillance—United States, 2003. *Morbidity and Mortality Weekly Report*, 54(SS02), 25–39/June 3, 2005. http://www.cdc.gov /mmwr/preview/mmwrhtml/ss5402a2.htm.
180. WHO, WHO reports on new, re-emerging diseases threatening world health, *The Nation's Health*, Vol. 24, November 1995.
181. Ewald, P.W., Guarding against the most dangerous emerging pathogens: Insights from evolutionary biology, *Emerging Infectious Diseases*, 2, 4, 1996.

8

Foodborne Illness

Introduction

Martin Molina was a health promoter in the small village of El Achiotal located on the Pacific coast of El Salvador. Like most evenings, all lights were off by 8 p.m. and most had settled quietly into their hammocks; a custom left over from the civil war that had just ended. One night, Martin was awoken by a knock at the door. He remained silent and listened, as the knock grew frantic and more rapid. Then a familiar voice called his name. It was Mrs. Cruz, a mother of six. Martin was relieved—he had heard stories of those who disappeared in the night after answering an unknown knock, either recruited into the military or disappeared at the hand of the infamous death squads. He went to answer the door.

Martin had fled his village to the United States when the civil war broke out, and like many of his country members, he had managed to return to El Salvador with enough money to buy property, build a house, and even finance a high school education. Martin did not know why he had returned to his small village with no potable water, no human waste facilities, and an electrical current so weak it could power little more than a light bulb, though he did enjoy the prestige of being the most educated person in the village. Plus, he found his work as a health worker extremely rewarding.

Outside his door, a distraught Mrs. Cruz explained to Martin that her son had been vomiting with profuse diarrhea for several hours, a scenario that was all too familiar to Martin. Martin grabbed his small bag that he constantly kept packed for emergencies and went to see little Santos Cruz, the fifth patient of the day with these symptoms.

When Martin reached Santos, he saw the boy was dry heaving into an old bucket. As Martin prepared a rehydration solution using a powder concentrate and water from a nearby well, Mrs. Cruz explained that she had been giving Santos the same solution that she had received at the village meeting about this sudden sickness. She said Santos had vomited it up and had bad diarrhea as well. Still, Martin insisted that Santos should try to drink the mix since he was severely dehydrated and this condition would only get worse without taking in fluids. But as soon as Santos swallowed a cup, he vomited on Martin's feet.

Martin looked up at the sky, aware that his village was experiencing an outbreak of some sort but that the cause of the outbreak would remain unknown. He also was aware that he was ill equipped to remediate an outbreak of this magnitude. Was it the food, the water he was using to mix the rehydrating solution, or something else? Martin returned his gaze to the sick child, he thought about the nice apartment in New York that he had shared with four others and wondered what life would be like if he had not returned.

– Christopher Landry, Peace Corps Volunteer in El Salvador, 1994–1996.

Worldwide Distribution of Foodborne Pathogens

In 2017, the World Health Organization (WHO) released the first-ever global estimates of foodborne illness. WHO found that each year, about 10 percent of the world's population suffers from a foodborne illness, more than 600 million people each year.[1] Young children, fewer than 5 years old, are particularly susceptible to the dangers of foodborne illness.

Overall, there were 420,000 deaths from foodborne disease across the globe in 2010.[1] Each year, 125,000 children die from a foodborne illness. Children under 5 account for almost one third of the total deaths, although they are only 9 percent of the population.

Foodborne agents causing diarrhea led to 230,000 of these deaths.[1] In fact, diarrheal diseases are responsible for more than half of the global burden of foodborne diseases, causing 550 million people to fall ill. Children are at particular risk of foodborne diarrheal diseases. Each year 220 million become ill and 96,000 die. Food-borne diarrhea may develop after eating undercooked meat, eggs, fresh produce, or dairy products contaminated by viral or bacterial agents.

Certain pathogens such as salmonella cause foodborne illness in countries all over the globe, regardless of a country's development. Other pathogens including foodborne cholera, the parasite helminths, and aflatoxin cause harmful but localized disease in developing regions but are not a health problem in developed countries.[2] Campylobacter is an important pathogen in high-income countries. Overall, the burden of foodborne illness is particularly bad in Africa and South-East Asia. But people in every country can be affected and can die from a serious foodborne illness. Unfortunately, lesser-developed countries (LDCs) lack the infrastructure and resources necessary to monitor many of the causative agents of foodborne disease. Further, it may be in the best interest of certain LDC governments to minimize reports on foodborne illness to protect the food export and tourism industries. Unfortunately, it is these LDCs that suffer most from foodborne illness.[2]

Reporting Accuracy

When estimating foodborne illness, certain principles or *rules of thumb* apply. When severe diseases such as botulism are reported, one can assume that the degree of under-reporting is small and reported numbers are considered close to the actual. When symptoms of disease are mild or short in duration, such as those caused by *Staphylococcus aureus*, investigators can expect a higher rate of under-reporting, which is factored in during the estimation of the disease. When there is little or no acceptable data available, researchers base their estimates on the incidence of disease in infants, or by extrapolating data from similar regions where such information may be available.[4,5]

In addition, differences in the prevalence of foodborne illness that may appear among the various regions of the world are often due to the following:

Climate: Climate influences lifecycles and growth patterns of foodborne pathogens. A colder climate may impede the growth of certain pathogens; a warmer climate will promote the spread of some pathogens.

Susceptible population: Populations with amplified numbers of elderly or infants, which are more susceptible to disease, will have higher rates of illness.

Poor nutrition: The nutritional status of a culture plays an integral part in determining the health status of a population. A population with a healthy diet will most likely possess a competent immune system and will be capable a fending off potential health threats.

Culture: Cultural aspects such as the knowledge, attitudes, stress (such as illness, loss of a significant other or poor living conditions), and practices of individuals may be the most important factor in determining the risk factors for disease. For example, African indigenous populations retain the custom of eating raw meat or pork, which puts them at high risk of consuming viable Trichinella spiralis (roundworm) or Taenia spp. (tapeworm) organisms.[6,7]

Reasons for Food Protection Programs

The implementation of programs to minimize foodborne diseases is important due to the morbidity, mortality, and economic loss.

Morbidity and Mortality due to Foodborne Disease

In the United States, the CDC estimates that each year 48 million people get sick from a foodborne illness, 128,000 are hospitalized, and 3,000 die.[3,8] Foodborne illness is a major health and economic problem in the United States and abroad. Consequently, surveillance systems have been put in place to monitor the incidence and impact of these important foodborne pathogens. The top five agents of foodborne illness in the United States are:[3]

- Norovirus
- Salmonella
- *Clostridium perfringens*
- Campylobacter
- *Staphylococcus aureus* (Staph)

Nine million of the cases of foodborne illness were caused by 31 known identifiable pathogens. The remaining cases were caused by unknown agents, perhaps viral, bacterial, or chemical. The number of deaths due to foodborne illness in the U.S. is around 3,000 annually.[8–10] This number may be low, as foodborne illnesses/deaths are not always duly reported as such, since people may not know their illness was food related. Further, some areas of the country lack the resources to investigate all foodborne illness cases.[9]

The symptoms from foodborne illness range from nausea and mild diarrhea to severe dysentery and death. Different organisms cause illnesses with a range of time to onset and duration (Table 8.1). The onset of illness after exposure can be as short as half an hour to 24–48 hours. Or it may take years, as heavy metals accumulate in our body until they reach toxic levels. Foodborne illness may be self-limiting, as from Cryptosporidium, or it may cause life-threatening complications, as seen with *E. coli* O157:H7.

United States: Morbidity and Mortality

Of the 9.4 million cases of foodborne illness in 2011, viruses caused 59 percent, bacteria 39 percent, and parasites the remaining 2 percent.[11] Salmonella alone caused an estimated 1.2 million illnesses and 19,000 hospitalizations.[12] Based on available records, it is

TABLE 8.1

Ten Foodborne Diseases, Symptoms

Top Five Foodborne Pathogens Resulting in Death: USA, Domestically Acquired Illness, 2010	
Pathogen	**Estimated Number of Deaths**
Salmonella, nontyphoidal	378
Toxoplasma gondii	327
Listeria monocytogenes	255
Norovirus	149
Campylobacter spp.	76

Sources: Symptoms and Sources of Ten Foodborne Germs, https://www.cdc.gov/foodsafety/symptoms.html.

estimated that annually, 1,351 deaths are caused by contaminated food eaten in the United States. Bacterial, viral, and parasitic agents were implicated, with 64 percent of the deaths caused by bacteria, 25 percent by parasites, and 12 percent by viruses. Table 8.2 lists the top five pathogens causing foodborne deaths, with nontyphoidal Salmonella causing the highest number of foodborne deaths, at 28 percent (Table 8.2). Next, T. gondii caused 24 percent, L. monocytogenes 19 percent, and norovirus 11 percent of foodborne death.[11] Listeria monocytogenes has become more prevalent in the United States and is now a listed agent. Nearly 1,600 cases of foodborne disease and 260 related deaths have been attributed to Listeria annually.[13]

European Union: Morbidity and Mortality

Campylobacteriosis was the most prevalent foodborne illness in the EU in 2016, with 461 outbreaks, accounting for 4606 related illnesses.[14] This was a sharp increase from 2015, when there were 399 outbreaks but only 1488 cases. In the EU, campylobacter has been the number one most reported gastrointestinal bacterial pathogen since 2005. Some countries experienced particular spikes in cases in 2016. In Sweden alone, there were 3,000 domestic Campylobacteriosis cases from eating contaminated poultry. By comparison, in the U.S. between 2010 through 2015, there were 209 foodborne Campylobacter outbreaks, accounting for 2,234 related illnesses.[3]

Although cases of Salmonella had been falling for several years, this is no longer the case.

In 2016, there were 94,530 human cases of salmonellosis reported in the EU. *S. Enteritidis*—the most widespread type of Salmonella—accounted for 59 percent of all salmonellosis cases originating in the EU and is mostly associated with the consumption of eggs, egg products, and poultry meat.[14] In 2016, there was an outbreak among 14 EU countries of *S. Enteritidis* from contaminated eggs from Poland. Some researchers believe this outbreak began in 2012.

Listeria appears to be an increasing challenge in the EU as well as in the USA. In 2016, listeria had the highest morbidity of foodborne illness in the EU, with 2536 listeria infections and 247 deaths.[15] In fact, it is estimated that there has been an active outbreak of invasive Listeria monocytogenes (L. monocytogenes) infections in various EU countries since 2015.[16]

Economic Consequences of Foodborne Illness

An outbreak of foodborne illness is almost always accompanied by a cost that reflects the amplitude of the outbreak. Medical costs incurred during the treatment of an illness and the loss of wages for those who must stay home from work are examples of expenses encountered by individuals inflicted with a foodborne illness. Upon identifying a food product as the origin of illness, the industries responsible for producing, processing, or distributing the foodstuff suffer losses reaching into the millions as they halt production, recall the contaminated product for disposal, lose sales, and increase marketing efforts to compensate for losses. Other costs result from investigations searching for the source of contamination, increased quality assurance measures to produce safer products, litigation fees from lawsuits, and the loss of trade (Figure 8.1).[17]

TABLE 8.2

Top Five Food Pathogens Causing Death in USA

Germ	Typical Time for Symptoms to Appear	Typical Signs and Symptoms	Common Food Sources
Campylobacter, Bacteria	2–5 days	Diarrhea (often bloody), stomach cramps/pain, fever	Raw or undercooked poultry Raw (unpasteurized) milk Contaminated water
Clostridium botulinum, Bacteria	18–36 hours	Double or blurred vision, drooping eyelids, slurred speech. Difficulty swallowing, breathing. Dry mouth. Muscle weakness and paralysis. Symptoms start in the head and move down as severity increases.	Improperly canned or fermented foods, usually homemade Prison-made illicit alcohol
Clostridium perfringens, Bacteria	8–12 hours. Rapid onset, 24 hour duration	Diarrhea, stomach cramps. Vomiting and fever are uncommon.	Beef Poultry Gravies Dried or precooked foods
Cyclospora, Parasite	1 week	Watery diarrhea, loss of appetite and weight loss. Stomach cramps/pain, bloating, gas, nausea, and fatigue. Vomiting, headache, fever, and flu-like symptoms.	Raw fruits, vegetables, and herbs
Escherichia coli, Shiga toxin-producing, such as O157, Bacteria	3–4 days after exposure, but may be as short as 1 day or long as 10 days	Severe stomach cramps, diarrhea (often bloody), and vomiting. Fever less than 101°F. 5-10% of those infected develop a hemolytic uremic syndrome, a type of kidney failure.	Raw or undercooked ground beef Unpasteurized (raw) milk and juice Raw vegetables, such as lettuce and sprouts Contaminated water
Listeria, Bacteria	1–4 weeks	Pregnant women experience fever and other flu-like symptoms. Infections may lead to miscarriage, stillbirth, premature delivery, or infection of the newborn. People other than pregnant women: headache, stiff neck, confusion, loss of balance and convulsions, plus fever and muscle aches.	Raw (unpasteurized) milk Soft cheeses made with unpasteurized milk, such as queso fresco, queso blanco, panela, brie, Camembert, blue-veined, or feta Raw sprouts Melons Hot dogs, pâtés, lunch meats, and cold cuts Smoked seafood
Norovirus, Virus	12–48 hours	Diarrhea, nausea/stomach pain, vomiting	Spread from an infected person via contaminated food or water or by touching contaminated surfaces. Leafy greens (such as lettuce), fresh fruits, shellfish (such as oysters)
Salmonella, Bacteria	12–72 hours	Diarrhea, fever, stomach cramps, vomiting	Eggs Raw or undercooked poultry Raw or undercooked meat Unpasteurized milk or juice Cheese Raw fruits and vegetables
Staphylococcus aureus (Staph), Bacteria	30 minutes to 6 hours	Diarrhea, nausea, stomach cramps, vomiting	Uncooked foods handled by people, such as sliced meat, puddings, pastries, and sandwiches. Also unpasteurized (raw) milk and cheeses made from it.
Vibrio, Bacteria	24 hours	Watery diarrhea, nausea. stomach cramps, vomiting, fever, chills	Raw or undercooked shellfish, particularly oysters

Source: Scallan, E. et al., *Foodborne, Emerg. Infect. Dis.,* 17, 7–15, 2011.

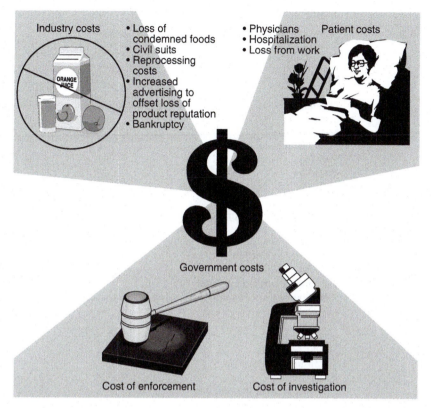

FIGURE 8.1
Economic losses associated with foodborne outbreaks.

Causative Agents of Foodborne Disease

Foodborne illness is defined as any illness incurred from the consumption of contaminated food. Food contaminants come from a wide variety of sources. Certain plants and animals produce substances that are toxic to humans, chemicals such as arsenic and lead may contaminate food through agricultural or industrial processes, and radionuclides from nuclear fallout are present in food products cultivated at or near testing sites. Parasites, viruses, fungi, and bacteria are all examples of living organisms that have the potential to contaminate a food supply, these agents are termed foodborne pathogens (Table 8.3).[7]

Radionuclides

Radiation is introduced into the food chain naturally from mineral deposits beneath the earth's surface or from the atmosphere in the form of ultraviolet and cosmic rays. Radiation from these sources is emitted at low levels tolerable by plant and animal life. Radionuclides, which are deposited in the environment accidentally or intentionally, as a direct result of human activities, are of much greater concern. The levels that result from

TABLE 8.3

Some Causative Agents of Foodborne Disease

Some Causative Agents of Foodborne Illness
Chemicals
Antimony
Cadmium
Lead
Mercury
Polychlorinated biphenyls
Packaging materials
Pesticides
Industrial processes
Food Additives
Gras (Generally Recognized as Safe)
Saccharin
Monosodium glutamate
Nitrates and nitrites
Color additives
Poisonous Plants and Animals
Plant Sources
Alkaloids
Lectins
Saponins
Glucosinolates
Mushroom poisoning
Animal Sources
Paralytic shellfish poisoning
Ciguatera poisoning
Pufferfish poisoning
Foodborne Pathogens
Parasites
Nematodes
Trichinella spiralis, Tapeworms
Protozoans
E. histolyitca, *G. lamblia*,
Cryptosporidium
Viruses
Hepatitis, Norwalk-type
Fungi
Aspergillus spp., *Penicillium* spp.,
Mucor spp.
Bacteria
Toxin producing
C. botulinum
Staphylococcus aureus
Vibrio cholerae
Infectious
Salmonella, Campylobacter, E. coli

these activities are tens to hundreds of times higher than those normally encountered by plant and animal life. Exposure to radiation at high levels has the potential to cause irreversible damage to whole organ systems, genetic mutations, or deformities in offspring, and if the central nervous system is damaged, symptoms of disorientation, convulsions, and lack of muscle coordination are experienced.

The fallout from the Chernobyl accident of 1986 deposited Cesium 137 ($137Cs$) and Iodine 131 (131_I) throughout the northern hemisphere. The isotopes, $137Cs$ and 131_I are absorbed by plants through the roots and foliage, which are in turn used as feed for farm animals. Humans ingest these radionuclides directly from the consumption of fruits and vegetables or indirectly by consuming meat and dairy products.[17] There is recent concern that the concrete vault meant to contain the Chernobyl radioactivity is crumbling and the danger of additional radioactive leakage is imminent. A similar scenario is put forth from nuclear weapons testing, and while we are aware of the dangers that arise from fallout, nuclear weapon testing continues to be a concern. India and Pakistan conducted a series of nuclear weapons tests in an attempt to achieve the upper hand in an arms race.[17] North Korea too conducts nuclear tests periodically. Although the nuclear testing sites are underground and usually in remote areas where there is little or no agricultural activity, radioactive plumes may travel across national boundaries where fallout could still pose a significant threat to human populations.

Chemicals

Ironically, humans are responsible for many chemical contaminants presently found in food. Between 80 percent and 90 percent of our exposure to potentially harmful chemicals is from food consumption.[5] Chemicals enter the food from packaging materials, agricultural applications of pesticides and fertilizers, as preservatives or colorings in foods, or by the release of industrial chemicals into the environment. Many chemicals have become ubiquitous. Table 8.4 lists some of the many chemicals that are found in foods. These chemicals largely originate from human processes.

TABLE 8.4

Some of the Many Chemicals Originating from Human Processes
That May Be Found in Food

Unintended Chemicals Found in Food
Insecticides
DDT, parathion, pyrethrum, arsenicals, others
Fungicides
Dithiocarbamates, mercurials, others
Herbicides
Carbamates, chlorophenoxy cpds (2, 4-D), Bipyridyls
Fertilizers
Nitrogen, others
Treatments and supplements
Food additives and veterinary drugs
Accidental and inadvertent
Mercury, PCBs, lead, dioxin, aluminum and cadmium from kitchenware
Migration from packaging
Plasticizers, stabilizers such as alkylphenols, printing inks, tin, and lead
Chemicals resulting from processing or preparation
Polycyclic aromatic hydrocarbons (PAHs), nitrosamines, mutagens

Packaging Materials

Chemical poisoning from certain toxic packaging materials may cause illness and death within 30 minutes. Substances such as plasticizers, stabilizers, and ink can migrate from the packaging to the food and cause illness when ingested.[7] Exposure to these substances is usually exceptionally low and the chemicals tend to be non-toxic. In the United States commercial packaging materials are tested for chemical migration and most materials can be considered safe. There are a few chemical contaminants of this nature that are worth mentioning.

Antimony

Antimony leaches into foods stored in chipped enamel containers. Acidic foods such as lemon juice will cause antimony to leach out into the product. The ingestion of antimony can cause complications of the gastrointestinal, cardiovascular, and hepatic systems.

Cadmium

Cadmium that is used as a plating material for trays and containers is also leached into food from containers by acidic conditions. Chronic exposure to cadmium can lead to kidney damage. Cadmium accumulates in both the liver and the kidneys, but in the liver, cadmium is tied up by metallothionein, whereas in the kidney, metallothionein is broken down by lysosomes, which allows the unbound cadmium to accumulate.

Lead

Aside from contaminating food through pesticide application, lead is present in the solder of metal containers and cans. Again, acidic foods will cause exposed lead to leach into the food. But in canning there is a coating that prevents the food from coming in contact with the metal. Lead consumption causes neurological complications, kidney failure, and weakens bone integrity. For this reason, lead is being phased out of use in the industry.

Industrial Processes

Traces of chemicals from industrial processes are often found in foods. Levels are usually extremely low and therefore, they go unnoticed. Occasionally, hazardous quantities are dumped into the environment by accident, carelessness, or with malicious intent. These chemicals can migrate into the food chain, causing foodborne illness.

Mercury

Mercury is a by-product of many industrial processes. Chronic exposure to mercury presents as an emotional or psychological disorder. Methyl mercury, however, is an acute toxin that causes tremors, neurological complications, kidney failure, and birth defects. Methyl mercury is the primary form of mercury found on foods.

In Minamata Bay, Japan, industries were dumping waste contaminated with inorganic mercury into the bay. Bacteria in the bay converted the inorganic mercury to methyl mercury that resulted in over 1,200 cases of food poisoning from the consumption of contaminated fish.

Mercury is also a component of fungicides used on animal feed. Animals consume this feed, then they are slaughtered for human consumption. Consuming meats and fish containing high levels of mercury can also cause blindness, paralysis, or death.[7]

Polychlorinated Biphenyls

Polychlorinated biphenyls (PCBs) are widely used in industry. They are extremely stable compounds that do not degrade easily, they are resistant to heat, and are also highly toxic. In Japan, in 1968, 1,000 people suffered food poisoning from PCB contamination in rice oil. During the rice oil process, a heating pipe with a small hole allowed PCBs to contaminate the rice oil. Symptoms included swelling of the eyes, rash, and gastrointestinal illness, which resulted in five deaths. Alternatives to PCBs are being researched, but they are so widespread today that we can expect to see traces of them for many years to come.

Dioxin

Another chemical that is found in foods is dioxin. In early 1999, high levels of dioxin were found in Belgian cattle. The government admitted that it does not inspect cattle that are brought up near industrial sites (a common source for dioxin along with incinerators), nor do they have requirements to measure dioxin levels in beef. Following this discovery, new efforts were made in Belgium and other industrial nations to keep dioxin-contaminated beef out of the global market.[18] Interestingly, the same year dioxins were also found in Belgian poultry and eggs, as well as poultry, eggs, and pork from several other European countries. Animal feed tainted with PCB-waste oil was identified as the culprit.[19] Nearly 10 years later, in 2008, pork meat and pork products from Ireland were recalled when high levels of dioxins were found in product samples. Contaminated feed was again the cause.

Pesticides

Pesticides are introduced into the environment through the air, water, and soil. Although they are used to control insects, fungi, and rodents, they may reduce crop yield. Organochlorine compounds (such as DDT and chlordane), organophosphates (such as parathion and malathion), and inorganic compounds (such as arsenics) have been applied to food in the form of a pesticide. DDT was found to interfere with the reproductive cycle of some birds and was banned in 1972. Since then, the EPA has banned aldrin, dieldrin, hepaclor and kepone. Yet traces of these compounds and their metabolites continue to be found in our food. Many Americans are eating imported foods that still contain these potentially dangerous residues. Many of these chemicals banned from use in the United States are sold to developing nations that use them extensively in producing crops for export to the American market.

The characteristics of certain pesticides make them especially dangerous. DDT and other chemicals of its class accumulate in the environment. This is why we still find trace amounts of residue in foods produced in the United States, especially shellfish. In response to this, pesticides were formulated to breakdown more readily. Because these pesticides spend less time in the environment, they tend to be more toxic. For example, parathion is a threat to farm workers because it can be absorbed through the skin. Carbamate compounds inhibit cholinesterase, an enzyme essential to transmitting nerve impulses at the synapses. The EPA has proposed tolerance levels for pesticides applied to food products. Tolerance levels are the maximum permitted amounts of pesticides that can be found on a raw product.

Although these efforts are well intentioned, only a small portion of imported products entering the United States is tested for pesticides.[18,20] Researchers analyzed data collected by the U.S. Department of Agriculture's Pesticide Data Program (PDP) to compare the relative amounts and toxicity of pesticide residues in different foods. They obtained pesticide residue data on over 27,000 food samples tested by the PDP from 1994 to 1997. The researchers weighed the amounts of residues present to account for differences in the toxicity of individual pesticide chemicals and computed a Toxicity Index (TI) for each food. The researchers calculated TI integrated measures of the frequency of pesticide detection, the levels of residues present and the relative toxicity of the detected residues, yielding an index of the relative toxicity loading of each food.[21] The researchers suggested that values greater than 100 on the TI scale showed comparatively high pesticide contamination, and values less than 10 indicated that those foods are comparatively quite "clean." Fresh peaches (both domestic and imported); frozen and fresh winter squash grown in the U.S.; domestic and imported apples, grapes, spinach, and pears; and U.S.-grown green beans were among the foods showing the highest TI values, and 11 of the 12 highest TI scores were for U.S.-grown foods.[22]

Unfortunately, pesticide residues in food appear to be on the rise. Dangerous pesticides are found in foods, including an herbicide known as 2,4-D and a dangerous insecticide called chlorpyrifos, which ranked fourth for prevalence in foods.[23] There has been a 13 percent increase in sampled domestic foods found with residues, now at 50 percent of domestic foods tested showing pesticide residue in 2015. Oddly, despite these increases, the number of samples the FDA is testing is decreasing, from 7,900 samples 10 years ago to 5,989 samples for the 2015 test. For comparison, it is worth noting that the U.S. Department of Agriculture tests for pesticide residue annually, looking at almost double the samples of the FDA. The 2015 USDA report found that 85 percent of samples contained pesticide residues.[23]

Food Additives

Food additives are used to alter taste, color, texture, nutritive value, appearance, and resistance to deterioration. There are thousands of food additives ranging from simple compounds such as salt to more complex compounds such as monosodium glutamate (MSG). Food additives are also considered to be the least hazardous source of foodborne illness, ranking behind pesticides, environmental contaminants, natural toxins, and microbial toxins.

In the early part of the twentieth century, increased knowledge of the toxic effects from long-term chemical exposure prompted the passage of the Food and Drug Act of 1906, to which several amendments have been added. In 1958, the Food Additive Amendment to the Food Drug and Cosmetic Act required the manufacturer of any new potential food additive to conduct testing and achieve FDA approval before marketing. Not included in this amendment are "generally recognized as safe" (GRAS) substances and color additives, which are covered under the Color Additive Amendment of 1950.[18]

Saccharin

Saccharin is a low-calorie sweetener that was approved by the FDA but still remains highly scrutinized. In laboratory animals, saccharin promotes bladder cancer after exposure to a bladder carcinogen. In 1977, the FDA proposed a ban on saccharin as a food additive. However, the ban was postponed by Congress pending additional studies to support the results. Saccharin was not removed under the Delaney Clause because various studies have reported conflicting results. A considerably high amount of saccharin is needed to produce the observed effects. It has been speculated that sodium, not saccharin, may be responsible for the bladder cancer.[18]

Monosodium Glutamate

MSG is used to enhance flavor in certain foods, and can cause headaches, possible nausea, and even lesions of the retina. An allowable daily intake (ADI) of 120 milligrams per kilogram has been established for individuals over one year of age. Manufacturers in the United States have voluntarily halted the use of MSG in infant foods.[30]

Nitrates and Nitrites

Nitrates and nitrites are used to cure ham, bacon, and other meats. They give cured meats their characteristic taste and color while preventing the growth of *Clostridium botulinum* spores. Nitrate and nitrites can also be found in vegetables, water, saliva, and tobacco. In the body, nitrates can be reduced to nitrites, which in turn oxidize hemoglobin into methemoglobin. If methemoglobin accumulates in the blood, oxygen cannot be transported and anoxia will result.

In food, nitrites react with amines to form nitrosamines. This is of concern because nitrosamines have caused cancer of the liver, kidney, bladder, stomach, and pancreas in laboratory animals. Because nitrosamines are not added to foods, they are not excluded by the Delaney Clause. The FDA has lowered the amount of nitrites added to meats and has suggested the addition of ascorbate or sodium erythorbate to prevent the formation of nitrosamines.

Generally Recognized as Safe Substances

GRAS substances are chemicals that had a history of safe use before the 1958 Food Additive Amendment passed. There are approximately 700 GRAS substances. Currently, the FDA is reviewing the safety of these substances and reclassifying them if deemed necessary. The Delaney clause is included in the Food Additive and Color Additive Amendments of the Food Drug and Cosmetic Act. This clause prohibits the use of any food additive found to induce cancer in laboratory animals under appropriate conditions. This clause has been challenged with much resistance from manufacturers due to controversy surrounding the extrapolation of results from lab animals to humans.[18]

Color Additives

Colors are added to foods such as candy, cakes, cereal, ice cream, and soft drinks to give them a color that appeals to the consumer. Natural colors such as grape skin extract, caramel, and fruit juices did not have to undergo certification after the Color Additive Amendment of 1960 passed. All synthetic colorings, however, required certification if there were any question as to their safety, even if the substance had been previously certified. There are seven synthetic colorings that are certified for unlimited use according to good manufacturing practices (GMPs). They are FD&C Blue No. 1 and 7, Red No. 3 and 40, Yellow No. 5 and 6, and Green No. 3. Other colors such as Green No. 1 and Red No. 2 were removed from use because of concerns over their carcinogenicity and other deleterious effects from chronic exposure.[18]

Poisonous Plants and Animals

Although there are numerous species of plants and animals that are poisonous, the majority is not part of our diet. By the process of trial and error, humans have identified plants that were either harmful to humans or possessed little nutritional value and excluded them from our diet. Some plants and animals known to be harmful have a significant nutritional value and are still part of our diet. Plants such as castor bean, alfalfa, radish, and some mushrooms produce substances that could be harmful if consumed. Currently, there are 1200 species of marine organisms that produce toxins. Fortunately, only a few are important in foodborne illness. Toxic plankton is consumed by shellfish and other marine animals, which in turn are harvested for human consumption. Likewise, a species of the puffer fish, which is important in China and Japan, also produces a toxin. In some instances, the toxic substance is inactivated by properly cooking the food (puffer fish, lectins). However, many plants and animals are served raw, leaving the consumer defenseless against any toxins that may be present in the food.

Plant Sources

Alkaloids

Alkaloids are part of a plant's natural defense. These nitrogen-containing organic compounds ward off predators such as insects and other herbivores that might consume a portion of the plant. Three major groups of alkaloids important to humans are:

1. The pyrrolizidine group of alkaloids is found in herbs such as coltsfoot, comfrey, and petasite. These are used in home remedies. Humans consume these herbs in the form of teas and other herbal concoctions, which are a common source of these alkaloids. Pyrrolizidine alkaloids cause extensive damage to the liver and lungs and have a high mortality rate among children who ingest them.
2. Solanum alkaloids cause diarrhea, headaches, vomiting, and neurological problems in humans who have consumed green, sprouted, stressed, or storage abused potatoes. Solanum can withstand normal preparatory procedures such as baking, boiling, or microwaving. It was once thought that solanum played a role in anencephaly-spina bifida (ASB), but research has refuted any such hypothesis.
3. Xanthine alkaloids are found in coffees, teas, and many other beverages in the form of caffeine. The caffeine found in 10 cups of coffee (0.5–1.0 gram) is enough to produce adverse effects in adults, and child mortality is possible after the ingestion of 5 gram of caffeine.

Lectins

Lectins are plant proteins that agglutinate red blood cells. More than 800 species of edible plants contain lectins, and over 600 of those species belong to the Leguminosae family. Ricin, found in the castor bean, is a highly toxic lectin. Effects in humans vary from stunted growth to fatality, depending on the source and dose of the lectin. Lectins are destroyed by cooking in wet heat.

Saponins

Saponins are glycosides that hemolyze red blood cells. When alfalfa, which has a high concentration of saponins, is fed to lab animals, they suffer from depressed growth rates and anemia. Saponins are currently of significant concern in the United States. As we shift to healthier foods such as alfalfa and soy-based products, we can also expect an increase of saponin intoxications.

Glucosinolates

Glucosinolates are found in broccoli, cabbage, radish, horseradish, and mustard greens and are thought to cause hypothyroidism (enlargement of the thyroid gland); hence, they are also termed goitrogens.

Mushroom Poisoning

There are approximately 100 species of mushrooms thought to be toxic and 12 species known to contain lethal toxins. Mushroom toxicity can be classified into two types of syndromes based on the latency time of the symptoms. Rapid onset syndrome is rarely severe and produces symptoms such as gastrointestinal illness, paresthesia, and/or CNS syndrome within three hours of consumption. Delayed onset syndrome presents symptoms as headaches, diarrhea, hepatic incompetence, muscle pains, and chills, six hours after consumption. Symptoms of mushroom poisoning vary with mushroom species. Mushroom-induced headache is common to European species, while profuse diarrhea and hepatic complications are more common from North American mushrooms.

Animal Sources

Paralytic Shellfish Poisoning

Paralytic shellfish poisoning occurs when humans consume shellfish that have fed on toxic plankton. Shellfish become toxic to humans when they feed on dinoflagellates such as *Gonyaulax catenella* in numbers greater than 200 per milliliter of water. When theses toxic organisms are present in numbers greater than 20,000 per milliliter they are visible to the naked eye. This phenomenon has been termed *red tide* for its crimson appearance. Saxitoxin, which is just one of the various toxins involved in paralytic shellfish poisoning, can be lethal after consuming only 4 milligrams. Symptoms include a tingling or burning sensation of the lips and gums, ataxia, and paralysis of the diaphragm. If treatment is sought early enough mortality can be avoided.

Ciguatera Poisoning

Species of barracuda, eels, jacks, groupers, red snapper, and sea bass specific to the Caribbean and South Pacific have been implicated in ciguatera poisoning. The ciguatera toxin, like the saxitoxin, is found in a dinoflagellate. The species of marine animals mentioned earlier do not feed directly on the dinoflagellate *Gamblerdiscus toxicus*, which contains the ciguatera toxin, but the small feed fish they thrive on consume the plankton as a primary component of their diet. Signs of ciguatera poisoning are a tingling or numbing sensation of the lips, tongue, and throat, which gives way to nausea, vomiting, and diarrhea. In extreme cases, muscle and joint pains arise making it difficult for the patient to walk.

Pufferfish Poisoning

The puffer fish, also known as the blowfish, is a common delicacy in China and Japan. While not all puffers are poisonous, those belonging to the genus Fugu (fugu fish) contain the lethal tetrodotoxin and are also a most sought-after delicacy. When the fish is prepared in acidic or alkaline conditions, the toxin is inactivated. If the fish are prepared incorrectly, victims experience symptoms similar to intoxication from saxitoxin, and death from respiratory paralysis can occur from the consumption of only 1–2 milligrams of the toxin.

Foodborne Pathogens

Until recently, when one thought of a foodborne outbreak, Salmonella was assumed to be the causative agent unless investigation proved otherwise. Since 1975, 25 new foodborne pathogens have been identified. More than 40 potential foodborne pathogens have been listed by CAST. A number of these are discussed in the following sections, along with fungi, an area not considered by CAST (Table 8.5).[5]

Coupled with these new foodborne pathogens are changes in dietary habits and food processing techniques, which are partly responsible for the surfacing of new pathogens and an increase in the number of illness due to traditional pathogens. Some of these changes are mentioned as follows:[20–29]

TABLE 8.5

Foodborne Pathogens Included in the CAST Report and Fungi

Foodborne Pathogens	
Viruses	**Parasites**
Hepatitis A	Anisakid nematodes
Norwalk	*Cryptosporidium parvum*
Norwalk-like	*Diphyllobothrium latum*
	Entamoeba histolytica
	Giardia lamblia
	Taenia saginata
	Taenia solium
	Toxoplasma gondii
	Trichinella spiralis
Bacteria	**Toxins**
Aeromonas hydrophila	Ciguatera toxin
Bacilus cereus	Diarrheic shellfish poisons
Brucella abortus	Domoic acid
Campylobacter jejuni	Histamine
Clostridium botulinum	Scromboid
Clostridium perfringens	Paralytic shellfish poisons
Coxiella burnetti	Tetrodotoxin
Escherichia coli	**Fungi**
Listeria monocytogenes	*Aspergillus* spp.
Mycobacterium bovis	*Peniclllium* spp.
Salmonella spp.	*Mucor* spp.
Shigella spp.	*Rhizopus* spp.
Staphylococcus aureus	*Candida* spp.
Vibrio cholerae	*Cryptococcus neoformans*
Vibrio parahemolyticus	
Vibrio vulnificus	
Yersinia enterocolitica	

Source: Motarjemi, Y. and Kaferstein, F.K., *World Health Org. Stat. Q.*, 50, 8, 1997.

- A decrease in lactic acid–producing bacteria as a result of processing milk increases the chances of survival of Salmonella or Campylobacter, as their competition has been removed.
- Water polluted with raw sewage or manure containing harmful numbers of microbes comes in contact with food, causing a variety of illness.
- Due to the abuse of antibiotics and sulfur-containing drugs in farm animals, some bacteria such as Salmonella have acquired resistance to antibiotics.
- A change in dietary preference. In an attempt to eat healthier, we are consuming foods that are lower in fat but also more susceptible to microbial growth.
- The increase in demand for packaged food with a longer shelf life prevents aerobic growth but supports anaerobic pathogens like Clostridia, which can be deadly.
- The growing popularity of ready-to-eat or take out foods that require only minimal heating before consumption raises the temperature of the food within the ideal growth range of many bacteria without destroying the pathogen.

The presence of these pathogens in foods and in the body can result in illness due to infection or intoxication. For example, infections such as that caused by *Giardia lamblia* arise when the organism colonizes on the intestine wall, reducing the absorption surface and therefore decreasing the intake of nutrients. Infections caused by *Salmonella typhi* or Hepatitis A have invasive properties and can spread beyond the gastrointestinal tract entering the bloodstream or liver, causing typhoid fever and liver disease, respectively. Intoxication results when an organism releases a toxin while growing on a food or while inhabiting the gastrointestinal tract of humans. Illness from toxigenic pathogens has a rapid onset and is usually more severe than that from an infection. In many cases, food products that contain a toxin have no altered appearance, taste, or smell, making it especially dangerous. In the following sections, we will examine parasitic, viral, fungal, and bacterial pathogens and the illnesses they cause.[22]

Parasitic Infections

Parasites are organisms that require another animal (host) to complete their life cycle. Organisms such as *Entamoeba histolytica* or *Cryptosporidium parvum* require only one host, they are shed in the feces, and infection occurs by consuming food or water contaminated with these feces. *Trichinella spiralis*, on the other hand, causes infection in those who eat meat infested with the cyst of the organism. In the case of Trichinella, the organism requires two hosts, the intermediate host (pig) that houses the cyst, and the definitive host (human), where the organism develops into an adult and is shed in the feces.

The Nematodes

Trichinella Spiralis

Species of *Trichinella spiralis* are found in warm-blooded animals such as bear, hogs, walrus, wild boar, and rats. This organism causes trichinosis when meat from warm-blooded animals containing viable larvae is consumed by humans (Figure 8.2). Back in the 1940s, there were an average of 400 cases of trichinosis per year in the United States. This number has decreased markedly to just 15 cases annually in recent years, as a result of improved cooking habits, greater access to home freezers, and legislation that regulates the quality of feed intended for commercial hogs. Trichinella organisms can be destroyed by cooking

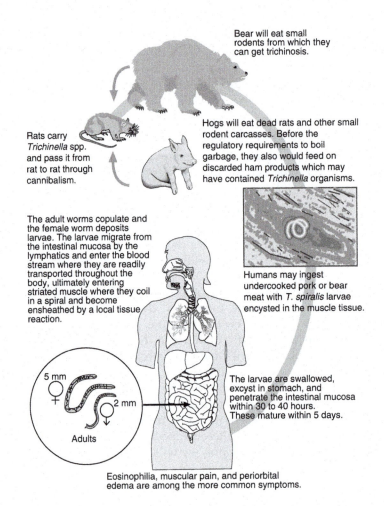

Bear will eat small rodents from which they can get trichinosis.

Rats carry *Trichinella* spp. and pass it from rat to rat through cannibalism.

Hogs will eat dead rats and other small rodent carcasses. Before the regulatory requirements to boil garbage, they also would feed on discarded ham products which may have contained *Trichinella* organisms.

The adult worms copulate and the female worm deposits larvae. The larvae migrate from the intestinal mucosa by the lymphatics and enter the blood stream where they are readily transported throughout the body, ultimately entering striated muscle where they coil in a spiral and become ensheathed by a local tissue reaction.

Humans may ingest undercooked pork or bear meat with *T. spiralis* larvae encysted in the muscle tissue.

5 mm ♀

2 mm ♂

Adults

The larvae are swallowed, excyst in stomach, and penetrate the intestinal mucosa within 30 to 40 hours. These mature within 5 days.

Eosinophilia, muscular pain, and periorbital edema are among the more common symptoms.

FIGURE 8.2
Life cycle of *Trichinella spiralis*.

meat until it turns a gray color or by freezing. In Idaho, in 1995, an outbreak occurred from the ingestion of cougar jerky. The outbreak was contained to ten people who displayed symptoms of eosinophilia, fever, rash, and overall weakness.[28]

Tapeworms

The beef tapeworm *Taenia saginata* and the pork tapeworm *Taenia solium* infest humans through the ingestion of undercooked beef or pork that contains viable larvae. *Taenia solium*, or pork tapeworm, occurs from the ingestion of undercooked pork products. The pig is an intermediate host for *T. solium*, and the pig becomes infected when it consumes food that is contaminated with human fecal material that contains the tapeworm eggs (Figure 8.3). The eggs enter the pig stomach and excyst where they penetrate through the mucosa into the lymphatics and blood stream, where they are then carried to the striated muscles, the brain, and possibly other organs. They become embedded in the muscle of the animal where they develop into cysterici (encysted larvae) that are thin-walled, fluid-filled bladders containing an inverted scolex or head of the tapeworm (Figure 8.3). If undercooked pork containing viable larvae are ingested by humans, the larvae excyst

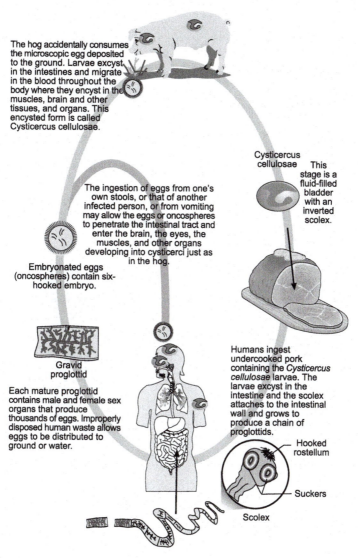

The hog accidentally consumes the microscopic egg deposited to the ground. Larvae excyst in the intestines and migrate in the blood throughout the body where they encyst in the muscles, brain and other tissues, and organs. This encysted form is called Cysticercus cellulosae.

Cysticercus cellulosae This stage is a fluid-filled bladder with an inverted scolex.

The ingestion of eggs from one's own stools, or that of another infected person, or from vomiting may allow the eggs or oncospheres to penetrate the intestinal tract and enter the brain, the eyes, the muscles, and other organs developing into cysticerci just as in the hog.

Embryonated eggs (oncospheres) contain six-hooked embryo.

Gravid proglottid

Each mature proglottid contains male and female sex organs that produce thousands of eggs. Improperly disposed human waste allows eggs to be distributed to ground or water.

Humans ingest undercooked pork containing the *Cysticercus cellulosae* larvae. The larvae excyst in the intestine and the scolex attaches to the intestinal wall and grows to produce a chain of proglottids.

Hooked rostellum

Suckers

Scolex

FIGURE 8.3
Life cycle of *Taenia solium*.

in the intestinal tract. The scolex of *Taenia solium* contains four suckers and two rows of hooks that protrude from the cystericus and attach to the intestinal wall, where the larva develops into an adult with the formation of proglottids. The proglottids mature to develop seven to 13 lateral uterine branches that produce tens of thousands of embryonated eggs. The most distal (and most mature) proglottids detach from the chain of proglottids and are either discharged from the intestine or migrate out. The adult tapeworm can grow up to seven meters in length and inhabit the intestine for 30 years, during which it sheds gravid proglottids (tapeworm segments containing eggs) in the feces of the primary host (human). If careless hygiene is practiced, person-to-person transmission is possible via direct contact or by food contaminated with fecal matter. There is also a cycle that may occur in humans whereby the *T. solium* eggs may be ingested from external sources, from

Cattle accidentally consume the microscopic eggs deposited to the ground. Larvae excyst in the intestines and migrate in the blood throughout the body where they encyst in the muscles, brain and other tissues and organs. This encysted form is called Cysticercus bovis.

Cysticercus bovis   This stage is a fluid-filled bladder with an inverted scolex.

Embryonated eggs (oncospheres) contain six-hooked embryo.

There are 15–20 lateral uterine branches of the uterus in this species.

Gravid proglottid

Each mature proglottid contains male and female sex organs that produce thousands of eggs. Improperly disposed human waste allows eggs to be distributed to ground or water.

Humans ingest undercooked beef containing the *Cysticercus bovis* larvae. The larvae excyst in the intestine and the scolex attaches to the intestinal wall and grows to produce a chain of proglottids.

Adult T. saginata may attain a length of 25 meters, but usually 12–13 meters

Suckers

Scolex

FIGURE 8.4
Life cycle of *Taenia saginata*.

self-contamination with fecal matter, or from reverse peristalsis when the contents of the intestines are back-flushed into the stomach from the process of vomiting. In this case, the eggs excyst and the larvae migrate to the human muscles, the brain, and possibly other organs such as the eye where they develop into cysticerci.

The cycle of the beef tapeworm, *Taenia saginata*, follows a similar route with the exception that re-infection by the egg stage of the tapeworm does not occur in humans (Figure 8.4). The consumption of undercooked or raw beef containing the cysticerci leads to the development of an adult beef tapeworm in the human intestinal tract that has the capacity of reaching lengths of 25 meters but is usually less than half that size.

Although tapeworm infections are rare in the United States, they are more prominent in areas where beef or pork is consumed raw. In healthy individuals symptoms often go unnoticed. In malnourished individuals multiple infections can occur, leading to blockage of the intestine. If the tapeworm leaves the intestine and infects organs such as the heart or brain, effects can be severe regardless of the health status of the individual. Tapeworms can be prevented by appropriately disposing of contaminated feces, cooking meat to an internal temperature of 57 degrees Celsius or freezing it at −10 degrees Celsius for 5 days.[29,30]

The Protozoans[29–31]

Entamoeba Histolytica

Amebiasis, which is caused by *Entamoeba histolytica*, affects about 10 percent of the world's population. Outbreaks occur where sanitation is poor, where risky sexual habits are practiced, and in institutional facilities to such settings, fecal material comes in contact with water and food products, increasing the likelihood of the organism being ingested. While in the cyst stage of its life cycle, the organisms can survive in the external environment where it is transmitted by water or the fecal-oral route (Figure 8.5). Once ingested, the

FIGURE 8.5
Life cycle of *Entamoeba histolytica*.

cysts are broken down by normal digestive action. Motile trophozoites are then released. Trophozoites multiply and invade the large intestine, causing symptoms that vary from mild to bloody diarrhea with a dozen explosive liquid stools a day. This latter case is known as amoebic dysentery. The incubation time of the disease varies from 3 days to a month. In the case of chronic infections, the patient can be asymptomatic while continuously shedding viable cysts, or symptomatic, suffering from bouts of diarrhea and fever until treatment is sought.

Giardia Lamblia

Giardia lamblia is a protozoan flagellate found in areas with poor sanitation and in unfiltered surface water supplies (Figure 8.6). Giardia is considered to be the primary agent for diarrheal outbreaks associated with contaminated water supplies, having a prevalence of 3–7 percent in developed countries and 20–30 percent in LDCs.[27] In the U.S., giardiasis is most common among those who travel to endemic areas and in child day care settings where constant diaper changing facilitates the spread from one child to another. The organism is also found in surface waters used for drinking and recreational purposes. The cyst stage of the organism resists chlorination and can, therefore, pass through municipal supplies that do not also have adequate filtration systems. The cysts reach the surface water supplies through the fecal deposits of beaver and muskrats. When ingested cysts reach the small intestine, trophozoites are released. The trophozoites attach themselves to the intestinal wall by adhesive disks, causing inflammation and irritation of the bowel. This inflammation prompts the influx of fluids, causing a loose stool. As a result, food is malabsorbed. The patient suffers from dehydration. Symptoms usually manifest within one week of eating the contaminated food and consist of nausea, explosive diarrhea (up to 10 movements per day), and fatigue. Without treatment, symptoms linger and re-infection is possible.[28]

Cryptosporidium

The first reported case of cryptosporidiosis, which is caused by the parasite *Cryptosporidium*, occurred in 1976. Cases were prevalent primarily among immunocompromised individuals whose immune systems were unable to protect the body from infection. In 1993, the city of Milwaukee experienced the most comprehensive outbreak to date, which involved 400,000 inhabitants of the city who drank from the contaminated public water supply. Primarily a water-borne pathogen, Cryptosporidium is transmitted via water contaminated with feces from human and agricultural origins. In 1994, the first documented case of foodborne cryptosporidiosis stemmed from fresh-pressed apple cider. Foodborne transmission of

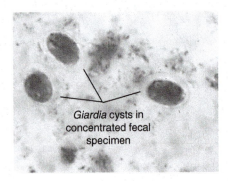

Giardia cysts in concentrated fecal specimen

FIGURE 8.6
Trophozoite and cyst stages of *Giardia lamblia*.

FIGURE 8.7
Life cycle of *Cryptosporidium parvum*.

Cryptosporidium occurs via the fecal-oral route, usually from careless food handlers shedding the hardy oocysts (see "life cycle," Figure 8.7) of the organism (Figure 8.7). In healthy individuals, symptoms appear as mild diarrhea, nausea, cramps, and a low-grade fever. Immunocompromised patients, such as infants, the elderly, or those with AIDS, experience high-volume diarrhea, weight loss, and severe abdominal cramps. Cryptosporidium is of particular concern because the oocysts remain viable in the chlorinated water of swimming pools and can survive the water treatment processes through which our drinking water passes. We can expect to see more outbreaks of cryptosporidiosis until researchers and regulators devise more adequate means to destroy the organism in our water supplies.[32]

Viruses[33–37]

Viruses are microscopic particles that usually contain a single strand of RNA. Although viral particles can exist outside a host cell, as during transmission, they require a host cell for replication to occur. After invading and replicating within a cell, death or damage to the host cell is common, but this is not always the case. The two most prominent foodborne viruses of present day are Hepatitis A and Norwalk-like virus.

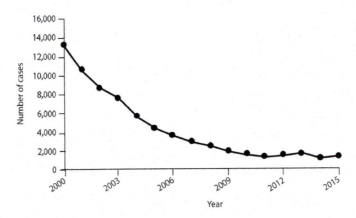

FIGURE 8.8
Hepatitis A Cases in USA, 2000–2015. (From CDC Viral Hepatitis Surveillance, United States, 2015.)

Hepatitis A

Hepatitis A is transmitted via the fecal-oral route and causes liver infection occasionally accompanied by jaundice. On average, symptoms occur 28 days after infection, but infected individuals shed the virus before symptoms manifest, producing an exceptionally risky situation as the infected person might handle food without added precaution. Contamination can occur by infected food workers handling foodstuffs, or from food products that have come in contact with water polluted with fecal matter, as is the case with shellfish harvested from sewage-infested waters. Hepatitis A can also be transmitted from patient contact or intravenous drug use. In each of the last three decades there have been sharp peaks in the number of Hepatitis A cases in the United States. The number of Hepatitis A cases in the United States has gone up and down over the past several years, with 1,390 cases seen in 2015. Overall, however, the incidence of illness has decreased markedly from the early 2000s, dropping 88 percent between the years 2000–2012 (Figure 8.8). In 2000, total case numbers were 10 times the current level. It helps that there are new techniques to determine the source of outbreaks. Genetic analysis is used to compare Hepatitis A samples from patients in a suspected outbreak area to determine if the patients' strains come from similar or different sources. This technology has been used to determine the source of Hepatitis A outbreaks in Michigan, Maine, and other states in which frozen strawberries in schools were responsible for the outbreaks.[36]

Norovirus

Norovirus, originally called Norwalk-like virus, is a small circular organism thought to contain RNA. Until very recently, it had been impossible to replicate in the laboratory. In the U.S., norovirus infection is the leading cause of food-related illness, causing 19–21 million cases of acute gastroenteritis each year.[3,40] Thousands of those afflicted are hospitalized. Annual U.S. norovirus-related deaths are between 570 and 800, mostly among the very young and the very old. The winter months see severe spikes of norovirus cases.

Salad greens and fresh fruit are common sources of norovirus contamination.[40] Food products such as creams and cream fillings are also efficient vehicles for viruses because they do not undergo any extensive heating before being served. Shellfish are the major vehicle of norovirus transmission. Norovirus has an attack rate near 50 percent following the consumption of contaminated shellfish or other foods. The onset of illness occurs between 18 and 36 hours after consuming the virus. Symptoms include diarrhea and nausea, lasting 24–36 hours.

A Case History: Almost a Tragedy   Colleen Tobin was a normal 13-year-old schoolgirl growing up in Northampton, Massachusetts, when she started to have odd symptoms: stomach aches, diarrhea, fatigue, and weight loss. She had stopped growing and her weight was decreasing quickly. Naturally, her parents had brought her to meet several specialists in the state, but the initial examinations turned up no answers or leads. Doctors did not consider a parasitic infection for two primary reasons. One, she had not traveled to any country where unsanitary conditions might facilitate the spread of parasitic disease, and two, she had not been on any outdoor excursions where she might have consumed untreated water. Unable to find any explanation for Colleen's illness, doctors suspected she was deliberately starving herself and suggested she be admitted to a psychiatric hospital. Appalled by this suggestion, Colleen's parents continued to seek new doctors, as Colleen had now lost 5.4 kilograms and had a body temperature of 33.3°C. The gastrointestinal specialists from Springfield, Massachusetts, ran several tests, all of which turned up negative. Colleen was blamed again, accused of taking laxatives. In her defense, Colleen mentioned that she did not know what a laxative was. Finally, Colleen was tested for parasites and the results returned negative, a phenomenon Colleen's parents learned was not unusual, even among infected individuals.

Nearly one year after the onset of her initial symptoms, Colleen spent 10 days at Massachusetts General Hospital. She now weighed only 40.8 kilograms, 13.6 kilograms less than when she developed symptoms. She had a feeding tube down her throat, her fingernails and mouth were turning blue, and her hair fell out in clumps. After having been seen by the hospital psychiatrist again, it was recommended that she be admitted to a psychiatric hospital.

Finally, after more than 2 years of illness, Colleen's family had received word of a similar case in Northampton were the illness had gone undiagnosed until they sought the help of a well-known parasitologist in New York City. Having exhausted every other resource, the Tobins went to New York City in hopes of gaining some insight into Colleen's condition.

Using a special rectal smear that the parasitologist had developed himself, three parasites were identified in Colleen's specimen within 2 hours. *Giardia lamblia*, *Entamoeba histolytica*, and *Enterobius vermicularis* parasites had stripped the lining of Colleen's intestine. So, although she was eating, her body was unable to absorb the necessary nutrients. Colleen was put on a regimen of 11 different antibiotics that initially turned her skin green and yellow. Twice weekly she had to receive an intravenous injection of vitamins, while taking pills to enhance the absorption in her intestines. Thankfully Colleen gained back the weight she had lost and began leading the life of a normal person. The origin of the infection remains unknown.[32]

Fungi[37–39]

Fungi, such as molds and yeasts, are single- and multi-celled plant-like organisms that grow on cereals, breads, fruits, vegetables, and cheeses. For years humans have used fungi for their beneficial properties. Antibiotics such as penicillin, and vitamins such as vitamin C, are derived from molds. Yeasts are used in making wine and bread They also serve as a protein source in foods. Aside from these beneficial effects, certain fungi possess properties that are harmful to humans. The potato blight in Ireland during the 1800s was caused by the fungus *Phytothora infestans*. Certain molds, such as *Penicillium* and *Aspergillus*, are toxigenic and can produce mycotoxins, such as aflatoxin. Mycotoxins are mold metabolites produced on food, which cause illness or death when ingested by humans or animals. Other molds such as *Mucor* and *Rhizopus* species are common spoilage organisms, which can also be pathogenic to humans. Fungi can grow in conditions that most bacteria could not survive. Molds can tolerate a pH as low as 2.0 and can thrive in low-moisture environments.

FIGURE 8.9
Morphology of molds and yeasts.

Most molds are aerobes and are therefore found on the surfaces of substrates where oxygen is present. In contrast to this, yeasts are facultative anaerobes that may be found within a substrate, require more moisture than molds, and prefer slightly acidic environments.

The body of a mold, called a colony or thallus (Figure 8.9), gives rise to long filaments known as hyphae. Hyphae that are divided by septae into cell-like compartments containing nuclei are called septate hyphae. If septae are not present, the hyphae are said to be coenocytic, with continuous cells, and nuclei throughout the appendage. As hyphae grow they form a tangled mesh called a mycelium, which can be divided into two parts, the vegetative mycelium, which is responsible for the uptake of nutrients, and the aerial mycelium where the reproductive spores can be found. The spores on the mycelium give color and texture to the molds, producing various shades of green, blue, brown, and even black, with a cottony, velvety, or waxy texture. When the molds are grown using a technique called "slide culture," the spores may be viewed microscopically in an undamaged state in relation to the hyphae. These structures are very useful in identifying the molds. Examples of three common molds, *Aspergillus*, *Penicillium*, and *Mucor*, are shown in Figure 8.9. Yeasts, on the other hand, are not made up of unicellular filaments, but usually come in the form of a sphere. Yeasts reproduce by mitosis, forming a colony of unicellular organisms that normally forms a glistening, smooth, pasty colony on agar. Yeast growth on foods commonly has a white powder-like appearance.[39]

Aspergillus spp.[37,39]

In England during the 1960s, Turkey X disease killed thousands of turkeys, ducklings, and baby chicks, which resulted in a huge economic loss to the area. An investigation revealed that the peanut meal feed was contaminated with *Aspergillus flavus* and a dangerous toxin that was produced by the fungi. Since this incident, extensive research has been conducted on fungi and the mycotoxins they produce. Aflatoxin, produced by strains of *Aspergillus flavus*, causes hemorrhaging, anemia, ataxia, hematosis, and cirrhosis of the liver and is a very potent carcinogen. There are four primary aflatoxins, B1, B2, G1, and G2, which are found in peanuts, corn, and cottonseed. In the United States, aflatoxin is of concern during the storage of corn and in milk from cows that have consumed the aflatoxin. Techniques such as treating milk with ultraviolet light or soaking corn in bisulfite during processing are ways that the food industries eliminate the toxin. Aflatoxin and other mycotoxins are a serious health threat in developing countries, particularly in Africa, Asia, and South America.[41] Fungi grow well in the warmth and humidity of these regions. Unfortunately, many developing countries lack the resources to prevent mycotoxin contamination or to treat contaminated foodstuffs prior to consumption. Further, in these countries, some contaminated products, like milk, are consumed fresh, without any treatment, increasing the risk of illness.

Another toxin produced by Aspergillus species is the ochratoxin, produced by *Aspergillus ochraceus*, found in peanuts, Brazil nuts, specific grains, red and black pepper, and citrus fruit. The ochratoxin causes swelling of the liver and necrosis of the kidney tubules, leading to blockage.

Penicillium spp.[38]

Rubratoxin, patulin, and yellow rice toxins are produced by members of the genus Penicillium. *Penicillium rubrum*, which produces the rubratoxin, can be found on plants such as peanuts, corn, legumes, and sunflower seeds. Patulin, which is produced primarily by *Penicillium expansum*, is found in rotting fruit due to the growth of organisms. Yellow rice toxins are produced by various species of penicillium growing on rice during storage; the toxin causes the rice to turn yellow, hence the name "yellow rice toxins." Citrinin and Citreoviridin are examples of such toxins that produce symptoms such as vomiting, difficulty breathing, low blood pressure and respiratory arrest.[38]

Mucor and Rhizopus spp.

Mucormycosis is the disease caused by fungi in the order Mucorales. Species of the genera Mucor and Rhizopus are the common species responsible for this disease. Found in the soil, Mucor and Rhizopus species are common spoilage organisms of bread and fruit. Pathogenic strains such as *Rhizopus oryzae*, are opportunistic organisms, infecting diabetics, leukemics, and those with suppressed immune systems. The onset of the disease is rapid and includes the invasion of blood vessels, causing embolisms and tissue necrosis. The disease can spread to whole organ systems, causing meningitis and death within 2–10 days.[37,39]

Yeasts

Yeast is sometimes difficult to distinguish from mold. Yeasts are used in industry for the manufacture of beer, ale, bread, and glycerin. Yeasts also produce infections and disease as a result of food contamination. True yeasts, such as *Saccharomyces cerevisiae*, which is used in making beer, produce ascospores after sexual mating. In the laboratory, many yeasts do not display this sexual stage and have been termed "false yeasts," or yeast-like fungi. Candida albicans and *Cryptococcus neoformans* are two examples of yeast-like fungi. They are the most common causes of these infections. Small numbers of *Candida albicans* may

Penicillium   Aspergillus   Mucor

FIGURE 8.10
Microscopic views of slide cultures of *Penicillium*, *Aspergillus*, and *Mucor*.

be found in healthy individuals, but infection can occur in infants, when taking broad-spectrum antibiotics, during pregnancy, or when the body is burdened with a malignant disease. Infections vary considerably in severity. Infection can occur between fingers and toes while causing a rash; in the mouth, vagina, and lungs; or systemic involvement of multiple organs. *Cryptococcus neoformans* is normally a ubiquitous saprophytic organism that is recovered in large numbers from the filth and debris associated with pigeon roosts. The organisms gain entry into the host through the lungs, normally producing a rapidly resolved pulmonary infection. In a few cases, it produces a serious and often fatal central nervous system infection, where the organisms appear in India ink preparations of the cerebrospinal fluid (Figure 8.10).[37,39]

Bacteria

Bacteria are the single-celled organisms that are responsible for more than 80 percent of foodborne illness.[7] Bacteria are much larger than viruses and can be viewed with a conventional microscope. Bacteria are categorized according to the physical and chemical properties they possess. Two broad groups of bacteria classifications are *gram-positive* and *gram-negative*, terms that refer to the amount of peptidoglycan (layer of sugar derivatives that confers strength to the cell wall) in the bacterial cell wall. The cell wall of gram-negative bacteria such as Salmonella and *Escherichia coli* contains only 10 percent peptidoglycan. In gram-positive bacteria such as Staphylococci and Clostridia organisms the cell wall contains up to 90 percent peptidoglycan. This cell wall difference allows scientists to distinguish these two bacterial groups by applying a special stain to the bacteria. Only gram-positive bacteria retain the violet crystal stain, which is taken up by the dense peptidoglycan material. After staining, examination under a microscope will reveal the different shapes of bacteria in the form of coccus, rods, spirilium, spirochete, and appendaged bacteria (Figure 8.11). Rod- and coccus-shaped bacteria are most commonly associated with foodborne illness.

Another important characteristic useful in identifying bacteria is the ability to grow in the presence or absence of oxygen. Aerobic bacteria require oxygen for growth, while the growth of anaerobes such as Clostridium is inhibited by the presence of oxygen (Figure 8.12). Some organisms, such as Campylobacter, are microaerophiles, which require low concentrations of oxygen for growth. If the anaerobe *Clostridium botulinum* is

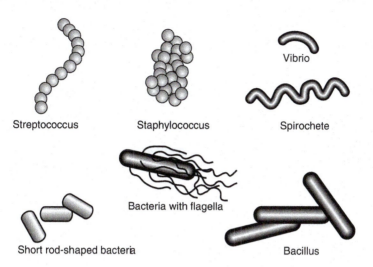

FIGURE 8.11
Examples of various bacterial forms.

FIGURE 8.12
Different types of microbial oxygen requirements.

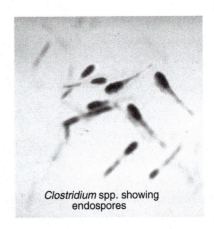

Bacterial cell

Bacterial chromosome (DNA)

Peptidoglycan layer

Spore coat

Plasma membrane

Terminal endospore

Clostridium spp. showing endospores

FIGURE 8.13
Endospores of *Clostridium* spp.

suspected, the investigator might search for endospores, which are structures produced during the life cycle of certain bacteria (Figure 8.13). Endospores can withstand extreme heat, harsh chemicals, and are very difficult to destroy. The dehydrated spore consists of bacterial DNA, a little RNA, some enzymes, and several layers around the spore including a thick protein spore coat that confers resistance to harsh chemicals. The spore cytoplasm also contains dipicolinic acid that forms a complex with calcium. The dipicolinic acid is a major factor in the heat resistance of spores and is essential for later resuming metabolic functions. Aside from the characteristics mentioned earlier, several other factors such as motility, lactose fermentation, or growth on a specific medium are properties that aid in the identification of bacteria.

Salmonella spp.[42–44]

Salmonellosis is the disease caused by the Salmonella organisms, gram-negative facultative anaerobes. Although there are approximately 40,000 confirmed cases each year, investigators estimated that the actual number of cases is between two and four million.[45] Salmonella species are part of the normal intestinal flora of humans, cattle, poultry, pigs, and many other animals. The disease is transmitted via food and water, putting every person at risk. There are three syndromes caused by Salmonella species: typhoid fever, enteric fever, and gastroenteritis.

Typhoid fever, caused by *Salmonella typhi*, accounts for less than 2.5 percent of all salmonellosis in the United States. Humans are the only reservoir for *S. typhi*, and transmission occurs through fecal contamination of food and water by an infected food worker or sewage contamination. Infection occurs when ingested organisms invade the epithelial cells of the small intestine. After penetrating the intestine wall, they invade the immune system, where they multiply and then are carried in the blood throughout the body causing systemic infection. Symptoms include septicemia, high fever, constipation, diarrhea, vomiting, spotting on the chest and trunk, and bleeding from the bowels and nose. Not all who harbor the organism show symptoms, even though they may shed the organism and infect others. These people are called carriers, and the most notorious of these is "Typhoid Mary." The term "Typhoid Mary" originated with Mary Mallon, an Irish immigrant who is thought to have infected hundreds of people by transferring the *Salmonella typhi* organisms she carried through her unsanitary cooking habits until she was placed in exile on an island off of Manhattan.[43]

Enteric fever is the second type of syndrome caused by *Salmonella paratyphi*. This illness comprises only 0.5 percent of the salmonellosis cases in the United States and is similar to typhoid fever, but less severe.

The third and most prominent syndrome of salmonellosis is gastroenteritis, caused by some 2300 serotypes of Salmonella. Only 150 of this number are actually known to cause human sickness, but all are assumed to be pathogenic. Gastroenteritis from Salmonella species does not include the invasive properties or systemic infection as seen with the typhi and paratyphi species. These organisms colonize in the small intestine, causing intestinal inflammation, resulting in diarrhea, abdominal cramps, chills, fever, and vomiting, which last 1–4 days. The severity of the infection varies with the involved Salmonella species, the health status of the individual before infection, and the implicated food. Many healthy individuals who become infected do not seek medical treatment, as symptoms are often mild, lasting 1–3 days. This is the primary reason why only 40,000 of the estimated 2–4 million cases are reported each year.

Salmonellosis stems from a wide variety of sources, including milk, eggs, ice cream, poultry, beef, and salads, infected with such species as *S. typhirium*, *S. enteriditis*, and *S. montevideo*, just to mention a few. Investigations of poultry products in the United States revealed that Salmonella species are present in 30 percent of the chicken available on the market. The numbers of bacteria are usually low, which is why we don't experience a greater number of outbreaks. Considerable efforts are being put forth to eradicate Salmonella in poultry, as it is implicated in one-third of the foodborne outbreaks in the U.S.

In 1997 the FDA approved PREEMPT, the first competitive exclusion product for use in poultry. The concept of competitive exclusion is rather simple. When chicks are hatched they are especially vulnerable to Salmonella infection because their natural intestinal flora, which compete with Salmonella and other bacteria for nutrients, have not yet been established. By spraying newly hatched chicks with a mixture of live bacteria, researchers were able to eliminate 99.9 percent of the Salmonella in poultry. Although this technique may eliminate Salmonella in poultry, there are other means by which an individual may become infected.[44] Salmonella is also spread by infected people who are shedding the organisms in their feces. Unsanitary habits such as not properly washing hands after using the toilet can facilitate the spread through a handshake or other contact. This mode of transmission is known as the person-to-person mode.

Domestic animals such as turtles and iguanas should be considered potential carriers of Salmonella and handled accordingly. Before 1975, domestic turtles accounted for 280,000 cases of Salmonella annually. These numbers prompted the FDA to put a ban on the interstate shipment of turtles, which resulted in an 18 percent reduction of salmonellosis in children ages 1–9.[45]

Salmonella species are commonly found in iguanas (*S. marina*), but the species involved rarely causes human disease. Even though iguanas are imported into the United States from many foreign countries, there are no quarantines or health inspections upon entering this country.

Treatment of gastroenteritis from Salmonella is limited to hydrating fluids and rest. Antibiotic treatment is offered only to severe cases for two reasons. First, non-selective antibiotics will eliminate the natural flora of the intestine leaving the body prone to other infections, and second, antibiotic treatment will not kill organisms that have acquired resistance to these drugs leaving only them to flourish and infect. This second reason is a primary argument of advocates lobbying against the use of antibiotics in meat and poultry.

Staphylococcus spp.[46]

Staphylococcus food poisoning, caused by the gram-positive cocci *Staphylococcus aureus*, is one of the most common foodborne illnesses in the United States. Staphylococci organisms can be found present in the nasal passages, pimples, throat, and wounds of humans. The organism can grow between 7°C and 48°C and produces an enterotoxin between 20°C and 37°C. Sickness is due to the consumption of the enterotoxin that is produced; therefore, Staphylococcus food poisoning is considered an intoxication as opposed to an infection. The intoxication is characterized by nausea, vomiting, and diarrhea occurring within 30 minutes to eight hours after consumption of the entero-toxin. Sickness is usually short in duration and self-limiting, but death has occurred from intoxication. The enterotoxins can survive prolonged boiling and are considered to be heat stable. Therefore, once the enterotoxins are produced in food, they have the capacity to produce illness even if the food is reheated and the Staphylococcus organisms are killed.

Contamination occurs through the preparation of foods by infected food handlers. Foods such as creams, cream pies, potato salad, and ham have all been implicated in outbreaks of Staphylococci food poisoning. Cream products were of major importance in Staphylococci food poisoning, but prevalence of infection has decreased as a result of proper refrigeration. Currently, ham is the most common vehicle of transmission for the Staphylococci organisms.

Staphylococci growth is inhibited by the presence of competitors, but it can survive in high salt and sugar concentrations that destroy other organisms. Pre-cooked ham contains salt in concentrations up to 3.5 percent, which kills off other organisms, leaving a perfect environment for Staphylococci to thrive. Several years ago, 18 people at a retirement party in Florida suffered from Staphylococci food poisoning after eating ham and rice. Samples of the leftover ham and rice pilaf were found to contain the enterotoxin produced by *Staphylococcus aureus*. Although investigators never found the source of contamination, it was likely due to the whole ham being placed immediately in the freezer after cooking, which would have prolonged the cooling time, allowing Staphylococcus organisms to multiply.[47]

Although the foods mentioned previously have been implicated in staphylococcal intoxications, virtually any food can serve as a vehicle if handled carelessly by infected food workers and left out at temperatures ideal for bacterial growth.

Clostridium spp.[48–50]

Clostridium perfringens is an anaerobic spore-forming bacteria found in soil and the intestinal tracts of humans and other animals. *C. perfringens* grows at temperatures between 20°C and 50°C. In optimum conditions (45°C) with the right media, *C. perfringens* can double its numbers in as little as 8–10 minutes, making it the fastest-growing bacteria known.[48]

In order for food poisoning to occur, numbers in excess of 10^8 bacteria must be ingested.[48] When the organism senses harsh conditions, such as a lack of nutrients or water, it sporulates and releases the enterotoxin responsible for the intoxication. Two types of spores that are classified by their sensitivity to heat are produced by various strains of *C. perfringens*. Heat resistance is measured by the D value (the amount of time and heat required to kill 90 percent of the population). Heat-resistant spores have a D90°C value of 15–45 minutes, which implies that 90 percent of the organism population is eliminated by heating to 90°C for 15–45 minutes. Heat-sensitive spores have a D90°C value of 3–5 minutes. The enterotoxin produced by *C. perfringens* is destroyed by heating to 60°C for 10 minutes. *C. perfringens* food poisoning usually occurs when cooked foods such as beef, turkey, or chicken are held for several hours at room temperature before serving or when dense foods are refrigerated after cooking. These conditions allow the vegetative cells to grow to dangerous levels in the food before it is consumed.

In March 1993, 156 persons suffered from food poisoning at a St. Patrick's Day meal. Investigation of the outbreak revealed that the delicatessen serving the food had prepared corned beef sandwiches early in the day and held them at room temperature until they were purchased. As mentioned earlier, holding foods at room temperature for extended periods of time is a dangerous practice that gives many pathogenic organisms the chance to proliferate.[49]

Food poisoning by *C. perfringens* is mild, usually limited to diarrhea, abdominal pain, and vomiting. Symptoms can manifest as soon as two hours after ingesting the contaminated food and last about 24 hours. Death from *C. perfringens* is rare but does occur among the elderly and those with an underlying condition.[49]

Clostridium botulinum

Clostridium botulinum, like *Clostridium perfringens*, is a spore-forming anaerobic bacterium found in soils throughout the world. Botulism is the illness that results when *C. botulinum* spores germinate and produce a toxin in the food that can be ingested. Outbreaks of botulism are rare in the United States, and although commercial foods have been implicated, home canned foods are the most common source. *C. botulinum* organisms are not a threat in food unless they can germinate into the toxin-producing vegetative cells. By destroying the spores in foods before canning or storing products, risk of botulism can be eliminated. The spores, however, are quite resistant to heat ($D_{90°C}$ is 30 minutes) and chemicals, making them difficult to destroy. The neurotoxins produced by the organism are fortunately heat labile and can be easily destroyed by boiling for 3 minutes or by heating to 80°C for 30 minutes.

C. botulinum organisms can grow in temperatures between 3.3°C and 45°C, but this varies with the type of organism. Aside from heat and chemicals, pH and oxygen concentration also affect growth. A pH of 4.6 inhibits the germination of spores, foods such as tomatoes may naturally have a pH as low as 4.6, while food of a higher pH may have acids added to them during processing to prevent spore germination. The oxygen concentration of the growth medium is a critical factor to *C. botulinum* growth. While the conditions on the surface of a food product may be aerobic, anaerobic conditions may exist within the food mass. In products that are smoked, such as fish and sausage, anaerobic conditions can develop, producing a particularly dangerous situation because the organism may be present in lethal numbers without having affected the appearance of the food. Fresh mushrooms were previously packaged in containers sealed with a cellophane wrap. If aerobic bacteria consumed all the oxygen present in the package, they would die, leaving anaerobic conditions without competitors, an ideal setting for an outbreak of botulism. The cellophane wrap now used is punched with holes to avoid this scenario.

Currently, there are seven types of *C. botulinum* listed, A-G, which are identified by the toxin they produce. Only types A, B, E, and F have been found to cause foodborne illness in humans. Types C and D cause botulism in cattle and other animals, while G has been reported to cause infant botulism not associated with food. Botulism caused by the A toxin is the most common in the United States, followed by B, E, and F. The A and B toxins have been isolated in fruits, vegetables, fish, condiments, beef, pork, and poultry. The E toxin has been isolated primarily in fish and the F toxin only in meat products.

Food poisoning in adults occurs within 12–36 hours of consuming the toxin. At the onset, symptoms such as nausea, vomiting, and diarrhea are present; then, as the condition develops, fatigue, blurred vision, difficulty speaking and swallowing are experienced. If the condition goes untreated, death can occur from respiratory paralysis. Infant botulism occurs in children less than one year old. Ingested honey has been the source of most food associated cases, but 50 percent of infant botulism occurs in children who were strictly breastfed, leaving air, dust, and soil as the likely sources of contamination.[50]

Campylobacter

Campylobacter is one of the leading causes of acute gastroenteritis in the United States, responsible for approximately 2 million cases annually. *Campylobacter* species are part of the normal flora of the gastrointestinal tract of warm-blooded animals. Of primary concern are cattle and poultry. Contamination occurs primarily during food processing when the intestinal tract is lacerated, allowing feces to contaminate the food product. Infected food handlers and raw milk are also significant sources of infection. *Campylobacter jejuni* is the organism most commonly implicated in Campylobacter infection.

Campylobacter jejuni is a microaerophile (requiring little oxygen), capable of growing in 3–15 percent oxygen. It has a narrow temperature range for growth (between 30°C–47°C) but more importantly, Campylobacter can survive for weeks in refrigeration at 4°C. Although no growth may occur during refrigeration, Campylobacter is still a threat because infection can occur from as little as a few hundred cells. Symptoms in the infected are usually mild, including nausea, vomiting, and bloody diarrhea. In severe infections, Guillain-Barré Syndrome develops, which causes neuromuscular paralysis.[51]

Escherichia coli[52–54]

E. coli organisms can be found among the normal bacterial flora in the intestine of humans and other warm-blooded animals. The O157:H7 strain of *E. coli* has caused several outbreaks of severe sickness and death, which has given the public the impression that all *E. coli* strains are harmful, when in fact the majority of strains are harmless and some strains can be found in the intestinal tracts of healthy, asymptomatic individuals.

E. coli organisms that are important to foodborne illness can be divided into four groups—enteroinvasive, enterotoxigenic, enteropathogenic, and enterohemorrhagic—to which the infamous *E. coli* O157:H7 strain belongs. Enteroinvasive *E. coli*, as the name implies, invade the epithelial cells of the intestine, resulting in fever, chills, and bloody diarrhea. In contrast, enterotoxigenic *E. coli* that are responsible for traveler's diarrhea exhibit cholera-like symptoms, where cells colonize the wall of the intestine, then produce a toxin that prompts watery diarrhea, vomiting, and abdominal cramps. This can result in dehydration. Enteropathogenic *E. coli* is most commonly found among infant nurseries in developing countries. During the 1940s and 1950s, enteropathogenic *E. coli* was a problem in the United States, but the development of more sanitary practices has diminished its prevalence in recent decades. However, enteropathogenic *E. coli* remains a major human health problem across the globe in developing countries, particularly causing diarrhea in children.[1]

Enterohemorrhagic *E. coli*, also known as *E. coli* O157:H7, has become one of the more problematic foodborne pathogens during the 1990s. Cattle are the major reservoir for the organism. Human infection is the result of consuming improperly cooked ground beef, raw milk, or unpasteurized apple cider. Symptoms generally include abdominal cramps, watery to bloody diarrhea, vomiting, and possibly a fever. *E. coli* O157:H7 is especially dangerous because it can be accompanied by hemolytic uremic syndrome (HUS), which is the primary cause of renal failure in children.[52] In January 1993, Washington State experienced an outbreak of *E. coli* O157:H7 due to improperly cooked hamburgers from a fast food chain. The outbreak consisted of 501 cases, 45 cases of HUS, and 3 deaths.[54] Outbreaks such as this can be avoided by properly cooking meat to an internal temperature of 68°C. Another outbreak of *E. coli* O157:H7 occurred in 1996 in Japan due to consumption of raw alfalfa sprouts. There were 9,000 illnesses and 17 deaths due to that outbreak. This outbreak and several smaller outbreaks in the United States prompted the FDA to put out a warning against consumption of raw alfalfa sprouts until new measures are put in place to ensure sprout safety.[53] There are also new techniques being developed to fight off *E. coli* O157:H7 in milk and other products. Cold pasteurization is a process by which an electron beam is directed toward the product and reacts with *E. coli* DNA, inactivating it. Thus far, this new technology has not been found to alter the foods in any significant way.[55]

Vibrio cholerae[56–58]

The disease cholera, caused by the organism *Vibrio cholerae*, is most prominent in LDCs where unsanitary conditions prevail. In the United States, most cholera cases are associated with international travelers who either become infected while abroad or who bring contaminated food items back into the United States where the infection may occur. In 1992, 102 cases of cholera were identified in the United States, the most reported to the CDC in a single year. Ninety-five of the cases were associated with travel, 75 of which occurred during an outbreak on an Argentinean airline.[56]

In Peru in 1991, an outbreak of cholera spread to 322,562 Peruvians. Following this initial outbreak, cholera spread from Peru throughout most of South America, north through Central America, Mexico, even reaching the U.S. Statistics from the Pan American Health Organization indicate that between 1991 and 1992 there were 645,030 cases with 5,694 deaths related to this epidemic.[57] A member of the genus Vibrio, this gram-negative rod can proliferate in temperatures between 15°C and 42°C, in a pH range of 6–10, and in a solution containing up to 6 percent sodium chloride. Hence it can be found in coastal waters contaminated with fecal matter. Crabs, shrimp, mussels, and claws can harbor the organism for long periods of time and have all been implicated as sources of infection.[56]

Vibrio cholerae does not invade the intestinal being. Instead it colonizes on the lining and produces the toxin choleragen, which causes the massive influx of fluids into the intestine. Symptoms precipitate 2–3 days after consuming the contaminated food and generally appear as abdominal pains, dehydration, and a characteristic diarrhea that has been termed "rice water stool" for its similar appearance to the water left behind after rice has been cooked. Complications associated with cholera include hypoglycemia, severe dehydration, and even renal failure. Cholera can have a case fatality rate of up to 50 percent in untreated patients, who usually die of complications surrounding the extreme loss of fluid from the body (up to one liter per hour in adults).

Although a vaccine exists, it is not practical for the inhabitants of endemic areas because it is effective for only 3–6 months. In those endemic areas, an increase in sanitation practices would be more practical. However, the vaccine is recommended for travelers spending short periods of time in endemic areas. Antibiotics can be given for treatment, although, as seen with Salmonella, many strains have acquired resistance to the traditional medicines. Rehydration solution can be used to treat mild cases that usually pass without any other treatment. In severe cases, rehydrating fluids are introduced intravenously to counter the rapid loss of fluid.[58]

Factors Frequently Cited in Foodborne Illness

During the food processing chain, many controls are put in place to prevent foodborne disease. To produce a food that is 100 percent free of contamination is not feasible due to the economic costs as well as the fact that such a rigorous sanitation process would alter the food so much that it would no longer appeal to the consumer. Instead, acceptable limits have been set that allow low numbers of certain bacteria to present in a food. If the food is treated properly, the presence of such low numbers of bacteria should not pose a threat. In many cases, however, food is abused. The result is often a foodborne outbreak.

Factors that contribute to foodborne illness can be broken down into three categories: (1) time and temperature, (2) poor hygiene practices by food handlers, and (3) cross contamination. Following factors within these categories are commonly found to be the cause of foodborne outbreaks, which are largely preventable.[59,60]

1. Improperly refrigerated food
2. Improperly heated or cooked food
3. Food handlers who practice poor hygiene
4. Lapse of a day or more between preparing and serving food
5. Introducing raw or contaminated materials to a food that will not undergo further cooking
6. Improper storage of foods at temperatures ideal for bacterial growth
7. Failure to properly heat previously cooked foods to temperatures that will kill bacteria
8. Cross contamination of ready-to-serve foods with raw foods, contaminated utensils or machinery, or the mishandling of foods

Some suggestions for reducing the hazards associated with the contamination of food are shown in Figure 8.14.

Hazard Analysis Critical Control Points[61]

Although there are several measures in place to prevent foodborne disease, pathogens such as *E. coli* and *Salmonella* still continue to pose a serious health threat. In response to this threat, the federal government has mandated the implementation of hazard analysis critical control points (HACCP) strategies in the seafood, poultry, and meat industries.

In contrast to the traditional end product testing methods, HACCP intervenes at various stages in the food processing/preparation chain to ensure a safer product. The objective of a HACCP plan is to identify hazard points within a certain industry that may lead to foodborne illness, implement strategies to minimize the identified hazards, and monitor these practices to ensure that all implemented plans are adhered to and are effective. There are seven key principles to the HACCP system: (1) assessing the hazard, (2) identifying critical control points (CCPs), (3) establishing standard procedures and critical limits, (4) monitoring procedures, (5) corrective actions, (6) record keeping, and (7) verification the system is working properly. By adhering to these principles, food establishments can reduce the risk of a foodborne illness to a minimum (Table 8.6).

Refrigerator/freezer

- Freezer should be kept at −17.7°C and refrigerator at 3.3–4.4°C, check temperature periodically with a reliable metal thermometer
- Store raw meats on bottom, and enclose in plastic wrap or baggies
- Large amounts of hot food should be divided into shallow pans for faster cooling, then covered and placed in refrigerator
- Clean refrigerator on a regular basis.
- Freeze meat, poultry, and fish products if you are not going to use them quickly

Time and temperature

- Serve or discard foods within 4 hours from time when it is removed from temperature control
- The danger zone for food is 4.4°C to 60°C

Foods

- Never mix different meats, poultry, or fish on counter or in storage
- Package meat and poultry in plastic storage bags to prevent cross contamination
- Do not allow high-risk foods to remain at room temperature for more than a couple of hours, refrigerate or cook immediately
- Do not partially cook foods.
- Cook poultry to 82.2°C, meat and ground meat to 71.1°C, and reheat leftovers to 73.8°C or until very hot

Counters/sinks

- Disinfect dishcloths frequently each day by placing in disinfectant soap, or heating in a cup of water in a microwave before cleaning countertops, washing dishcloths in a dishwasher is also good practice
- Wash hands thoroughly before food preparation
- Clean countertops with hot, soapy water between placement of high-risk foods on the same surface
- Thaw high-risk foods such as poultry in the refrigerator or run under cold water

Microwave cooking

- Use thermometer or meat probe to insure proper cooking temperature
- Be sure to let the food stand as instructed since further cooking will take place during this time
- Rotate or stir to assure even cooking throughout

Cutting boards

- Avoid preparing food on wooden cutting surfaces where deep grooves permit high numbers of bacteria to grow
- Use plastic or ceramic cutting surfaces that can be easily sanitized between uses or between the preparation of high-risk foods, fruits, and vegetables

FIGURE 8.14

Recommendations for reducing hazards associated with the contamination of food.

TABLE 8.6

The Seven Key Principles of HACCP

Hazard Critical Control Point (HACCP) System

Assess the Hazards
- Identify potentially hazardous foods
- Follow the flow of food to assess
  hazards at receiving, storing, preparing, cooking, holding, serving, cooling, and reheating
- Estimate risks

Identify Critical Control Points (CCPs)
- Develop procedure and flowcharts showing the flow of food and all of the CCPs

Set Up Procedures and Standards for CCPs
- Standards must be met at each CCP and should be: measurable, based on fact, correct for the recipe, clear directions with specific actions

Monitor CCPs
- Check to see if standards are met; employees should be involved in process; standards must be met

Take Corrective Actions
- If standard not met, correct it
- Have specific steps for correction

Set Up Record Keeping System
- Blank forms near equipment where they are to be used
- Notebooks to write down actions
- Flowcharts and recipes near work areas

Verify That the System Works
- Identify and assess all hazards
- CCPs selected
- Standards set with monitoring and schedules
- Corrective actions in place
- Monitoring being done
- Flaws or omissions corrected
- Monitoring equipment calibrated

Are you interested in learning more about foodborne disease? Visit <https://people .umass.edu/~envhl565/> and click on one or more of the following sites under Chapter 8.
General Links to Foodborne Disease

- *FDA/CFSAN bad bug book*: Introduction to Foodborne Pathogenic Microorganisms and Natural Toxins
- FDA Center for Food Safety & Applied Nutrition
- HACCP Web site
- Food Safety Education
- Foodborne Illness and Death in the United States
- U.S. FDA/Center for Food Safety and Applied Nutrition (CFSAN)
- FDA Food Safety and Modernization Act
- European Union Food Safety
- International Guidance on Food Safety in Trade

Assessing the Hazard

Adequately assessing the hazards in a food process is essential to an effective HACCP plan. The person or persons conducting the hazard assessment must review the foods on site to determine which are potentially hazardous, employee activities should be supervised for dangerous practices, and the design of the facility should be examined for potential flaws.

Identifying Critical Control Points

Once the hazards have been assessed, CCPs are identified to control the potential hazards. For example, if raw chicken is being handled, Salmonella would be an identified hazard, and the cooking process would be the CCP, because it is there that Salmonella can be eliminated. Depending on the hazard, other CCPs could be refrigeration to prevent microbial growth or fluctuation of the pH to inactivate a toxin.

Establishing Standard Procedures and Setting the Critical Limits

Procedures such as washing hands, sanitizing machinery, and food preparation should be established and well defined to ensure the production of a safe product. Critical limits such as time, temperature, pH, and water activity are parameters that ensure the production of a safe product. Operating outside the critical limits could result in foodborne illness. In the case of raw chicken, the critical limits during cooking would be heating to an internal temperature of at least 73.9°C. By not meeting these critical limits, Salmonella organisms can remain viable in numbers high enough to cause infection or possibly contaminate another food item.

Monitoring Procedures

The periodic monitoring of procedures is necessary to track the operation of the facility, indicate when critical values have not been followed, and provide documentation in the event of an inspection or other situation that may require verification. For instance, while cooking our chicken we do periodic checks on the oven temperature, inspection reveals that the oven is operating at 74°C, the time, date, and temperature are recorded, and the supervisor might request that the temperature be raised slightly. By monitoring the CCPs, improperly cooking the chicken has been avoided, preventing a possible outbreak of Salmonella.

Corrective Actions

Corrective actions must be taken when a CCP has operated outside of the critical limits. In some cases, it may be necessary to discard a product, while other situations may simply require bringing the CCP back within the critical limits. In either case, the deviation from the critical limit should be documented. Any corrective action should be approved by the proper regulatory bodies before implementing HACCP.

Record Keeping

If a facility is required to have an HACCP plan, a written copy of the plan must be kept on-site. In the event government inspectors visit a facility, documentation regarding CCPs, critical limits, and deviations from those limits should be available.

Verification the HACCP System Is Working Correctly

The purpose of verification is to check how effective the plan is in reducing the identified hazards, and possibly reveal any hazards that might not have been previously identified. Verification may include reviewing current documentation for patterns in errors or corrective actions, which may reveal false reporting, ineffective CCPs, or other flaw in the HACCP plan. Detection methods such as those discussed previously, confirm the presence or absence of identified microbes, and should reveal if the HACCP plan has resulted in a decreased number of those pathogens.[60]

United States Regulatory Efforts with Regard to Food Protection

On December 18, 1997, the FDA required that all seafood processors, domestic and those importing to the United States, carry out a hazard analysis of their products and processes. In the event hazards are recognized, the facility is required to have a written HACCP plan to address the problem. The ruling also identified eight sanitation concerns that must be monitored and controlled. A formal program of sanitation standard operating procedures (SOPs) is not required by the FDA; however, the FDA does require that records on sanitation SOPs be filed.

The United States imports more than 50 percent of the seafood that Americans eat, and there are an estimated 6,000 firms in 172 different countries that are involved in the seafood industry. Any importers of seafood will have to verify that their product was produced in accordance with the HACCP ruling. HACCP plans will undergo yearly evaluations to compensate for any changes in manufacturing practices and to add more controls if additional hazards are identified. As can be expected, this ruling was met with resistance from the seafood industry. Many importers will have to find new producers who can comply with the ruling, while domestic producers may have to change their current practices in order to meet the new requirements. The meat and poultry industries are currently experiencing similar rulings.

On January 27, 1997, the USDA required meat and poultry slaughterers and processing facilities to have sanitation SOPs in place and that they also conduct generic *E. coli* testing. Large facilities employing 500 or more workers must also adhere to reduction performance standards for Salmonella organisms, to ensure the HACCP plan is reducing the risk of foodborne illness. From 1999, facilities harboring 10–499 workers were subject to this regulation, and by January 25, 2000 over 6,200 meat and poultry plants were operating under this ruling.[61]

Fast forward to 2018 and new regulations called the FSMA Produce Safety Rule are now in place for produce in the United States.[62] Both small and large farms are affected by the regulations. These regulations cover a spectrum of processes and standards around produce, including worker training, health and hygiene, and the cleanliness and upkeep of equipment, tools, and buildings. Microbial contaminant limits also apply to the produce.

Surveillance Efforts

Outside of the food industry, considerable efforts are also being put forth to control and better understand foodborne pathogens. The Foodborne Diseases Active Surveillance Network (FoodNet), which is part of the Centers for Disease Control and Prevention (CDC) emerging infections program, is the latest surveillance strategy aimed at detecting foodborne illness that would normally go unnoticed under the traditional system of passive surveillance.[63]

Previously, foodborne disease was monitored by a passive approach of reporting cases that had been confirmed in a clinical lab. This method only accounted for a fraction of the actual cases of foodborne illness, underestimating the impact foodborne disease. The FoodNet is an active approach of tracking disease, where health officials contact laboratories to request data on new cases of foodborne illness. The health official in turn, reports the information to the CDC via a computer network. Surveillance sites have been established in several states where surveys are conducted among clinical laboratories, physicians, and the general population. The purpose of these surveys is to more accurately assess the incidence of foodborne illness, and to monitor the procedures of physicians and laboratory personnel that might influence the number of pathogens that are isolated and reported.

Case control studies are another component of the FoodNet that provide information on which food items are most commonly associated with these diseases. Isolates from these studies are sent to the CDC where they are phagotyped and tested for antibiotic resistance. This information alerts health officials to the presence of new pathogens that have not been previously isolated, or to strains that have become resistant to antibiotics used to reduce their numbers in animals or to combat infections in the human body. The FoodNet has been up and running since January 1, 1996; since then it has identified outbreaks of Campylobacter in California, Salmonella in Oregon, and two outbreaks of *E. coli* O157:H7 in Connecticut.[63]

Another agency recently set up in the United Kingdom is the Food Standards Agency. This agency will be used for creating policies regarding food safety and reporting of food poisoning. There is some controversy over this agency however, because this agency is looking into issues that are normally covered by the Ministry of Agriculture, Fisheries and Food and local governments. Many in England felt that these agencies did not do a good job with the recent outbreaks of both *E. coli* and salmonella and pushed for a new agency that will actually investigate the cases and make new regulations, something the other agencies have only advised on.[64]

Along with sources like FoodNet and agencies like the Food Standards Agency, there are temperature indicators that consumers can buy to test their foods. These indicators are adhesive strips that signal when foods have changed temperature and are therefore dangerous. The strips can also be used in restaurants and grocery stores to determine freshness.[65]

Safety of Food for Import and Export

The international food trade is governed by a set of international food standards called the Codex Alimentarius.[66] The Codex provides guidance on food safety and quality across country borders. This codex was established in 1963, set forth by the Codex Alimentarius Commission. This commission currently consists of 189 Member Countries across the globe, as well as the European Union. It exists under the auspices of the Food and Agriculture Organization of the UN (FAO) and WHO. Countries that produce food that is exported to other countries, including meat, eggs, grains, follow the guidelines for food that is exported. The Codex provides guidance for a wide variety of foods that may be processed, partly processed, or raw. Each specific foodstuff, such as canned applesauce or sweetened condensed milk, has its own written standard. The standard provides clear guidance on the essential components of a product, contaminant levels, production, labeling, methods of sampling and analysis.

Food Safety in South America

The international food trade drives food safety improvements of food export from countries across the globe. The $200 billion U.S. dollar international food trade incents countries to higher standards for their exported foodstuffs than exist for food sales within a country's own borders.[67] This is the case for many countries in South America, where meat and grain are major exports to the U.S. and other developed countries but foodborne illness remains a risk intra-country due to a lack of political drive for higher internal food safety standards. Further, lack of awareness among local consumers about foodborne disease compounds this food safety problem in many Latin American countries.[66]

Food Safety in the EU

Just as the United States Food and Drug Administration monitors food safety and the CDC tracks foodborne disease outbreaks, the European Union has similar regulatory bodies. The European Commission and the European Food Safety Authority promulgate the General Food Law (GFL) as the cornerstone of the EU food regulatory framework. This law covers the food sector, "from farm to fork."[68] The EU has a food safety monitoring system in place that alerts when food exceeds levels of regulated substances or exceeds food safety rules in any way. This allows entire batches of food to be recalled quickly, as soon as an alert is received. The EU also has a robust method of food tracing, so it may track a foodborne disease outbreak up or down the chain of production and processing.[69]

Food Safety in China

China has experienced a number of significant crises around food safety, with some problems due to the country's highly fragmented food chain. The country has 500,000 food production and processing companies. More than 2/3 of these are very small, with fewer than 10 employees.[70]

In China, the Food Safety Commission is in charge of leading China's food safety problems. This has included regulations issued to food producers about traceability and record keeping, especially for high-volume food producers and food produced for high-risk groups like babies.[71]

Despite new regulations, China has experienced some high-profile food safety violations. The meat scandal in 2014 revealed that Shanghai Husi Food, a subsidiary of U.S.-based food supply giant OSI Group, was repackaging and selling expired beef and chicken products with new expiration dates.[72] Further, journalists revealed that a popular chain restaurant in China had serious worker hygiene problems. Video showed that at the Chinese hotpot chain Haidilao, dishes and cutlery were being washed with materials used to clean the restaurant floors.[73]

Food Fraud

An additional challenge beyond foodborne illness is food fraud. In the United States, food fraud often consists of adulterated food. For example, olive oil may be diluted with less expensive oil, like peanut oil, but sold as 100 percent olive oil. Unfortunately, if this product is consumed by someone with a peanut allergy, he or she may become ill or die.[74] Other foodstuffs in the U.S. that have been found to be fraudulent are honey, apple juice, milk, orange juice and coffee.

Food fraud is often done for financial gain, substituting a cheaper alternative while charging the higher price for the legitimate expensive product.

There have been large-scale international investigations of food fraud. One such operation, known as Operation Opson IV, took place across 47 countries between

December 2014 and January 2015.[75] It led to the discovery and confiscation of thousands of kilograms of 2,500 fraudulent foods including:

- Previously frozen fish, sold as fresh, but thawed and doused with chemicals to make it appear fresh
- Mozzarella cheese sold as fresh, but sold after expiration date and treated with chemicals to make the cheese appear fresh
- Counterfeit alcohol sold as brand name vodka, whisky, and beer

Future Challenges in Foodborne Illness

New challenges to food safety will continue to emerge, largely due to:[76]

- Increased amounts of imported foods, especially fresh fruit and vegetables
- Increased food contamination
- Better detection of outbreaks
- Emerging disease
- Increasing antibiotic resistance
- Changes in consumer preferences and habits

References

1. WHO, Estimates of the Global Burden of Foodborne Diseases, Foodborne Disease Burden Epidemiology Reference Group 2007–2015. http://www.who.int/mediacentre/news/releases/2015/foodborne-disease-estimates/en/ (accessed April 21, 2018).
2. Hoffmann, S., Devleesschauwer, B., Aspinall, W.et al., Attribution of global foodborne disease to specific foods: Findings from a World Health Organization structured expert elicitation, *PLoS ONE* 12(9): e0183641, 2017.
3. Estimates of Foodborne Illness in the United States, Burden of Foodborne Illness. https://www.cdc.gov/foodborneburden/2011-foodborne-estimates.html (accessed April 21, 2018).
4. Kaferstein, F.K., Food safety: A commonly underestimated public health issue, *World Health Organization Statistics Quarterly*, 50, 3, 1997.
5. Motarjemi, Y., and Kaferstein, F.K., Global Estimation of foodborne disease, *World Health Organization Statistics Quarterly*, 50, 8, 1997.
6. Buzby, J.C., and Roberts, T., Economic costs and trade impacts of microbial foodborne illness, *World Health Organization Statistics Quarterly*, 50, 57, 1997.
7. Theis, M, Case problem: Balancing the risk of foodborne illness with patient rights and quality of life. *Journal of the American Dietetic Association*, 99, 9, 1999.
8. Centers for Disease Control (CDC), *Foodborne and Diarrheal Diseases Branch (FDDB)*, June 13, 2006. http://www.cdc.gov/foodborne/ (accessed April 21, 2018).
9. Stuart, D. Study puts U.S. food poisoning at 76 million yearly, *The New York Times*, September 17, 1, 1999.
10. Anonymous, Food poisoning: The real extent of food-borne disease, *Consultant*, 39, 3, 1999.
11. Scallan, E., Hoekstra, R.M., Angulo, F.J., Tauxe, R.V., Widdowson, M., Roy, S.L., Griffin, P.M., Foodborne illness acquired in the United States—Major pathogens, *Emerging Infectious Diseases*, 17(1), 7–15, 2011.
12. USDA Serotypes Profile of Salmonella Isolates from Meat and Poultry Products January 1998 through December 2014, United States Department of Agriculture, Washington, DC. https://www.fsis.usda.gov (accessed April 20,2018).

13. National Enteric Disease Surveillance: The Listeria Initiative, February 2016. http://www.cdc .gov (accessed April 21, 2018).
14. EU summary report on zoonoses, zoonotic agents and food-borne outbreaks 2016, *European Food Safety Authority, European Centre for Disease Prevention and Control*, November 13, 2017.
15. *Salmonella Cases No Longer Falling in the EU*, December 12, 2017. https://www.efsa.europa.eu /en/press/news/171212 (accessed April 21, 2018).
16. *Multi-country outbreak of Listeria monocytogenessero group IVb, multi-locus sequence type 6, infections probably linked to frozen corn*, European Food Safety Authority European Centre for Disease Prevention and Control.
17. Koren, H. and Bisesi, M., *Handbook of Environmental Health and Safety*, Lewis Publishers, Boca Raton, FL, 1996, Chap. 2.
18. McGeary, J., Nukes...They're back, *Time*, p. 34, May 25, 1998.
19. Dioxins and their effects on human health fact sheet, October 2016. http://www.who.int /mediacentre/factsheets/fs225/en/ (accessed April 21, 2018).
20. Ennever, F.K., *Metals, Principles and Methods of Toxicology*, Raven Press, New York, 1994, Chap. 12.
21. Groth, E., Benbrook, C.M., and Lutz, K., Do you know what you are eating? An analysis of U.S. Government data on pesticide residues in foods, Consumers Union of United States, Public Service Projects Department, Yonkers, NY, Technical Division, February 1999.
22. Taylor, S.L., *Chemical Intoxications, Foodborne Diseases*, Academic Press, San Diego, CA, 1990, Chap. 10.
23. Gillem, C., Hold the plum pudding: US food sampling shows troubling pesticide residues, December 21, 2017, Environmental Health News. http://www.ehn.org/what-foods-have-the -most-pesticides-on-them-2518891617.html (accessed April 21, 2018).
24. Castle, S., Dioxin in meat is Europe's new food scare, *The Independent* (London), September 22, 2, 1999.
25. Reddy, C.S., and Hayes, A.W., *Food-Borne Intoxicants, Principles and Methods of Toxicology*, Raven Press, New York, 1994, Chap. 9.
26. Hunter, B., Overlooked threats of foodbourne illness, *Consumer's Research Magazine*, 78, October 1, 1995.
27. Halliez, M.C., and Buret, A.G., Extra-intestinal and long term consequences of Giardia duode-nalis infections, *World Journal of Gastroenterology: WJG*, 19(47), 8974–8985, 2013..
28. CDC, Outbreak of Trichinellosis Associated with Eating Cougar Jerky-Idaho, *Morbidity and Mortality Weekly Report*, 45(10), 205, 1995.
29. Cliver, D.O., *Parasites, Foodborne Diseases*, Academic Press, San Diego, CA, 1990, Chap. 21.
30. Markel, E., Voge, M., and John, D., *Medical Parasitology*, 7th ed., W.B. Saunders Company, Philadelphia, PA, 1992, Chap. 3.
31. Eckman, M., and Priff, N., *Diseases*, Second ed., Springhouse Corporation, Springhouse, PA, 1997.
32. Cameron, J., Parasites, a "select epidemic," *Hampshire Daily Gazette*, 23, September 12, 1995.
33. Juranek, D.D., Cryptosporidiosis: Sources of infection and guidelines for prevention, *Clinical Infectious Diseases*, 2(1), S57, 1996.
34. Hutin, Y.J.F., Pool, V., Cramer, E.H. et al. A multistate foodborne outbreak of Hepatitis A, *New England Journal of Medicine*, 340, 8, 1999.
35. Cliver, D.O., *Viruses, Foodborne Diseases*, Academic Press, San Diego, CA, 1990, Chap. 20.
36. CDC, 1998. http://www.cdc.gov/ncidod/diseases/hepatitis/h96trend.htm (accessed April 21, 2018).
37. Tortora, G.J., Funke, B.R., and Case, C.L., *Microbiology, An Introduction*, 5th ed., The Benjamin /Cummings Publishing Company, Redwood City, CA, 1995, Chap. 12.
38. Marth, E.H., *Mycotoxins, Foodborne Diseases*, Academic Press, San Diego, CA, 1990, Chap. 8.
39. Moore, G.S. and Jaciow, D.M., *Mycology for the Clinical Laboratory*, Reston Publishing, Reston, VA, 1979.
40. U.S. Trends and outbreaks, Norovirus. https://www.cdc.gov/norovirus/trends-outbreaks.html (accessed April 21, 2018).

41. Sowley, E., Aflatoxins: A silent threat in developing countries, *African Journal of Biotechnology*, 15(35), 1864–1870.

42. CDC, Salmonella serotype montevideo infections associated with chicks—Idaho, Washington, and Oregon, spring 1995 and 1996, *Morbidity and Mortality Weekly Report*, 46(11), 237, 1997.

43. Huckstep, R.L., *Typhoid Fever and other Salmonella Infections*, E. & S. Livingstone, London, UK, 1962, p. 227.

44. Murphy, J., FDA approves new anti-Salmonella spray developed by ARS, *Food Chemical News*, 26, March 23, 1998.

45. Doyle, M.P. and Cliver, D.O., *Salmonella, Foodborne Diseases*, Academic Press, San Diego, CA, 1990, Chap. 11.

46. Cliver, D.O., *Transmission of Disease Via Foods, Foodborne Diseases*, Academic Press, San Diego, CA, 1990, Chap. 1.

47. CDC, Outbreak of staphylococcal food poisoning associated with precooked ham—Florida, 1997, *Morbidity and Mortality Weekly Report*, 46(50), 1190, 1997.

48. Johnson, E., *Clostridium Perfringens Food Poisoning, Foodborne Diseases*, Academic Press, San Diego, CA, 1990, Chap. 16.

49. CDC, *Clostridium perfringens* Gastroenteritis associated with corned beef served at a St. Patrick's Day meals—Ohio and Virginia, *Morbidity and Mortality Weekly Report*, 43(8), 137, 1993.

50. Sugiyama, H., *Clostridium Botulinum*, Foodborne Diseases, Academic Press, San Diego, CA, 1990, Chap. 6.

51. CDC, Outbreak of *Campylobacter* enteritis associated with cross-contamination of food Oklahoma, *MMWR Weekly*, 47(7), 129–131, 1996.

52. Doyle, M.P, and Cliver D.O., *Escherichia Coli*, Foodborne Diseases, Academic Press, San Diego, CA, 1990, Chap. 13.

53. Kurtzweil, P. Questions keep sprouting about sprouts, *FDA Consumer*, 33, 18, 1999.

54. Bell, B.P., Goldoft, M., Griffin, P.M. et al. A multi-state outbreak of *Escherichia coli* O157:H7, *Journal of the American Medical Association*, 272, 1349, 1994.

55. Earth Vision Reports, New Pasteurization Technology makes foods safer, Missouri, 1999. http://www.earthvision.net/ColdFusion/News_Page1.cfm?NewsID=8790 (accessed April 21, 2018).

56. CDC, Cholera associated with international travel, *Morbidity and Mortality Weekly Report*, 41(36), 664, 1992.

57. CDC, Update: Cholera-Western Hemisphere, 1992, *Morbidity and Mortality Weekly Report*, 41(36), 667, 1992.

58. Doyle, M.P., and Cliver, D.O., *Vibrio, Foodborne Diseases*, Academic Press, San Diego, CA, 1990, Chap. 17.

59. The National Restaurant Association, *Applied Foodservice Sanitation*, 4th ed., Educational Foundation of the Restaurant Association, Chicago, IL, 1992.

60. The National Restaurant Association, *Serving Safe Food*, The Educational Foundation of the National Restaurant Association, Kendall/Hunt publishing company, Dubuque, IA., 1995.

61. Reisman, M., Regulated HACCP moves in, *Food Quality*, 5(1), 21, 1998.

62. FDA, What to Expect Now That Larger Farms Must Comply with the FSMA Produce Safety Rule, February 1, 2018. www.fda.gov/Food/GuidanceRegulation/FSMA/ucm594712.htm (accessed April 21, 2018).

63. CDC, CDC's Emerging Infectious Program, CDC/USDA/FDA Foodborne Diseases Active Surveillance Network (FoodNet), November 4, 1997. http://www.cdc.gov.ncidod/foodnet /foodnet.htm (accessed April 21, 2018).

64. Concentrated drive to end food poisoning, BBC Online, January 28, 1999. http://new2this.bbc .co.uk/hi/end...Fsafety/newsid%5f263000/263873.stm (accessed April 21, 2018).

65. Carter, J., New weapon aids food safety, *Springfield Sunday Republican*, 11, February 14, 1999.

66. "About Codex Alimentarius", Codex Alimentarius International Food Standards, http:// www. fao.org/fao-who-codexalimentarius/about-codex/en/#c453333 (accessed April 21, 2018).

67. Leake, L., *Food Safety in South America*, August 12, 2015. http://www.foodqualityandsafety .com/article/food-safety-in-south-america/ (accessed April 21, 2018).
68. A Modern and Sustainable Food Law in the EU FactSheet. https://ec.europa.eu/food/safety /general_food_law/transparency-and-sustainability-eu-risk-assessment-food-chain_en (accessed April 21, 2018).
69. Food safety in the EU. europa.eu/european-union/topics/food-safety_en.
70. Hunt, K., Why Chinese food safety is so bad, January 16, 2015. www.cnn.com/2015/01/16 /world/china-food-safety/index.html.
71. Regulations on the Implementation of the Food Safety Law (Revised Draft Notified as SPS 1055), September 27, 2017. https://www.fas.usda.gov/data/china-regulations-implementation -food-safety-law.
72. Li, Z., *China's Tainted Meat Scandal Explained*, July 30, 2014. ed.cnn.com/2014/07/29/world /asia/explainer-china-meat-scandal/index.html.
73. *Chinese Hotpot Chain Haidilao Reopens Beijing Branches After Food Hygiene Scandal*, October 6, 2017. http://www.scmp.com/news/china/society/article/2114291/chinese-hotpot-chain-haidilao -reopens-beijing-branches-after-food.
74. Smith, G., *What is Food Fraud*, December 13, 2016. http://fsns.com/news/what-is-food-fraud.
75. Record Seizure of fake food and drink in Interpol-Europol Operation, February 16, 2015, Press Release. https://www.europol.europa.eu/newsroom/news/record-seizures-of-fake -food-and-drink-in-interpol-europol-operation.
76. Centers for Disease Control, Challenges in Food Safety. https://www.cdc.gov/foodsafety /challenges/index.html.

9

Water and Wastewater

OBJECTIVES

A student reading this chapter will be able to:

1. Describe the different stages of the hydrological cycle and the relative amounts of freshwater on the planet.

2. List the main consumers of freshwater, noting each consumer's major use of water.

3. Discuss water scarcity, pointing out the areas of the world at risk of water shortage and potential associated conflict. Describe some methods of water management.

4. List several sources of freshwater.

5. Describe the types of wells used to pump groundwater.

6. Describe groundwater formation, including a discussion of contamination, recharge, water mining, fossil water, and some problems associated with overuse.

7. List the various types of pollution threatening water supplies, noting the difference between point and non-point sources.

8. Briefly outline the Clean Water Act (CWA) and the Safe Drinking Water Act (SDWA), explaining the purpose of each.

9. Describe the process and purpose of wastewater treatment.

10. List the components of a septic system and the components of a typical wastewater treatment plant. Define an.d describe BOD and sag curve.

Introduction

Water is powerful, a force of change. It is a life-sustaining force, growing crops for food, sustaining life for man and beast, and washing away all manner of wastes. It knows no boundaries and can exist in the sky as water vapor, on land in rivers and seas, and underground in aquifers. While water plays a crucial and positive role in our roles, water can also be destructive. It contributes to most of the world's major natural disasters, from massive floods to tsunamis to droughts.

As important as it is, water is not distributed evenly across the globe. Some countries are water rich while others are in near constant drought. Clean fresh water is a scarce resource. Across the globe, nearly a third of the human population, 2.1 billion people, lack access to safe, readily-available freshwater.[1] 844 million people do not even have a basic drinking water service. This includes 263 million people who have to travel more than 30 minutes to collect safe water and 160 million who still drink untreated water from surface water sources, such as streams or lakes. Of the 160 million people using untreated surface water (from lakes, rivers, or irrigation channels), 150 million live in rural areas. Even for those with good water access, the source of that water may be at risk, as several of the world's largest aquifers, vast groundwater reserves, in Africa, Eurasia, and the Americas are being overused, tapped into at unsustainable rates.[2]

Globally, water usage increased six-fold over the twentieth century—twice the rate of world population growth.[3]

Water is not always as pure or safe as we think. In countries throughout the world, lakes, rivers, and oceans are used as dumping grounds for all kinds of waste. All too often, that water is contaminated with waterborne disease, due to lack of proper purification or water treatment systems. This situation results in thousands of cases of waterborne illness every year in countries with inadequate clean water. An awareness of the problem and a strong sense of responsibility are essential for proper management and preservation of a safe, plentiful water supply. However, different cultures and different countries treat water according to their customs and traditions. Not everyone shares the same level of concern about water supply and water pollution. These differences in beliefs about the importance of water quality, water scarcity, and water use can cause conflict and hinder a collective effort to protect the world's water supplies.

The Properties of Water

Water is a unique compound. It is composed of one hydrogen molecule and two oxygen molecules and has a bent configuration. The oxygen carries a partial negative charge and the two hydrogen atoms carry partial positive charge (Figure 9.1). Because of this shape and charge configuration, water exerts a major influence on the earth's environments, dissolving polar molecules easily. Due to water's high boiling point and a high melting point, large amounts of energy are required to vaporize water or to melt ice. Clouds, rain, and snowmelt in winter all occur from the absorption of heat from the sun to transform water into different states. This heat absorption leads to the hydrological cycle.[4]

Hydrological Cycle

Consider all the different forms of water. Water exists in the atmosphere as clouds, rain, snow, sleet, and fog; on the ground as rivers, lakes, and oceans; and underneath the earth's surface as groundwater. These different states of water comprise the hydrological cycle.

FIGURE 9.1
Configuration of water. (Adapted from Turk, J., and Mirk, A, *Environmental Science*, 4th ed., Saunders College Publishing, Philadelphia, PA, 1988, Chapter 17.)

The hydrological cycle is a process involving the sun, the atmosphere, the earth, and water (Figure 9.2).[5] The hydrological cycle consists of evaporation, condensation, transportation, transpiration, precipitation, and runoff. When surface water absorbs heat from the sun, it evaporates or vaporizes into the atmosphere. Clouds or water vapor transport water through the atmosphere. Plants also add water to the atmosphere, in a process called transpiration or evapotranspiration. When water condenses to rain, sleet, or snow, it falls to the earth as precipitation. The precipitation that is not absorbed by the soil becomes runoff, flowing over the land.[6]

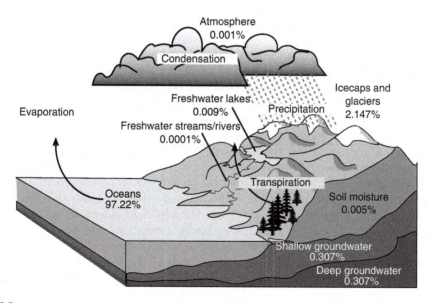

FIGURE 9.2
The hydrological cycle and percentages of water in various Earth compartments. (Adapted from Grigg, N.S., *Water Resources Management*, McGraw-Hill, New York, 1996, Chapter 1.)

Water Resources

A view of the ocean would convince even a desert dweller that vast amounts of water cover large regions of this planet. Water covers 71 percent of the earth's surface. The approximate amount of water on the earth is 1.3 billion cubic kilometers. Ninety-seven percent of this volume is salt water contained within the oceans and seas. However, humans and animals require freshwater for consumption, which makes up only 3 percent of the total amount of water. Much of this 3 percent of freshwater exists as icebergs or deep groundwater. Less than 1 percent of the earth's water exists as accessible freshwater, available for human use.[7] Thus, despite the appearance of vast amounts of available water, fresh water sources are limited. In addition, most of the easily accessible sources of fresh water are currently in use. To obtain new sources would require significant expense and effort.[8] An increasing world demand for water, coupled with the threat of water pollution, is creating a growing crisis of water supply. As the demand for water continues to grow, new sources or new water management strategies will be required to meet people's needs.

Water and Health

Water comprises approximately 67 percent of the human body and is a major component of blood and lymph flowing through our bodies. Humans must drink water to survive; without water, a human will live only 3 days. Water also plays a major role in hygiene, as simple hand washing can prevent the spread of disease. Access to clean water is critical to human well-being and survival.

More than 850 million people throughout the world lack basic water sources and have only limited access to fresh water.[9] This means they must travel for longer than 30 minutes to access fresh water from an improved (i.e., protected) source or they use water from unprotected wells or surface water. In developing countries, the poor living on the urban outskirts must purchase water from vendors at rates 40 times the municipal rate.[10] Unfortunately, some people have no other option but to use polluted water for drinking, as it is the only water source available.

In an effort to provide all people with clean water and sanitation facilities, the United Nations declared the 1980s the "International Drinking Water Supply and Sanitation Decade." The World Health Organization and UNICEF have carried this vast challenge into the present. In 1990, they established a global monitoring program focused on the United Nation's Sustainable Development Goal (SDG) 6, which drives toward providing universal access to basic water services by 2030. Basic water service includes safe water sources or improved water sources within 30 minutes of the home (Figure 9.3). Progress has been made against this goal, but there are still nearly one billion people lacking basic service.

Water plays an important role in sanitation and hygiene as well. We shower, bathe, wash our hands. But across the globe, more than three quarters of all human-created wastewater is released, untreated, into waterways.[11] The Ganges River in India demonstrates this problematic practice. Revered as a goddess by Hindus, the Ganges serves as a source of water for drinking and bathing, religious rituals, cremation, and sewage outflow. Yet people still drink, cook, and bathe in this holy but polluted river.[12] The waste flushed into the Ganges each year includes 1.3 billion liters of domestic waste and 260 million liters of industrial waste.[13] It also acts as a catchment for rainwater run off of the millions of megagrams of fertilizers and thousands of megagrams of pesticides used on adjacent land. This water pollution problem extends beyond one river. Fewer than 10 percent of India's 3,119 towns and cities have wastewater sewers and treatment facilities. This lack of

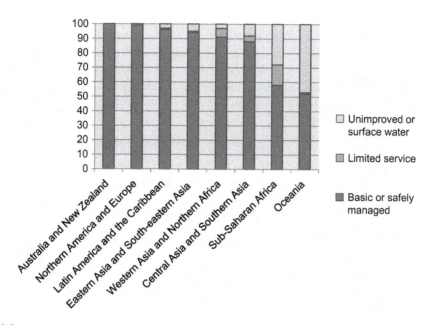

FIGURE 9.3
Means of access to drinking water, percentage of population. (Adapted from Progress on Drinking Water, Sanitation and Hygiene: 2017 Update and SDG Baselines. Geneva, Switzerland: World Health Organization (WHO) and the United Nations Children's Fund (UNICEF), 2017. License: CC BY-NC-SA 3.0 IGO.)

wastewater and sewage treatment infrastructure is repeated in developing countries across the globe. Consequently, in addition to clean drinking water, WHO/UNICEF also monitor improved sanitation and hygiene practices, as both are directly related to water's role in human health.[9] Together, these three components—water, sanitation, and hygiene—are known as WASH.

Interestingly, while modern technology may help clean up polluted waters and improve public health, it is not always welcome. A water supply project in Quiescapa, Bolivia initially stalled because of villager resistance to abandoning unsanitary traditions, such as allowing animals free access to the water source. Teaching the villagers about water purity helped elicit local cooperation towards completion of the project.[14] Clearly, the challenges to ensuring universal access to clean water are myriad and many. Drought, military conflicts, inadequate government support, resistance to change, and insufficient funds all contribute to the problem of supplying clean water to the world's population. Rural areas are especially hard-hit. Worldwide, estimates of investments in water and sanitation projects find that 80 percent is allocated to urban needs and only 20 percent to rural facilities. Unfortunately, in developing countries, water facilities in rural areas are often in disrepair. The lack of simple spare parts can render a water supply such as a hand-pump useless.[15]

Water Shortage and Scarcity

Water on this planet is not evenly distributed. Some areas of the world have plentiful water supplies; others have little or none. Nearly 700 million people in 43 countries experience water scarcity.[16] Further, two thirds of the world's population live in regions of the world that experience water scarcity for at least one month a year. These water shortages threaten the public health by causing dehydration, forcing the use of polluted water, and limiting access to proper sanitation facilities.[17] People living in regions of water scarcity often must devote large amounts of time and energy to obtain water, sometimes walking several kilometers to reach a source of drinking water. To complicate the challenge, these water shortages occur in areas with high rates of population growth, such as Central Asia and Sub-Saharan Africa. Since increased population means an increasing demand for water, the problems will only escalate. Parts of China, India, and South America are also experiencing water shortages.[18]

Globally, the demand for water has been increasing at about 2.3 percent annually, with a doubling of demand occurring every 21 years. Such a demand for water cannot be met by our current water supplies. In addition to the pressure of increased demand for water, water pollution also threatens the available water supply. More than 75 percent of sewage throughout the world and an increasing amount of industrial waste is discharged untreated into the world's rivers.[18] At some point in the future, worldwide water use will be limited by physical, economic, and environmental limitations.[8]

Water Rights and Conflicts

With clean freshwater in demand, conflicts over water rights and resources are inevitable. Many major water sources cross national boundaries, ensuring disputes between countries staking their claim to this valuable resource. Most of the undeveloped sources of freshwater traverse two or more countries. These international waters are actual and potential sources of contention.[19] Asia and the Middle East suffer from water-related disagreements. Such turmoil is not new: In 1950, a quarrel over water occurred between Pakistan and India, bordering on war.[20] Other more recent disputes have involved several other countries. Bangladesh and India have clashed overuse of the Ganges and other shared rivers.[21] Turkey's water projects have met opposition from Syria and Iraq. Egypt opposed Ethiopia's plans to use additional water from the Blue Nile for irrigation.[22] The Jordan River basin has increased tension between Israel and Palestine. Laos's plans to dam the Mekong have created tension with Cambodia. In the United States, opposing factions also bristle over water rights. NYC's attempts to control the upstate New York watersheds that supply the city's water angered upstate residents.[23] Water disputes have also occurred in the western United States, where water is scarce and demand high. The Colorado River serves many cities and agricultural areas, all of which want their fair share of the river.[24] In such cases, choices must sometimes be made regarding water allocation. Clearly, water scarcity and unclear water rights can cause or exacerbate national and international conflict.

These examples of water conflict highlight the importance of water as a pressing global issue. Interestingly, water scarcity has not yet garnered the level of attention global warming or ozone depletion have received, even though it poses a more immediate health threat to billions of people. The countries most affected by water scarcity are developing countries, where a lack of freshwater could seriously limit or hinder their growth. The amount and level of water conflict will increase with the level of water scarcity. Proper water management and international cooperation could halt some of the disagreements over water allocation and even prolong the use of scarce water resources.[25]

Water Consumption and Management

Water is an integral, necessary part of our daily lives. We use water for food production, power generation, washing, and recreation. While most people agree that water is a valuable resource, it does not appear to be properly valued. Water abuse and overuse demonstrate that it is not. Many factors impact the amount and the way a country uses water, including the economy, available technology, level of industry, and agriculture, culture, and climate. For example, as standards of living increase in developing countries throughout the world, water use will increase as more people use flush toilets and use dishwashers and washing machines.[5] Effective water management would improve water efficiency, decrease consumption, and help preserve remaining resources.

Water management encompasses a number of techniques to preserve water resources, such as adequate water pricing, community education, and leak prevention. Part of the problem with excess consumption and abuse of water through pollution has to do with inadequate water pricing. The consumer's price of water rarely reflects the actual cost of storage, treatment, and delivery. In the United States, water is generally inexpensive, which contributes to an attitude of excess. A higher water price that includes the true cost would stimulate conservation, as consumers would bear the cost of over-consumption. If water has no or little value, then it will be wasted without a second thought. As the value of water increases, so does the attention to its effective use. For example, Germany has the highest price for water in all of Western Europe and North America. It also has a low overall average water use per person. On the other hand, the United States, with the least expensive water, has a very high per-capita use.[26] One drawback to accurate water pricing is the unfair burden it places on the poor. Urban poor often pay more for water than non-urban consumers. Increasing water prices to truly reflect its costs while ensuring that basic water needs are met would stimulate water conservation without detriment to anyone.[27]

Education is another key component to a successful water management plan. Informing consumers about the impact of their water use will engender a greater sense of responsibility in the user and hopefully aid in conservation efforts. In addition, large quantities of water are lost through leakage. Stopping this loss would save tremendous amounts of water. All of these techniques encourage sustainable water use. Sustainable water use implies that current needs are met without diminishing the resource for future generations' use and at no expense to environmental need.[27] As the world's water demands increase, water management might be an increasingly valuable tool to combat scarcity and pollution.

One new aspect of water management is the idea of *Virtual Water*, a concept coined by Professor Tony Allan. Virtual water is defined as the amount of water that is used to produce the things we buy and use in our daily life, including food.[28] As an example, beef production requires 15,400 liters of water for every 1 kilo of beef produced, via the grain a cow eats, the water it drinks, and water used to process the beef. Even more water is required to produce 1 kilogram of chocolate, 24,000 liters of water. Awareness of the water used in production can help people make more educated choices.

Water Use

Overview

Water use changes over time as countries and economies develop, standards of living rise, populations grow, and technology expands the realms of water use. Water use even varies with time of day and time of year, with the greatest uses occurring during the evening and the summer, respectively. The three major water consumers in the world are agriculture, industry, and households or individual use (Table 9.1). Total annual water use is greatest in Eastern and Southern Asia, at 2107×10^9 cubic meters per year, with the bulk of that from agricultural use.

TABLE 9.1

Water Withdrawals by Type and Region

	Municipal Water Withdrawal (10^9 m³/year)	Industry Water Withdrawal (10^9 m³/year)	Agriculture Water Withdrawal (10^9 m³/year)	Total Water Withdrawal (10^9 m³/year)
North America	123	321	415	859
Central America and Caribbean	8	6	20	34
Southern America	26	26	154	216
Africa	33	9	184	227
Middle East	25	20	231	276
Central Asia	7	10	128	145
Eastern and Southeastern Asia	206	216	1703	2107
Europe	69	181	84	334
Oceania	5	4	16	25

Source: Adapted from Food and Agriculture Organization of the United Nations, www. FAO.org, http://www. fao.org/nr /water/aquastat/countries_regions/profile_segments/.

Note: Based on latest values available for each country.

The highest industrial water use occurs in North America, at 321×10^9 cubic meters per year. Around the globe, water use varies, often dependent on the amount of available water, the intensity of agriculture, and the extent of industry. Per capita water use also varies significantly across the globe (Table 9.2). This selection of countries shows a lower per capita usage in drier countries, such as those in Africa and higher per capita water usage where water sources are plentiful. These amounts are only a snapshot and do not reflect the stress on many of these countries' water sources.

Agriculture

Agriculture consumes the largest portion of the freshwater supply, with over two thirds of the world's water demand used for irrigation. Agriculture includes not only growing crops but also livestock and aquaculture. These are addressed in the following section. Sixty percent of water used for irrigation is lost to evaporation or runoff. Such an inefficient system cannot continue as water scarcity escalates in scope. As water resources become scarce, agriculture will need to develop a more efficient system of water usage, using less water to grow more food.[29] In addition, agricultural water use tends to be heavily subsidized, both in the United States and worldwide. Water subsidies provide little incentive to economize on water use and may in fact promote wasteful practices. Droughts can wreak havoc on agriculture since they rely so heavily on water withdrawals.

Although agriculture on average demands about 70 percent of global water withdrawals, in some countries, that percentage is even higher. For instance, in Africa, the Sudano-Sahelian region uses 95 percent of its water withdrawal for agriculture. In Asia, the numbers vary widely. Vietnam uses 95 percent of its water withdrawal for agriculture while Malaysia uses only 50 percent for agriculture. The average agricultural withdrawal for Southern and Eastern Asia is higher than the global average, at 82 percent of their water withdrawals.

Livestock

Globally, the livestock sector is the fastest-growing agricultural sector, which suggests it will require a larger share of available fresh water in the future. Currently, it is estimated

TABLE 9.2

Per Capita Water Withdrawals,[a] Select Countries

	Water Withdrawal per Capita (m³/person/year)
Dem Republic of the Congo	11
Rwanda	17
Mozambique	46
Kenya	72
Singapore	82
Paraguay	88
Denmark	118
United Kingdom	129
Tanzania	145
South Africa	270
Brazil	306
China	410
Malaysia	414
Russia	455
France	512
India	605
Mexico	674
Spain	694
Japan	714
Afghanistan	823
Argentina	865
Saudi Arabia	928
Vietnam	965
Egypt	973
Pakistan	1020
Canada	1113
United States	1543
Chile	1603
Uzbekistan	2113
Turkmenistan	5752

Source: Food and Agriculture Organization of the United Nations, www. FAO.org, http://www.fao.org/nr/water/aquastat/data/query/index.html.

Note: Total water withdrawal per capita = (Total country water withdrawal) × 1000000/ (Total population).

[a] Annual quantity of water withdrawn per capita for agricultural, industrial, and municipal purposes. It can include water from primary renewable and secondary freshwater resources, as well as water from groundwater, direct use of agricultural drainage water, direct use of (treated) wastewater, and desalinated water. It does not include in-stream uses, such as recreation, navigation, hydropower, and inland fisheries.

that 8 percent of freshwater is used in livestock production, with 7 percent of that used to grow the feed crops and only 1 percent to service the animals.[31] As developing countries see economies improve and people shift from a primarily plant-based diet to a higher protein animal-based diet, water usage for livestock production will increase. There is a possibility that future livestock water demand could directly conflict with water for human consumption and use. Because water is not distributed evenly in countries, and livestock need similar quality water as humans, drier areas may be the first to experience conflicts about water usage between livestock production and human consumption.

Aquaculture

Aquaculture is the fastest growing food industry in the world. Almost 50 percent of fish consumed globally is now farm-raised.[32] Aquaculture is a water-based industry, but there are differing views on calculating the actual water footprint of aquaculture. These views diverge on whether such footprint calculations should include the freshwater used for processing fish for sale as well as freshwater water used to grow food for the fish. All calculations account for the water used in the farm pens and raceways where the fish live. But this is negligible, as this water is generally returned to its source. Regardless, freshwater aquaculture on average consumes less freshwater versus livestock production. In the United States, aquaculture water withdrawals were 3 percent of total withdrawals compared with 8 percent for livestock.[33]

Industry

The industrial sector also uses large quantities of water for numerous purposes. Industrial water withdrawals are used for creating, processing, washing, diluting, cooling, or transporting a product; adding water to a product; or sanitation needs within the manufacturing facility. Some industries that use large amounts of water make a variety of products, such as food, paper, chemicals, oil, or primary metals.[33] Water is also used for the cooling and condensation by power plants and waste disposal. Approximate industrial water use in the United States is 757 billion liters per day, equivalent to 0.757 cubic kilometer.[30,34] Annual industrial water use in the United States has stayed consistent for the past 10 years at or under 300 cubic kilometers. Worldwide, industrial demand for water has increased in some countries.[35] For instance, China's annual industrial water use nearly tripled in 30 years, from 46 cubic kilometers in 1980, to 142 cubic kilometers in 2012. Most recently, in 2015, it was 133 cubic kilometers per year.

Domestic

U.S. domestic water use includes internal household use such as cooking, showering, flushing toilets, and washing dishes and laundry, and outdoor use such as washing cars and watering the lawn. Water is also used extensively for recreation. In the United States, household water use is extensive at 522 liters per household per day. Technology allows for large volumes of water to be used, in dishwashers, washing machines and flush toilets. In fact, flush toilets consume the largest proportion of water used in single-family homes, at 24 percent[36] (Figure 9.4). Next largest proportion of water consumption comes from shower usage and faucet usage. In the United States, domestic water use has decreased 22 percent over the past 15 years, with the greatest decrease in water use seen in clothes washers and flush toilets. Faucet, bathtub, and shower use remained the same, with the average shower time of 7.8 minutes per shower. Public supply usage comprises 87 percent of the annual U.S. Domestic usage, with the other 13 percent of domestic water coming from private wells and cisterns.[33]

Domestic water usage in developing countries differs greatly from that in developed countries. In LDCs, water serves to meet basic domestic needs such as cooking and drinking. But lack of plumbing for flush toilets, and lack of large electric appliances like dishwashers and washing machines, limit domestic water use in many developing countries.

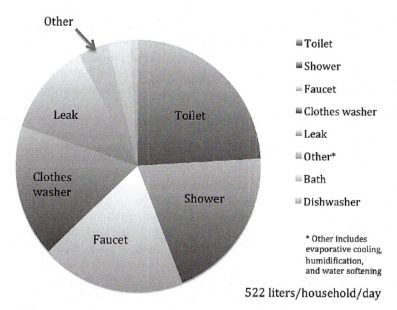

FIGURE 9.4
Domestic indoor water usage for a single-family household in U.S./Canada. (Adapted from Residential End Uses of Water, Version 2 EXECUTIVE REPORT, Water Research Foundation, April 2016. http://www.waterrf.org.)

Sources of Drinking Water

In 2015, 5.2 billion people, 71 percent of the population, used safely managed drinking water services.[9] Included in this number are one third of the populations of the least-developed countries. Seventeen percent of the population used an improved water source, which is considered a basic water service by the World Health Organization. "Improved" is defined as:

> Improved sources are those that have the potential to deliver safe water by nature of their design and construction. These include piped supplies (such as households with tap water in their dwelling, yard or plot; or public standposts) and non-piped supplies (such as boreholes, protected wells and springs, rainwater and packaged or delivered water). (WHO)

If freshwater is untreated or from an unprotected or polluted source, it may not be safe to drink. The following section outlines the range of drinking water sources.

Surface Water

The amount of surface water on the planet is approximately 230,000 cubic kilometers, or about 0.017 percent of the total volume of water.[5] The United States has vast amounts of surface water, with 5.6 million kilometers of rivers and streams and 16.5 million hectares of lakes. This surface water serves as drinking water to a large number of Americans. Sources of surface water develop through rainfall or snowmelt. Precipitation that is not absorbed into the soil flows along the ground as runoff, entering streams, rivers, and lakes. These waterways are sources from which water may be drawn off for human use. The areas surrounding rivers and lakes from which water flows into the waterways are called watersheds. Surface water is accessed through intakes, concrete structures on or below the surface of the water that divert or draw water from the source. Due to its exposed state, surface water is prone to contamination from a number of sources, including diffuse pollution such as agricultural runoff. The Food and FAO estimate that each year about 450 cubic kilometers of wastewater are discharged into rivers, streams, and lakes.[37] Surface water sources and water sheds require protection and management to limit or prevent contamination. Watershed fencing, limited recreational access, and public education can help protect surface waters against pollution.[38]

In the United States, an estimated 268 million people use a public water supply for their household water.[33] The majority of that water, 63 percent, comes from surface sources, such as lakes and streams. Groundwater comprises the remaining 37 percent.

Groundwater

Groundwater is formed in the following way: When rain falls on the earth, some of the water percolates downward through the spaces in the soil, pulled by gravity. At a certain point, the water reaches an impermeable layer of rock. At this layer, the water stops moving (Figure 9.5). As additional water flows downward, it fills all of the spaces in between the rocks and sand that make up the soil above the impervious layer. This area of saturated soil is the zone of saturation. The top of the zone of saturation is the water table. The water table can rise and fall depending on the amount of rainfall and outflow of water from the zone of saturation into a river or a lake. Groundwater sources able to economically deliver water are referred to as aquifers. Some groundwater sources have a layer of impermeable rock above and below the zone of saturation. This type of water formation, called an Artesian aquifer, is under pressure greater than atmospheric

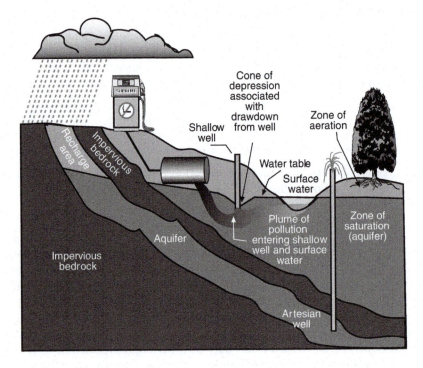

FIGURE 9.5

Groundwater configuration. (Adapted from Koren, H. and Bisesi, M., *Handbook of Environmental Health and Safety*, 3rd ed., Lewis Publishers, Boca Raton, FL, 1996, p. 273.)

pressure, due to the weight of the water.[4] Aquifers around the globe contain 30 percent of the world's available freshwater.[2] Water may remain in an aquifer for hundreds of years or even thousands or longer.[40]

Groundwater is currently the primary source of freshwater for approximately two billion people.[39] Groundwater volume worldwide is estimated to be 8.5 million cubic kilometers or 0.62 percent of the total water volume.[5] Groundwater sources supply drinking water to 50 percent of the people living in the United States and to 90 percent of people living in U.S. rural areas.

Aquifers can act as a backup drinking water sources in times of drought. But the world's water aquifers are under stress from overuse. The top three most-stressed aquifers are[39]

1. The Arabian Aquifer System
2. The Indus Basin aquifer in northwest India and Pakistan
3. The Murzuk-Djado Basin in northern Africa

Aquifer stress and depletion develops as a consequence of drought, poor management, leaks in municipal water pipes, aging infrastructure, lack of technology, population demand, and agriculture demand.[39] In the United States, the Ogallala Aquifer has been depleted by 30 percent, leaving some parts of the Aquifer dry. In other areas, the water table has lowered by more than 91.4 meters.[40]

Wells

Wells provide drinking water for millions of people throughout the world. In some areas of the United States, no access to a public water supply exists, so wells serve as the central water supply. Wells may be private, serving a single user or municipal, serving many people. A well site requires careful consideration of several factors: the source of the well water; its purity; the location of the well; and well water protection, both above and below the ground. Several types of wells exist, ranging from the crude to the sophisticated. Some wells are dug (Figure 9.6) or bored; others are driven or drilled (Figure 9.7). Wells vary in their susceptibility to contamination. Open hand-dug wells are frequently contaminated by animal feces, exposing the user to potential waterborne disease. Such wells tend to be shallow and less than 7.6 meters in depth. The shallow nature of these wells makes them prone to drought and contamination. Covering a dug well decreases the likelihood of contamination. However, if the groundwater itself is contaminated, the well water should be treated before it is consumed. More than 90 percent of residential wells are drilled. These wells are much deeper, reaching depths of up to 106.6 meters. They resist contamination and yield 7.5–18.9 liters per minute more than 98 percent of the time. Septic tanks and farm runoff pose a significant pollution threat to well water.[41,42]

Groundwater Contamination

Although groundwater may appear protected, it is not immune from contamination. As far back as the late 1960s, synthetic organic chemicals (SOCs) were found in groundwater sources. Additional instances of groundwater contamination occurred in the 1970s and 1980s. Despite its seemingly inaccessible location underground, groundwater is vulnerable to several means of pollution. Rainfall may transport surface pollutants deep into the soil, down to the water table. Polluted surface water

FIGURE 9.6
Dug well with two-pipe jet pump installation. (Adapted from USEPA, *Manual of Individual Water Supply Systems*, U.S. Government Printing Office, Washington, DC, 1973, p. 34.)

FIGURE 9.7
Drilled well and submersible pump. (Adapted from USEPA, *Manual of Individual Water Supply Systems*, U.S. Government Printing Office, Washington, DC, 1973, p. 34.)

TABLE 9.3

Sources of Groundwater Pollution

Planned Activities			
Surface	Waste disposal Landfills NP Open dumps NP Surface impoundments P Waste tailings P Land application of sewage sludge P	Storage or transport Material stockpiles P	Agricultural pesticides NP Fertilizers NP Irrigation practices NP Animal feeding P	Other Road de-icing NP
Sub Surface	Waste disposal Injection wells Septic tanks P Cesspools P Other wells P Sewage pipelines NP	Storage or transport Underground storage tanks P Pipelines NP		
Unplanned Activities			
Surface	Urban runoff NP Acid deposition NP Accidental spills P Natural leachates NP			
Sub Surface	Saltwater intrusion NP	Groundwater/surface water interaction NP		

NP = Nonpoint source and P = Point source.

may infiltrate a groundwater source. Improperly disposed hazardous waste can seep into groundwater. A staggering number of potential pollution sources exist in the United States including (1) over 23 million septic systems, (2) between 5 and 6 million underground storage tanks, (3) millions of megagrams of pesticides and fertilizers, (4) municipal landfills, and (5) abandoned hazardous waste sites. Agricultural runoff, accidental spills, stormwater, and wastewater are all potential sources of groundwater contamination. Sources of groundwater pollution may be divided into planned activities and unplanned, involving surface and subsurface pollution from point and non-point sources (Table 9.3). A point source refers to pollutants entering the environment from a specific point such as a pipe or a specific source such as a factory or treatment plant. Non-point source refers to pollutants entering the environment from a broad area and may include scattered sources. The types of pollutants found in groundwater include nitrates, inorganic ions, SOCs, and pathogens. Many of these contaminants pose significant health risks.[43–45]

Recharge and Water Mining

Recharge is the replacement of groundwater by natural processes. Recharge of aquifers can take hundreds and possibly thousands of years. The rate of groundwater use in the United States presently exceeds the rate of recharge of these aquifers, which may lead to water shortages in the future. Overuse of some water systems in the United States has caused severe water stress. Water stress is an actual term of art defined as "the ratio of water withdrawal to water availability." Water stress falls into four categories: low, moderate, medium-high, and high. It occurs when more than 10 percent of the available freshwater is being used.[46] As mentioned, the Ogallala Aquifer, which underlies eight states in the

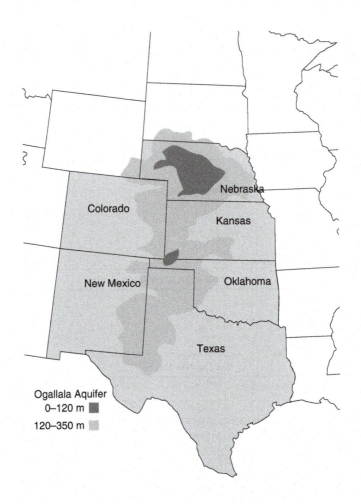

FIGURE 9.8
Geographic location of the Ogallala Aquifer. (Adapted from Turk, J. and Mirk, A., *Environmental Science*, 4th ed., Saunders College Publishing, Philadelphia, PA, 1988, Chap. 17.)

western and mid western United States and water systems in Florida, is stressed from the large quantities of water being pumped out of it (Figure 9.8). In addition, certain sources of groundwater are considered nonrenewable resources, like coal or oil, because of the length of time they took to develop. The overuse of these "fossil" waters is called water mining because the resource is being permanently depleted. An example of "fossil" water is the Ogallala Aquifer. The large quantities of this water used in irrigation greatly exceed the rate of replenishment. Unfortunately, rainfall in this region is low compared to other parts of the country: only 25–100 centimeters per year, compared to the eastern United States, which receives 100–150 centimeters per year.[4,44]

Though the process is controversial, aquifer recharge can also be a human-controlled process. In Western Australia, engineers have been injecting desalinated water into the aquifer that the city of Perth uses for drinking water, in an effort to recharge the overused aquifer.[2]

Subsidence and Salination

Aquifer depletion can result in subsidence and salt-water intrusion. The first, subsidence, involves a settling of the soil as the water is pumped out. The rocks and dirt settle more tightly together, and the volume of the soil decreases. This decreased volume of earth cannot hold as much water as it did previously. The second problem with aquifer depletion is saltwater intrusion, which occurs in most coastal areas. Fresh water is pumped out of the aquifer, and the zone of saturation decreases at both the upper and lower levels. Saltwater can seep into the aquifer at the lower level, polluting the freshwater.[4]

Groundwater Protection

In the United States, groundwater is protected directly and indirectly by various laws including the Resource Conservation and Recovery Act (RCRA), the Comprehensive Environmental Response, Compensation, and Liability Act (CERCLA), the SDWA, the Federal Insecticide, Fungicide, and Rodenticide Act (FIFRA), the Toxic Substances and Control Act (TSCA), and the Pollution Prevention Act of 1990. By regulating the use, transportation, and disposal of thousands upon thousands of chemicals, these laws are an attempt to preserve the quality and purity of U.S. groundwater. However, a national comprehensive groundwater protection plan has not yet been established by the EPA.[45]

Do you want to learn more about water and wastewater? Point your browser to <https://people.umass.edu/~envhl565/> and click on any of the Chapter 9 Web links associated with water and wastewater including:

- Envirofacts Querying Capability
- Water Quality: USEPA
- Water Environment Federation
- WHO/UNICEF Water Monitoring Program
- United Nations FAO Global Water Data
- Water Conflict Across the Globe

Other Sources

Desalination

Desalination is a technique of removing the salt from ocean water to provide drinking water. First, filters remove sediments, suspended solids, and salt from the feed water. With the use of pressure, this water is then pushed through a semi-permeable reverse osmosis filter, resulting in water that is 99 percent free of salts, minerals, and other ions. The product water then passes through a charcoal filter, removing unwanted tastes and odors. In the last step, the filtration process is completed by an ultraviolet sterilizer where 99.8 percent of all micro-organisms, including viruses and bacteria, are destroyed.[42] While this process currently carries a high price tag, it does offer a means to access plentiful ocean water. This practice is currently in use in some areas of water scarcity, such as the Middle East and Western Australia.

Bottled Water

Even though Americans have easy access to tap water, as a country, we consume large quantities of bottled water. In 2016, Americans consumed 48.4 billion liters of bottled water, equal to 148.7 liters per person.[47] The highest per capita bottled water consumption is Mexico, at 254.3 liters per capita in a year. Many people believe that bottled water is safer and cleaner than tap water. However, bottled water is sometimes no better and can be worse than regular tap water. Tests by government agencies have discovered arsenic, bacteria, and concentrated chemicals in bottled water. In addition, many of the claims on bottled water, such as "pure" and "natural," are meaningless, as these terms are not regulated by any government agency. While many people visualize crystal-clear streams filling their bottles of water, the source could be anywhere, including a kitchen faucet. Despite these drawbacks, bottled water can serve as an excellent substitute for tapwater that may be unsafe for drinking. More responsible bottlers take additional precautions in purifying their water products. Specific wording on the product reveals information about the level of purity. A water labeled "Artesian" is well water from a contained underground body of water. Such water is less likely to be contaminated by chemicals from the surface. Bottled water that has been "purified" has been treated by one of several means of purification, such as distillation or disinfection with ozone or filtration through carbon or by reverse-osmosis. Both Artesian and purified water are lead free.[48]

Dams

Dams have long been used as a means of controlling and storing water for human purposes. Reservoirs created by dams allow access to large quantities of water that would otherwise have flowed downstream. In addition to providing means to meet water demand, dams provide a method to manage river basins. Despite these positives, dams are not popular, as they may disrupt the natural ecosystem. Despite the immediate benefits of increased power production and better navigation, the Egyptian built Aswan Dam was one such unpopular national project. The Aswan Dam succeeded in the dislocation of 60,000 indigenous people, the flooding of archeological sites, a decline in overall fish populations, and the destruction of many natural bank ecosystems.[49] By the same token, large-scale dams such as the Three Gorges Dam in China have generated tremendous international environmental controversy and protest. Nevertheless, dams continue to provide water access to areas around the world.[50] Dams may also may serve as a source of energy: Hydropower produces 16 percent of the total electricity generated worldwide.[51]

Water Re-use

The possibility of re-using water has become a reality in some countries as freshwater resources become depleted. Re-use typically involves the use of treated wastewater for non-consumptive purposes, such as agricultural irrigation, landscaping, and industrial uses. Due to the high purity of treated effluents, water re-use does not appear to pose a health threat when used for nonpotable purposes. An excellent means of water conservation, water re-use will continue to expand into areas of the world with increasingly limited water supplies.[52]

Re-using wastewater for human consumption is a more controversial, perhaps less appealing, option. However, it is a reality in some U.S. cities, such as Los Angeles and El Paso, Texas. In these cities, treated wastewater is injected directly into aquifers, in a process called artificial groundwater recharge. This technology has attracted some criticism, as the safety of this practice is still in question.[53] Interestingly, some areas of the world already use wastewater for drinking and bathing, albeit indirectly. Large cities at the downstream end of major rivers, such as London and New Orleans, draw drinking water from the river. By the time the river reaches these cities, the water has been used several times over for various purposes upstream, including many polluting uses.[54]

Water Pollution

Overview

Water in its natural state is not pure. Water naturally picks up dirt, minerals, and particles as it flows over the ground. Much of the pollution that threatens our water supply today is anthropogenic in that it is generated by humans and not part of a natural process. Industry, agriculture, and overpopulation have all contributed to pollution of the world's water supply. In this country, the media reports about beaches closed due to dangerously high bacterial counts or about wells contaminated with carcinogenic chemicals. Many different sources of contamination and pollution exist, from microscopic organisms to organic chemicals. In the United States, laws and regulations exist to try to maintain pure water supplies. Worldwide, water protection and water purity vary from country to country. Some strategies to protect water supplies include protection of areas near sources of drinking water, limitations on pollutant discharges into our waterways, and the processes of chlorination and filtration.[55] However, these strategies are not always effective. Sometimes a disinfection method, such as chlorination, might not kill all the microorganisms, which might lead to waterborne illness. Other times, a contaminant might go undetected in a public well for years, exposing people to increased risk of cancer from their drinking water. A highly publicized example of well contamination occurred in Woburn, Massachusetts, where town wells were discovered to contain trichloroethylene, a suspected carcinogen. Many feared this contamination resulted in a cluster of childhood leukemias in the vicinity of the accused industries.

The potential for water pollution varies in type and scope throughout the United States and the world. A country's level of industry, agriculture, population, socio-economic conditions, water delivery systems, and regulation all impact the degree of water contamination. Historically, water pollution in this country did not garner significant national attention until the 60s and 70s. Early legislation in the 1940s, 1950s, and 1960s allocated and expanded federal funding for municipal wastewater treatment plants. But it was not until the passage of the Federal Water Pollution Control Act in 1972 that the United States began a serious effort to mend its damaged and abused waterways. The Federal Water Pollution Control Act, renamed the Clean Water Act in 1977, established national standards for the nation's waterways and set limitations on allowed pollutant discharges. The Act's intent was to clean up the nation's waterways, including a return of navigable waters to a "fishable and swimmable" condition by July 1983 and a halt of pollutant discharges into waterways by 1985.[55] Subsequent amendments in 1977 and 1987 strengthened this Act, tightening regulation on pollutants. Despite this well-intentioned legislation, many rivers and lakes remain dangerously polluted.[56] The EPA estimates that the public and private cost for water pollution treatment is $64 million/year.[57]

Water Quality

Water quality encompasses various characteristics of water, from taste and color to temperature and purity. Different levels of water quality are found regionally and throughout the world, based on available technology and economic resources. Water quality can vary, depending on its intended use: high quality is needed for drinking water; lower quality is sometimes acceptable for irrigation purposes, as in wastewater reuse. Water use may also affect water quality, degrading its purity, through pollution or a pH change. With pollution threatening the U.S. water supply, drinking water must be evaluated against certain standards to determine its safety and acceptability. Several different types of pollutants may contaminate a water supply. These can be categorized as physical, chemical, biological, and radioactive contaminants. Each different category poses different types and levels of hazards to the consumer when they are present in the water supply. Consequently, in the United States, public water supplies water undergo testing to ensure its purity and safety to the consumer.

Routine biological testing involves the culturing of samples of 100 milliliters (ml) of tap water for the presence of indicator bacteria known as coliforms. The coliforms inhabit the intestines of warm-blooded animals including wildlife, birds, and humans. Their presence in the water supply indicates the possibility of fecal contamination and the potential for the presence of pathogens. The coliform test is used because the pathogens are difficult to grow. The absence of the indicator organisms does not assure the freedom of water from pathogens. Chlorination may kill the coliforms while permitting the resistant cysts of parasites such as *Cryptosporidium* or *G. lamblia* to survive. However, the coliform test has value as an indicator of potential contamination. The standard for water supplies is less than one coliform per 100 milliliters water. Supplies exceeding this level require additional testing. If the supply continues to exceed the standard, then additional chlorine is usually added. In the event that the coliform remains high even after increasing the chlorine level, the town may issue a boil order requiring consumers to boil their water before consumption.

Tests for physical contaminants include odor, taste, color, turbidity, and total dissolved solids, referring to bicarbonates, sulfates, calcium, and other mineral and matter. These aesthetic water quality criteria impact the consumer's acceptance of the water. Foamability and excessive heat are also considered pollutants. Turbidity occurs when the water contains suspended clay or silt and may indicate other pollutants and stimulate bacterial growth. Color in the water results from organic matter, algae, metals, or colored industrial waste. Taste and odor signify organic material in the water and may also indicate the presence of pathogens or other pollutants. Foamability exists when detergents are present in the water supply. While not a health hazard, foam may indicate the presence of contamination.

Chemical contamination, either organic or inorganic, is tested using a spectrum of methods. Presently, more than 83 chemicals are the basis for testing in municipal water supplies that are defined under the SDWA as having 15 service connections used by year-round residents or regularly serving 25 or more people. The EPA has established standard or maximum contaminant levels (MCLs) for these chemicals. The SDWA was reauthorized in 1986 also requiring filtration for most public supplies having surface water sources in order to remove pathogens such as *Cryptosporidium* and *Giardia lamblia*. Most of the public water supplies serving more than 10,000 people may require filtration under these requirements, and most of these systems have insufficient funding to carry out these federal mandates. Underfunding may also be the reason why a majority of small water supplies have not been adequately checked and may be operated by persons lacking training or qualifications resulting in a large share of waterborne diseases in these communities. More than 60 percent of the nation's 60,000 community water supplies that serve fewer than 500 people are at increased risk to chemical and microbial contamination.[58] The contaminants of concern in drinking water include microbes, volatile organic chemicals, SOCs (including pesticides), inorganic chemicals (lead, mercury, and arsenic), and radionuclides.[29]

Right to Know

In the United States, consumers concerned about their tap water now have easier access to water quality information, thanks to the finalization by the EPA of a law that addresses this. This law requires public water systems to inform their consumers about the quality of their drinking water, any contaminants, and related health hazards. Water suppliers that serve the same people year-round must prepare annual water quality reports (consumer confidence reports or drinking water quality reports) for their customers. Since two hundred million Americans receive their water from water utilities affected by this law, its impact will be significant. Utilities are required to send out written reports detailing the quality of the water. As these utilities already report this information to state Departments of Environmental Protection, no additional water testing is necessary. This legislation empowers consumers through access to vital public health information.[59] EPA also has a local drinking water information web page that offers links to many Internet-accessible water quality reports (https://www.epa.gov/ground-water-and-drinking-water/drinking -water-data-and-reports). Additionally, the public can use the site to find specific information about its local drinking water supply, including information about the state's drinking water program and source water protection program.

Types of Pollution

Inorganic Compounds

Inorganic compounds of many types may be present in the water systems. Of particular concern are lead, cadmium, mercury, arsenic, and copper. These substances can cause serious acute and chronic health problems. Lead in particular has been linked with brain damage and nervous system disorders. Mercury has also been implicated in central nervous system problems.[60] Cadmium has been known to cause kidney damage.

Synthetic Organic Chemicals

Organic compounds in the water systems can present a serious health threat. The word "organic" refers to a substance containing the element carbon. Carbon atoms can bond with four other atoms, which allows it tremendous variety in the types of compounds it forms. The types of SOCs in existence today number in the tens of thousands. Organic compounds can occur naturally in water, in the form of decaying vegetation and microorganisms. These natural organics may cause taste and odor problems in water. Of greater concern are the tremendous numbers of synthetic organic compounds that may pollute waterways. Sources of synthetic organic compounds include industry, agriculture, even households. Each time a toilet is flushed or sewage sludge and manure is applied to the fields, large quantities of drugs like antibiotics, hormones, and painkillers, etc. are entering the environment and seeping into our waterways.[60] Disinfection with chlorine may also create organics in a water supply. Synthetic organic compounds can be classified as volatile organic compounds (VOCs) or SOCs. VOCs, as their name implies, are volatile and vaporize easily. VOCs migrate quickly in the ground and travel far from their site of introduction. VOCs present a threat to groundwater, where they are less able to vaporize and can accumulate. A dangerous VOC, carbon tetrachloride, has been found in well water. SOCs threaten surface waters, through both accidental and purposeful discharges into waterways. Atrazine, chlordane, lindane and toluene are several dangerous SOCs in use today.[61,62] Polychlorinated biphenyls or PCBs are other hazardous organic chemicals that have been found in some U.S. water supplies, even though their U.S. production ceased in 1977.

Radioactive Material

Radioactive materials from natural and manufactured sources can migrate into water systems. The radiation emitted from decaying radionuclides can have dangerous, even fatal health consequences. The most common radioactive substances in water are radium, uranium, radon, and certain manufactured radionuclides. While naturally occurring radionuclides appear mainly in groundwater, surface waters are more likely to contain artificial radionuclides from atmospheric fallout.[63]

Microbiological Pollution

Microbiological pollution of natural waters can occur from microorganisms found in sewage water or runoff from fields with grazing animals or manure spreading.[64] In addition, wildlife, domestic animals, and people may cause microbiological water pollution. The most common microbial contaminants are *E. coli*, *Giardia lamblia* and *Cryptosporidium parvum*.

Emerging Pollution Risks

Plastic is an emerging pollution problem, especially microplastics. Microplastics are generally defined as being less than 5 millimeters in size.[64] Health risks of such microplastics remain largely unknown.

Sources of Pollution

Overview

Pollution sources are labeled "point" sources or "non-point" sources. A point source of pollution stems from a single discrete location, such as an outflow pipe. This type of source can be monitored, measured, and controlled. Industrial discharges are generally considered point sources, as they enter waterways at a single location. The United States Environmental Protection Agency regulates point source discharges through a permit system. A non-point source of pollution has a wide range of entry into a waterway and is difficult to measure and manage. Agricultural and livestock farming activities are considered non-point sources of pollution.[65] Non-point sources of pollution have now surpassed point sources as the main culprits of water pollution in the United States.[66]

Underground Injection Wells

Underground injection wells (UIW) may pollute water. An UIW is any hole in the ground that is deeper than it is wide, into which liquids are injected. Five types of UIWs exist, Class I–Class V, each used for a different purpose. The most troubling are the Class V wells, which are shallow injection wells. These simply constructed wells may contain various pollutants, including industrial waste, agricultural drainage waste oil, and wastewater. All of these may threaten sources of drinking water. The likelihood of contamination depends on the proximity of the well to an aquifer, the quality of the well construction, and the type and amounts of fluid injected.[45]

Industrial Discharges

Unfortunately, part of creating a product includes creating waste. In the past, the liquid portion of this waste was dumped directly into our waterways or into poorly constructed landfills that leaked dangerous substances into the water. As knowledge about the degree and the risks of water pollution increased, public concern over the health effects of chemicals in our water systems also increased. Many people worried that chemical pollutants in our water system would cause cancer and other diseases. Today, restrictions are placed on what and how much can be dumped into waterways. The government has enacted laws, such as the SDWA and the CWA, which require monitoring for specific chemical contaminants in water. The National Pollutant Discharge Elimination System (NPDES) regulates the discharge of wastewater into U.S. waterways. A NPDES permit stipulates the allowable levels of contaminants that may be discharged in an effluent. This level is called an effluent limitation. The EPA has set standard effluent limitations for specific industries, such as paper mills and chemical manufacturers. The permit also specifies the required tracking of discharged chemicals.[65] Despite the permit system, chemical contamination is still a growing problem. The vast number and volume of chemicals used by industry make it impossible to monitor water systems for each specific chemical. Even with the limited number of chemicals monitored, each year thousands of public water systems violate one or more of the thresholds, due to the high cost of monitoring. Adding new chemicals to the hit list every few years increases the financial strain.[65-67]

Agriculture

Agriculture demands an average of 70 percent of the world's fresh surface waters, making it the largest consumer of freshwater. Not surprisingly, much of the pollution entering our waterways comes from agricultural sources. During rainfall, fertilizers and pesticides applied to crops can flow as runoff into surface waters or percolate through the ground to groundwater. In September of 1999, North Carolina was flooded as a result of Hurricane Floyd. The floodwaters managed to shut down one third of the state, an area or 46,619 square kilometers affecting 2.1 million people. The floodwaters also succeeded in causing huge amounts of hog feces and urine to flow into rivers and enter the groundwater there. Drinking water was tainted by such overflow of manufactured lagoons causing an increase of fecal-coliform bacteria levels to be three times the normal average.[68] Increasing amounts of fertilizer and pesticides used for agricultural purposes also increases the opportunity for water pollution. Information on agricultural pollution in developing countries is limited. However, we do know that pesticide use in developing countries continues to increase, suggesting that incidents of water contamination will occur more frequently.[69] Another area of the world that is affected by agricultural runoff is an area in the Gulf of Mexico known to farmers and fishers as the "dead zone." At last estimation this area was at about 12,437 kilometers. The source of the dead zone lies inland, stemming from the agricultural runoff of farms, which is then carried down the Mississippi River and out into the Gulf of Mexico. The principle cause of this "dead zone" is as a result of nitrogen fertilizer and hog farm surface runoff. When the nitrogen reaches the Gulf, it stimulates growth just like it would on land, causing major algae growth. As the algae die and decay, they remove oxygen from the water, which results in the death of plant and animal species.[70]

Pesticides

Although pesticides have improved the world's food supply, they do pose potential health risks if they enter the water supply. Some pesticides have been linked to an increased risk of cancer. The United States uses 499 million kilograms of pesticides annually. Pesticides have been found in stormwater runoff from both agricultural areas and urban areas.[56] Pesticide contamination of surface waters occurs when rainfall carries the chemicals in runoff into the waterway. Pesticides may also enter surface waters directly from the atmosphere. Pesticides have been found in U.S. surface waters since the commencement of water tests in the 1950s. Surface water monitoring studies conducted throughout the United States have detected over 100 pesticides or related compounds. The types of pesticides found in surface waters change from region to region, seasonally, and over time. Organochlorines, most of which the United States banned in the 1970s, have been seen in decreasing concentrations in the water supply. Herbicide use has increased, as has the concentrations of herbicides in the water supply, which regularly surpass the MCL standard following applications.[70]

Fertilizer

Worldwide, fertilizers are increasingly found in water sources, both surface and groundwaters. Nitrate, a major component in fertilizer, has been detected in wells in many rural areas of the U.S., where fertilizer use is greatest. In fact, 9 percent of wells in agricultural regions of the U.S. exceed the 10 milligram per liters nitrate limit. Nitrates pose a health risk to pregnant women and infants and young children. Babies are especially at risk from a condition called methemoglobinemia or "blue baby" syndrome, in which nitrates act to reduce the ability of red blood cells to transport oxygen. This condition can be fatal.[71–73] Fertilizers in water can also cause eutrophication, which is nutrient-overload of a water source. Eutrophication promotes algae growth, disrupts the ecosystem, deoxygenates water, and may result in fish kills. Nitrate contamination of groundwater poses a significant problem, as they are difficult to clean up.[66]

Stormwater

Defined as "stormwater runoff, snow melt runoff, and surface runoff and drainage," stormwater poses a threat to drinking water sources. It may contain pathogens, nutrients, such as nitrates, SOCs, heavy metals and salts. Groundwater contamination can occur when stormwater transports pollutants through the soil into the groundwater. Impervious surfaces, such as roads and parking lots, increase both the amount of stormwater and the level of contamination. Currently, stormwater discharges from industrial, municipal, and other small facilities are regulated under the Water Quality Act of 1987. Permits are required for industrial and municipal stormwater discharges, to lessen the possibility of water pollution.[74]

Acid Mine Drainage

Water sources may be contaminated by drainage from mines. Acid mine drainage generally comes from active or abandoned underground mines. This acidic drainage enters streams and rivers, acidifying the water and harming the aquatic environment. Heavy metals may also be leached into the water during drainage formation. Sealing off abandoned mines can alleviate this pollution threat. Reducing water flow from the mines also can limit the amount of acid mine drainage. Treatment of drainage is a third option to reduce its impact on waterways.[75]

Fecal Contamination

Microbial contamination of water occurs when a source of freshwater is polluted by raw sewage or animal feces. This problem occurs when an undetected contamination source enters the water delivery system of a city or town, polluting the water. Contaminated water may cause a host of gastrointestinal diseases and dehydration from severe vomiting or diarrhea could be fatal in children, the elderly, and people with weakened immune systems.[76]

Over the past 15 years, the increased incidence of waterborne disease has forced many countries to examine their water purification methods. Even when water is chlorinated, it might still harbor disease-causing microorganisms. Certain microorganisms, in particular, *Giardia lamblia* and *Cryptosporidium parvum*, are very hardy and can tolerate the concentrations of chlorine used to treat drinking water. Unfortunately, when ingested, these organisms can cause disease and even death. A major U.S. outbreak of Cryptosporidiosis occurred in Milwaukee, Wisconsin in 1993, when over 400,000 people developed gastrointestinal illness from drinking water contaminated with Cryptosporidium. One hundred people died from the illness during this outbreak. As Cryptosporidium lives in the water supplies of approximately 45 million Americans, it represents a significant public health threat. Removal of this dangerous parasite from drinking water requires expensive filtration systems, as the organism can survive for 10 days in chlorinated water.[77,78] Further, Cryptosporidium health outbreaks are on the rise: there were 32 swimming pool-related outbreaks in the United States in 2016, compared to only 16 outbreaks in 2014. Swallowing only a mouthful of contaminated pool water can cause illness.[78]

In developing countries, the likelihood of water source fecal contamination is highest when people use an unimproved water source that is open to the elements and in areas where people still practice open defecation.[9]

Water Treatment

Municipal Water Treatment

In the United States, most of the drinking water comes from 200,000 community water supply systems. These systems consist of approximately 140,000 small-scale suppliers and 60,000 municipal supply systems, supply water to 241 million Americans.[79] Most of this water requires treatment to remove contaminants before it can safely be used for drinking water. As mentioned previously, surface water is exposed to many sources of pollution and requires extensive treatment. Due to its protected nature, groundwater may require less treatment, though increasing reports of groundwater contamination may suggest additional purification.

Water treatment proceeds as a series of steps, each intended to remove additional unwanted material from the water and improve the water quality. The main steps of treatment are sedimentation, coagulation-flocculation, filtration, and disinfection. Other treatment includes softening, fluoridation, taste and odor removal, algae control, and aeration. After this treatment, the water should be free of almost all contaminants. Sedimentation involves the settling out of large or weighty suspended objects from the water. A settling tank holds the water for 24 hours while the objects settle out. The next step in the process, sedimentation-flocculation, involves the addition of chemicals such as hydrated aluminum sulfate to facilitate large particle formation. These particles, or floc, settle out of the water when the water is allowed to sit. Filtration also removes suspended material but involves passage of the water through beds of sand or other porous material. While filtration is generally used in large water systems, it can also be used for individual water supplies: filtration units may be attached directly to the faucet. Disinfection, discussed in the following section, is the next crucial step. Other types of water treatment include aeration and water softening. Aeration is a technique used to remove VOCs from a water supply. An upward stream of air intersects with a downward flow of water. As the air rises through the water, it pulls the VOCs into the air stream and out of the water. Water softening involves removing the minerals calcium and magnesium from a water supply to prevent the buildup of mineral deposits on cooking equipment, washing machines, and water tanks.[73]

Disinfection

The most critical step in water treatment, disinfection, should destroy all organisms in the water supply. To ensure proper disinfection, organic matter and other material must be removed prior to disinfection. An important characteristic of an effective disinfectant is that it leaves a residual to prevent re-growth of microorganisms. This residual maintains the purity of the water as it travels to the consumer. Chlorine is the major disinfectant used in United States water systems today. It provides an inexpensive, relatively effective method of killing most waterborne microorganisms and protecting water as it travels through a delivery system. However, chlorine does have some drawbacks. When added to water containing certain organic matter, chlorine can combine with these pollutants to form dangerous disinfectant by-products called trihalomethanes or THM. These compounds have been linked with an incidence of bladder and rectal cancer. Despite the increased risk of cancer from trihalomethanes, chlorinated water is safer than drinking unchlorinated water, due to the threat of waterborne disease. In the many countries throughout the world that lack purified chlorinated water, waterborne disease is a tremendous public health problem. The World Health Organization estimates that several million people die annually worldwide from disease caused by waterborne pathogens. Not chlorinating water to avoid the risks from trihalomethanes would increase the incidence of waterborne disease.[80,81]

Although chlorine is the most widely used disinfectant, other disinfection methods exist, including ozonation, ultraviolet light exposure, and chloramination. Ozone can disinfect water 3100 times faster than chlorine and leaves no residual by-products. However, because ozone does not provide a long-lasting residual, ozonated water is treated with a small dose of chlorine. Chloramination, which involves mixing chlorine and ammonia in the water to create chloramines, also disinfects the water and does not form trihalomethanes. But because chloramination lacks the disinfectant strength of chlorine, the water usually requires a second dose of chemical.[63]

Home Water Treatment

In some areas of the United States, the available water might not meet purity or safety standards. In such situations, people have the option of buying home water treatment systems for their individual water supply. The water treatment market contains many options for the water wary consumer, from the simplest of systems to the complex (and expensive). A relatively inexpensive option, activated carbon filters, can remove mercury, VOCs, trihalomethanes, and unpleasant tastes and odors. However, if these filters are improperly maintained, they may act as breeding grounds for bacteria. More expensive home treatment units, such as "reverse osmosis" or "distillation" units, remove most heavy metals and bacteria.[77]

Regulations

United States

Safe Drinking Water Act

In the United States, drinking water received from "Public Drinking Water Systems" is regulated by the SDWA. The SDWA allows the United States EPA to set MCLs for water pollutants to protect the public health. An MCL is determined by multiplying the greatest dose of a chemical that generates no negative health effects in animal studies. Chemicals with a MCL also have a Maximum Contaminant Level Goal, which is the target level of a chemical in the water. While MCLs are required, MCLGs are not. MCLs have not been established for all the potential chemical contaminants in treated water. Enforcement of the SDWA is left to the individual states, with oversight provided by the Office of Groundwater and Drinking Water, a division of the USEPA. State standards for public water systems must be as strict as EPA standards. All public water systems in the U.S. must comply with the National Primary Drinking Water Regulations, which contain the MCLs. These water supply systems must test the water for specific contamination, including various inorganic and organic compounds, trihalomethanes (a specific organic), turbidity, coliform bacteria, and natural and synthetic radionuclides.[82]

The SDWA has evolved considerably since its inception in 1974. Presently, every 5 years the EPA administrator selects at least five chemicals from a list of contaminants not currently monitored but which are known to be in the water supply. The administrator must decide whether to regulate for these chemicals. A substance may also be added to the list if it is a CERCLA hazardous substance or a FIFRA pesticide. While new requirements provide improved water quality, these requirements place a financial strain on a community's water budget. To properly implement the current goals of the SDWA will require tens of millions of dollars. The EPA estimates that $138.4 billion will be required over the next 20 years to overhaul the nation's 55,000 aging community water systems.[83]

European Union

Water Industry Directives

In the EU, three key pieces of legislation protect freshwater: the Bathing Water Directive, the Drinking Water Directive, and the Urban Waste Water Treatment Directive (UWWTD).[84] Together, these are called the "Water Industry Directives." In 2000, the release of the Water Framework Directive offered a more comprehensive approach to water management, with a focus on interrelationships of water and pollution sources. The EU continues to strive for a holistic approach to protecting water sources. This is a challenge given the diversity of its membership, with member countries along a broad spectrum of water supply infrastructure.

China

Environmental Protection Law

In the past few years, China's interest in protecting its water sources has increased significantly. To that end, China amended its Environmental Protection Law in 2015 for the first time in 25 years.[85] This along with the 'Water Pollution Prevention and Control Action Plan' ("Water Ten") has created inroads on a water management system that is rife with siloes, local laws at conflict with national water goals, and overall inefficiency.[86] The country continues its focus on water with dozens of new water conservation projects starting each year, with fifteen projects started in 2017 and 24 in 2016.[85] These include accountability for river water quality at an individual level.

Wastewater Disposal and Treatment

Sewage

In many developing countries around the world, human waste pollutes the land and the water. In a recent study, WHO found that more than 2 billion people lack access to basic sanitation, such as a simple covered privy.[9] In some countries, due to this lack of facilities, people openly defecate on the ground, leaving the fecal matter to contaminate soil and water sources. Despite WHO's efforts to reduce this practice, nearly 900 million people still practice open defecation, about 12 percent of the human population (Figure 9.9). This practice is on the increase in sub-Saharan Africa and Oceania. Another 850 million use pit latrines or bucket latrines, which are considered unimproved facilities. Population growth puts pressure on reducing some of these unsafe practices. Improved sanitation services that are shared across multiple households are used by 600 million people (Figure 9.9).

The developed world generally has ready access to sanitation in the form of septic systems or municipal sewers. In fact, 39 percent of the global population, or 2.9 billion people, use a

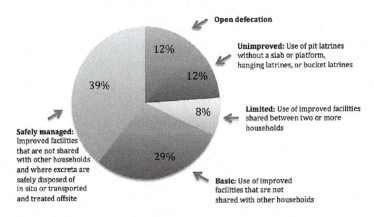

Sanitation Service Levels: Global Usage

FIGURE 9.9
Sanitation service levels: Global usage. (Adapted from: Progress on Drinking Water, Sanitation and Hygiene: 2017 Update and SDG Baselines. Geneva, Switzerland: World Health Organization (WHO) and the United Nations Children's Fund (UNICEF), 2017. Licence: CC BY-NC-SA 3.0 IGO.)

safely managed sanitation service, in which feces are either disposed of onsite or treated off-site.[9] In the United States, wastewater from household water use is generally routed away from the house through pipes to treatment facilities or septic systems. This wastewater contains human wastes, food, detergents, and household chemicals. Sewage or wastewater is 99.9 percent water. Organic matter and impurities comprise the remaining 0.1 percent. Upon entering the ground or a waterway, this 0.1 percent of matter can drastically impact the quality of the ground water or surface water. This organic material can serve as food for organisms living in the water.

Biological Oxygen Demand

As microorganisms decompose organic material in surface water, they use oxygen dissolved in the water. When a waterway has plentiful dissolved oxygen (DO) to meet the biological oxygen demand (BOD) of the organic matter, then the water can still support aquatic life. However, if a waterway is overloaded with biodegradable organic pollutants, this decomposition process can deplete the supply of DO. This deoxygenation diminishes the water's capacity to support aquatic life and can result in fish kills and algal blooms. The BOD of waste can be determined through a simple BOD test. Water reaerates through a natural process over time and distance from the waste input. The deoxygenation and reaeration of water can be presented graphically as a sag curve. A sag curve demonstrates the level of DO over time, showing the critical level where aquatic life dies (Figure 9.10).[7]

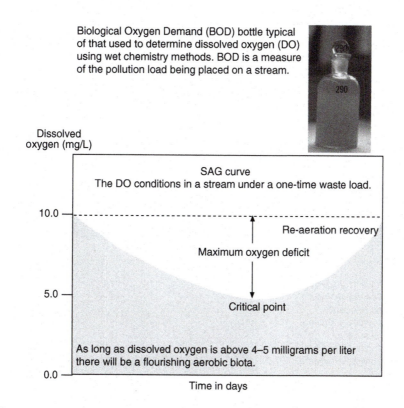

FIGURE 9.10

A typical sag curve resulting from the deposit of high BOD materials into a receiving body of water.

Types of Disposal

Pit Privies

In many areas of the world, human excreta and wastewater is dumped directly into waterways, for lack of proper sanitation facilities. Such practices contaminate the water supply and increase the spread of waterborne disease. Rudimentary facilities such as pit privies can protect water supplies from fecal contamination. A pit privy is simply a manually dug hole, with some type of seat placed over it and surrounding walls for privacy (Figure 9.11). Used mainly only in developing countries, the pit privy isolates human waste from the water supply. However, if not properly maintained, the privy itself can become a source of contamination.[87]

Septic Systems

Septic systems provide a method of wastewater disposal to many people in the United States, especially in rural areas or where no access to municipal sewage treatment exists. A septic system is a two-stage system, made up of the septic tank, usually a pre-cast or reinforced concrete tank, and an absorption field, where the effluent flows into the soil. Sewage flows from the house into the septic tank, where it is held for 24–72 hours (Figure 9.12). During this time, the heavier materials settle to the bottom, where they form a layer of

Ventilation opening

Floor

Mound draining away from base

Base

Pit is 1.5–1.8 meters deep. The waste should be frequently covered with lime. Must be a minimum of 15 m from water source.

FIGURE 9.11
Typical pit privy construction. (Adapted from Chanlett, E.T., *Environmental Protection*, 2, McGraw-Hill Book Company, New York 1979, Chapter 3.)

Failed septic system showing leakage at manhole cover

Septic system being pumped showing scum layer.

Septic tank
At least 15 meter from water sources.

Distribution box

Absorption (leach) field in homogeneous, loamy soil. 30 meter from water sources.

Perforated pipe

FIGURE 9.12
A typical septic system construction. (Adapted from Chanlett, E.T., *Environmental Protection*, 2, McGraw-Hill Book Company, New York 1979, Chapter 3.)

sludge. Lighter particles migrate to the top of the wastewater, forming a scum layer. The middle layer, consisting of wastewater, flows out of the tank into a series of perforated pipes that lie in the drainage field. The effluent flows out of the pipes and percolates through the soil. The soil in which the effluent flows must allow for adequate percolation. This capacity is crucial for an effective septic system. Soil permeability is determined through a perc test. Filtration through the soil further purifies the effluent as soil organisms break down the organic material into simpler inorganic forms. The minimum septic tank capacity is 2839 liters, extending up to 7570 liters for large households. Tanks require periodic cleaning to remove the scum and sludge layer (Figure 9.13). A view of a typical installation of a septic system including the septic tank and leach filed is shown in Figure 9.14.[63]

FIGURE 9.13
A cross-section schematic of a septic tank.

FIGURE 9.14
A schematic of a typical municipal wastewater treatment plant.

Municipal Sewage Treatment

Sewage treatment provides a method to speed up and control water's natural process of purification, through bio-oxidation, filtration, and settling. The end product of sewage treatment, called an effluent, should have significantly reduced BOD and pose little or no harm to the waterway or human health. Achieving these goals requires several stages of treatment, including primary, secondary, and tertiary treatments and sludge disposal (Figure 9.15).

FIGURE 9.15
An overview of a septic tank process.

Primary Treatment

Primary treatment is largely a mechanical process, concerned with the removal of solids. First, screens are used to catch large materials as the sewage flows through them. Next, the comminuter, a grinder or a shredder, reduces the size of the solids in the water and removes any coarse objects (Figure 9.16). The sewage flows into a grit chamber, where heavy matter such as stones, sand, and glass particles can settle out. This prevents such materials from entering the pumps and destroying the mechanisms by abrasion. From the grit chamber, the sewage flows to the primary settling tank, called the primary clarifier.[82] Here the flow is slowed, the settleable solids descend to the bottom where they form sludge that is pumped out for processing and drying. A skimmer removes floating debris, and the liquid portion of the sewage overflows into a receiving channel around the clarifier (Figure 9.17). At this stage, the effluent still contains substantial organic matter causing a high BOD and numerous microbes. Many municipal systems throughout the United States still stop treatment at this primary stage. Chlorine is usually added at this point and the effluent is pumped to a receiving body of water where the pollution load from this effluent will be high.

Secondary Treatment

Secondary treatment employs biological oxidation to break down pollutants into more stable inorganic or oxidized organic forms that resist further metabolism and so do not contribute to the biological oxygen demand when released into a body of water. Trickling filters and activated sludge treatment employ bacteria to breakdown and digest organic material in the sewage. Trickling filters consist of a bed of coarse stones or plastic media, covered with bio-mass consisting of bacteria, protozoa, and fungi. When wastewater is applied to the stones, dissolved organic matter is adsorbed by the slime and digested by the microorganisms.

FIGURE 9.16
Bar screen (left) and comminuter (right) at a municipal wastewater treatment plant.

FIGURE 9.17
A primary clarifier at a municipal treatment plant.

Paddles and baffles used to create turbulence and entrain oxygen into activated sludge tank.

FIGURE 9.18
An activated sludge unit seen full and empty.

Activated sludge treatment occurs in an aeration tank, where sewage is mixed with "active" or bacteria-rich sludge to create "mixed liquor." This mixture is aerated for several hours while the sludge bacteria digest the organic material in the sewage (Figure 9.18). This mixed liquor is transferred to a settling tank, where the activated sludge settles out, most of which is routed back to the aeration tank. The effluent flows to a sedimentation tank, known as a secondary clarifier. An advantage of activated sludge treatment is that it can remove nitrates and phosphorus from sewage, whereas trickling filters do not.[88,89]

Tertiary Treatment

In some cases, additional treatment of the effluent is necessary to further reduce the BOD and suspended solids. A number of tertiary treatments or advanced wastewater treatment methods accomplish this goal, including air stripping by ammonia and rapid granular filtration.[76] At this stage the effluent may or may not be disinfected before it is released into receiving waters. Controversy exists about disinfecting wastewater with chlorine. While the chlorine kills disease-causing organisms, it also may kill aquatic life when treated effluent is discharged into a waterway. Disinfection practice varies from state to state and from country to country.[6]

Sludge Treatment and Disposal

Sludge refers to the solids and liquids separated out of wastewater during sewage treatment. Wastewater treatment in the United States generates approximately 7.3 million megagrams of sludge annually. Sludge is generally composed of organic matter, pathogens, metals, and chemicals removed from wastewater. Treatment of sludge entails the removal of liquids and digestion of the organic matter. The treatment process includes thickening to increase solid density, dewatering to remove liquid, stabilization to digest or neutralize the sludge, conditioning to promote larger particle formation, and disinfection. Sludge disinfection is a crucial step, as it destroys pathogens in the sludge to prevent the spread of disease. Digested sludge may be air-dried (Figure 9.19).[83,87]

In the past, sludge was either dumped in the ocean or buried in a landfill. Today, sludge disposal poses an increasing problem since ocean disposal is no longer legal and landfill space is increasingly scarce. One solution to sludge disposal is to use it as fertilizer (Figure 9.19). However, some sludge may contain heavy metals. These should not be used for agricultural purposes. Such sludge may pose a risk to groundwater sources if improperly disposed in a landfill, as the metals or toxins may leach into the soil and water. Incineration presents another disposal option, although it generates air pollution, requires significant energy input, and creates ash waste.[88,90]

FIGURE 9.19
A sludge-drying apparatus and a sludge composting effort.

Water Pollution and Health

In both developing and developed countries, the reality of ingesting polluted water can be both disruptive and devastating. Health risks range from gastrointestinal upset to chronic disease. While bacterial infection can be treated with antibiotics, exposure to toxic chemicals in water is often unknown and may have a more significant and longer-term impact. Only time can reveal the true dangers of contaminants such as SOCs in our waters, as little is known about the long-term effects of exposure to various water pollutants. In Massachusetts, wells contaminated with the cleaning solvents trichloroethylene and chlorobenzene forced homeowners to limit their exposure to their well water through drinking and cooking with bottled water, limiting shower time, and wearing gloves to wash dishes.[91] Consumers armed with information about their water sources can protect themselves from exposure to polluted water. Worldwide, water pollution continues to plague many developing countries, inflicting immediate and long-term damage on the public's health.

Waterborne Disease

Historically, waterborne disease has been a significant threat to human health. Presently, 3.3 billion people in 127 countries of the developing world suffer water-related diseases. There are almost 6 million deaths from water-related diseases each year, and 50 percent of people on earth lack adequate sanitation.[9] In the developing world, waterborne illness causes 900 million cases of diarrhea and claims the lives of several million children annually.[77] Each year, an estimated 4 billion episodes of diarrhea globally result in an estimated 2 million deaths, mostly among children. Waterborne bacterial infections may account for as many as half of these episodes and deaths.[9] Before methods of water purification were established in this country, diseases like cholera and polio were common health problems. With the development of chlorine treatments and other purification methods in the early part of the twentieth century, the incidence of waterborne disease in this country decreased significantly in the U.S. However, despite extensive water treatment in the United States, water may still carry waterborne disease. Over the past 10 years, several outbreaks of waterborne disease have occurred. In fact, in the United States, waterborne illness causes over 900 deaths annually and more than one million cases of gastrointestinal illness. Two of the reasons for the increase in waterborne disease are inadequate purification and water contamination.[79]

Every day, across the globe, due to a lack of access to clean water and good sanitation and hygiene, more than 800 children under age five die from diarrhea-related disease.[92] It is not surprising then that water-related diseases are one of the top killers of children in this age group. In an effort to diminish the challenge of water-related disease, the World Health Organization and UNICEF continue to promote global progress on the three key aspects of water safety: basic freshwater service, good sanitation, and hygiene.[9] Together, these three stand for the acronym WASH: Water, Sanitation, and Hygiene. Part of the focus of WASH is on education—teaching people the importance of safe WASH practices. But education is only part of the human health equation. Access to basic services is critical.

Future Outlook

To adequately meet water demand in the future, we must recognize the value of water resources today. Increasing population, increased pollution, and competition for water challenge the goal of satisfying everyone's basic needs for water. Conflicting user groups, such as industry and agriculture, will vie for water rights in the face of scarcity. Effective water management, including conservation, accurate pricing, and education can aid efforts to use water efficiently. Throughout the world, the problems of pollution and scarcity will increase over time. The need to protect water from pollution will increase as freshwater supplies dwindle. The notion of sustainable water use will continue to gain momentum in the face of a water crisis, encouraging conservation and preservation of our water resources. On a planet covered with water, yet plagued by waterborne disease, drought, and water waste and mismanagement, we must promote conservation, efficiency, cooperation, and frugality. Every drop counts.

References

1. WHO Press Release, 2.1 billion people lack safe drinking water at home, more than twice as many lack safe sanitation, July 12, 2017. https://washdata.org/ (accessed April 26, 2018).
2. Parker, L., What you need to know about the world's water wars, *National Geographic*, July 14, 2016. https://news.nationalgeographic.com/2016/07/world-aquifers-water-wars/ (accessed April 26, 2018).

3. China Issues New Regulation on Water Management, Sets Fees for Usage. http://www .worldwatch.org/china-issues-new-regulation-water-management-sets-fees-usage (accessed April 26, 2018).

4. Turk, J. and Mirk A., *Environmental Science*, 4th ed., Saunders College Publishing, Philadelphia, PA, 1988, Chap. 17.

5. Grigg, N.S., *Water Resources Management*, McGraw-Hill, New York, 1996, Chap. 1.

6. Nadakavukaren, A., *Our Global Environment*, Waveland Press, Prospect Heights, IL, 1995, Chap. 14.

7. Chanlett, E.T., *Environmental Protection*, 2, McGraw-Hill Book Company, New York, 1979, Chap. 3.

8. Biswas, A., Water development and environment. *Water Resources*, Biswas, A. (Ed.), McGraw-Hill, New York, 1997, Chap. 1.

9. Progress on drinking water, sanitation and hygiene: 2017 update and SDG baselines. World Health Organization (WHO) and the United Nations Children's Fund (UNICEF), Geneva, Switzerland, 2017. Licence: CC BY-NC-SA 3.0 IGO.

10. Johnson, B.K., and Donahue, J.M., *Water Culture and Power*, Island Press, Washington, DC, 1998, Chap. 1.

11. United Nations Goal 6: Ensure access to water and sanitation for all. https://www.un.org /sustainabledevelopment/water-and-sanitation/ (accessed April 26, 2018).

12. Stille, A., The Ganges' next life, *The New Yorker*, 58, January 19, 1998.

13. Dakkak, A., *Water Pollution Worries in Developing World*, February 27, 2016, EcoMENA. https://www.ecomena.org/water-pollution/ (accessed April 26, 2018).

14. Yates, L.D., A water supply development project in Quiescapa, Bolivia, *Journal of Environmental Health*, 3, 20, 1997.

15. World Health Organization, Implementation of the global strategy for health for all by the year 2000, 8th Report on World Health, 1993, Chap. 6.

16. Mekonnen, M., and Hoekstra, A., Four billion people facing severe water scarcity, *Science Advances*, February 12, 2016: e1500323.

17. Anonymous, Thirsty planet, *Environmental Health Perspectives*, 104(l), 12, January 1996.

18. Hutchings, C., Down to the last drop, *Geographical Magazine*, 71, 24, August 1997.

19. Pinkayan, S., *Asian International Waters*, in Asian International Waters: From Ganges-Brahmaputra to Mekong, A.K. Biswas, and T. Hashimoto (Eds.), Oxford University Press, New York, 1996, p. 1.

20. Verghese, B.G., Towards an eastern Himalayan rivers concord, *Asian International Waters*, A.K. Biswas and T. Hashimoto (Eds.), Oxford University Press, 1996, p. 29.

21. Nishat, A., Impact of Ganges water dispute on Bangladesh, *Asian International Waters*, A.K. Biswas and T. Hashimoto, Oxford University Press, 1996, p. 60.

22. Hinrichson, D., The world's water woes, *International Wildlife*, 22, July/August 1996.

23. Revkin, A.C., Chasing a deal on water with a few pitchers of beer, *New York Times*, 41, November 5, 1995.

24. Grigg, N.S., *Water Resources Management*, McGraw-Hill, New York, 1996, p. 493.

25. Biswas, A.K., *Water for Sustainable Development of South and Southeast Asia in the Twenty-First Century*, A.K. Biswas and T. Hashimoto (Eds.), Oxford University Press, 1996, Chapter 2.

26. Lam, B., Finding the right price for water, *The Atlantic*, March 24, 2015, www.theatlantic.com (accessed July 14, 2018).

27. Organization for Economic Co-operation and Development (OECD), *Water Consumption and Sustainable Water Resource Management*, OECD, Paris, France, 1998, p. 23.

28. The World Counts, *Water Water Everywhere and Not a Drop to Drink*, April 14, 2014. http://www .theworldcounts.com/stories/average-daily-water-usage (accessed April 26, 2018).

29. Von Huben, H., *Water Sources*, 2nd ed., American Water Works Association, Denver, CO, 1995, Chap. 5.

30. FAO, Water withdrawal, Country and Region Profiles. http://www.fao.org/nr/water/aquastat /countries_regions/profile_segments/ (accessed April 26, 2018).

31. Schlink, A.C., Nguyen, M.L., and Viljoen, G.J., Water requirements for livestock production: A global perspective, *Revue Scientifique et Technique*, 29(3), 603–619.

32. Troell, M., Metian, M., Beveridge, M., Verdegem, M., and Deutsch, L., Comment on "Water footprint of marine protein consumption—Aquaculture's link to agriculture, *Environmental Research Letters*, 9(109001), 4, 2014.

33. Maupin, M., Kenny, J., Hutson, S., Lovelace, J., Barber, N., and Linsey, K., *Estimated Use of Water in the United States in 2010*, Circular 1405 U.S. Department of the Interior, U.S. Geological Survey, Reston, VA, 2014.

34. Easter, K.W., Becker, N., and Tsur, Y., Economic mechanisms for managing water resources: Pricing, permits, and markets, *Water Resources*, A.K. Biswas (Ed.), McGraw-Hill, New York, 1997, p. 589.

35. FAO Aquastat database, China water use query results. http://www.fao.org/nr/water /aquastat/data/query/results.html (accessed April 26, 2018).

36. Residential End Uses of Water, Version 2 Executive Report, Water Research Foundation, April 2016. http://www.waterrf.org (accessed April 26, 2018).

37. Squandering the Planet's Lifeblood, *International Wildlife*, 24, September/October 1999.

38. Von Huben, H., *Water Sources*, 2nd ed., American Water Works Association, Denver, CO, 1995, Chaps. 3 and 7.

39. Richey, A.S., Thomas, B.F., Lo, M.-H., Reager, J.T., Famiglietti, J.S., Voss, K., Swenson, S., and Rodell, M., Quantifying renewable groundwater stress with GRACE, *Water Resources Research*, 51, 5217–5238, 2015.

40. Oskin, B., *Aquifers: Underground Stores of Freshwater*, January 14, 2015. https://www.livescience .com/39625-aquifers.html (accessed April 26, 2018).

41. Whelan, E.M., *Toxic Terror: The Truth Behind the Cancer Scares*, Prometheus Books, Buffalo, NY, 1993, Chap. 14.

42. USEPA, *The Quality of our Nation's Water*, Office of Water. http://www.epa.gov/enviro (accessed April 26, 2018).

43. USEPA, *Manual of Individual Water Supply Systems*, U.S. Government Printing Office, Washington, DC, 1973, p. 34.

44. Kahn, J., *Economic Approach to Environmental and Natural Resources*, Harcourt and Brace, New York, 1995, Chap. 14.

45. USEPA, Is someone contaminating your drinking water? *Groundwater Protection Council*, Oklahoma City, OK, 1997.

46. Lindeburg, M.R., *Civil Engineering Reference Manual*, 6, Professional Publications, Belmont, CA, 1997, p. 7.

47. *Per Capita Consumption of Bottled Water Worldwide in 2016, by Leading Countries (in gallons)*, Statista 2018. https://www.statista.com/statistics/183388/per-capita-consumption-of-bottled-water -worldwide-in-2009/ (accessed April 26, 2018).

48. Uehling, M.D., How safe is your drinking water? *Popular Science*, 10, 63, 1996.

49. White, F.G., The environmental effects of the high Aswan Dam, *Environment: Science and Policy for Sustainable Development*, 30(7), 4–40, 1988.

50. Grigg, N.S., *Water Resources Management*, McGraw-Hill, New York, 1996, Chap. 13.

51. United Nations Sustainable Development Goals: Goal 6: Ensure access to water and sanitation for all. https://www.un.org/sustainabledevelopment/water-and-sanitation/ (accessed April 26, 2018).

52. Geyer-Allely, E., *Water Consumption and Sustainable Water Resources*, Organization for Economic Co-operation and Development, Paris, France, 1998, p. 28.

53. Pinholster, G., Drinking recycled wastewater, *Environmental Science and Technology*, 29, 174, 1995.

54. Asano, T., Wastewater reuse, *Water Resources*, McGraw-Hill, Biswas, AX, 1997, p. 383.

55. Frederick, K.D., Changing water resources institutions, *Sustaining Our Water Resources*, K.D. Frederick (Ed.), National Academy Press, Washington, DC, 1993, p. 73.

56. Grigg, N.S., *Water Resources Management*, McGraw-Hill, New York 1996, Chap. 14.

57. Palmer, T., *Lifelines: The Case for River Conservation*, Island Press, Washington, DC, 1994, p. 117.

58. Nadakavukaren, A., *Our Global Environment*, Waveland Press, Prospect Heights, IL, 1995, Chap. 15.

59. Pfarrer, S., Water quality mandate praised, *Daily Hampshire Gazette*, Debra Scherban (Ed.), Northampton, MA, August 13, 1, 1998.

60. Von Huben, H., *Water Quality*, American Water Works Association, Denver, CO, 1995, Chap. 6.

61. Montague, P., Drugs in Water, September 3, 1998, Environmental Research Foundation, www.rachel.org (accessed July 14, 2018).

62. Von Huben, H., *Water Quality*, American Water Works Association, Denver, CO, 1995, Chap. 7.

63. Koren, H., and Bisesi, M., *Handbook of Environmental Health and Safety*, 3rd ed., Lewis Publishers, New York, 1996, Chap. 3.

64. European water policies and human health, EEA Report. https://www.eea.europa.eu /publications/public-health-and-environmental-protection (accessed April 26, 2018).

65. Ongley, E.D., *Control of Water Pollution from Agriculture*, 55, Food and Agriculture Organization of the United Nations, Rome, Italy, 1996, Chap. 1.

66. Gallagher, L.M., *Clean Water Act, Environmental Law Handbook*, T.F.P. Sullivan (Ed.), Government Institutes, Rockville, MD, 1997, p. 114.

67. Tibbets, J., Water pressure, rewriting the safe drinking water act, *Environmental Health Perspectives*, 104, 930, 1996.

68. Kilborn, P.T., Storm highlights flaws in farm law in North Carolina; A threat to water supply, *The New York Times*, October 17, 1999.

69. Ongley, E.D., *Control of Water Pollution from Agriculture*, 55, Food and Agriculture Organization of the United Nations, Rome, Italy, 1996, Chap. 3.

70. Annin, P., Down in the dead zone, *Newsweek*, 60, October 18, 1999.

71. Larson, S.J., Capel, P.D., and Majewski, M.S., *Pesticides in Surface Waters*, 3rd ed., Ann Arbor Press, Chelsea, MI, 97, Chaps. 1, 3, 4, 7.

72. Mitchell T.J. and Harding, A.K., Who is drinking nitrate in their well water? *Environmental Health*, 11, 14, 1996.

73. Holton, C., Nitrate elimination, *Environmental Health Perspective*, 104, 3(6), 1996.

74. Pitt, R., Clark, S., Parmer, Y., and Field, R., *Groundwater Contamination from Stormwater Infiltration*, Ann Arbor Press, Chelsea, MI, 1996, Chap. 1 and 2.

75. Kelly, M., *Mining and the Freshwater Environment*, Elsevier Science Publishers, Essex, UK, 1988, Chaps. 1 and 4.

76. Thompson, E., *Rotting Carcasses Sewage Create Mounting Health in N.C.*, The Associated Press, New York, September 20, 1999.

77. Riley, M.B., Murphy, A.D., and Rosado, M.A.M., A reversal of tides, *Water Culture, and Power*, J.M. Donahue and B.R. Johnston (Eds.), Island Press, Washington, DC, 1998, p. 237.

78. Welch, A., "Crypto" outbreaks in swimming pools on the rise, May 18, 2017, *CBS News*. https://www.cbsnews.com/news/crypto-outbreaks-swimming-pools-cryptosporidium-parasite/ (accessed April 26, 2018).

79. Wasik, J., Special report: How safe is your water? *Consumers Digest*, 35, 63, May, 1996.

80. Whelan, E.M., *Toxic Terror*, Prometheus Books, Buffalo, NY, 1993, Chap. 11.

81. Pilkington, N.H., Disinfection: An overview, *Modem Techniques in Wastewater Treatment*, L.O. Kolarik and A.J. Priestly, CSIRO Publishing, East Melbourne, Australia, 1995, p. 75.

82. Von Huben, H., *Water Quality*, 2, American Water Works Association, Denver, CO, 1995, Chap. 1.

83. Williams, S.E., Safe drinking water act, *Environmental Law Handbook*, 14, T.F.P. Sullivan, Government Institutes, Rockville, MD, 1997, Chap. 6.

84. European water policies and human health, EEA Report. https://www.eea.europa.eu /publications/public-health-and-environmental-protection (accessed April 26, 2018).

85. McGregor, D., Key Water Policies 2016–2017, March 16, 2017, China Water Risk. http://chinawaterrisk.org/resources/analysis-reviews/key-water-policies-2016-2017/ (accessed April 26, 2018).

86. Interview with Liu Feiqin, Environmental Law: 1 Year On, October 19, 2016. http:// chinawaterrisk.org/interviews/environmental-law-1-year-on/ (accessed April 26, 2018).

87. McGarry, M.C., *Waste Collection in Hot Climates: A Technical and Economic Appraisal in Water Wastes and Health in Hot Climates*, John Wiley & Sons, New York, 1977, p. 243.

88. Barnes, D., Aerobic wastewater treatment systems, *Modern Techniques in Water and Wastewater Treatment*, L.O. Kolarik and A.J. Priestley (Eds.), CSIRO Publishing, East Melbourne, Australia, 1995, p. 92.

89. Casey, T.J., *Unit Treatment Processes in Water and Wastewater Engineering*, John Wiley & Sons, New York, 1997, Chap. 16.

90. Cheremisinoff, P., *Sludge Management and Disposal*, PTR Prentice Hall, Englewood Cliffs, NJ, 1994, Chap. 1.

91. Humphrey, P., Water fears contaminate westport, *The Boston Globe*, 41, February 26, 1995.

92. WASH in a Changing World, UNICEF. https://www.unicef.org/wash/3942_statistics.html (accessed April 26, 2018).

10

Air, Noise, and Radiation

OBJECTIVES

A student reading this chapter will be able to:

1. Explain the reasons why air pollution is considered a national and global threat.
2. Discuss the chemical and physical components of the atmosphere and explain the mechanisms of dispersion.
3. Describe the regulatory efforts in the U.S. with emphasis on titles of the 1990 CAAA.
4. Discuss the issues behind stratospheric ozone depletion and global warming.
5. List the nature, sources, and health and welfare effects of the criteria pollutants.
6. Describe the major sources of indoor air pollution, including health effects and methods of control.
7. Define noise pollution and radiation.
8. List the major sources and known health effects of noise and radiation.

Introduction

Did you think air pollution is a very recent occurrence? Did modern air pollution all start with the Industrial Revolution? According to Joseph Stromberg, air pollution has been a problem dating back to ancient Rome. Scientists pulling ice core samples in Greenland can look at gas samples trapped in the 1000ds of meter of ice dating back 115,000 years. Their research looked at samples dating back 2,100 years. The samples revealed the chemical signature of methane dating back to the time of ancient Rome. The methane concentrations did not follow the various warming periods during this time but did correspond to human activity with increases in metallurgy and agriculture starting around 100 B.C. It was during this time that the Han dynasty in China also expanded rice fields that allowed for the growth of large amounts of methane-producing bacteria. More wood was being burned to create weapons. These situations were enough to show a global increase in methane production from this time.[1] Metal forging in China, Rome, and Greece appeared to be an important turning point in the development of increased air pollution levels outside of the home.[2]

As cities began to grow in frequency and size throughout the centuries, the burning of coal and wood along with increased agriculture, use of horses and cattle, and other livestock all contributed to air pollution. Coal was so prevalent in the thirteenth and fourteenth centuries along the shoreline in England[3] that people would fill wheelbarrows with it and haul it back to their homes for burning. Burning this sea coal created huge air pollution issues at that time in England . The obnoxious stench of burning sea coal was so annoying to Edward I and Edward II of Great Britain, that those caught burning coal were at risk of losing their lives. In fact, a man was hanged during the reign of Edward II for burning coal and fouling the air.[4] Such rigorous penalties haven't been witnessed since then. However, air pollution has continued to be a problem that worsened with the combined effects of an increasing population and the Industrial Revolution. Interestingly, the introduction of the automobile that began to replace horses was initially heralded as "miracle of cleanliness." That interpretation began to wane in the 1940s when smog from automobiles became a major issue in Los Angeles.[5] Even so, air pollution was generally considered to be only an annoyance, resulting in respiratory irritation, deposits of dust and filth on surfaces, and reduced visibility. However, the potentially severe consequences of air pollution were thrust into the vision of the public eye in documented episodes that have become "classics" in the air pollution literature. These episodes began in the first week of December in 1930 in the Meuse River Valley, Belgium. This deep river valley was highly industrialized with manufacturing plants in steel, glass, lime, fertilizers, and sulfuric acid. That week in December, the air became stilled, and the continuously emitted pollutants in the valley were trapped. The residents complained of irritated throats, coughing, and tightness in the chest. More than 60 deaths were attributed to the episode, primarily among the elderly and those with pre-existing cardio-respiratory disorders. Nearly 18 years later in Donora, Pennsylvania, similar conditions prevailed with steel manufacturing concentrated in a river valley. A stable, slow-moving air mass resulted in the formation of a dense, cool layer of air near the earth with warmer air above in a phenomenon known as an inversion. The noxious pollutants of industry were trapped below the interface of the warm air above and the cool air below, and began accumulating on Wednesday, October 28th. The first death from this episode occurred on that Saturday and nearly 20 people died by Sunday evening. Six thousand people, or 40 percent of the population, developed symptoms associated with the episode including nausea, vomiting, irritation to the eyes, nose and throat, headaches, and muscular aches and pains. The severest symptoms and the most deaths were among the elderly and those with pre-existing respiratory and cardiovascular conditions. The lungs of those who died showed evidence of capillary dilation, hemorrhaging, purulent bronchitis, and edema.[6,7]

Excess deaths were also demonstrated for London, England during several air pollution episodes associated with stagnant air. These episodes occurred in 1952, 1956, 1957, and 1962. London's worst episode was in 1952 during which a thick, yellow smog settled over the city for 5 days, resulting in the deaths of 4,000 more people than the number expected in the absence of air pollution. The deaths were mostly among the elderly and those with pre-existing respiratory or cardiac problems. The excess deaths spiked with the air pollution episode and subsided when the stagnant air was eventually swept away by more turbulent conditions.

Episodes of air pollution resulting in measurable levels of excess deaths have also occurred in Los Angeles and New York, primarily in the 1950s and 1960s. The air pollution disasters focused public attention on the severity of air pollution. It became clear that exposures to concentrated levels of ambient pollutants could cause severe symptoms and even death. Such awareness prompted greater attention to studying the factors associated with air pollution and then establishing regulations to control and limit the pollutants. Such efforts have made it unlikely that similar air pollution episodes will occur in industrialized countries. However, countries such as China, Mexico, and Eastern Europe are facing substantial air pollution problems as they strive for industrial growth in the absence of vigorous air pollution control policies.

Although progress in controlling air pollution in the developed nations has been made, problems of global concern remain. Government leaders and scientists are building toward a consensus in the midst of controversy that industrial emissions such as carbon dioxide, and other greenhouse gases, are contributing to global warming with potentially severe consequences. Emissions of nitrogen and sulfur oxides are thought to be major contributors to acid deposition, which is being associated with the disappearance of fish and other biota in tens of thousands of lakes and streams worldwide, and the decimation of high-altitude forested areas. Chlorofluorocarbons (CFCs), once widely used in air conditioners, in electrical component cleaning, and in the production of foam cups, have been linked to the catalytic destruction of stratospheric ozone in a process known as ozone depletion. Such depletion is thought to result in higher levels of damaging ultraviolet radiation reaching the earth. These atmospheric effects of air pollution have mobilized governments internationally to meet and attempt to resolve these issues. Actions that may be taken are likely to have far-reaching effects on economies, standards of living, and the livelihood of hundreds of millions of people. The quality of our lives, the stability of our environment, and the survivability of the planet have been brought forward as arguments in these controversial issues. Is there a global warming problem? Is stratospheric ozone being depleted? Is acid deposition killing forests internationally and destroying aquatic life? These issues are presented in this chapter in a scientific and objective framework for the reader to draw an independent conclusion.

Air pollution not only represents a perceived threat to global ecological damage, but also exerts both acute and chronic threats to human health at exposures within the normal range of pollutants in our cities and towns.[8] These threats include respiratory disorders such as asthma, bronchitis, and emphysema. There are also increased risks of cardiovascular disease, cancer, and respiratory infections. Air pollutants have been implicated in direct injury to living plants, and to materials including (1) corrosion of metal; (2) soiling of buildings; and (3) degradation of paints, leather, paper, textiles, and dyes. These effects result in significant economic losses approaching several billion dollars annually.[9] Based on the overview given earlier, the arguments for controlling air pollution are substantial. The student of air pollution requires an understanding of the mechanisms of air pollution, which must begin with a discussion of the atmosphere and methods of pollution dispersion.

The Atmosphere and Methods of Dispersion

Chemical Characteristics

The earth's atmosphere consists of a mixture of gases, water vapor, and both solid and liquid particles. It is a dynamic system with most components leaving and entering on a continuing basis and with a wide variety of chemicals being created in the atmosphere in the presence of sunlight, catalysts, and/or moisture. In spite of this continuing exchange of atmospheric ingredients, some of the gases remain relatively constant in their proportions. Nitrogen (N_2) represents a constant 78 percent of the 453 billion megagrams of air surrounding the planet, while oxygen (O_2) remains steady at 21 percent, and argon (Ar) at 0.9 percent. These three gases collectively represent 99.9 percent of the atmosphere. The remainder of gases that are in constant proportion consist of traces of neon (Ne), helium (He), hydrogen (H_2), xenon (Xe), and krypton (Kr) (Figure 10.1). The atmosphere also contains substances that tend to be variable in their concentration and include water vapor, carbon dioxide, carbon monoxide, methane, ozone, ammonia, and nitrogen oxides.

FIGURE 10.1
Atmospheric components.

Many of these latter compounds enter the atmosphere from both anthropogenic and natural sources, and include substances such as carbon monoxide, ammonia, and nitrogen oxides, which are regulated as pollutants. Additionally, many other substances enter the atmosphere from human activities, and those that have the capacity to cause human disease and welfare effects resulting in disturbances to plant life and physical structures are collectively known as air pollutants.

Although the main components of the atmosphere, nitrogen and oxygen, have remained relatively constant for millions of years, current scientific thought suggests this was not always the case. The primitive atmosphere is thought to have been rich in carbon dioxide with little free oxygen. Nearly 2 billion years ago, green plants evolved that used carbon dioxide to grow by photosynthesis, producing energy-rich sugars and a waste by-product now known as oxygen. Organisms adapted to the presence of this waste oxygen and developed into the aerobic species of today, including humans.[10] The earth's atmosphere and climate have achieved a dynamic balance through the removal and recycling of carbon, nitrogen, oxygen, and other substances that have sustained plant and animal life for millions of years. Human technology and explosive populations could potentially alter this balance causing changes in the earth-atmosphere system that jeopardizes our sustainability.

Physical Characteristics

Solar Radiation

The life on Earth requires a continuing source of energy. That energy is supplied by the sun, which radiates enormous electromagnetic energy at effective temperature of 6,000°F and travels in less than 10 minutes to the earth at 300,000 kilometers per second.[11] More than 99 percent of the energy from the sun is within the spectral range of 150–4,000 nanometers (0.15–4.0 µm). Fifty percent of the energy striking the earth is in the infrared spectrum exceeding 0.7 µm (700 nanometers), while slightly more than 40 percent is in the visible spectrum of 400–700 nanometers (0.4–0.7 µm) and about 9.0 percent is in the energetic and shorter wavelengths of 150–400 nanometers (0.15–0.4 µm) (Figure 10.2).[11,12] The solar radiation reaching the earth's atmosphere is absorbed, scattered, and reflected so that an average of less than 50 percent reaches the earth's surface. The upper and lower atmosphere absorbs and scatters 15–20 percent of the total incident solar radiation. When skies are

FIGURE 10.2
Electromagnetic spectrum and components of solar radiation.

clear, carbon dioxide and water vapor strongly absorb infrared radiation, while clouds and dust particles in the lower atmosphere scatter and diffuse visible light wave. Clouds cover a major part of the earth at most times causing significant scattering and reflection of incoming solar radiation. Solar radiation is also reflected back to space by ground surfaces including oceans and lands. The combined reflective ability of cloud cover and ground surfaces are known as the earth's albedo. About 30–40 percent of the solar radiation is reflected by this albedo, with most of it attributed to cloud cover. The amount of solar energy absorbed must ultimately be radiated back to space in order to maintain a global radiation balance, otherwise the earth would soon be overheated. This is not normally a uniform process, but on average, the annual global temperature varies less than one degree centigrade. Solar energy that is absorbed by ground surfaces is radiated back as heat in longer, lower energy, infrared wavelengths. You can observe this effect if you look up a long stretch of blacktop roadway on a hot summer day when the absorbed energy from the sun heats the air above the pavement causing air density changes that distort the scenery in the distance. Air density changes cause shifts in optical refraction that appear as "waves" when they distort distant views. The blacktop absorbed the higher energy, wavelengths of the visible range, and radiated back longer wavelength, infrared radiation in the form of heat. If this reflected heat energy (infrared) is absorbed by infrared-absorbing gases such as carbon dioxide, CFCs, methane, and water vapor, then it traps the warmth and reflects it back to the earth's atmosphere in a process known as the greenhouse effect (Figure 10.3).

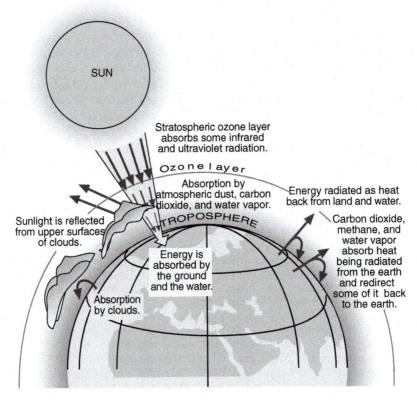

FIGURE 10.3
Mechanisms of reflection and absorption of solar radiation.

If the greenhouse gases increase in concentration, it is logical that more heat energy will be absorbed and the average annual global temperature may rise causing a global warming trend. This is the focus of world attention, and a matter of discussion later in this chapter.

Vertical Temperature Differences and Atmospheric Regions

Imagine yourself in the basket of a hot air balloon lifting off the ground with a thermometer in your hand. You would normally experience a declining temperature as you gained altitude at a rate of about −6.5°C/km, or a loss of about 65°C over the 0–10 kilometers altitude range of this region known as the troposphere (Figure 10.4). Of course, when you entered levels of 6 kilometers or higher, the air has become so thin that you've entered the "killing zone" where supplemental oxygen is required. The thickness of the troposphere is greater over the equator (12 kilometers), and lesser over the poles (8 kilometers).

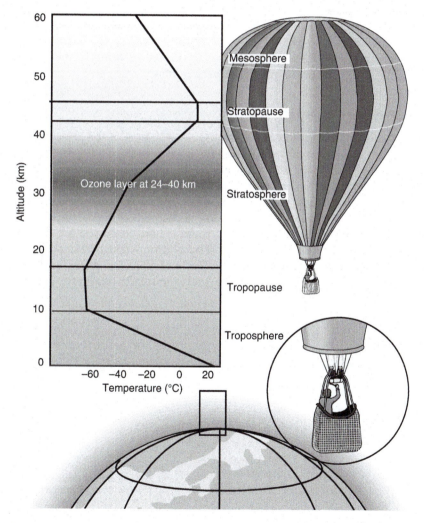

FIGURE 10.4

Atmospheric regions and temperature profile. (Adapted from Chambers, L.A., Classification and extent of air pollution problems, In *Air Pollution*, 3rd ed., Stern, A.C. (Ed.), Academic Press, New York, 1976, Chapter 1.)

The troposphere is where life activities occur and where the changes and manifestations in weather and climate are driven by variations in thermal energy, friction, and global rotation (Coriolis force). The end of the troposphere is marked by stationary or isothermal temperature as you increase in altitude from 10 to 20 kilometers in a region known as the tropopause. The temperature begins to increase to 0°C beyond this region to an altitude of about 50 kilometers. This area is known as the stratosphere and contains concentrations of ozone approaching 200 parts per billion (ppb) in the densest region at 24–40 kilometers. Ozone (O_3) is an isotope of oxygen that absorbs infrared radiation causing the elevated temperature of the stratosphere. Ozone also strongly absorbs ultraviolet radiation in the UV-B region of 290–320 nanometers thus restricting damaging levels of UV-B from reaching the earth and causing increased skin cancer in humans. The stratospheric ozone concentrations vary geographically and seasonally. Peak ozone levels are found between autumn and spring and are normally most dense over the poles. Levels above the stratosphere include the stratopause, mesophere, mesopause, and thermosphere. These regions are in the outermost atmosphere where light molecules are lost to space, and the relatively few molecules of N_2 and O_2 are widely separated and may become photoionized. The region at 90–400 kilometers is also known as the ionosphere where the earth's magnetic field captures protons and electrons emitted by solar storms resulting in the Aurora Borealis (northern lights) and Aurora Australis (southern lights).[11]

Atmospheric Pressure and Density

Gravity pulls the atmosphere to earth with the densest part of the atmosphere closest to the surface. About 99 percent of the atmospheric mass is below 30 kilometers, 90 percent is below 12 kilometers, and 75 percent of the atmosphere is below 10 kilometers (Figure 10.5). A column of air exerts a pressure on an object at sea level of 43 Kg.cm^2. This pressure can be demonstrated by creating a vacuum in a plastic bottle. The external atmospheric pressure will cause the bottle to collapse. Pressure is defined as the force per unit area and is related to the density of the air such that there is both low pressure and low density at higher altitudes. This can be demonstrated by boiling water first at sea level and then at high altitude. The temperature remains steady at 100°C as gases are released from the boiling water. Boiling the water atop a high mountain peak will occur at a lower

FIGURE 10.5
Atmospheric density as a function of altitude.

FIGURE 10.6
The demonstration of atmospheric pressure at low and high altitudes.

temperature of perhaps 90°C, since the pressure is lower and gaseous vapors can escape more readily under lower pressure (Figure 10.6). Atmospheric pressure is measured by a barometer and normally expressed in millimeters or inches of mercury. The pressure at sea level is equal to 760 millimeters mercury (Hg), 29.92 inches of mercury, or 1010 millibars. Pressure and atmospheric density in the troposphere is not constant and varies in response to changes in thermal radiation and air movement. Air that is warmed rises in contrast to the cooler air around it. A hot air balloon uses a propane burner to heat the air inside a fabric envelope causing the balloon and its passengers to rise. The energy driving air movement in the atmosphere is sunlight, which strikes the earth at varying

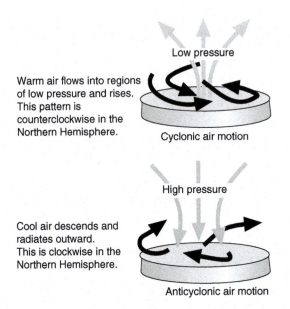

FIGURE 10.7
Cyclonic motion in a low-pressure system and anticyclonic air motion in a high-pressure system in the Northern Hemisphere.

angles, warming the air unevenly. Warmer air becomes less dense, expands, and rises, thereby creating a column of air that is lower in pressure. Cooler air is denser, settles, and is characterized by higher pressure. Differences in pressure cause air to move in both horizontal and vertical patterns. Air tends to move from regions of high pressure to low pressure. Therefore, as sun heats a body of air and it rises, a region of low pressure is created into which flows cooler air. The greater the temperature differences, the larger the pressure differences will be, and the more rapid or violent will be the flow of air. If the air flows horizontally, this is known as wind. The direction of the wind is influenced by the rotation of the earth (Coriolis force—Chapter 1), friction, and the differential warming provided by the sun. Such forces cause air to flow into regions of low pressure in a cyclonic motion and then to rise. This cyclonic motion is counterclockwise in the northern hemisphere and clockwise in the southern hemisphere (Figure 10.7). When cool air descends in the northern hemisphere it is radiated outward in a clockwise motion known as an anticyclone (Figure 10.7). The direction is reversed to a counterclockwise motion in the southern hemisphere. These represent very large-scale air movements and play important roles in the dispersion of air pollutants. Low-pressure migrating cyclones usually are associated with inclement weather including precipitation, clouds, and windy conditions. High-pressure systems normally signal cooler dry air with sunny conditions.

Atmospheric Inversions

There is an ample supply of naturally occurring and anthropogenic pollutants in the atmosphere. They are normally removed by chemical reactions in the atmosphere, dry or wet deposition, and microbial conversions known collectively as sinks. Additionally, such pollutants are normally dispersed by wind and turbulence to very low concentrations that are unlikely to produce adverse health effects. However, sometimes atmospheric conditions become very stable and slow-moving, with the effect of causing the pollutants to build in concentration. The concentration of pollutants is markedly influenced by meteorological conditions. The stability of the air directly impacts the mixing and transport of pollutants along with chemical reactions in the atmosphere that form new categories of pollutants. A stable air mass is slow-moving with little turbulence and permits the accumulation of pollutants. Certain meteorological conditions promote stability. Under normal conditions in the troposphere, air that is warmed rises as long as the air above and around it is cooler. The warmer the air mass in relation to its surroundings, the more rapidly it will rise. As the air rises, it will lose heat from such mechanisms as radiative cooling at an average of −6.5°C/km, which is termed the normal lapse rate. Under theoretical conditions, a parcel of warm, dry air would cool at the rate of −10°C/km and this would be referred to as the dry adiabatic lapse rate. The actual change of temperature with altitude is referred to as the environmental lapse rate and it is a direct measure of atmospheric stability. This stability can be measured by placing a thermometer at ground level and another several meters in the air to determine the environmental lapse rate (Figure 10.8). If the air cools with altitude at a rate faster than the dry adiabatic lapse rate of −10°C/km, then the vertical motion of the air will be quite rapid, permitting cooler air to sweep in and replace it, thereby creating turbulent conditions with good dispersion of pollutants. Atmospheric turbulence promotes pollution dispersion and results from mechanical obstruction and atmospheric heating. The higher a column of air can rise, the greater the maximum mixing depth (MMD) will be. The MMD varies with the time of day, the season, cloud cover, geography, and the presence of anticyclones. Mixing depths tend to be greater during the daytime, in summer, and in geographical areas where there is little ground cover or water to absorb the heat such as deserts or the Great Plains. However, if the air temperature does not cool sufficiently with increasing altitude, or even gets warmer at higher elevations, the upward movement of air is prevented, stable conditions predominate, and dispersion is poor (Figure 10.8).[7,13] The warming of air with increasing elevation is a reversal of the normal tropospheric cooling process and is known as atmospheric inversion. It is the most common atmospheric condition associated with stability. The cooler, denser air near the ground is unable to rise through the warm layers above it and is therefore trapped until turbulent conditions can be reestablished. There are two primary types of inversions.

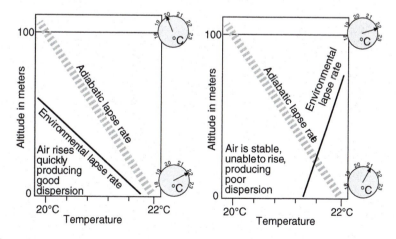

FIGURE 10.8
Dispersion characteristics of air having environmental lapse rates that are cooler (good dispersion) or warmer (poor dispersion) that the adiabatic lapse rate.

Radiation inversion: this type of inversion normally occurs at night when heat absorbed by the earth during the day radiates out quickly at night leaving the air nearest the ground cooler than the radiated warmer air above. These inversions are normally not persistent and are easily broken up when the sun rises in the morning to cause convective currents. Such short-lived inversions seldom present a health hazard since there is little time to accumulate pollutants.

Subsidence inversion: Subsidence refers to the descent of air masses. Cool air masses (anticyclones) associated with Hadley cells descend causing air molecules to compress in a layer above the ground. This layer becomes warmed by compression while the air at ground level remains unchanged and often cooler than the air above, producing an inversion layer at distances of 500–1000 meters above the ground (Figure 10.9). Such inversions may occur frequently along coastal areas such as California, which has documented inversions more than 300 days of the year. Such subsidence inversions occur most often during the fall and winter months and are particularly troublesome since they may persist for days and permit the buildup of potentially dangerous levels of pollution.

Nocturnal inversions also occur when the cool dense air along mountain tops descends into the river valleys during the evening hours and accumulates under a layer of warm air above. The shaded valleys are often covered with mists or clouds, resisting penetration by sunlight until mid-day when the sun is directly overhead where it can cause convection currents to break up the inversion layer. When such conditions combine with migratory anticyclones the inversions become very persistent and hazardous. The stagnation of air produced by such migrating anticyclones was responsible for the classic air pollution disasters of Meuse Valley Belgium in 1930; Donora, Pennsylvania in 1948; and others occurring in England, New York, and Los Angeles.[7,13,14]

Inversions are not alone in meteorological events that influence air pollution. Sunlight and humidity are important in forming chemical reactions in the atmosphere, producing secondary pollutants such as ozone created from nitrogen oxides in the presence of sunlight. Nitrogen oxides are considered primary pollutants since they are derived from identifiable sources such as smokestacks or automobile exhausts. Rain, sleet, and snow serve as sinks to remove pollutants from air through absorption or adsorption, carrying pollutants

FIGURE 10.9
Environmental lapse rate conditions associated with subsidence inversion, and nocturnal inversion occurring in a river valley.

to the ground or bodies of water where they may cause harmful effects. Pollutants emitted into an air mass may not widely disperse, and a migrating air mass may accumulate pollutants as it moves great distances in a process known as long-range transport. Much of the acidity from atmospheric deposition falling on the northeastern United States has been attributed to industries and power plants in the Mid west burning high-sulfur coal.

The History of Air Pollution Control in the United States

The control of air pollution was seen as local control issue up until the 1950s when a federal role began to slowly evolve. It was in 1955 that Congress authorized the Public Health Service in the Department of Health, Education and Welfare (DHEW) for the purpose of researching air pollution and conducting training programs. However, air quality continued to grow worse and Congress passed the Clean Air Act of 1963 which gave federal authority to (1) develop air quality criteria for protecting the public health; (2) conduct research on sulfur dioxide pollution; (3) provide grants to establish state air pollution control agencies; and (4) to provide more research, training, and technical assistance with regard to air pollution. Amendments to this Act in 1965 established the National Air Pollution Control Agency (NAPCA) and set emission standards for 1968 model light-auto motor vehicles. In spite of these regulatory initiatives, air pollution continued to worsen, and the 1963 Clean Air Act proved inadequate to deal with the problem. The response of Congress was to pass the Comprehensive Air Quality Act of 1967, which represented the first attempt to develop a regional approach for the control of air pollution through the designation of Air Quality Control Regions (AQCRs). However, states still kept the main responsibility for enforcement and were expected to develop air quality standards and plans for implementing them. Progress proved to be slow, and as the Americans celebrated "Earth Day" for the first time on April 22, 1970, public attention to environmental concerns reached a high level, stimulating Congress to strengthen air pollution control initiatives. This time Congress passed the 1970 Clean Air Act Amendments (CAAA), which (1) dissolved NAPCA; (2) transferred air pollution control activities to the U.S. Environmental Protection Agency (USEPA), which the President created by executive order; (3) increased federal enforcement authority; (4) established uniform National Ambient Air Quality Standards (NAAQS) for six criteria pollutants; (5) required states to develop plans for implementing programs to achieve NAAQS in the states and submit the State Implementation Plans (SIPs) to USEPA; and (6) gave American citizens the right to sue private and government entities to enforce air pollution requirements. Additional amendments were made in 1977 to the 1970 CAAA that postponed and extended many compliance deadlines for air-quality standards and auto emission standards.[15]

However, the 1977 CAAA did authorize USEPA to regulate chemicals suspected of destroying stratospheric ozone, and also provided for the protection of clean air areas by preventing further air pollution through the concept Prevention of Significant Deterioration (PSD).[11] Despite these efforts and some success under pollution control, three major threats to the nation's environments and to the health of millions of Americans defied these regulatory initiatives. These threats included acid rain, toxic air emissions, and urban air pollution. Additionally, the threats to global environment persisted in the form of global warming by greenhouse gases, and the continued destruction of stratospheric ozone by ozone-depleting chemicals. President George Bush proposed major revisions to the Clean

Air Act in June of 1989 that were passed by Congress and signed into law on November 15, 1990. These revisions are known as the Clean Air Act Amendments of 1990. Many of the provisions of this Act are still coming on line as there are many progressive and creative new directions included in the Amendments.[16] There are several titles to the Act that are intended to produce a healthy, productive environment combined with a sustainable energy policy and economic growth.

Titles of the Clean Air Act

Title I: Provisions for Attainment and Maintenance of the NAAQS

Building upon the 1970 and 1977 CAAA, the 1990 CAAA attempt to strengthen the provisions protecting the public against seven of the most widespread and common pollutants designated as criteria pollutants (Table 10.1). The criteria pollutants are those for which standards have been set establishing maximum concentrations allowed in the ambient or outdoor air environment and are known as NAAQS. USEPA promulgated standards for six classes of pollutants in 1971, and an air-quality standard for lead was promulgated in 1978. These standards are based upon Air Quality Criteria, which document the relationship of these pollutants to effects on health and welfare therefore providing a scientific basis for the standards.[15]

Over 100 million Americans live in cities that fail to meet the NAAQS for ozone, and many people continue to be exposed to unacceptable levels of carbon monoxide and particulates. The 1990 CAAA established these as non-attainment areas, which were first described in the 1970 CAAA and refers to those AQCRs as designated by USEPA that exceed the levels of one or more of the criteria pollutants during the year. USEPA ranked

TABLE 10.1

National Ambient Air Quality Standards

Pollutant	Averaging Time	Primary Standard	Main Sources
Carbon monoxide	8 hours	9.0 ppm	Transportation
Hydrocarbons (corrected for methane)	3 hours (6–9 a.m.)	160 µg/m³	Transportation, industrial processes
Nitrogen dioxide	Annual average	0.05 ppm	Stationary source fossil fuel combustion, transportation
Sulfur dioxide	Annual average	0.03 ppm	Stationary source fossil fuel combustion
Particulates PM10 PM2.5[a]	Annual arithmetic mean Annual arithmetic mean 24 hours average	50 µg/m³ 15 µg/m³ 65 µg/m³	Multiple sources including stationary source fuel combustion, industrial processes, and transportation; fine particles are associated with photochemicals products of fossil fuel combustion
Ozone	1 hour 8 hour[a]	0.12 ppm 0.08 ppm[a]	Secondary pollutant formed in presence of sunlight, NOx, and hydrocarbons; fossil fuel combustion is a major contributor
Lead	3 months	1.5 µg/m³	Food, dust, older houses with lead paint

[a] July 1997 amendments to the 1990 CAAA.

these areas according to the severity of an area's air pollution problem. A non-attainment status carries serious implications because new industries or sources that might contribute to the pollution would be prohibited and therefore limit economic development. This was economically very confining. USEPA, therefore, developed an offset policy that allowed an industry to build a new facility if it achieved agreements with nearby sources to reduce their emissions so no net increase in pollution levels occurred. Alternatively, new sources could be developed in a non-attainment region if the pollution emission rates could be reduced maximally by the best available control technology (BACT) as designated by USEPA. Under the 1990 CAAA, USEPA classified non-attainment areas for ozone, carbon monoxide, and particulates based on the extent to which the NAAQS is exceeded. The 1990 CAAA further establishes specific pollution controls and attainment dates for each control. Carbon monoxide and PM10 are classified as either moderate or serious non-attainment areas, while ozone has five non-attainment classifications ranging from marginal (Albany, NY) to extreme (Los Angeles, CA). The more severe the non-attainment problems, the more controls the area needs to apply and the more time it is allotted to do so.

Title II: Provisions Relating to Mobile Sources

Most automobiles emit up to 80 percent less pollution than in 1960, but they still account for the greatest combined amount of criteria pollutants including carbon monoxide, hydrocarbons, and nitrogen-oxides. The main reason for this is the proliferation of vehicles on the nation's highways. More than one in every five households in the United States has three or more cars and the kilometers traveled are projected to grow by 2 percent each year as Americans no longer just travel from home to work, but link their trips to shopping, after-school activities, day-care, and many other activities. Efforts by government agencies to discourage the use of single-occupant vehicles (SOVs) has been generally discouraging since Americans value their travel independence and continue to enjoy the luxury of inexpensive fuel, affordable cars, and free highways.[17] To combat this problem, the 1990 CAAA reduced the allowable emissions for carbon monoxide, hydrocarbons, and nitrogen-oxides from vehicle tailpipes through tighter inspection and maintenance (I/M) programs. The amount of evaporation of gasoline while refueling at the pumps has also been reduced through the use of specially designed recapture nozzles. Programs requiring cleaner reformulated gasoline and/or oxygenated fuels are also being phased in throughout areas of the country in non-attainment areas for ozone and carbon monoxide, respectively.

Methyl-tertiary-butyl ether (MTBE) is the biggest volume oxygenate used in gasoline and contains at least the 2.7% O_2 by weight required to reduce CO emissions. The additives must be used during the winter months of November through December when levels of CO peak. The amount of MTBE has increased from the 6.8 billion liters used in 1992 because the program to use reformulated gasoline to combat ozone depletion began in 1995.[18] The law also required California to phase-in the use of 150,000 clean fuel cars (i.e., propane or electric cars) by 1996 and 300,000 by the year 1999. Additionally, 26 of the country's most polluted areas will have to further regulate companies and agencies with centrally-fueled fleets of 10 or more vehicles such that vehicle emissions are further reduced. This process began in 1998. The use of MBTE has decreased in the U.S. significantly since it was discovered to be a potent groundwater contaminant that could cost from 1 to 30 billion dollars to mitigate.[19] There has been a switchover to using ethanol as a gasoline additive since there is no liability protection for the use of MBTE.[20]

Title III: Air Toxics

Congress recognized that very hazardous chemicals other than the criteria pollutants were being emitted into the nation's air and directed USEPA under the 1970 CAAA to establish a list of such chemicals along with standards and control techniques. Unfortunately, USEPA acted very slowly over 20 years and listed only eight hazardous air pollutants (HAPs) including arsenic, asbestos, benzene, beryllium, mercury, radionuclides, vinyl chloride, and emissions from coke ovens. Congress was aware that more than 1.2 billion kilograms of toxic air pollutants were being emitted annually in the United States and that exposure to such substances may be causing up to 3000 cancer deaths annually.[16] They obtained the information that industry was required to supply under the Superfund "Right to Know" rule (SARA, Section 313). This rule was passed in 1986 in response to the deaths of nearly 2,000 people in 1984 at Bhopal, India, because of the accidental release of toxic chemicals from a chemical manufacturing facility. This heightened concern prompted Congress to list 189 toxic air pollutants for which emissions must be reduced. USEPA was required to publish a list of source categories that omit certain levels of these pollutants and to issue standards based on the best demonstrated control technology within the regulated industry known as the Maximum Achievable Control Technology (MACT). The source categories were to be controlled by the year 2000, although companies voluntarily reducing emissions could obtain a 6-year extension.

Title IV: Acid Deposition Control

Emissions of nitrogen and sulfur oxides are partially converted in the atmosphere to nitric and sulfuric acids, which return to the earth in rain, snow, fog, and on dry particles to cause acidic deposition. Such acidic deposition harms non-buffered lakes and streams by killing fish and other aquatic organisms, causing forest destruction, corroding steel, damaging concrete and marble structures, reducing visibility, and even contributing to adverse health effects.[16] Despite progress in reducing emissions under the 1970 CAAA, nearly 18 megagrams of SO_2 were being emitted annually in 1990, mostly from combustion of fossil fuels by electrical generating facilities. The goal of the 1990 CAAA is to reduce the total SO_2 emissions by half to 9,071 megagrams per year through a phased program requiring power plants to reduce emissions. The 1990 CAAA is the first air regulation to encourage the use of market-based principles, such as emission banking and trading through the use of allowances. It works as follows: coal-burning electric utilities and other industries will be allowed to emit nearly 8.1 million megagrams of SO_2, while close to 226.8 thousand megagrams theoretical were set aside for auction. If a utility expected it could not reduce its SO_2 emissions to meet the new federal standard, it could purchase an "allowance" from other utilities whose emissions were lower than that required under federal regulations. An allowance is defined under 1990 CAAA as the right to emit 0.91 megagram of sulfur dioxide. This is a market incentive approach where allowance, or the right to pollute, could be traded or banked on the market. This concept was originally proposed by members of the Environmental Defense Fund (EDF) as part of the effort to reduce SO_2 emissions 50 percent by the year 2010. According to EDF representatives, the program has been very successful with emissions reductions of 3.08 million megagrams tons in 1 year alone at unexpectedly low program costs.[15] However, the market for the sale of allowances has significantly diminished, and auction prices for a 0.91 megagrams allowance has dropped from the 1990 expected price of $1,000 to $250 in 1992, $140 in

1995 and \$68 in 1996. Such inexpensive prices worry some people that utilities will find it less expensive to buy allowances and so pollute considerably more.[21]

Title V: Permits

The 1990 CAAA also introduced an operating permits program similar to that used on other federal legislation designed to protect water such as the Federal National Pollution Elimination Discharge System (NPDES). This updated the Clean Air Act to make it more consistent with other federal statutes. Under the permit program, air pollution sources regulated by the Act must obtain an operating permit through their state environmental program: State programs must be approved by USEPA if they are to administer the permit system. The permit requires that all rules and regulations regarding air pollution control be listed on the one permit, making enforcement a much easier success. Additionally, the permit will list the permissible levels of pollutants and the required control measures. The source is required to file periodic reports showing its level of compliance with the permit obligations. They must also pay a fee for the permit, which is intended to cover costs by the state in administering the program.

Title VI: Stratospheric Ozone and Global Climate Protection

Stratospheric ozone is most concentrated in the region of 24–40 kilometers although it is among the least concentrated of naturally occurring atmospheric gasses. Ozone is formed when ultraviolet rays split oxygen molecules to release an oxygen atom that combines with an oxygen molecule to form ozone (O_3). Although created by UV light, ozone also absorbs UV energy in the range of 200–300 nanometers, protecting plant and animal life from the harmful effects of intense ultraviolet radiation. In the mid-1970s, concern about the integrity of the ozone layer grew when the scientists Rowland and Molina from the University of California reported that CFCs were attacking and destroying the ozone layer. The CFCs were widely used as refrigerants, aerosol propellants, and foam-blowing agents in styrofoam containers. The scientists began their work in the 1970s and showed that CFCs could be photolytically destroyed in the atmosphere to release chlorine atoms that could then catalytically destroy O_3 molecules in the stratosphere (Figure 10.10). Mario Molina, Sherwood Rowland, and Max Planck received the Nobel Prize for chemistry in 1995 for their work in establishing that CFCs were destroying the ozone layer.[22] The work of Rowland and Molina erupted from the narrow view of a few esoteric scientists to international attention in 1985 when British scientists discovered a large hole in the ozone over the Antarctic during September to mid-October. This was subsequently confirmed by National Aeronautics and Space Administration (NASA).[23] The Antarctic hole has continued to grow since then, while also lasting longer. In the early 1990s, NASA scientists reported that very low concentrations of stratospheric ozone were appearing over the middle latitudes of 20–60 degrees. While some ozone loss was partially due to volcanic dust and gasses hurled into the atmosphere by the eruption of Mount Pinatubo in the Philippines, losses in ozone were being observed over the United States, Japan, Europe, Russia, and China before and since then.[24] The World Meteorological Organization reports the ozone hole over Antarctica as being approximately as large as the combined area of Canada and the United States, peaking at 12.4 million kilometers and lasting for 50 days. Scientists predict the ozone hole will likely grow larger over the next 15–20 years. Despite global efforts to halt the production of ozone-depleting CFCs, there remain substantial amounts of CFCs to be released, and CFCs may persist in the stratosphere for many years (Figure 10.11).[25]

Free chlorine is then available to attack more ozone molecules. This can be repeated thousands of times for each chlorine atom.

Chlorine monoxide combines with an oxygen atom to form molecular oxygen and a free chlorine atom.

CL

O O
Oxygen molecule

Oxygen atom O

O O

Chlorine CL monoxide

OZONE

Free chlorine atoms impact ozone to create O_2 and chlorine monoxide.

UV light from the sun

SUN

CL

UV light breaks the chlorine free from the CFC molecule.

CL CL CFCs

Chlorofluorocarbons (CFCs) are released from air conditioners, refrigerators, and some industrial uses. The CFCs rise to the stratosphere.

FIGURE 10.10
The catalytic destruction of ozone in the atmosphere by CFCs according to the Rowland-Molina hypothesis.

The efforts to control and halt ozone-depleting chemicals have been controversial. However, representatives from 29 nations met in Montreal, Canada and reached international agreement (Montreal Protocol) to freeze CFC production at 1988 levels by 1990, and then to cut production in half by the year 2000. Mounting evidence that the ozone layer was being depleted resulted in a subsequent conference in London in 1990 and the Montreal Protocol was amended to hasten the phase-out of CFCs production by the year 2000. The amendments also called for a phase-out of ozone-depleting halons (used in fire extinguishers) and CCL_4 by the year 2000. Methyl chloroform is to be phased out completely by the year 2005. More than 80 countries signed the 1990 agreement in London. The 1990 CAAA required that Class I ozone-depleting chemicals be phased out on a schedule specified in the 1987 Montreal Protocol. Class II chemicals (hydrofluorocarbons or HCFCs) were to be phased out by 2030. However, as stratospheric ozone depletion

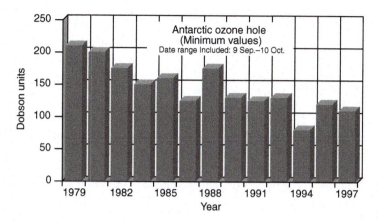

FIGURE 10.11

Antarctic ozone hole minimum values, 1979–1997. (Adapted from USEPA, *The Size and Depth of the Ozone Hole*, written by USEPAs Stratospheric Protection Division, June 10, 1998, http://www.usepa.gov/ozone/science/hoe /size.html.)

became progressively worse and became apparent over North America and northern Europe, representatives of more than 100 nations met in Denmark and agreed to speed up the phase-out of CFC production that was to have been completed by January 1, 1996, and Halon by 1994.[12,15] According to NASA, measurements from satellites in 2017 showed the hole in the earth's ozone layer that forms over Antarctica each September was the smallest observed since 1988. They reported the hole will continue to shrink, but that the smaller ozone hole in 2017 had more to do with weather conditions than human intervention.[26]

The efforts to find substitute chemicals have usually resulted in inferior performance and the requirement to modify existing equipment to make use of the alternatives such as HCFCs or chlorine-free HFC-134a (CF_3CFH_2). These substitutes are not expected to damage the ozone layer but are still infrared absorbers and contribute to the greenhouse effect. Therefore, the amendments to the Montreal Protocol target many CFC substitutes for phase-out by 2040 through voluntary efforts.[15] However, there was concern that many LDCs would be unable to afford the expense of modifying equipment to accept the substitute refrigerants. The Montreal Protocol allowed for continued production of CFCs in developed nations for use in developing countries in order to prevent them from building their own CFC-producing factories. American factories may produce 48,534 megagrams each year for export until 2005. Unfortunately, there has been evidence that CFCs have been returning to this country illegally where a 150-kilogram cylinder that costs $70 (51 Euros) in Europe is sold for $242 (206.8 Euros) in the United States. This is still less costly than the CFCs recycled in the United States. Officials believe as much as 9071 megagrams entered the United States in 1995 from Russia and other countries in the former Soviet Union.[27,28] Despite these troublesome and controversial events overall production of CFCs is down markedly and the National Oceanic and Atmospheric Administration (NOAA) has reported a reduction in the rate at which CFCs are accumulating in the atmosphere.[29] A graph is shown featuring the lowest recorded column ozone during September/ October of the years 1979–1997 as measured by satellite using total ozone mapping spectrometers (TOMS) (Figure 10.14). The minimum value of 88 Dobson units was recorded in 1994.[30] Although there appears to be a slight increase in stratospheric ozone compared to 1994, the recovery of stratospheric ozone is expected to take decades or up to 100 years.[29]

Title VII: Provisions Relating to Enforcement

Since the start of the clean act movements in the middle of the twentieth century the laws and regulations have changed with relationship to ongoing research and public policy. In view of the fact that the laws have changed it is also important to keep the enforcement ability up to date. Included in the 1990 Clean Air Act there is a broad array of authorities to make the law more readily enforceable, thus bringing it up to date with the other major environmental statutes. Through the Clean Air Act of 1990, the EPA has gained authority to issue administrative penalty orders up to $20,000 and field citations up to $5,000 for lesser infractions. Contained in the Clean Air Act of 1990 is a clause that actually enhanced civil judicial penalties. Criminal penalties for knowing violations are upgraded from misdemeanors to felonies, and new criminal authorities for knowing and negligent endangerment will be established.

In addition, polluting sources must remain within compliance and annually certify their compliance with the EPA. The EPA has also gained the ability to administer subpoenas for compliance data. EPA will also be authorized to issue compliance orders with compliance schedules of up to 1 year.

Health and Welfare Impacts of Ozone Depletion

If the stratospheric ozone that protects the earth from UV-B radiation (280–320 mm) were brought to sea level pressures, it would be about 3.1 mm thick. The destruction of this very delicate layer could lead to increased UV-B radiation, which is thought likely to result in (1) increases in basal and squamous cell skin cancer (such cancers now affect 500,000 people annually along with 25,000 new cases each year of malignant melanoma); (2) suppression or weakening of the human immune response system; (3) damage to the cornea and conjunctiva of the eye; (4) reduction in plant leaf size, total dry weight, and stunting of plant growth; and (5) decreased amounts of phytoplankton and zooplankton. Studies showed that phytoplankton production was reduced by 6–12 percent under the ozone hole. This lack of productivity can have several negative impacts. Since phytoplankton photosynthesis is reduced, less carbon dioxide will be removed from the atmosphere, further enhancing the greenhouse effect. Also, since phytoplankton are at the base of the food chain, a reduction could threaten the food source of the world's oceans.[31,32] It needs to be mentioned that pilot studies in Southern Chile, where people have been periodically exposed to high levels of UV-B, found no convincing evidence to support the reported anecdotal acute adverse health effects due to ozone thinning.[31] The authors conclude that exposure to UV-B at current levels in Punta Arenas, South America, were not likely to produce adverse health effects, but cautioned that increased levels may be of concern if they should occur.

The reduction of CFCs is important in the protection of stratospheric ozone. CFCs also contribute to global warming. The greenhouse gases contribute the following amounts to global warming: CO_2 (55%), CFCs (7%–24%), methane (15%), and nitrous oxide (6%).[12] The CFCs are strong infrared absorbers and are up to 20,000 times as efficient as carbon dioxide in producing warming effects. Efforts at reducing CFCs are doubly important to protect the stratospheric ozone layer and to reduce the impact on global warming.

Other Titles to the 1990 CAAA

The CAAA contains many provisions with regard to tougher enforcement, new research, and development in air monitoring. There are also unemployment benefits through the Job Training Partnership Act for those workers who may have lost their jobs because industries had to comply with the new provisions of the 1990 CAAA. Efforts under the Act also attempt to reduce the problem of pollution-created haze areas at National Parks and other parts of the country.

Revised Ozone and Particulate Standards

On July 17, 1997, USEPA announced National ambient air quality standards (NAAQS) for ground-level ozone and particulates.[33] Ground-level ozone is a major component of smog that is photochemically produced as a secondary pollutant of the stratosphere from the interaction of sunlight, nitrogen oxides, and hydrocarbons. Ground-level ozone will be discussed in more detail later in the chapter. The previous ozone standard was last revised in 1979 and set at 0.12 part per million for 1 hour. However, based on more than 3,000 new studies on the ecological and health effects of ozone published since 1980, it's become clear that the standard did not adequately protect the public health, and USEPA replaced the previous standard with an 8-hour standard set at 0.08 parts per million.[26]

Similarly, USEPA announced a new standard for particulate matter (PM) under the NAAQS based on the review of hundreds of scientific studies. USEPA has added a new annual PM 2.5 standard set at 15 micrograms per cubic meter ($\mu g/m^3$) and a new 24-hour PM 2.5 standard set at 65 micrograms per cubic meter. USEPA is retaining the current annual PM10 standard set at 50 micrograms per meter[7,34] Health effects of particulates are discussed in detail in a following section.

These changes to the standard were met with significant opposition from businesses such as the auto industry, oil corporations, and from some state governors. Areas of the country producing more smog and soot than considered healthy under the proposed standards would face economically-strangling restrictions including controls on truck operations, mandates to reduce pollution-causing auto emissions to zero, and requirements that vehicles be run on fuels other than gasoline, strict controls on power plants and factories, and increased emission checks to autos. The new standards are considerably tougher than the one's developed under the 1970 CAAA and 1990 CAAA, which are still unmet by many states across the country.[35]

Bush Clear Skies Initiative

The Bush administration developed a plan called the Clear Skies initiative and submitted it to Congress in February 2003 as a proposal to amend the Clean Air Act, which is the primary federal law governing air quality. According to USEPA, the emissions reductions from Clear Skies would help to alleviate our nation's major air pollution–related health and environmental problems including fine particles, ozone, mercury, acid rain, nitrogen deposition, and visibility impairment. Clear Skies would:

- *Cut sulfur dioxide (SO_2) emissions by 73 percent*: from year 2000 emissions of 9.98 million megagrams to a cap of 4.08 million megagrams in 2010 and to a cap of 2.7 million megagrams in 2018.

- *Cut emissions of nitrogen oxides (NO$_x$) by 67 percent*: from year 2000 emissions of 5 million tons to a cap of 1.9 million megagrams in 2008 and to a cap of 1.5 million megagrams in 2018.
- *Cut mercury emissions by 69 percent*: the first-ever national cap on mercury emissions. Emissions would be cut from 1999 emissions of 43.5 megagrams to a cap of 23.6 megagrams in 2010 and to a cap of 13.6 megagrams in 2018.[36]

The President's "Clear Skies" bill has been stagnated in both the Senate and the House since it was first introduced in late 2002. "Just a few months ago, the Senate rightfully rejected the president's bill. We hope that the House will do the same and protect public health for all Americans."[37] Some argue that if Congress passes the Clear Skies bill, the result will be to weaken and delay health protections already required under the law. The Clear Skies legislation sets new targets for emissions of sulfur dioxide, mercury, and nitrogen oxides from U.S. power plants. But some argue that the targets are weaker than those that would be put in place if the Bush administration simply implemented and enforced the existing law. According to these authors, when compared to current law, the Clear Skies plan would allow three times more toxic mercury emissions, 50 percent more sulfur emissions, and hundreds of thousands more megagrams of smog-forming nitrogen oxides. The authors also suggest that the plan would delay cleaning up this pollution by up to 10 years compared to current law and force residents of heavily polluted areas to wait years longer for clean air compared to the existing Clean Air Act.[38] In early March 2005, the initiative failed to move out of committee when members were deadlocked 9–9. Seven Democrats, James Jeffords (I) of Vermont and Lincoln Chafee (R) of Rhode Island, voted against the bill; nine Republicans supported it. Within days, the Bush Administration moved to implement key measures, such as the NOx, SO$_2$ and mercury trading provisions of the bill, administratively through EPA. It still remains to be seen how resistant these changes will be to court challenges.[39]

The Issue of Global Warming

The Hot Air Treaty, Kyoto, Japan

The global warming treaty completed in December 1997 (Kyoto, Japan) asked Western nations to reduce greenhouse gases to pre-1990 levels by 2010. The United States, if it were to comply, would need to reduce greenhouse emissions to seven percent below the level of the year 1990 or by about 33 percent, one third of its otherwise projected use in the year 2010.[40] Western European nations and Japan are assigned similar goals. According to the agreement, reductions would not start taking effect until 2008. Even if the nations complied, total world production of CO$_2$ would be reduced by only 0.997 billion megagrams by the year 2010. The treaty opened for signature on March 16, 1998 and closed on March 15, 1999. The agreement came into force on February 16, 2005 following ratification by Russia on November 18, 2004. As of April 2006, a total of 163 countries had ratified the agreement (representing over 61.6% of emissions from Annex I countries). The United States and Australia were exceptions. Other countries, like India and China, which have ratified the protocol, are not required to reduce carbon emissions under the present agreement.[41] The LDCs have the urgent problem of providing for exponentially increasing populations, and further reducing use of fossil fuels is unlikely. Inexpensive labor and unregulated industry could attract business from the West thereby shifting jobs, machinery, and CO_2 production to the LDCs with no net overall decrease in greenhouse gases, while greatly disturbing Western economies. The U.S. Congress has already voted down any treaty that does not include the LDCs, and if the accord is not ratified, global emissions of greenhouse gases could become double their pre-industrial level by the year 2050.[40] As an example, China, with its heavy reliance on coal, will be the largest single producer of climate warming carbon dioxide by 2010–2030. Despite rising health problems among its citizens because of increased air pollution, there appears to be no slowing in China's coal burning practices as it races to feed its billions of people and enter the world's industrial market place.[42]

The Paris Agreement

This is also known as the Paris Climate Accord. It is an agreement within the United Nations Framework Convention on Climate Change (UNFCCC) directed towards reducing greenhouse gas emissions, mitigation, adaptation, and finance starting in 2020. Representatives of 195 parties signed the agreement and adopted a consensus on December 12, 2015.[43] The aims of the agreement are to keep global temperature rise below 2 degrees Celsius above pre-industrial levels and to make effort to limit the temperature increase further to 1.5 degrees Celsius.[44] Based on this agreement, each country would determines plans and report on its own contribution it intends to make in reducing global warming. There is no regulatory or financial punishment for each country in setting a specific date or target. It's an expectation that participating countries will exceed previously set targets.[45] U.S. President Trump announced in June 2017 that he would withdraw the United States from this agreement. Under the current agreement, the earliest possible withdrawal for the U.S. would take place in November 2020, just before the end of Trump's first term. There is both domestic and international discord over this decision. This Paris agreement differs from the Kyoto protocol in that it does not set commitment targets with legal force, but emphasizes consensus-building with voluntary targets established by nations.[46] This is not a treaty but an executive agreement with no legal mitigation or finance targets. This Paris Accord also does not sharply differentiate between developed and lesser developed nations in that all agreeing parties will be required to submit plans for reducing emissions. There is the realization that some nations will have differing capacities to respond. So how effective would this Paris Accord be if all nations were fully compliant? MIT used their Integrated Global System Modeling (IGSM) to predict temperature increases based on a wide array of scenarios from no efforts to full compliance with the Paris Accord past 2030. They concluded that with all the scenarios there would be only a 0.1°C change by 2050, and that the goal provided by the Paris Agreement could not be met in the present situation.[47]

Global Warming: The Controversy

The question of global warming has been raised by climatologists since the early 1970s. It has been argued that human activities have upset the balance of atmospheric carbon dioxide through (1) the combustion of fossil fuels, which releases carbon oxides; (2) the clearing of forests, which produces CO_2 and removes a vital consumer of CO_2; and (3) the destruction of phytoplankton by pollution of the oceans. Since the pre-industrial age, a steady in increase of CO_2 in the atmosphere from 280 to 370 parts per million has prompted and alarmed specialists on what to do with excess CO_2.

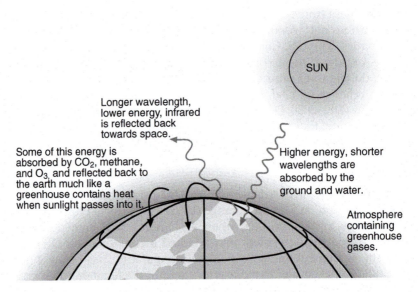

FIGURE 10.12
The absorption of reflected infrared energy to produce the greenhouse effect.

Greenhouse gases such as carbon dioxide, methane, ozone, CFCs, and nitrous oxide are strong absorbers of infrared radiation. When the sun's energy reaches the earth, much of it is absorbed by ground cover and water and reflected back as lower energy infrared radiation. An increasing blanket of carbon dioxide around the planet absorbs some of this infrared energy, trapping it and causing the earth to warm in a process known as the greenhouse effect (Figure 10.12). This concept of global warming caused by greenhouse gases is undergoing tremendous scientific debate, and has come to public attention starting in 1995 because (1) icebergs the size of small states have broken off the Antarctic ice shelf; (2) the annual average global temperature has risen by about 0.5°C (1°F) since the nineteenth century;[48] (3) U.S. and French scientists report that satellite measurements show the global sea level has risen faster (3 millimeters per year) in recent years than in previous decades;[49] (4) climate scientists reconstructed average annual temperatures back to year 1400 and found that 1990, 1995, and 1997 were the warmest in 600 years;[50] and (5) mountain glaciers around the world are rapidly retreating such as the one at Austria's highest mountain peak, Gross Glockner (Figure 10.13), while those at Glacier National Park in Montana have shrunk 50 percent since 1850.

There have been various measures proposed for combating greenhouse gas increases. Alternatives methods include (1) featuring other sources of energy including wind, solar, and nuclear power; (2) promoting increased numbers of trees and woody plants in deforested areas to serve as natural CO_2 sinks; or (3) implementing the newest method of carbon sequestration that takes CO_2 from stationary sources, such as a chemical factory or power plant, and storing it for thousands of years either underground or in deep oceans. The ocean contains at least 50 times more carbon than the atmosphere does, making it a more efficient reservoir for carbon dioxide, and yet having less effect in altering the chemistry of the water. It will take hundreds of years for the CO_2 to reach the surface of the ocean, rendering it ineffective as a greenhouse gas.[51]

FIGURE 10.13

Glaciers at Gross Glockner in Austria have receded rapidly in the last few years. This is a view of the visitors' center overlooking the receding glacier.

Researchers at the Max-Planck Institute for Meteorology in Hamburg, Germany, say that it is unlikely that natural forces (less than five percent chance) caused the temperature rise.[48] Other reputable scientists are not ready to point the finger to anthropogenic sources such as greenhouse gases as the source of global climate change. In fact, while international leaders battle over treaties to resolve a perceived global threat from greenhouse gases, scientists still line up along a mark in the sand that support or reject the role of human-induced global warming. Recent developments favor the theory that greenhouse gases play a role in increasing global temperatures, but uncertainties remain in the studies aimed at showing this, and a definitive answer may not be possible.[48] The scientific community will likely never depend on a single study or observation to accept the concept of greenhouse warming, but will gradually accept the concept when there is a consistent accumulation of evidence.[48] The combined search for such proof encouraged paleoclimatologists at the University of Massachusetts to reconstruct temperatures over the past 600 years using a method of indirect indicators (proxy climate indicators) including ice core samples, coral records, tree rings, and historical temperature measurements. They compared estimated temperatures to factors that affect climate including the atmosphere's carbon dioxide supply and discovered CO_2 to be the dominant factor over the past several decades.[50] Many factors influence climate as mentioned in Chapter 1, and global climate is likely influenced by many complex factors much greater in their effect than anthropogenic greenhouse gases. Climate is affected by (1) increases in atmospheric gases that absorb energy, (2) changes in the earth's orbital geometry, (3) changes in ocean temperature, (4) volcanic activity with increased atmospheric and reduced sunlight penetration, and (5) variations in solar radiation.

Are you interested in learning more about air, noise, and radiation? Visit <https://people.umass.edu/~envhl565/> and click on one or more of the following sites under Chapter 10.

General Links to Air, Noise and Radiation

- *EPA Region 1*: Ground-Level Ozone Information
- *Air pollution*: U.S. EPA Office of Air and Radiation
- Radiation Protection Division Home Page
- Kyoto Protocol
- Climate Change Indicators
- USEPA Montreal Protocol Site
- Montreal Protocol
- Bush "Clear Skies" Initiative-National Resource Defense Council (NRDC)
- USEPA-Clear Skies Initiative
- The Paris Agreement (Accord)

While the meaning of the rise in global temperature and effects such as increasing ocean levels continue to be disputed, the spread of certain diseases has been clearly documented. Those diseases have been found at latitudes where they have not been previously identified. The increased incidence of hanta virus, malaria, cholera, and toxic algal blooms suggest a change in ecological balance and predator-prey relationships that can influence the risk of disease in humans. Regardless of whether there is a proven relationship between global warming, climate changes, and the emergence of new patterns of infectious disease, the public health community is obligated to investigate these new trends and determine what additional analysis and intervention are necessary.

Factors Affecting Global Climate Change

Orbital Geometry as a Factor Affecting Climate

Some may recall the frigid weather of 1994, which plummeted temperatures to all-time lows of −20°F to−30°F in cities around the United States. This caused people to wonder if the greenhouse effect was real, and some considered that glaciers may be returning. In the 1840s the idea that glaciers covered much of the Northern Hemisphere was very controversial, but scientists have shown abundant evidence that thick ice sheets covered much of the North American continent, gouging out lakes, taming mountain ranges, and recreating the landscape. However, recent discoveries are even more surprising. Using core samples from sea-floor sediments and ice core samples from Greenland and Antarctica, scientists have shown that (1) carbon dioxide rises in positive, almost perfect, correlation with rising global temperatures; and (2) climate changes are normally abrupt and often extreme over the history of the planet.[51] The records show that mean global temperatures fluctuated widely with transitions from warm to cold often measured in decades (Figure 10.14). The interglacial period we are now in has lasted 10,000 years and this stable period of warmth is very unlike the unstable climate history of the planet. It is probably no coincidence that human population marked its beginning just about 10,000 years ago with the climate stability. Humans were certainly not sufficient in number or technology to affect global climate over the previous 160,000 years, so what events caused such drastic changes? Among the more widely accepted theories is that of Milankovitch, a Yugoslavian astronomer, who theorized such climate shifts were the result of subtle changes in the earth's orbit around the sun (eccentricity) and slight variations in the earth's tilt (obliquity), which produce

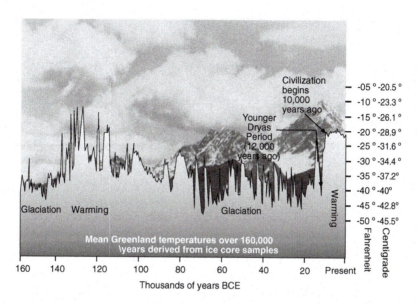

FIGURE 10.14

Record of average annual temperatures in Greenland derived from ice core samples. (Adapted from Lemonick, M.D., *Time*, January 31, 79, 1994.)

climatic cycles (ice ages) of earth at 23,000, 41,200, and 100,000 years. The Milankovitch theories gained support from examination of oxygen isotopes in ocean-floor sediments revealing climate conditions over eons of time. Heavy oxygen isotopes indicate warmer weather while lighter ones predominate in cooler weather.[23,51] The Milankovitch theory does not explain all ice ages, nor the rapid shifts in global temperature evidenced by the Greenland ice core samples.

Changes in Ocean Temperature

Wallace Brodkerad and George Denton from the University of Maine have advanced one of the more interesting theories about climate changes involving ocean currents. They propose that a very salty, dense stream of water flows northward under the Gulf Stream carrying heat from the tropics. It rises to the surface in the far north as water above is swept away by harsh weather and deposits its heat to warm the arctic, cools, sinks, and returns to the tropic. This continuing process keeps the arctic free of glaciation. Disturbances to this process could lead to a new period of glacier formation. This was thought to occur during the period of glaciation known as the Younger Dryas period, about 12,000 years ago. During this period, huge quantities of fresh water from the North American continent emptied into the North Atlantic for over 100 years.[51]

Some of greatest extremes in weather have been attributed to ocean currents known as El Niño ("the little boy"). El Niño is a change or shift in ocean temperatures along with atmospheric conditions in the tropical Pacific that changes weather patterns all around the world (Figure 10.15). Scientists believe that high atmospheric pressure in the eastern Pacific sends trade winds westward pushing Pacific water ahead of them. Consequently, the water on the Asian and Australian coasts is about 0.3–1.5 meters higher than of the Americas. When the pressure drops, the water moves east again forcing downward a cool layer of water underneath

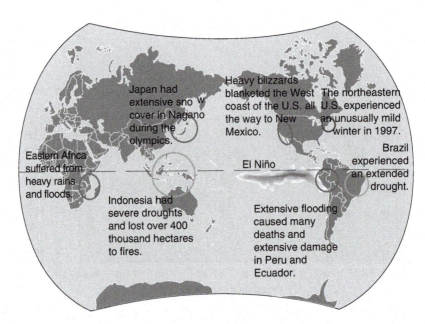

FIGURE 10.15
Global effects of El Niño. (Adapted from Noshi, J.M., *Time*, February 16, 67, 1998.)

called a thermocline, while permitting the warmer, unmixed water on top to carry its heat and moisture towards the Americas, and away from Indonesia.[52] Because of this process, Indonesia faced its worse drought in 50 years with fires that consumed more than 400,000 hectares of forest and threatened hunger for many, while western coasts of the Americas suffered heavy precipitation, powerful storms, and incredible winter blizzards. Torrential rain fell in East Africa, powerful hurricanes swirled in the Pacific, flash-floods killed hundreds in Peru and Ecuador wiping out entire villages, and typhoons smashed into Japan. The ocean currents associated with El Niño produced remarkable global effects, and scientists are concerned that El Niño is changing dramatically from its 2-million-year-old cycle, although somewhat chaotic, to recur far more frequently and persist much longer. Some climatologists argue that the increase in severity and frequency of El Niño is a result of global warming.[53] When the waves associated with El Niño bounce off the Americas and return, then bounce back off of Asia, they lift up the thermocline, causing a cooling effect that ends the El Niño effect. This causes shifts in weather that include a cold and wet north, and dry, hot south that is opposite the effects of El Niño. This shift in Pacific Ocean temperature is known as La Niña.

Volcanic Activity

Volcanic eruptions in the modern era may have extreme localized effects on land and may cause short-term global changes in weather patterns as sunlight is inhibited by a layer of particles thrust into the atmosphere. However, ancient volcanoes are thought to have erupted in Siberia some 250 million years ago and lasted for a million years. This coincided with the Great Dying, occurring at the end of the Permian period, which killed 95 percent of the life in the oceans and more than 70 percent of the vertebrates on land. Ocean levels dropped by 91.4 meters and acid rain enveloped the earth. Some scientists theorize that this prolonged volcanism blocked out the sun causing extensive glaciation and poisoning the atmosphere with sulfur oxides.

Scientists claim that other factors may have contributed to this mass extinction, which developed the events leading to the extinction of the dinosaurs some 65 million years ago, but accumulating evidence points towards extended volcanic activity as having dramatic effects on the earth's climate and its inhabitants.[54]

Solar Radiation

Scientists have recently discovered that the sun is not constant in its solar output, but varies with changes in the sun's luminosity, which is in turn associated with sunspot activity. Reduced sunspot activity appears to result in a cooling effect whereas increased activity increases warming of the earth. The sunspots show cycles of 11 and 22 years that correlate with nearly half of the global warming evidenced over the last 100 years.[23]

The Criteria Pollutants

Introduction

USEPA has reported substantial progress in improving the air quality in many states including the heavily polluted areas of Southern California, Hawaii, and regions in Arizona and Nevada. Despite major increases in population and acute travel over the past decade, air pollution has decreased by about one-third overall in these regions. The greatest reductions have been recorded for lead (93%), followed by carbon monoxide (35%) and PM (26%). While the majority of pollutants are being steadily reduced, nitrogen oxides, ground-level ozone and fine particulates remain a problem in many areas.[55–57]

The total emissions of criteria pollutants fell 32 percent since 1970 in the face of a 29 percent increase in the U.S. population, a 121 percent escalation in vehicle kilometers traveled, and a gross domestic product increase of 104 percent.[55] Still, nearly 46 million people live in counties that fail to meet the air quality standards for one or more of the criteria pollutants (Table 10.2).[55] Fine particulates from power plants, motor vehicles, and photochemical

TABLE 10.2

Percent Changes in Ambient Concentration of Air Pollutants from 1970 to 1996

Criteria Pollutants	Percent Change in Ambient Concentrations from 1970 to 1996
Carbon monoxide	−31%
Nitrogen dioxide	+8%
Sulfur dioxide	−39%
Particulates (PM_{10})	−73%
Ozone	−38%
Lead	−98%

Source: USEPA, National Air Quality and Emission Trends Report, USEPA Office of Air Quality Planning & Standards, USEPA document 454/R-97-013 1/13/98, www. USEPA.gov/oar /aqtmd96/trends.html, 1996.

TABLE 10.3

Sources and Health and Welfare Effects of Criteria Pollutants

Criteria Pollutants and Sources	Health Effects	Welfare Effects
Carbon monoxide Incomplete combustion of fossil fuels as in vehicles, kerosene heaters, boilers, and furnaces; cigarette smoking, forest fires, and biological decomposition	Interferes with oxygen transport in blood by binding with hemoglobin; causes headaches, fatigue, cardiovascular disease, and central nervous system disorders	Effects on plants or materials are not evident
Nitrogen dioxide Emitted from the combustion of fossil fuels in vehicles, industrial boilers, and electric generating utilities	Causes increased risk of respiratory infections and aggravates symptoms in persons with asthma and chronic bronchitis	Produces a reddish-brown haze over cities which reduces horizon visibility, causes leaves to yellow, and is a precursor to acid deposition and tropospheric ozone
Sulfur dioxide Fossil fuel combustion especially in coal-burning electric power utilities, metal smelters, oil refineries, and industrial boilers	Causes irritation of the throat and lungs and aggravates symptoms in persons with asthma and chronic bronchitis	Causes corrosion and deterioration of metals, brittleness of paper, paint discoloration, damages textiles and leaves of plants, and is a precursor to acid deposition
Particulates PM_{10} and $PM_{2.5}$ Fossil fuel combustion emissions, industrial processes, photochemical reactions in atmosphere, mechanical abrasion	Aggravates asthma, heart disease, and chronic lung disease; alters lung's natural cleansing mechanisms; smaller particles associated with the most severe symptoms	Causes soiling of materials, grime deposits and reduced visibility; particulates from volcanic eruptions can reduce solar energy and produce temporary cooling effect
Ozone A product of NOx emissions from motor vehicles, power utilities, and industries burning fossil fuels, combined with hydrocarbons and sunlight in the atmosphere	Causes breathing difficulty, irritation to mucous membranes, and increases risk to respiratory infections; acute exposures causes respiratory pain, bronchoconstriction, lung edema, and abnormal lung development	Corrodes rubber, paint; weakens fabrics, rubber, and produces leaf damage and retardation of plant growth
Lead Historically emitted from vehicles burning leaded gasoline; emissions have been reduced by 98% since 1974; most lead exposures in USA today are not airborne	Damage to nervous system, blood forming tissues, kidneys; evidence of neurobehavioral disorders including learning disabilities and antisocial behavior	No known effect on vegetation or materials

reactions in the atmosphere kill as many as 64,000 Americans each year from lung and heart disease.[58] Similarly, the American Lung Association released a survey of 13 U.S. cities linking exposure to ground-level ozone (smog) to as many 50,000 emergency room visits and 15,000 hospital admissions.[54] The health and welfare problems associated with criteria pollutants have not been completely resolved and many issues remain. A summary of criteria pollutant sources and health and welfare effects is presented in Table 10.3.

Particulate Matter

Particulate pollutants include airborne particles in liquid solid form that range in size from visible fly ash greater than 100 micrometers to particles 0.005 micrometer in size

TABLE 10.4

Sources and Characteristics of Coarse and Fine Particulates

Particulate Size	Particulate Sources	Chemistry
Coarse Particles These are coarse particles from 1–100 micrometer including the course fraction PM_{10} (2.5–10 micrometer)	Industrial and mechanical processes such as fragmentation of matter and atomization of liquids, agricultural and forestry activities, and dust from unpaved roadways, mold spores, and wood ash	Silicon, aluminum, iron, potassium, and calcium are common components; the coarse particle samples tend to be alkaline
Fine Particles These are fine particles less than 1 micrometer, but including much of the mass of the $PM_{2.5}$ fraction; the $PM_{2.5}$ fraction consists of fine particles less than 1.0 micrometer and some coarse particles in the 1–2.5 micrometer range	Industrial and residential combustion of fossil fuels; secondary particles produced by direct, catalytic, and photochemical oxidation of nitrogen and sulfur compounds, and volatile hydrocarbons to produce sulfates, nitrates, and oxyhydrocarbons	Elemental and organic carbon such as from fuel combustion (soot), sulfates, nitrates, condensed organic compounds, oxyhydrocarbons, and trace metals; the fine particle samples tend to be acidic

(Table 10.4). Particles may be produced naturally, such as pollen or sea spray, or by human activities, such as industrial processes, agricultural activities, fossil fuel combustion, and traffic. Particulates include dust, smoke, soot (carbon), sulfates, nitrates, trace metals, and condensed organic compounds. Particulates produce a number of effects adverse to human interests including (1) respiratory and cardiac health hazards to humans; (2) the deposit of grime and soot on buildings; (3) the reduction in sunlight, thereby causing a regional or global coding effect; and (4) reduced visibility in areas of extensive smoke or particulate pollution.[59]

The characteristics for coarse particles (2.5–10 micrometers) are very different from fine particles (<2.5 micrometers) and relate to the adverse effects they produce. Coarse particles originate from wind-blown dust coming from deserts, unpaved roads over which vehicles travel, and agricultural fields, pollen, mold spores, and plant debris. These particles tend to be mostly minerals consisting of aluminum, silicon, potassium, and calcium, which are chemically basic. Fine particles (<2.5 micrometers) are mostly emitted from fossil fuel combustion in industry, residences, and motor vehicles. They are also created in the atmosphere from gases of nitrogen and sulfur oxides and volatile organic compounds forming sulfates, nitrates, aliphatic and aromatic hydrocarbons including aldehydes, ketones, phenols, esters, terpenes, and phenylacetic acids.[34,60] Coarse particles can build up in the respiratory system causing respiratory disorders such as asthma in women. Smaller particles are more apt to be associated with (1) increasing lung function in persons who have pre-existing conditions such as asthma, (2) increasing premature death among the elderly and those with cardiopulmonary disease, (3) deterioration of respiratory defense mechanisms and adverse changes in lung tissue and structure, and (4) increased respiratory disease and symptoms.[34,56,61] Community studies provided support for issuing revised standards for particulates in order to protect the public health and welfare. The standard of PM 2.5 is set at 15 micrograms per cubic meter annual arithmetic mean, and 6.5 micrograms per cubic meter for a 24-hour average since these fine particles are more closely linked to mortality and morbidity effects than the previous PM_{10} standard.

Ozone and the Photochemical Oxidants

There are two types of ozone, which are known in the popular media as "good" ozone and "bad" ozone. Good ozone is that layer of ozone in the stratosphere previously discussed that absorbs UV-B radiation and protects the earth from excess amounts of these damaging rays. The "bad" ozone is formed on the troposphere (nose-level) by a complex series of reactions including sunlight, nitrogen dioxide, and volatile hydrocarbons. The reaction follows that outlined in Figure 10.16.

The photochemical oxidants resulting from such reactions include primarily ozone, nitrogen oxides, and alkyl peroxy radicals (RO_2). The alkyl peroxy radicals are produced by oxidation of hydrocarbons. An oxidant is a substance that readily gives up an oxygen atom or removes hydrogen from a compound. Photochemical refers to the initiation of these reactions by sunlight. This mixture of photochemical oxidants is often referred to as photochemical smog and is most concentrated in areas with heavy traffic, intense sunlight, and stable air conditions. Such conditions are especially prominent in Southern California and to a lesser degree along the northeast coast, especially during May through September. Ozone is the primary indicator of smog and in concentrated form is a dense (1.6 times as heavy as air), violet blue gas with a sharp metallic odor. It is a powerful oxidant capable of breaking molecular bonds and rapidly degrading many structured materials, as well as plant and human tissues.

The ozone standard was revised in 1979 to be 0.12 part per million for 1 hour. However, more than 3,000 studies since the 1980s show evidence that adverse health effects occur at lower exposures than the previous standard, and that these symptoms are aggravated by longer exposures. Exposures to ozone are associated with increased hospital admissions from asthma, and also accounted for up 20 percent of summertime admissions of respiratory distress. Respiratory infections and inflammation are aggravated by ozone as well. Children and adults who play and work outside during summer months are at elevated risk to ozone exposure and may experience coughing, chest pain, and a reduction in lung function. Exposure to ozone over long periods has been shown to cause inflammation

FIGURE 10.16
Proposed mechanism for the production and maintenance of tropospheric ozone. (Adapted from Godish, T., *Air Quality*, 3rd ed., CRC/Lewis Publishers, Boca Raton, FL, 1997, Chapter 2.)

of the lungs, impairment of lung defense mechanisms, premature aging of lung tissue through irreversible damage, chronic bronchitis, and emphysema. Consequently, USEPA has set a new 8-hour standard at 0.08 parts per million based on the decision that this level will protect the public health.[62]

Carbon Monoxide

Carbon monoxide is a colorless, odorless, tasteless gas that is produced from the incomplete combustion of fossil fuels with the greatest contribution by far being motor vehicles. In 1993, motor vehicles accounted for nearly 75 percent of anthropogenic carbon monoxide emissions.[60] Between 1986 and 1996 carbon monoxide concentrations in ambient air have decreased 37 percent while overall emissions decreased by nearly 20 percent. These improvements were made while vehicle kilometers traveled increased 28 percent during the same period.[63] Carbon monoxide enters the blood streams through the lungs and combines with hemoglobin of red blood cells to form carboxyhemoglobin. Carbon monoxide has a stronger affinity for hemoglobin than oxygen, therefore as levels of carboxyhemoglobin rise, the adverse effects associated with oxygen deficiency increase. The health threat from chronic low-level exposure is most serious for people who have pre-existing cardiovascular disease.[51] Inhalation of slightly elevated levels of carbon monoxide can produce symptoms ranging from dizziness and headaches to visual impairment, reduced work capacity, poor learning ability, lowered manual dexterity, and difficulty in performing complex tasks. Higher levels of CO inhalation can lead to paralysis of motor function and death.[63]

Lead

Lead is a systemic heavy metal poison that enters the human body through inhalation of contaminated air and ingestion of lead in food, water, soil, or dust. Lead accumulates in the bones, blood, and soft tissues with target organs being the central nervous system, the blood-forming tissues, and the gastrointestinal tract. Lead exposures have been associated with neurological impairments, mental retardation, and behavioral disorders. Even at very low doses, children and fetuses suffer from central nervous system damage. The association of lead with behavioral problems and reduced intellectual ability caused lead to be placed on the list of criteria pollutants in 1977 when the Clean Air Act was re-authorized. Since automobiles burning leaded gasoline were the largest contributor to lead exposure, the phase-out of leaded gasoline has been the predominant control strategy. Lead emissions from highways have decreased 99 percent since 1987 while overall ambient lead concentrations decreased 95 percent since 1977, and 75 percent since 1987.[55]

Sulfur or Oxides

Health and Welfare Effects

Sulfur, phosphorous, carbon, oxygen, hydrogen, and nitrogen are the main components of most living things, and consequently are found in fossilized forms including the fossil fuels such as coal and oil. The combustion of fossil fuels results in the oxidation of sulfur to produce sulfur dioxides, which react with moisture, oxidizing agents and reactive hydrocarbons and other substances in the atmosphere to produce sulfates, sulfites, and sulfuric acid. The primary source of emissions is electric utilities. These compounds may produce a variety of adverse health effects. Sulfur oxides are also involved in corroding metals; disintegrating marble, limestone, and dolomite; attacking fabrics; reducing visibility; and impairing the growth of plants and forests. The most prominent health concerns associated with SO_2 include respiratory illness, effects on breathing, a reduction in lung defenses, and aggravation of existing cardiovascular disease. Persons with asthma, cardiovascular disease, and chronic lung disease are most sensitive to the effects of sulfur dioxide. Children and the elderly are also at increased health risks from the inhalation of sulfur oxides.[63] Sulfur oxides are among the main precursors to acid deposition, with nitrogen oxides being the second greatest contribution.

Acid Deposition

Acid deposition is a term that represents a more inclusive understanding of the acidification process than the term acid rain, since acidity may be found in rain, sleet, snow, fog, clouds, and adsorbed to particles. Consequently, the term acid rain is being replaced by the term acid deposition although the two are presently used synonymously. The problems associated with acid deposition were first recognized about 30 years ago in Sweden when changes in the plant and animal life of freshwater lakes was associated with increased acidity from acidic deposition.[17] The revelation of acidic deposition was not widely accepted by the world scientific community initially, but as measurements of rainfall pH demonstrated greatly expanding areas of increasing acidity, the problem became the subject of domestic and international debates over acid rain controls. The United States passed the 1990 CAAA with major provisions for sulfur and nitrogen oxide controls although atmospheric acidity still presents a major environmental problem.

Rain, snow, clouds, and sleet are normally somewhat acidic because of the presence of CO_2 in the atmosphere producing carbonic acid with pH levels of nearly 5.0 in the absence of pollutants. The pH scale is logarithmic and is based on the negative log of the hydrogen ion concentration. Consequently, a pH of 3.0, as has been found in precipitation in the Ohio Valley, is 100 times the acidity of rain with a pH of 5.0. About 70 percent of sulfur oxide emissions originates from large electric utilities burning high sulfur coal while other major contributors to NOx and SOx include industrial boilers, metal smelters, and automobiles.[23] Sulfur acids have contributed about 65 percent to the acid deposition problem while nitrogen oxides contributed 35 percent. However, as regulatory controls become more effective, the levels of sulfur oxides in the atmosphere have decreased by 39 percent since 1970 while ambient concentrations of NO_2 increased eight percent over the same period.[63] Consequently, the emissions of NOx from automobiles and electric power generators is taking on greater importance.

Effects of Acid Deposition on Ecology

Acid deposition is a product of fossil fuels with most SOx occurring from electric utilities burning coal or oil. Many of these plants employ the strategy of building tall stacks so the discharged gases will be carried out of the vicinity. Although locally effective, the gas streams are carried by prevailing winds great distances to be deposited far from the origin in a process known as long distance transport. Heavily industrialized areas of the Ohio River Valley, upper Midwest and areas of Illinois, Pennsylvania, West Virginia, and Indiana are often considered to be the source of significant acid plumes that travel eastward to deposit all along the Northeast coast off the United States and parts of Canada. The heavily industrialized areas of Great Britain and Central Europe and parts of China create similar conditions for polluting distant locations. Sweden and Norway point to central Europe as a major cause of the acidity falling on them. Although long distance transport contributes significantly to acidic deposition, local sources are also important contributors and just how much each contributes to the problem has been strongly contested by those allegedly producing the pollutants, and those who receive them.

The New England areas including upstate New York, the mountainous areas of West Virginia, Pennsylvania, Virginia, Kentucky, Wisconsin, Minnesota, parts of Canada, and Western mountainous areas of North America are sensitive to acidity because these areas are deficient in magnesium and calcium carbonates. Therefore, the soils and water have low acid neutralizing capacity and are referred to as acid-sensitive ecosystems (Figure 10.17).[23]

Aquatic Ecosystems

When lakes and streams receive acid deposition in acid-sensitive areas they are unable to neutralize or buffer the acidity. Consequently, aquatic systems in the Adirondacks or

FIGURE 10.17
Acid-sensitive regions of North America and Canada. (Adapted from Godish, T., *Air Quality*, 3rd ed., CRC/Lewis Publishers, Boca Raton, FL, 1997, Chapter 4.)

New York, New England, and other sensitive areas become chronically and progressively acidified or receive sudden massive doses of acidity in spring thaws resulting in a process known as shock loading. As the acidity of a body of water increases there is a reduction in the diversity of species and a shift in species composition. The smallest of the organisms, unicellular phytoplankton, disappear followed by the benthic invertebrates. These organisms serve as important food sources and nutrient recyclers (decomposers) so that large aquatic creatures are adversely affected by their disappearance. Acidification can lead to mobilization of aluminum and possibly other toxic metals, which in combination with the increased acidity, leads to reductions in fish populations. At pH levels below 5.5, fish populations decline with the most sensitive, juvenile fish first disappearing. The eggs, larvae, and juvenile fish fail to thrive in a process called recruitment failure often leaving acidified streams and lakes with a diminishing population of older, larger fish. Eventually, entire lakes and streams become devoid of fish, such as in the Adirondacks, mid-Atlantic mountains, New England, Ontario, Quebec, Sweden, and Norway.

Effects on Forest and Plants

The causal role of acid deposition in the decline of forested areas in North America, Germany, and Central Europe has been a matter of considerable investigation and debate. Acidic deposition at levels of pH 4.0–5.0, which are most common, does not appear to cause widespread adverse effects on forest ecosystems.[23] However, conifer forests such as Red Spruce on mountaintops in New Hampshire, Vermont and in the Appalachians have been more than 80 percent decimated at the cloud line. Severe damage is also evidenced in Central Europe in such places as the Czech Republic and Poland where 60–70 percent of forests show evidence of damage associated with sulfur and nitrogen disposition. The mechanisms to such destruction are not immediately devious but may be attributed highly to combinations of ozone and acid clouds (as low as pH 2.2), which (1) directly damage leaves; (2) mobilize toxic metals in soil such as aluminum, which adversely affect roots; (3) leaching nutrients from soil; and (4) over-stimulate plants from excess nitrates, which aggravates deficiencies of other nutrients. These factors may combine to increase forest susceptibility to insect and fungal pathogens.[64]

Current Directions in SOx Control

Ambient concentration of SO_2 decreased 37 percent since 1987 while emissions decreased by more than 14 percent. In October 1999 the EPA conducted a study to determine whether emission regulations impacted the recovery of lakes and streams, in North America and Northern Europe, which had suffered years of acid deposition. The study concluded significant declines of 1–6% sulfate per year in many lakes and streams. However, reduced emissions do not automatically translate to immediate improvement in streams lakes and rivers. For example, some waters in Maine continue to acidify even though sulfuric acid concentrations in rain have decreased. The scientists concluded that several factors contributed to continued lake acidity: (1) variations in climate, (2) increasing levels of nitrogen compounds in precipitation such as nitric acid, (3) declines in the ability of a watershed to neutralize an acid, and (4) the short length of time that data have been collected.[65]

Reductions in SO_2 emissions are due mostly to controls implemented under USEPA's Acid Rain Program.[63] This program features a number of options including (1) switching to low-sulfur fuel; (2) using scrubbers that remove SO_2 from the stack emissions; (3) washing coal, which removes up to 50 percent of the sulfur content; and (4) using advanced combustion technologies such as fluidized-bed combustion to remove sulfur and nitrogen oxides.

Nitrogen Oxides

Nitrogen and oxygen are combined in the high temperature confines of industrial boilers and automotive engines to produce nitric oxide. This is oxidized in the atmosphere to produce nitrogen dioxide, an orange-yellow to reddish-brown gas that can often be seen from a distance, enveloping cities. Nitrogen oxides are an important precursor for ozone and acidic deposition (nitric acid). Nitrates arising from NOx emissions fall from the sky to stimulate algae growth (i.e., red tide), causing toxic conditions for aquatic life in bays and estuaries. Nitrogenous compounds also contribute significantly to the formation of fine particulates. Nitrogen dioxide irritates lung tissues and reduces resistance to respiratory infections, while possibly increasing the incidence of acute respiratory disease in children.[63]

Health Implications of Air Pollutants

The criteria air pollutants have been implicated in a number of adverse health problems including reduced lung functions, exacerbating problems such as asthma, increasing risk of chronic bronchitis and emphysema, and aggravating acute respiratory infections. Carbon monoxide, sulfur dioxide, and particulates have also been shown to contribute to cardiovascular disease.[56,58,61,57] It is calculated that 125 million Americans breathe unhealthy air resulting in approximately 70,000 premature deaths in the United States alone.[66] Fine particulates from motor vehicles and power plants are reported to kill some 64,000 Americans a year and may be a major contributor to the epidemic of childhood asthma sweeping the country and developing nations worldwide.[58]

Asthma occurs in persons with hyperactive airways when the bronchial tubes respond to allergens, pollution, irritants, cold air, and even exercise by small muscle contraction, dilation of blood vessels, and mucous secretion. The result is suffocating paroxysms where the asthma sufferer struggles for each breath through closed airways clogged with mucous (Figure 10.18).[63] Asthma has increased by 42 percent since the mid-1950s. In 2015 alone, 24.6 million people suffered from severe asthma, 6.1 million of which were children (<18 years of age).[66,68] According to the Centers for Disease Control and Prevention (CDC), asthma rates among children below 17 years of age increased 72% between 1982 and 1994.[69] In reality there has been a two-fold increase in the number of children who have asthma since 1980. This statistic alone has spurred concern among public health researchers.[70] It has become more severe and research shows that children and other minorities and urban poor are at greatest risk. Asthma now tends to be concentrated in urban pockets where children live under poor conditions and are frequently exposed to allergens such as mites and cockroaches,[71,72] and to episodes of air pollution involving fine particulates and ozone

Asthma attack
Beginning signs
• Wheezing
• Coughing
• Difficulty in breathing

Extended attack
• Sweating
• Pulse becomes rapid
• Skin turns blue
• Arms and legs become chilled or cold

Bronchiole

Muscles contract and membranes swell around bronchioles.
Mucus forms in airways.

Narrowed airway makes breathing difficult.

Alveoli

Normal airway   Airway in spasm

FIGURE 10.18
Symptoms and effects of asthma.

(Figure 10.19).[73] Air pollutants, particularly ozone and fine particulates, were found to be consistently associated with asthma exacerbation in a prospective cohort study of children with asthma.[74] However, this link to specific air pollutants remains controversial. The prevalence of asthma is rising in both industrialized and developing countries, and the two countries with the greatest increase in asthma prevalence are Australia and New Zealand, neither of which has extensive air pollution.[75] Serious questions remain to be answered on this controversy. Have all the responsible gases been identified? Is there an interaction of gases and particulates? What are the mechanisms by which pollutants exacerbate asthma?

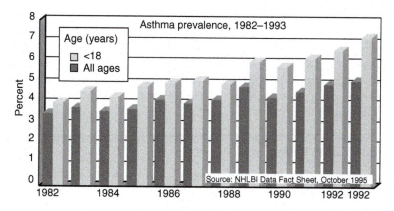

FIGURE 10.19
Trends in the prevalence of asthma from 1982 to 1996. (Adapted from Friebek, E., *Environ. Health Perspect.*, 104, 22, 1996.)

Main Mechanisms of Air Pollutant Effects on Respiratory System

Air is generally laden with dust, microbes, and gaseous emissions that may have adverse effects on human health. However, the human respiratory system has a number of defense mechanisms to ward off such attacks. The first line of defense includes nasal hairs that trap larger particles. A combination of mucous and cilia line the bronchi and bronchioles, which trap and move smaller particles out of the respiratory tree. Irritating dusts and gases may cause the smooth muscles of the bronchioles to constrict, restricting heavier exposure in the short term. However, prolonged or exaggerated constriction is often combined with mucous secretions leading to difficulty in breathing. Particles that reach the deeper portions of the lungs, such as the alveoli, may be confronted by the alveolar macrophages that can ingest, destroy, and discard many microbes. Pollutants may produce their adverse effects by (1) inhibiting and inactivating mucociliary streaming; (2) killing or neutralizing alveolar macrophages; (3) constricting airways; (4) causing vasodilation and excess mucous secretion; or (5) causing changes in alveolar cell wall structure through abscesses and thickening, which causes scar formation. The formation of scar tissue is characteristic of emphysema in which the walls of alveoli disintegrate and the compartments become unnaturally enlarged. Elasticity of the living tissue diminishes and the lungs are unable to recoil and expel air reducing the efficiency of the lungs in enhancing oxygen and carbon dioxide. Victims ultimately may become wheelchair bound, gasping for each breath, and unable even to expel sufficient air to blow out a candle. There is no cure for emphysema.

Indoor Air Pollution

Imagine suffering through episode after episode of pneumonia within a few short months, suffering an attack of breath-restricting asthma every time you enter the basement of your home, or developing nausea and violent headaches upon entering your place of employment each day. These are examples of afflictions associated with indoor pollution in private homes and public places.[76] At one time, most people were aware of the potential threats of outdoor air pollution, but today many experts are cautioning against the effects of indoor pollutants. There has been a growing body of scientific literature in the last several years that points to indoor air as being more polluted and potentially hazardous than outdoor or ambient air.[77] Additionally, people spend an average of 90 percent of their time indoors while some at-risk subgroups such as the elderly, very young, and chronically ill may spend nearly all their time indoors.

Sources of Indoor Air Pollution

Sources of indoor pollution come from several major categories that include combustion sources such as: (1) acid, coal, gas, oil and kerosene; (2) household cleaning products; (3) furnishings such as cabinetry or furniture made of pressed wood, and newly installed carpeting that emits volatile organic carbons (VOC) from adhesives; (4) building materials and paints that off-gas volatile compounds; (5) products for personal care and hobbies; (6) cooling and humidification systems; (7) external sources such as radon and tobacco smoke; and (8) moisture, which supports the growth of molds and other biological contaminants (Figure 10.20). Sources such as air fresheners, paneling, cabinets, and rugs may emit pollutants at a fairly continuous rate for some time, while other pollutant sources or activities release pollutants intermittently such as cigarette smoking, use of household cleaners, or pesticide sprays. How long pollutants remain in the air and their relative concentration depends on several factors. Elevated heat and humidity increase the release of some pollutants into the air such as formaldehyde. Concentrations are greatly influenced by the amount of ventilation as well.

Air enters a house in three ways:

1. *Natural ventilation*: Air enters through open windows and doors, ridge vents, on the roof, ventilation pipes, or other portals designed for air access.
2. *Infiltration*: Air is drawn into the house through cracks around doors and windows, spaces in construction joints, or spaces or cracks in the foundation.
3. *Mechanical ventilation*: Air maybe drawn out of the building at home through externally vented fans or into the structure by whole-house fans in air handling systems.

The rate at which air is replaced in the structure by external air is termed the air exchange rate. This average rate for an American home is 0.7–1.0 air changes per hour. Houses that are tightly sealed without provisions for an exchange may experience as little as 0.2 air changes per hour while drafty houses may approach up to 10 times that amount.[78] More recently constructed office buildings and nonresidential structures may have sealed windows that restrict air infiltration. While ventilation and cooling systems are intended to maintain comfort, air that is recirculated in a building without sufficient outside makeup air may exhibit increased levels of CO_2 and pollutants. Prior to 1973, at least 0.4 cubic meters per minute (CFM) per occupant was required for makeup air. However, the standard was reduced after 1970 to 5.0 CFM, but this level may be insufficient in providing a safe and healthier environment.

Potential sources of indoor air pollutants

1. Moisture
2. Pressed-wood furniture
3. Contaminated humidifier
4. Moth repellants
5. Personal care products
6. Room air fresheners
7. Chemical cleaners and disinfectants
8. Pressed wood cabinets
9. Unvented gas stove
10. Household chemicals
11. Tobacco smoke
12. Wood stove
13. Wood paneling
14. Asbestos pipe insulation
15. Unvented dryer
16. Radon infiltration
17. Pesticides
18. Auto exhaust
19. Auto cleaners and additives
20. Paints, thinners, and stains

FIGURE 10.20

Sources of potential indoor air pollutants. (Adapted from Anonymous, *Environ. Health Perspect.*, 104, 21, 138, 1996.)

Signs of Indoor Pollution

Signs of indoor air pollution may include physical or health signs or both. Physical symptoms may include (1) heating or cooling equipment that is dirty and/or moldy, (2) moisture condensation on walls and windows, (3) air that has a stuffy or has an unpleasant odor, and (4) signs of water leakage anywhere in the building with the growth of molds. Health indicators of indoor air pollution may include immediate or acute effects such as eye irritation, dry throat, headaches, fatigue, sinus congestion, sun irritation, shortness of breath, cough dizziness, nausea, sneezing, and nose irritation. These symptoms are vague and may originate from many sources. Some people recognize that the symptoms occur when they enter a structure and diminish when they leave. In many cases, however, there is not a readily identifiable pattern or illness. Additionally, chronic illnesses such as emphysema, cancer, and heart disease may not appear for years, and any association with exposure to substances in a particular building is usually not recognized. When a number of occupants of a building display acute symptoms without a particular pattern and the varied symptoms cannot be associated with a particular source, the phenomenon is often referred as sick building syndrome (SBS). Further, maladies may not be the cause of indoor air pollution, but may be related to psychological stress, inadequate lighting, excessive heat or cold, or even noise.[79]

When well-defined illnesses occur in a building and they can be traced to specific building problems, the illnesses are referred to as building-related illnesses (BRIs). The well-defined illnesses normally include asthma, hypersensitivity pneumonitis, humidifier fever, or Legionnaire's disease. The problems associated with SBS or BRI have apparently increased as concerns about energy conversation have escalated resulting in the construction of air-tight buildings with thicker insulation, magnetically sealed doors, sealed windows (office buildings) or triple-glazed windows that all serve to limit the exchange of air with the outside. Office buildings are at risk to indoor air pollution from (1) poorly designed, or improperly used or maintained ventilation systems; (2) the presence of pollution sources such as harsh chemicals, improperly vented photocopiers, or biological contaminants from dirty ventilation systems; and (3) uses of the building for which the building was not designed such as a paint shop or automotive repair service in the basement of an office building.

Common Sources of Indoor Air Pollution

The most common sources of indoor pollution include environmental tobacco smoke (ETS), radon, biologicals, nitrogen dioxide, carbon monoxide, organic gases, formaldehyde, respirable particles, and pesticides.

Environmental Tobacco Smoke and Other Combusted Materials

Although smoking has been under vigorous attack in the United States, and the opportunities for smoking in public buildings has greatly diminished, smoking still contributes to nearly 500,000 deaths each year in the United States.[80] Now we are seeing a renewed interest and increase in smoking. Sixteen percent of adult New Yorkers smoked in 2013, up from 14 percent in 2010. Even so, this is still less than the national average of 18 percent.[81] We are also seeing an increase in teenaged smokers in the UK and the United States.[82] The smoker is exposed to nearly 4700 compounds in main-stream smoke through direct inhalation. The side-stream smoke that originates from the burning end of the cigarette and exhaled smoke make up ETS, to which non-smokers may be exposed in a form referred to as "passive" or "second hand" smoke. Such involuntary exposures have prompted numerous legal actions further restricting the public places where smoking is allowed. However, people continue to be involuntarily exposed to ETS in the home and many commercial establishments increasing the risk of respiratory disorders. Second hand smoke has been linked the rise of lung cancer risk by an average of 30 percent in non-smoking spouses.[76] Other sources of combustion products include gas and kerosene space heaters that are unvented, and improperly vented wood stoves, fireplaces, and gas stoves. Pollutants from these sources include carbon monoxide, nitrogen oxides, particulates, and a variety of aromatic hydrocarbons and acid aerosols. Unintended deaths occur every year from the buildup of CO in the home because of improper ventilation of home space heaters or furnaces. Lesser concentrations may lead to headaches, dizziness, and aggravate pre-existing heart disorders. Nitrogen oxide exposure can increase the risk of respiratory infection, emphysema, and may aggravate respiratory disorders such as asthma.

Radon

Radon is a colorless, odorless gas that occurs naturally by the decay of radium-226, a product of uranium decay. It is found most everywhere but is highest in concentration in uranium and phosphate ores. Areas with lesser concentrations of radon are found in limestone, shale, and granite. Radium-226 is ubiquitous in low concentrations and doesn't present a health issue until it is trapped and concentrated in areas such as the basement of a building or house built in soils where uranium is in the soil or rock.

As the uranium naturally radioactively decays it releases radon gas that further decays into short-lived, radon daughters and gamma rays (Figure 10.21). The radon daughters posing the greatest risk are the alpha-emitters polonium-218 and 214. The radon progeny quickly

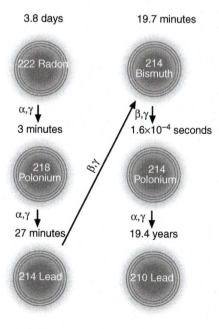

FIGURE 10.21
Radioactive decay of products of radon.

attack particles in the air which may then be inhaled to lodge in sensitive areas of the upper respiratory tree. Once lodged in human tissue, the radioactive materials increase the risk of lung cancer causing from 5,000 to 20,000 excess cancer deaths a year in the United States.[77]

Evidence released by the National Research Council points to radon as a significant public health problem causing as many as 21,800 deaths per year with the greatest risk being to people in homes where cigarettes are smoked. Radon is considered to be the second-leading cause of lung cancer exceeded only by cigarette smoking.[83] Radon in soil normally enters structures under slight negative pressure through cracks in concrete walls and floors, sumps, floor drains, dirt floors, joints, and crawl spaces. Radon may also be present in well water or even foundation materials. It is estimated by USEPA that as many as eight million homes in the United States have elevated radon levels. Although some regions are at greater risk than others, buildings and households should be checked since even adjacent houses may show remarkably different levels of radon. Most hardware stores have radon detection kits for sale. The kits are either alpha track detectors or chemical canisters. The kit instructions usually suggest you hang the detector in a closed cellar at .9 to 1.2 meters above the ground for several days to a month. The kits are returned by mail to the manufacturer for analysis and a report is usually returned within 2 to 3 weeks to the homeowner or building manager. Radon concentrations are reported as picocuries per liter (pCi/L). A picocurie is the equivalent to one trillionth of a curie representing slightly more than 2.0 radioactive disintegrations per minute. The average concentration of radon and its decay products in ambient air is about 0.25 picocuries per liter. The federal action guidelines recommend that corrective actions be taken at 4 picocuries per liter or above.

Some of the highest levels of radon have been measured in a region called the Reading Prong located in an area covering Southeastern Pennsylvania, and parts of Connecticut and New Jersey.[84] The highest level ever recorded in a household (2600 picocuries per liter) was identified in Southeastern Pennsylvania. The northeast has higher mean levels than most other sections of the country, but the presence of radon in a home is unpredictable. Households must be tested to determine if they exceed the recommended action level of 4 picocuries per liter. Surveys have shown that one in every 15 homes in the U.S. exceeds the action guideline. At this level, the estimated number of lung cancer deaths to radon exposure would be 13–50 per 1000 for an average lifetime exposure. This risk is equivalent to smoking five cigarettes a day.[77,84]

The mitigation of radon may involve a number of actions such as installing sub-slab or basement ventilation; covering sump holes; sealing cracks and other openings in floors, joints, and walls; ventilating crawl spaces; installing air-to-air heat exchanges; or installing a drain tile suction system around the outer foundation walls. In some cases, more than one method may be used.

Biological Contaminants

The most common biological contaminants include molds, mildew, viruses, bacteria, dust mites, cockroaches, pollen, animal dander, and cat saliva. The sources for these agents are numerous including people-transmitted viruses and bacteria, household pets, contaminated air handling or ventilation systems, moist surfaces on which organisms can grow, dried fecal material from cockroaches and other insects, and soil and plant debris. The major threat to the biological contaminant of the home is moisture. Water may enter through a porous basement floor or walls, a leaky roof, or remain in unventilated bathrooms or as a reservoir in dirty humidifiers. In these situations, water stimulates the growth of mold, dust mites and other infectious or allergenic organisms.[85]

Surveys have shown that 30–50 percent of homes surveyed in the northern U.S. and in Canada have problems with moisture while structures in the south may be at higher risk. Wet surfaces, moist carpeting (often associated with concrete floors), water damaged materials, and standing water can all support the growth of molds, mildew, bacteria, and insects such as house and dust mites. Dust mites are microscopic insects that feed on shed human skin, thrive in moist environments, and are powerful allergens. Molds are also potent allergens, and they can provide nutrients to bacteria, spread viruses, emit chemicals such as cancer-causing phenols and trichloroethylene, increase susceptibility to a host of diseases by weakening the immune system, and aggravate asthma.

Possible symptoms of illness caused by biological contaminants include running nose, colds, flu-like symptoms, headaches, unexplained fatigue, and digestive problems. Allergic rhinitis, hypersensitivity pneumonitis, asthma and other serious allergies are common. Some biological contaminants transmit infectious illnesses such as influenza, measles, and chicken pox while molds and mildews release disease-producing toxins.[77,85] Homes that have water stains or other signs of moisture, leaking plumbing, mold, mildew, or dust that remains after cleaning are signs of a possible water problem.

Reducing Exposure

Since moisture always migrates to colder surfaces using ample steps such as using fans, warming basements, ventilating attics and bathrooms, and insulating exterior walls can check the unwanted movement of water. Gutters may be used to drain water away from buildings. Other methods of controlling biological vectors include washing bedding frequently in hot water, maintaining humidity between 30 and 50 percent, using a high-efficiency particulate filtering vacuum cleaner, enclosing pet litter boxes, discarding wet carpeting and other water damaged materials, and practicing good housekeeping procedures.

Carbon Monoxide, Nitrogen Oxides, and Respirable Particles

These substances are manually emitted from unvented kerosene and gas heaters, wood stoves, fireplaces, environmental tobacco smoke, automobile exhausts in attached garages, and other combustion sources. These have been discussed previously under combustion sources.

Organic Gases, Pesticides

Most households and many businesses have a wide variety of products that can emit potentially hazardous materials including paints, strippers, disinfectants, cleaners, repellants, automotive products, hobby supplies, volatile office supplies, and pesticides. Such substances often exceed levels found outside the home by a factor of up to five times while some activities such as paint stripping may produce concentrations of 1000 times background outdoor levels.[77,86] The potential health consequences are numerous based on the extraordinary variety of chemicals available in the home and business. These symptoms range from initiation to mucous membranes, multiple organ damage, headaches, nausea, cancer, and many other maladies. Exposures to such substances can be limited by (1) properly disposing of unneeded products, (2) following instructions on a label and using it in a well-ventilated area if so advised, and (3) using non-toxic alternatives where possible.

Formaldehyde (HCHO)

Formaldehyde is found in pressed wood products such as cabinets and furniture made from plywood, particleboard, wall paneling, and fiberboard. It is also found in urea formaldehyde foam insulation (UFFI), some textiles and environmental tobacco smoke. However, formaldehyde-emitting products are now being manufactured with much less HCHO emission potential, and emissions from pressed wood products are presently 10 percent of those used 20 years.[87] Since older products, and structures, have off-gassed most of the volatile HCHO, and new products have very little HCHO, the significance of formaldehyde as a public health issue has been markedly reduced. Exposures still can occur, which is aggravated in the presence of a confined or poorly ventilated space, combined with high heat and humidity. Adverse health effects include irritation to the mucous membranes, severe allergic reactions, fatigue, wheezing, and coughing.

Noise

Introduction

People spend an average of 90 percent of their lives indoors and are exposed not only to chemicals and biologics, but also to sounds that enter their homes in the form of heavy traffic, descending or disembarking airliners, trains, and loud radios. Outside, they are confronted with the additional sounds of snowmobiles, jet-skis, and motorcycles in areas previously thought to be reserves of quiet. Sound itself is not a pollutant, but when it interferes with tasks, when it distracts, annoys, or disturbs, or when it causes losses in hearing or alters physiology in negative way then it becomes unwanted sound, or noise. Noise may be considered a pollutant whose primary medium of transmission is air and so it is considered in this chapter as another form of air pollutant. Organizations such as Noise Network have been created for people who cannot endure sources of noise disturbance any longer. The organizations will help clients relocate from an area of high noise to one of a lower frequency of occurrence. Tragedies have been the result of normally noisy areas. For example, a 78-year-old farmer was driven to the unthinkable end after months of enduring noisy neighbors. He reached his breaking point, shooting two of the neighbors, killing one and wounding the other. He was later convicted of manslaughter and is presently serving two consecutive life sentences. This is an uncommon effect of noise pollution; however, it is true and could be a cause for alarm down the noisy road.[88] The intensity or loudness of noise combined with its duration and frequency (high or low) are important to determining the hazardous effects from that noise. We know that high and low pitch noise causes more damage than noise in the middle spectrum, and mixed (white) noise is less damaging than noise at a specific frequency or pitch. Most people exposed to high energy noise on a daily basis can experience hearing loss. Such hearing loss may occur permanently. More than 30 million people in the U.S. were exposed to noise levels above 85 decibels, which is considered hazardous.[89] Compare this with the 9 million people exposed to the same level of noise in 1981. According to the World Health Organization about 12–15 percent of all employed people are exposed to noise levels of 85 decibels or more in developed countries.[90]

The Physics of Sound

Sound is a form of energy that is produced by the vibration of objects that compress and expand air, water, or solids to produce waves. When the vibration occurs in the ear, variations are produced in the normal atmospheric pressure that are carried to the eardrums and cause the membranes to vibrate. These vibrations are passed to the inner ear and ultimately to the brain which interprets these vibrations of acoustical energy as sound. Because sound depends on the movement of molecules causing positive and negative pressure waves, the source cannot travel in a vacuum. As an object vibrates, waves radiate outward in an expanding sphere much like the ripples produced when a stone is dropped in water. Waves are produced that consist of alternating peaks and troughs that may vary in their length, peak, and amplitude. These wave characteristics produce changes in frequency (or pitch) and amplitude, which translates to loudness. Frequency is directly related to the speed of vibration. The faster the vibration, the more frequent the waves are per unit time and the higher the pitch (Figure 10.22). Frequency is expressed as cycles per second and the standard unit of expression is hertz (Hz), which is equivalent to one wave per second passing over a particular point. The normal range of human hearing is in the frequencies of 50–20,000 Hz. Sound waves also vary in the height or amplitude and reflect the amount of energy formed in the sound wave (Figure 10.23). Soft sounds have a low amplitude while loudness is characterized by large amplitudes. This amplitude intensity

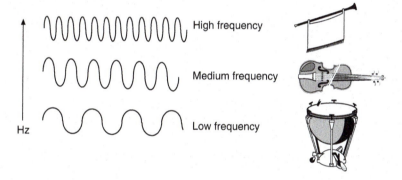

FIGURE 10.22
Frequency differences of sound. Higher frequencies are associated with more cycles per second (hertz) and are perceived as higher pitch.

FIGURE 10.23
Amplitude differences of sound. Higher amplitudes or wave heights are associated with greater energy and perceived as loudness.

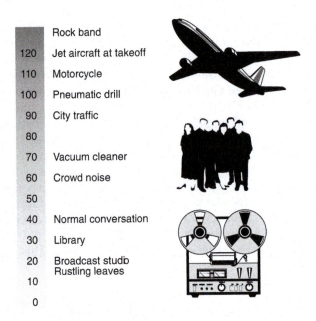

	Rock band
120	Jet aircraft at takeoff
110	Motorcycle
100	Pneumatic drill
90	City traffic
80	
70	Vacuum cleaner
60	Crowd noise
50	
40	Normal conversation
30	Library
20	Broadcast studio
	Rustling leaves
10	
0	

FIGURE 10.24
Sound levels of various activities.

or loudness is measured in decibels (dB). The scale of dB ranges from zero (0) dB, which is the threshold for hearing among the normal population, to 194, which is the threshold for pure tones.[91,92]

The perception of sound (hearing) is not uniform across frequencies (Hz) and amplitudes (dB). Human hearing is most acute on the range of human speech, which is about 400–4000 hertz. At frequencies on either side of the range, hearing is less sensitive and greater amplitudes are required to hear these frequencies. At amplitudes below zero dB, humans cannot detect sound pressures although they exist, and as sound increases above 70 decibels it becomes increasingly difficult to participate in conversation while sounds above 90 decibels may lead to hearing loss. The pain threshold for noise begins at 120–130 decibels for most people. Since the perception of loudness doubles with each 10-decibel increase, this is not a linear relationship and small differences in dB readings are heard as very large changes.[92,93] The relationship of decibels to common sounds is shown in Figure 10.24.

Physiology of Sound and Health Effects

Sound compression waves are directed by the external auditory canal (meatus) where they strike the tympanic membrane causing it to vibrate in concert with the waves. The tympanic membrane is connected to a series of three very small bones in the middle ear known as the malleus, incus, and stapes, which transmit the vibration to the oval window of small-shaped structure called the cochlea (Figure 10.25). The transmitted vibration causes corresponding waves in the liquid (perilymph), which under appropriate conditions cause the basilar membrane to vibrate. This in turn causes hair cells of the spiral organ (organ of Corti) to move against the tectorial membrane. The hair cells convert a mechanical force (pressure) into a nerve impulse (electrical signal) that is ultimately transmitted to

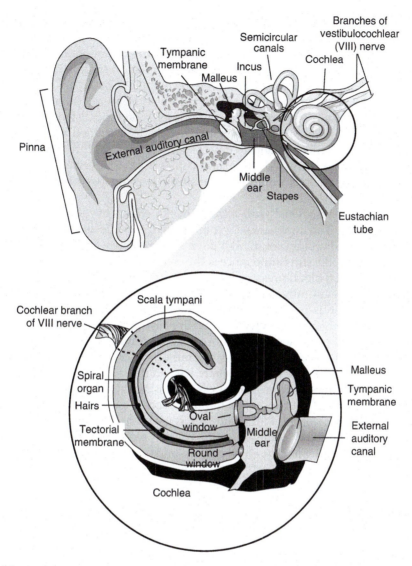

FIGURE 10.25
Structure of the ear and hearing apparatus.

the auditory area of the temporal lobe of the cerebral cortex.[93] Excessive sound pressure (loud noises) can destroy the delicate hairs in the spiral organ. This process usually begins with the hairs at the base of the cochlea most sensitive to high pitched sounds and with continued exposure to high dB sound, the damage extends to other hairs and corresponding lower pitches. Such continued exposure can lead to increased damage of hair cells until 30–50 percent of the hair cells are destroyed at which time measurable hearing loss becomes evident. Such damage resulting in loss of hair cells produces permanent hearing loss and is known as permanent threshold shift (PTS). Permanent hearing loss can result from prolonged exposures to sound levels of 70–85 decibels, which are at or above the level of vacuum cleaners or city traffic. Higher sound intensities require much less time to cause damage and being on the landing field without ear protection where jet planes are disembarking could cause measurable hearing losses in less than an hour. Noise can also result in a temporary threshold shift (TTS), which is a reversible hearing loss normally lasting from moments to hours, although longer periods of temporary hearing loss are possible. Very loud explosions or gunfire can cause acoustical trauma that can rupture or damage eardrums, middle ear bones, and even the organ of Corti.

The measurement of hearing loss is usually performed on an instrument called an audiometer, whereby a person is placed in a soundproof room and listens on earphones to sounds produced at different pitches and dB (decibels) by a trained audiologist. Persons with hearing problems will be unable to detect sounds at normal threshold levels, and the intensity will have to be increased. If normal values are 0–25 decibels, then hearing the same sounds at 26–40 decibels would indicate a mild hearing loss whereas a severe hearing loss would be 71–90 decibels.

Noise has also been associated with adverse health effects beyond hearing loss. Noise may be an annoyance. The annoyance level appears to escalate as a function of intensity (loudness) and pitch. Therefore, loud, high pitch sounds (above 5,000 hertz) tend to be the most annoying.[93] Related to this annoyance is the creation of stress and associated physical effects such as increased blood pressure, elevated heart rate, headaches, colitis, and ulcers. Such responses may seem unreasonable but in the presumably quieter realm of historic man, sudden loud sounds prompted immediate physiological responses to the presence of danger as a mechanism of survival. Unfortunately, prolonged exposure resulting in continued exaggerated physiological response leads to damaging health consequences.

Noise can lead to interference with speech communication; most of us have been in the presence of a loud band or noisy crowd where we had difficulty hearing a person yelling at us within a few centimeters from our ears. Noise ratings have been developed that measure its ability to interfere with speech communication or a scale known as the preferred octave speech-interference level (PSIL). The PSIL is useful in determining what changes in PSIL are necessary to making communication easier.[57] Noise levels have also been implicated in decreased learning ability, work performance, and increased safety hazards and sleep disruption.[87,91]

Regulation of Noise

The regulation of sound requires that it be measured according to a standard. The measuring device most frequently used is a sound pressure level meter that converts sound captured by a microphone to electrical signals that can deflect a needle on an indicating meter (Figure 10.26). Most instruments contain A, B, and C scales. The A-weighing network is biased to high-frequency noise and best correlates with the annoyance effects and hearing damage in human. Thus, the sound reading on this scale would be expressed dBA units. This is the scale most frequently used in assessing noise exposure. Exposures are regulated for the occupational environment by the Occupational Safety and Health Administration (OSHA) Department of Labor with a permissible exposure limit of 90dBA for an 8-hour day, 40-hour workweek. Permissible exposure times decrease half for each 5-dB increase.

Federal agencies that regulate noise include the Federal Highway Administration (FHA), Housing and Urban Development (HUD), OSHA, Federal Aviation Agency (FAA), and USEPA. Each of these agencies is responsible for the regulation and control of noise to specific operations or segments of society under their direction. OSHA regulations are designed to protect workers from occupational noise induced hearing losses. USEPA is mandated under the Federal Noise Control Act of 1972 and Quiet Communities Act of 1978 to abate noise in the ambient environment and in communities through (1) identifying levels of environmental noise considered protective of human health, (2) identifying major noise sources requiring regulation, and (3) providing financial assistance to state and local governments to investigate noise problems and develop control measures in the communities. Many programs were initiated in the 1970s, but the elimination of federal funding and associated programs in the 1980s all but eliminated state and local noise control programs throughout the country. Many federal regulations under the NSAD 1972 noise abatement were never promulgated and it is unlikely that significant regulatory attention will be paid to issues of noise until there is an increasing public outcry focusing on the increasing noise pollution in our environment.[92]

The 1990 CAAA added a new title IV, relating to acid deposition control, *without repealing the existing Title IV, relating to noise pollution*. The U.S. Code designates the original title IV (noise pollution) as subchapter IV and the new title IV (acid deposition control) as subchapter IV-A. The EPA administrator established the Office of Noise Abatement and Control (ONAC) under the Clean Air Act to investigate and study the effects of noise on the public health and welfare. In 1981 the Administration concluded that noise issues were best handled at the state and local level. However, the EPA retained authority to investigate and study the health effects associated with noise along with the effectiveness of existing regulations for protecting the public health and welfare.[94]

Radiation

Introduction

Marie and Pierre Curie termed the mysterious phenomenon "radioactivity" and isolated two radioactive substances, polonium and radium, from uranium ore samples. These discoveries, along with the discovery of "X-rays," quickly led to the birth of a whole new field of scientific inquiry with far-reaching implications in physics, biology, medicine and, unfortunately, warfare.

Ionizing Radiation

Early investigations of radioactivity revealed that the observed emissions were of several different kinds, consisting of subatomic particles—protons, neutrons, or electrons—that are released when atoms decay, later becoming known as particulate radiation. The particulate radiation includes alpha and beta particles. Atoms are the basic units of elements and consist of a small dense center called a nucleus surrounded by a cloud of negatively charged electrons (Figure 10.26). The positive charge in the atom is attributed to the protons or atomic number. The protons account for approximately half of the mass of the atom, with the other half composed of neutrons that carry no net electrical charge and are therefore, neutral. The total number of protons and neutrons is equal to the mass number of the atom. A particular element may have more than one mass number. Lead may have mass numbers of 214 or 210, although they have the same atomic number. Such atoms are called isotopes and the difference in mass number results from different numbers of neutrons. Many elements have stable and unstable isotopes. Unstable isotopes or radioisotopes (Figure 10.26) spontaneously emit energy in the form of radiation that was originally detected by their ability to expose photographic film in the dark. When a radioisotope

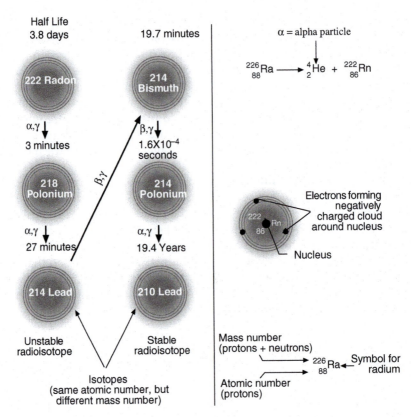

FIGURE 10.26
Radioactive decay of products of nuclear composition.

decomposes, it releases energy in the form of electromagnetic radiation (g or X-rays), and energy of motion from particles (a or b). Naturally occurring gamma rays (g) consist of highly energetic short wavelengths of electromagnetic radiation, a form of energy that also includes ultraviolet light, visible light, infrared waves, and microwaves.

Radioisotopes

Ionizing radiation includes alpha (a) and beta (b) particles, X-rays, and gamma rays because the particles or rays involved are sufficiently energetic to dislodge electrons from the atoms or molecules that they encounter. Ionizing radiation's ability to destroy chemical bonds gives it special significance. An atom that is missing one or more electrons is referred to as an ion, and energetic radiation capable of doing this is called ionizing radiation. X-rays and gamma rays are included in the category and are capable of breaking chemical bonds and ionizing molecules. When critical molecules are disrupted, the cell will undergo degenerative changes or even die. A critical molecule in human and animal tissue is DNA. Injury results from the ionization of water molecules in tissue that form ion pairs and free radicals that instantly react with the cell components to produce cross-linking of macromolecules causing changes in the tertiary or three-dimensional structure of the molecule. These changes may disrupt the function of proteins, deoxyribonucleic acid (DNA) and other molecules, causing mild reactions to cellular death. The disruption of the DNA molecule may cause genetic errors to occur that will be evidenced in abnormal growth and maintenance of the affected cells. Death of the cells is often the ultimate outcome.

Radiation Exposure

Every individual comes into contact with ionizing radiation from three general types of sources: (1) naturally occurring, (2) naturally occurring but enhanced by human actions, and (3) human generated.

Natural Sources

The earth, air, water, and food all contain traces of radioactive materials that constitute a source of "background" radiation. The following are sources of "background" radiation:

Cosmic Radiation

High energy particles composed primarily of protons and electrons that stream toward the earth both from outer space and form the sun following episodes of solar flares. Exposure to cosmic radiation is considerably less at sea levels than at high altitudes (for this reason, residents of Denver receive about twice as much cosmic radiation as the residents of LA or Miami).

Radioactive Minerals on the Earth's Crust

Uranium, thorium, potassium, and radium are all found in soils and rocks in many parts of the world. Inhabitants of areas such as the Rocky Mountains are exposed to higher levels than those on the East Coast, where radioactive minerals are in much less abundance.

Radionuclides in the Body

A number of radioactive substances enter the body by ingestion of food, milk, water, or by inhalation and are incorporated into body tissues where their concentration may be sustained at a steady rate or may increase with increasing age.

Enhanced Natural Sources

The impact of human activity on levels of exposure would be much lower under other conditions. Examples of enhanced natural sources include uranium mill tailings, phosphate mining, and jet airline travel (due to the increased exposure to cosmic radiation at high altitudes).

Human-Generated Sources

These include medical applications, nuclear weapons fallout, nuclear power plant emissions, and consumer products. Medical applications include medical X-rays and radiopharmaceuticals for both diagnostic and therapeutic purposes. Nuclear weapons fallout from the first atomic bomb to the end of atmospheric testing of nuclear weapons by several major powers of the world, primarily the U.S. and the former Soviet Union, resulted in significant amounts of radioactive fallout worldwide. Very small amounts of exposure are received from home smoke detectors, static eliminators, airport security checks, TV receivers, and tobacco products.

Health Impacts on Ionizing Radiation

During the early years of work with radioactivity was the medical community itself, particularly radiologists whose enthusiastic use of their new machines exposed their patients and themselves to large amounts of radiation on a daily basis, with a much-increased incidence of leukemia.

Dosage

Since ionizing radiation can neither be seen nor felt, human exposure is measured in terms of the amount of tissue damage it causes. Under the International System of Units, the term gray (Gy) is the unit of absorbed dose, used to quantify the amount of energy form ionizing radiation absorbed per unit mass of material. There is little evidence that present average levels of exposure to ionizing radiation are having a serious adverse effect on the health of the general public. However, higher doses can be injurious to living tissues. The International Congress of Radiation Research has been involved in a long-running argument over what is considered a safe dose of radiation. United States radiation experts from the University of California (UC) at Davis have stated that there are safe values of exposure to radiation (i.e., threshold) that the body can tolerate without the development of cancer. Conversely the International Commission on Radiological Protection stands behind the linear no-threshold theory, stating that the indicated safe dose proposed by the UC Davis researchers does not exist. These arguments about the validity of the linear no-threshold theories are important because regulators use the theory when setting radiation exposure standards.[94,95]

Dose Rate

The time span during which a given amount of radiation is delivered, which is thought to be very significant in determining the extent of tissue damage incurred, perhaps even more important than the total dose.

Radiation-Induced Mutations

Exposure to low levels of ionizing radiation can result in gene mutations. There is a linear relationship between increasing radiation dose and increased number of mutations. Nuclear power plant workers and medical personnel who receive small doses of radiation on a daily basis have been shown to exhibit a marked increase in the frequency of mutations in their chromosomes.

Radiation and Birth Defects

A number of congenital abnormalities, as distinguished from defects caused by mutant parental genes, are known to result from X-ray exposure in utero. Sensitivity to radiation is greater during fetal development.

Radiation-Induced Cancer

The potential for health damage as a result of radiation exposure was first recognized in connection with the development of skin cancers among early radiologists, of whom nearly 100 fell victim to this disease within 15 years following the discovery of X-rays. Leukemia is the classic example.

Radiation and Nuclear Power Generation

Most attention is focused on the nuclear power plants, due to their perceived threat. Nuclear power production involves a number of steps, referred to as the nuclear fuel cycle. These steps include mining the uranium, crushing and processing or milling the uranium, converting yellowcake to gaseous uranium hexafluoride, and enriching uranium to increase the concentration of the desired isotope from its original concentration.

Although highly controversial, nuclear power is the cleanest fuel known to man. The controversial problems with nuclear power do not deal with the toxic gas or water production as other fuels sources do, but to the production of the waste. The waste for a nuclear reactor, depending on the isotope used for production, has varying half-lives. These half-lives correspond to the time it takes for the isotope to decay to one half of its original limit. Nuclear power is used throughout the United States mainly for the production of public electricity. Another major source of nuclear power is through the naval defense departments of the world. International navies harness nuclear energy for use in submarines and other sea-going vessels. The use of nuclear power allows submarines to stay submerged for longer period of time, hence allowing a more efficient use of time and federal money.

In 1982, Congress enacted the Nuclear Waste Policy Act, which delegated responsibility for high-level radioactive waste management to the federal government and designated the U.S. Department of Energy as the lead agency to coordinate the effort to site, construct, and operate the nation's first permanent repository for such wastes.

Ultraviolet Radiation

Wavelengths of the electromagnetic spectrum ranging between 40 and 400 nanometers in length are categorized as ultraviolet (UV) light. Certain portions of this range are strongly absorbed by living tissues, particularly by DNA, which constitutes the major target of UV damage. Injury to the hereditary material of cells is the reason for the lethal or mutational effects that excess UV exposure can provoke in living organisms. Light in the ultraviolet range does not penetrate beyond the upper epidermis, but the lack of melanin in people without color places them at increased risk to skin cancers.[91]

Three major types of skin cancer account for over 700,000 new cases of the disease diagnosed in the U.S. each year. Basal cell carcinoma and squamous cell carcinoma represent the vast majority of skin cancers, accounting for nearly one-third of all cancers in the U.S. today. Chronic exposure to sunlight (especially the UV-B component of sunlight) is recognized as the cause of over 90 percent of these two cancers. Malignant melanoma is much less common but far deadlier. Americans born in the 1990s are at 12 times the risk of developing melanoma as those born 50 years ago and twice as likely as those born just 10 years ago. Unlike the non-melanoma skin cancers, malignant melanoma appears to develop as a result of occasional severe sunburn, rather than prolonged low-dose exposure to sunlight.

The beneficial effects of ultraviolet light include the production of vitamin D that prevents the formation of bone deforming rickets, and the germicidal properties of UV that are found to reduce bacterial and viral infections. This has been demonstrated in operating rooms and in the treatment of bacterial skin diseases including acne and boils.[91]

References

1. Stromberg, J., Air pollution has been a problem since the days of ancient rome, *Smithsonian Magazine*, February 2013. https://www.smithsonianmag.com/history/air-pollution-has-been -a-problem-since-the-days-of-ancient-rome-3950678/.
2. Hong, S., et al., History of ancient copper smelting pollution during roman and medieval times recorded in Greenland Ice, *Science*, 272(5259), 246–249, 1996. doi:10.1126/science.272.5259.246.
3. Urbinato, D., London's historic "Pea-Soupers," United States Environmental Protection Agency, Summer 1994. Retrieved August 2, 2006.
4. Chambers, L.A., Classification and extent of air pollution problems. In *Air Pollution*, 3rd ed., Stern, A.C., (Ed.), Academic Press, New York, 1976, Chap 1.
5. Jeffry, M.D., and Kurkpatrick, D., City, *Country, Empire: Landscapes in Environmental History*, University of Pittsburgh Press, Pittsburgh, PA, pp. 44–49, Summer 1994.
6. Morgan, M.T., Gordon, L., Walker, B. et al., *Environmental Health*, 2nd ed., Morton Publishing, Englewood, CO, 1997.
7. Whelan, E.M., *Toxic Terror*, Prometheus Books, Buffalo, New York, 1993, Chap. 10.
8. Godish, T., *Air Quality*, 3rd ed., CRC/Lewis Publishers, Boca Raton, FL, 1997, Chap. 5.
9. Liu, B. and Xu, E., *Physical and Economic Damage Functions for Air Pollutants by Receptors*, U.S. Environmental Protection Agency, Washington, DC, USEPA/600/6-76-011, 1976.
10. Gribbon, J., Carbon dioxide, ammonia-and life, *New Scientists*, 94(1305), 413, 1982.
11. Godish, T., *Air Quality*, CRC/Lewis Publishers, Boca Raton, FL, 1997, Chap. 1.
12. Nadakavukaren, A., *Our Global Environment: A Health Perspective*, Waveland Press, Prospect Heights, IL, 1995, Chap. 11.
13. Godish, T., *Air Quality*, CRC/Lewis Publishers, Boca Raton, FL, 1997, Chap. 3.
14. Nadakavukaren, A., *Our Global Environment: A Health Perspective*, Waveland Press, Prospect Heights, IL, 1995, Chap. 12.
15. Godish, T., *Air Quality*, CRC/Lewis Publishers, Boca Raton, FL, 1997, Chap. 8.
16. USEPA, *The Clean Air Act Amendments of 1990: Summary Materials*, USEPA Office of Air and Radiation, Reproduced by Library of Congress, Congressional Research Service, Washington, DC, November 15, 1990.
17. Kessler, J. and Schroder, W., *Meeting Mobility and Air Quality Goals: Strategies that Work*, USEPA, Office of Policy Analyses, Washington, DC, October 13, 1993.
18. Anderson, E.V., *Health Studies Indicate MTBE Is Safe Gasoline Additive*, C&EN, Washington, DC, September 20, 9, 1993.
19. Long Island Utility Fighting to Defeat MTBE Safe Harbor, Napoli, Kaiser, Bern & Associates, March 16, 2004, archived from the original on October 20, 2007.
20. CCL 2 List and Regulatory Determinations: Chemical Contaminant Candidates, United States Environmental Protection Agency.
21. Bradley, J., Buying high, selling low, *E/The Environmental Magazine*, July/August 1996. http://www.emagazine.com/2curr3.
22. Chandler, D.L., MIT scientist shares Nobel for identifying ozone damage, *The Boston Globe*, 101V, October 12, 1995.
23. Godish, T., *Air Quality*, CRC/Lewis Publishers, Boca Raton, FL, 1997, Chap. 4.
24. Associated Press (AP), Ozone hole grows to record size, *CNN Interactive*, November 2, 1996.
25. National Oceanic and Atmospheric Administration (NOAA), Stratospheric Ozone: Monitoring and Research in NOAA, Last Update March 3, 2006. http://www.ozonelayer.noaa.gov/index.htm.
26. Blumberg, S., NASA, Warm Air Helped make 2017 Ozone Smallest Since 1988, November 2, 2017. https://www.nasa.gov/feature/goddard/2017/warm-air-helped-make-2017-ozone-hole -smallest-since-1988.
27. Wald, M.L., Smuggling of Polluting Chemicals is Resolved, *NY Times National*, September 17, 1, 1995.

28. Begley, S., Holes in the ozone treaty, *Newsweek*, September 25, 70, 1995.
29. Stevens, W.K., Scientists report an easing in ozone-killing chemicals, *New York Times*, August 26, 1993.
30. USEPA, *The Size and Depth of the Ozone Hole*, written by USEPA's Stratospheric Protection Division, June 10, 1998. http://www.usepa.gov/ozone/science/hole/size.html.
31. Suitil, K., Holey war, *Discover*, January, 75, 1993.
32. Schein, O., Vincencio, C., Munoz, B. et al., Ocular and dermatologic health effects of ultraviolet radiation exposure from the ozone hole in Southern Chile. *American Journal of Public Health*, 85(4), 546, April, 1995.
33. USEPA, *USEPA's Revised Ozone Standard, Fact Sheet*, USEPA, Office of Air and Radiation, Office of Air Quality Planning and Standards, July 17, 1997. http://ttnwww.rtpnc.usepa.gov.
34. USEPA, *USERA's Revised Particulate Standard, FACT sheet*, USEPA Office of Air & Radiation, Office of Air Quality Planning & Standards, July 17, 1997. http://ttnwww.rtpnc.usepa.gov.
35. Gerstenzank, J., Clean-air plans fuels backstage U.S. fight, *Los Angeles Times: Science*, April 26, 1997. http://www.latimes.com/home/news/science/environ/ x000035587.html.
36. USEPA, *Clear Skies*, last revised, March 2, 2006. http://www.epa.gov/clearskies/.
37. Ledford, A. *On Today's House Subcommittee on "Clear Skies,"* Statement of Angela Ledford, Director of Clear the Air, A National Campaign Against Global Warming, Press release, May 26, 2005.
38. Natural Resources Defense Council (NRDC), *Dirty Skies: The Bush Administration's Air Pollution Plan*, last revised September 5, 2003. http://www.nrdc.org/air/pollution/qbushplan .asp#clearskies.
39. Bochert, L., *EPA in the District Court*, American College of Environmental Lawyers, September 23, 2008.
40. Easterbrook, G., Hot air treaty, *US News & World Report*, December 22, 46, 1997.
41. Wikipedia, The Free Encyclopedia, Kyoto Protocol: *The Kyoto Protocol to the United Nations Framework Convention on Climate Change*, last modified June 6, 2006. http://en.wikipedia.org /wiki/Kyoto_Protocol
42. Tyler, P., China's inevitable dilemma: Coal equals growth, *New York Times*, A I, November 29, 1995.
43. Sutter, J.D., and Berlinger, J., Final draft of climate deal formally accepted in Paris, CNN. Cable News Network, Turner Broadcasting System. Retrieved December 12, 2015.
44. Change, United Nations Framework Convention on Climate, The Paris Agreement—Main page, *UNFCCC.int*. Retrieved November 17, 2017.
45. Mark, K., COP21: What does the Paris climate agreement mean for me? *BBC News*. BBC. Retrieved December 14, 2015.
46. Taraska, G., *The Paris Climate Agreement* (PDF), Center for American Progress, December 15, 2015.
47. Dwortzan, M., How much of a difference will the Paris Agreement make? April 22, 2016. Retrieved December 4, 2017. http://news.mit.edu/2016/how-much-difference-will-paris -agreement-make-0422.
48. Monastersky, R., Dusting the climate for fingerprints: Has greenhouse warming arrived? Will we ever know? *Science News*, 147, 362, June 10, 1995.
49. Monastersky, R., Satellite detects a global sea rise, *Science News*, 146, December, 388, 1994.
50. Mann, M.E., Bradley, R.S., and Hughes, M.C., Global scale temperature patterns and climate forcing over the past six centuries, *Nature*, 392, 779, April 23, 1995.
51. Lemonick, M.D., The ice age cometh? *Time*, 79, January 31, 1994.
52. Noshi, J.M., The fury of El Nino, *Time*, 67, February 16, 1998.
53. Pfarrer, S., Expert fears El Nino global warming link, *Daily Hampshire Gazette*, April 17, 9, 1998.
54. Nash, M., When life nearly died, *Time*, 95, September 18, 1995.
55. USEPA, *National Air Quality and Emission Trends Report*, 1996, USEPA Office of Air Quality Planning & Standards, USEPA document 454/R-97-013 1/13/98. http://www.usepa.gov/oar /aqtmd96/trendsfs.html.

56. Suplee, C., Dirty air can shorten your life, study says. *Washington Post*, March 6, Al, 1995.

57. News and Analyses, Air pollution in the U.S., *Scientific American*, April 27, 1997.

58. Allen, S., More dying of dirty air than in cars, study finds, *Boston Globe*, 14, May 9, 1996.

59. Washington (CNN), America's dirty little secret: Smog still a health problem, *CNN Interactive*, June 21, 1996.

60. Godish, T., *Air Quality*, 3rd ed., CRC/Lewis Publishers, Boca Raton, FL, 1997, Chap. 2.

61. Raloff, J., Heart-y risks from breathing fine dust, *Science News*, 148, 5, July 1, 1995.

62. USEPA, *USEPA's Revised Ozone Standard, FACT Sheet*, USEPA Office of Air Quality Planning and Standards, July 17, 1997. http://ttnwww.rtpnc.usepa.gov.

63. USEPA, *National Air Quality and Emission Trends Report, FACT Sheet*, USEPA, Document Number 454/e—97-013, USEPA Office of Air Quality Planning and Standards, 1, 23, 1998. http://www.usepa.gov/out/aqtmd96/trendsfs.html.

64. Flynn, J., The following forest, *Amicus Journal*, 15(4), 1, 1994.

65. Shaw, R., *Environmental News Network, Recovery begins From Acid Rain Damage*, CNN.com October 18, 1999 Retrieved July 8, 2018. www.CNN.com/NATURE/9910/18/Acid.Rain.CNN/.

66. Reese, A., Blue Skies, *E/the Environmental Magazine*, November-December 1999.

67. National Center for Environmental Health, Center for Disease Control and Prevention, Most Recent Asthma Data, Last reviewed February 13, 2018. https://www.cdc.gov/asthma/most_recent_data.htm.

68. Stolberg, S.G., Poor are fighting baffling surge in asthma, *The New York Times*, October 18, 1999.

69. Cole, C.L., High rate of asthma inspires new study, *Boston Sunday Globe*, 1, November 24, 1999.

70. Shell, E.R., Does civilization cause asthma? *The Atlantic Monthly*, 285, 90, 2000.

71. American Lung Association, *Life and Breath: Answer and Questions*, item, #1245, 15, January 1990.

72. Rosenstreich, D.L., Eggleston, P., Kaltan, M. et al., The role of cockroach allergy and exposure to cockroach allergen in causing morbidity among inter-city children with asthma, *The New England Journal of Medicine*, 333, 1356, 1997.

73. Friebek, E., The attack of asthma, *Environmental Health Perspectives*, 104(1), 22, 1996.

74. Thurston, G.G., Lippman, M., Scott, M.B., and Fine, J.M., Summertime haze air pollution and children with asthma, *American Journal of Respiratory Critical Care Medicine*, 155, 154, 1997.

75. Barner, R.I., Air pollution and asthma, *Postgraduate Medical Journal*, 70, 319, 1994.

76. Lyons, B., Something in the air, *Ladies Home Journal*, 102, February 1995.

77. USEPA, *The Inside Story: A Guide to Indoor Air Quality*, USEPA Office of Air and Radiation, and the Consumer Product Safety Commission, US Government Printing Office, LM/4uc/1-88 /004,1988, 32 pp.

78. Anonymous, Carbon monoxide-heart failure link, *Environmental Health Perspectives*, 104(21), 138, 1996.

79. Menzies, D., Tumbly, R.M., Nunes, R. et al., Exposure to varying levels of contaminants and symptoms among workers in the office buildings, *American Journal of Public Health*, 86, 1629, 1996.

80. Godish, T., *Air Quality*, CRC/Lewis Publishers, Boca Raton, FL, 1997, Chap. 6.

81. The Associated Press, Smokers in NYC Increase to 1 Million, September 15, 2014. https://www.cbsnews.com/news/smokers-in-nyc-increase-to-1-million.

82. Walton, A., Why teens smoking's on the rise, *Salon*, November 9, 2013. https://www.salon.com/2013/11/09/teens_have_started_smoking_again_because_they_think_its_cool_partner.

83. Leary, W., Research ties radon to as many as 21,800 deaths each year, *The New York Times National*, A13, February 20, 1998.

84. USEPA, *A Citizens Guide to Radon*, US Government Printing Office, USEPA-86-CC4, Boca Raton, FL, 1986.

85. Hager, M.G., Mites and molds, *Self Magazine*, 167, May 1997.

86. Holusha, J.A., Study calls household materials especially toxic, *The New York Times*, March 15, 1995.

87. Godish, T., *Air Quality*, 3rd ed., CRC/Lewis Publishers, Boca Raton, FL, 1997, Chap. 11.

88. Geary, J., Mad about the noise, *Time*, July 27, 1998.
89. "Pollution, Noise." International Encyclopedia of the Social Sciences. *Encyclopedia.com*, February 15, 2018. http://www.encyclopedia.com.
90. World Health Organization, 2001, Fact Sheet No. 258, WHO Press, Geneva, Switzerland. http://www.who.int/mediacentre/factsheets/fs258/en/.
91. Nadakavukaren, A., *Our Global Environment: A Health Perspective*, Waveland Press, Prospect Heights, IL, 1995, Chap. 13.
92. Godish, T., *Air Quality*, 3rd ed., CRC/Lewis Publishers, Boca Raton, FL, 1997, Chap. 12.
93. Toroza, G.J. and Anagnostakos, N.R., *Principles of Anatomy and Physiology*, 6th ed., Harper and Row Publishers, New York, 1990, Chap. 17.
94. United States Environmental Protection Agency. Clean Air Act Title IV-Noise Pollution, Retrieved February 15, 2018.https://www.epa.gov/ clean-air-act-overview /clean-air-act-title-iv-noise-pollution.
95. Birchard, K. Experts still arguing over radiation doses, *Lancet*, 354, 400, 1999.

11

Solid and Hazardous Waste

OBJECTIVES

A student reading this chapter will be able to:

1. Explain the consequences of improper solid waste disposal.
2. Describe the typical municipal solid waste stream.
3. Discuss the methods of reducing the solid waste stream through reuse and recycling efforts.
4. Describe the methods of collection and disposal of municipal solid wastes including the benefits and problems associated with landfills and incinerators.
5. Differentiate types of hazardous waste.
6. Give examples of why proper hazardous waste disposal is important.
7. List different methods of hazardous waste control, emphasizing waste reduction, volume or hazard reduction, and long-term storage and disposal options.
8. Discuss the positive and negative aspects of clean up efforts under "Superfund" and some of the major concerns regarding its future operation.

Introduction

More than 3.2 million megagrams of solid waste are generated every day across the globe. In most areas of the world, including the United States, Asia, and Africa, that waste goes to garbage dumps and landfills.[1] For many of us, garbage is nothing more than a weekly chore: Put the garbage can by the curb and a truck comes by to collect it. But not all the waste generated is collected by garage trucks. It is estimated that four billion people throughout the globe live without regular garbage collection. Poorer countries often lack the financial resources to invest in a government-run waste collection program. Even developed counties struggle with how to manage all the waste that is generated. The largest landfill in the world by mass is located in Mexico City, the Bordo Poniente Landfill, which receives 4 million megatons of garbage each year.[2] The top twelve landfills by mass are found all across the globe (Table 11.1). Garbage, refuse or waste—whatever term best applies—is a global unifier.

Waste is a daily challenge, no matter where you live, because every person creates some form of waste, nearly every day. The more densely populated an area, as in large cities like New York, Shanghai, and Mumbai, the more municipal waste is generated. Some cities are developing programs to encourage citizens to reduce, reuse, and recycle, but these programs have a long way to go to impact the vast quantities of waste generated daily.

Cities have historically been centers for filth and disease. Cities in England and Western Europe were disease-ridden centers of filth up until the nineteenth century. It was the norm to cast garbage, rubbish, and body wastes out the windows onto the streets below or leave it on the floor about the living space. The consequence of increasing refuse, filth, insects, and rodents resulted in disease outbreaks of enormous consequences such as the bubonic plague or "black death" that ravaged Europe in the thirteenth and fourteenth centuries. The relationship between disease outbreaks and accumulated filth gradually clarified in the eyes of municipal leaders so that by the early 1900s, efforts were undertaken in Europe and North America to improve sanitation. One of the major efforts was directed at promptly collecting urban wastes and disposing of them at some

TABLE 11.1

Top 12 Largest MSW Landfills

Landfill Name	Location	Hectares	Megagrams/ Day	Megagrams/ Year (Millions)
Bordo Poniente Landfill	Nezahualcoyotl, Mexico (Mexico City)	375	10,886	4.0
Apex Regional Landfill	Las Vegas, Nevada	890	9,525	3.4
Sudokwon Landfill	Incheon, South Korea	231	16,329–18,144	6.3
Puente Hills Landfill	Los Angeles, California	255	9,344	3.3
Laogang Landfill	Laogang Landfill	405	5,443–9,071	2.7
Lagos Dumpsites	Lagos, Nigeria	40	8,165	3.0
Xingfeng Landfill	Guangzhou, China	–	5,443–7,257	2.3
Sao Joao Landfill	São Paulo, Brazil	61	6,350	2.3
Delhi Landfills	Delhi/New Delhi, India	202	5,443	2.0
West New Territories Landfill	Hong Kong	–	5,625	–
Malagrotta Landfill	Rome, Italy	275	5,629	2.1
Mumbai Landfills	Mumbai, India	–	3628–6350	1.8

Source: Worlds Largest Landfills With Photos and Statistics, November 28, 2017, https://owlcation.com /stem/15-of-the-Worlds-Largest-Landfills.

distance from the population to reduce the probability of disease. This normally meant dumping the collected refuse on the outskirts of town. Unfortunately, much of the world continues to follow this pattern of refuse disposal, which (1) invites the proliferation of rodents and insects, (2) becomes a source of contamination to groundwater, (3) pollutes ambient air when combusted, (4) facilitates the spread of debris around the dumping site, (5) lowers property values about the site, and (6) encourages the spread of disease from microorganisms and toxic chemicals. It has only been within recent years that the United States has adopted waste management policies that recognize the need for safer and more effective methods of collecting, storing, transporting, and disposing of the unusable or unwanted materials generated by society. This is happening in other countries around the globe as well, to varying degrees.

Municipal solid waste (MSW) management varies from country to country, as countries have different infrastructures to handle trash collection, sorting, and disposal. These processes differ most significantly between developed countries and developing countries. In developed countries, historically, there has been an evolution of regulations to improve management of MSW, starting in the 1960s. The history of MSW regulation in the United States reflects changes that occurred in other countries across the globe in the same time frame.

In the U.S. in the 1960s and early 1970s, a large percentage of MSW was thrown into open dumps, with very little recycling occurring. Some communities with limited open land for dumping built incinerators to burn the garbage. More than 300 municipal incinerators were operating in the U.S. during the 1960s. The introduction of strict air quality regulations in the 1970s caused many of the incinerators to close as they represented major sources of air pollution. These same regulations imposed restrictions on the burning at open dumps. During the mid-1970s, incineration capacity declined, open dumps disappeared, and the generation of MSW increased dramatically. During the same period, the recovery of materials, such as from recycling, grew very slowly.[3] The U.S. government passed regulations in 1976 that forbade open dumping while introducing the concept and use of the sanitary landfill. This law is known as the Resource Conservation and Recovery Act (RCRA) (pronounced "rickra") of 1976 (P.L. 94-580).

There are three distinct programs under RCRA. Subtitle D encouraged states to develop comprehensive plans for the management of solid wastes with emphasis on those of a non-hazardous nature, such as household wastes. Subtitle C was designed to control the improper disposal of hazardous waste through a manifest process known as the "cradle-to-grave" approach. Subtitle I was designed to minimize the contamination of groundwater from underground storage tanks through leak detection requirements, mitigation and prevention of leaks, and new performance standards for underground storage tanks. Subtitle D was soon found to be inadequate despite regulations, which dictated the conditions for the proper siting, construction, and management of sanitary landfills. Nearly 94 percent of 17,000 land disposal sites surveyed in the mid-1970s failed to meet minimum requirements.[4] The consequences included substantial odors and debris, contamination of groundwater, proliferation of pests, and other environmental pollution problems. Seeing the need for additional control over the disposal of solid wastes, Congress passed the Hazardous and Solid Waste Amendments Act of 1984 (P.L. 98-616). This provided strict requirements for the proper siting of landfills to minimize surface and groundwater pollution. However, in an effort to achieve an improved and sustainable waste management program, the USEPA promulgated new RCRA Subtitle D landfill requirements that went into effect in October of 1993. The landfills were to be constructed with double liners, systems for collecting the liquids percolating through the wastes (leachate), groundwater monitoring wells, and methane venting and detection systems. The costs for siting and operating landfills escalated dramatically as the number of landfills diminished in the face of rapidly growing amounts of MSW.

Globally, sanitary landfilling continues to be among the most widely used method of disposal. However, in the United States, the USEPA's Agenda for Action endorsed the concept of integrated waste management by which MSW is reduced or managed through several different practices that include:

1. Source reduction (including reuse of products and backyard composting of yard trimmings)
2. Materials recycling (including composting)
3. Waste combustion (preferably with energy recovery) and landfilling[3]

The concepts of MSW and MSW management are discussed in the following sections.

The problem of diminishing refuse disposal capacity is often further exaggerated by the community opposition to landfill siting that has become known as the "not in my backyard" (NIMBY) syndrome. The cost for a truck to unload its waste at a landfill (tipping fee) has approached $62 per megagram in the Northeast United States making other disposal alternatives more attractive.[5]

Definition and Characterization of Municipal Solid Waste

Definition of Municipal Solid Waste

In the United States, MSW includes paper and paperboard, yard trimmings, food wastes, plastics, glass, metal, and wood wastes. The USEPA's Office of Solid Waste uses a materials-flow methodology to estimate the amounts of MSW generated. This methodology is based on production data (by weight) for both materials and products in the waste stream and is adjusted for imports and exports. MSW may also be defined by the products generated, such as durable goods (e.g., appliances), non-durable goods (e.g., newspapers), containers and packaging, food wastes and yard trimmings, and miscellaneous organic wastes. Such wastes generally come from residential, commercial, institutional, and industrial sources. Examples of the types of MSW in each of these categories are listed in Table 11.2.

TABLE 11.2

Sources of MSW and Example of Products from Each Source

Sources	Example of Products
Residential, including single and multifamily houses	Nondurable paper items such as magazines, newspapers, and advertising flyers, plastic and glass bottles, aluminum and steel cans, packaging, food wastes, yard wastes
Institutional, including schools, hospitals, prisons, and nursing homes	Food wastes, papers from classrooms and offices, disposable tableware and napkins, paper towels from restrooms, and yard trimmings
Commercial, including restaurants, office buildings, and stores (wholesale and retail)	Food wastes, paper products from offices, restrooms, and serving tables, disposable tableware, corrugated and paperboard products, and yard wastes
Commercial, including restaurants, office buildings, and stores (wholesale and retail)	Wooden pallets, office paper, corrugated and paperboard, plastic film, and food wastes (from cafeterias)

The composition of a country's MSW may change over time, as a country's wealth increases. For instance, in China, food and other goods used to be sold loose, without any exterior packaging. Much of the waste generated was discarded organic matter. But with the rise of the middle class in China, packaging has become more prevalent, as consumerism has increased. The composition of waste changes from organic matter, which decomposes, to plastic packaging, broken electronics, and discarded toys, which persist in a landfill.[6]

Characterization of Municipal Solid Waste

Contrary to popular belief, MSW does not include everything that is discarded or everything that is landfilled in Subtitle D landfills (RCRA Subtitle D). It does not include municipal sludge, industrial non-hazardous waste, construction and demolition waste, agricultural waste, oil and gas waste, and mining wastes (Figure 11.1). The term "generation" refers to the weight of materials and products as they enter the waste management system from residential, commercial, institutional, and industrial sources and before materials are recovered or combusted.

The total annual amount of MSW materials generated (millions of megagrams) from 1960 to 2014 in the United States is shown in Figure 11.2. Total MSW generation has increased steadily since 1960, from 80 million megagrams to 182 million megagrams in 1990 to 234 million megagrams in 2014.[7] A person in the United States today generates more than 2.1 kilograms of waste per day, an increase of 1.2 kilograms from 1960. This exceeds 700 kilograms per person per year, more than in any other country.[8] In comparison, the waste generation rate in the EU is 1.5 kilograms per person per day. In developing countries, the average MSW generation rate is about half, at 0.759 kilograms per person per day.[9]

Paper and paperboard materials represented the largest component of that 2014 amount at 26.6 percent, followed by food wastes at 19.9 percent and yard trimmings at 13.3 percent (Figure 11.3). The amount of MSW generated in 2014, categorized as products, is shown

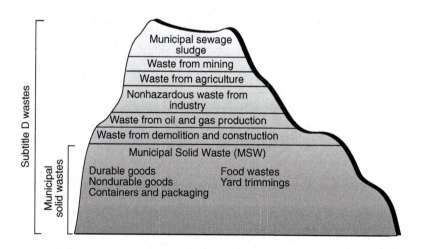

FIGURE 11.1
Municipal solid wastes as a component of RCRA Subtitle D wastes. (Adapted from United States Environmental Protection Agency, *Characterization of Municipal Solid Wastes in the United States: 1997*, Update, prepared for USEPA by Franklin Association, Prairie Village, KS, Report No. EPA 530-R-98-Oct-May, 1998, Chap. 1.)

FIGURE 11.2
Total amount of MSW generated (millions of megagrams/year) 1960-2014. (Adapted from United States Environmental Protection Agency, USEPA, Office of Land and Emergency Management, *2014 Advancing Sustainable Materials Management*, Washington, DC, November 2016.)

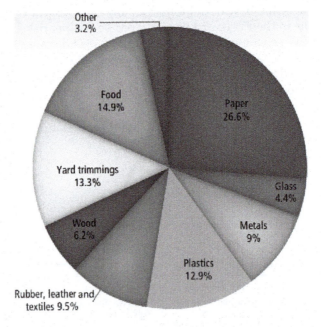

FIGURE 11.3
Major material components of MSW by percent (234 million megagrams), 2014. (Adapted from United States Environmental Protection Agency, USEPA, Office of Land and Emergency Management, *2014 Advancing Sustainable Materials Management*, Washington, DC, November 2016.)

2014 U.S. MSW Products Generated
234 Million Megagrams

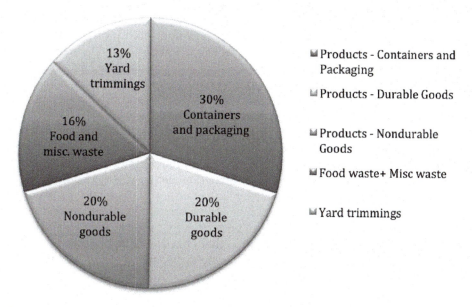

Products - Containers and Packaging

Products - Durable Goods

Products - Nondurable Goods

Food waste+ Misc waste

Yard trimmings

FIGURE 11.4

Products generated in MSW by weight 2014. (Adapted from United States Environmental Protection Agency, 2014 Advancing Sustainable Materials Management, November 2016.)

as Figure 11.4. The majority of generated waste is categorized as containers and packaging representing 30 percent or 67.86 million megagrams. Non-durable goods represented the second largest amount (20.4 percent), followed by durable goods at 20.2 percent.[7]

By comparison, in 2014, the total waste generated in the EU-28 (the 28 countries that comprise the EU) by all households amounted to 203.5 million megagrams, a decrease from the 2010 high of 214.4 million megagrams.[10] Where this household waste is sent for disposal depends on the EU country. Some EU countries have more advanced waste management infrastructure than others.

In China, the most populous country in the world, annual production of MSW in 2014 was around 300 million tons.[11] In 2014 in China, a total of 604 sanitary landfills, 188 incineration plants, and 26 other units were used for MSW management.[6] Disposal of MSW was as follows: 60.16 percent landfilling, 29.84 percent incineration, 8.21 percent untreated discharge, and 1.79 percent other treatments. China lags behind the United States and the EU in MSW disposal infrastructure and needs to focus on improvements, particularly around leachate treatment and landfill location and management.[6]

Collection and Disposal of Solid Waste

Collection of Municipal Solid Waste

Whether collected by private companies under contract or by municipal employees, the removal of solid waste is generally considered the responsibility of government, which should develop and enforce regulations that protect the public health by proper collection and disposal of municipal waste. The collection of MSW consumes 7.5 billion dollars of the 10 billion-dollar annual costs for the collection and disposal of solid waste. This effort requires nearly 170,000 employees distributed among public and private sectors.[11] Factors that must be considered in collection include (1) whether the waste will be collected by the municipality or under private contract, (2) the frequency of collection, (3) what materials will be salvaged, (4) the types and location of collection containers or bags, (4) what materials will be forbidden from routine disposal (i.e., tires, automobile batteries, mercury-containing batteries, fluorescent lights, durable white goods), (5) whether wastes should be separated prior to collection as required by certain recycling efforts, and (6) what type of equipment may be required. Collection vehicles often include manually loading compacting bodies, which increase the size of the load that can be carried in a vehicle and facilitate emptying at the disposal site (Figure 11.5). Communities that offer recycling programs and curbside collection will usually employ specialized trucks that have several compartments or sections necessary for the segregation of collected recyclables (Figure 11.6).

Many rural communities do not offer waste collection service but require the residents to bring their refuse and/or recyclables to a transfer station. A transfer station is a site where MSW is gathered from surrounding towns. There, many small loads of MSW are combined into one single large load before transfer to a processing facility or a sanitary landfill, allowing for fewer trips in bigger transport vehicles (Figure 11.7). The waste may

FIGURE 11.5
Waste-compacting collection vehicle.

FIGURE 11.6
Compartmentalized truck assigned for pickup of recycled materials.

FIGURE 11.7
Transfer station where waste compaction and separation take place.

be concentrated as well. Concentration most often involves compaction by placing the waste into a metal channel where a ram compresses it into a roll-off collection container (Figure 11.8). The container may then be hailed onto a tractor bed and transferred for processing or landfilling. Some transfer stations also feature recycling bins where separated items such as metal cans, plastic milk bottles made of high-density polyethylene (HDPE), paper and cardboard, and glass containers may be collected and periodically removed for further processing and resale (Figure 11.9).

FIGURE 11.8
Close-up view of a compaction ram at a transfer station.

FIGURE 11.9
Recycling bins at a transfer station.

Management of Municipal Solid Waste

Once waste is collected, it must be processed in some way for final disposal. Presently, most states in the U.S. use some combination of waste management that involves one or more components of landfilling, combustion, or recycling. Table 11.3 shows the changes in these activities over time, from 1960 to 2014. The amounts of MSW generated have been consistent since 2000, while amounts of MSW recycled and composted continue to increase. Combustion with energy recovery has been stable since about 2000. Nationally, in 2014, more than half (52.6 percent) of the total MSW collected was landfilled, 32.6 percent

TABLE 11.3

MSW by activity type, in millions of megagrams, 1960–2014

Activity	1960	1970	1980	1990	2000	2005	2010	2012	2013	2014
Generation	79.9	109.9	137.5	189.0	220.9	230.2	227.8	228.4	231.4	234.5
Recycling	5.1	7.3	14.5	26.3	48.1	53.7	59.2	59.5	59.1	60.2
Compositing*	neg.	neg.	neg.	3.8	15.0	18.7	18.3	19.3	20.3	20.9
Combustion with energy recovery[†]	0	0.4	2.5	27.0	30.6	28.8	26.6	29.5	30.1	30.0
Landfilling and other disposal[‡]	67.9	102.1	121.8	131.8	127.3	129.0	123.6	120.1	121.8	123.4

Source: Adapted from United States Environmental Protection Agency, USEPA, Office of Land and Emergency Management, 2014 *Advancing Sustainable Materials Management*, Washington, DC, November 2016.

* Composting of yard trimmings, food and other MSW organic material. Does not include backyard composting.

[†] Includes combustion of MSW in mass burn or refuse-derived fuel form, and combustion with energy recovery of source separated materials in MSW (e.g., wood pallets, tire-derived fuel).

[‡] Landfilling after recycling, composting and combustion with energy recovery. Includes combustion without energy recovery. Details might not add to totals due to rounding.

neg. Negligible = Fewer than 4,535 Megagrams or 0.05%.

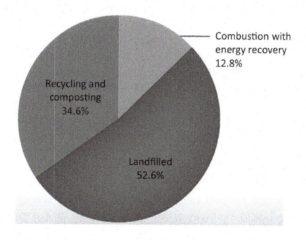

FIGURE 11.10

How MSW is managed in the U.S. (Adapted from United States Environmental Protection Agency, USEPA, Office of Land and Emergency Management, 2014 *Advancing Sustainable Materials Management*, Washington, DC, November 2016.)

was recovered by recycling (or composted), and 12.8 percent was combusted, with most combustion systems employing energy recovery (Figure 11.10). The percentage of MSW going to a landfill has decreased significantly over nearly four decades, from 89 percent in 1980 to less than 53 percent in 2014.[7]

Landfills

Design

Landfills have historically been sources of (1) pollution to streams and groundwater; (2) producers of methane gas, which can leak into basements in potentially explosive amount; (3) foul odors; and (4) breeding grounds for insects, rodents, and gulls. Stricter requirements for the siting and construction of landfills under RCRA Subtitle D mandate that all new landfills or ones that are to be expanded must be constructed with double liners consisting of impervious layers, leachate collection systems, monitoring wells to detect potential migration of pollutants from the landfill site, and methane detection systems also designed to detect lateral migration of methane.

The reason for such extensive design concerns is that a landfill is a depression in the ground (or constructed on the surface to resemble a stadium) that is designed to hold accumulated wastes. Its purpose is to contain wastes and prevent pollutants from reaching surface or groundwater. Landfills have been described as bathtubs that can leak from the bottom or spill over the top.[12] Landfills with two liners are like a bathtub within a bathtub and are designed to further restrict the migration of pollutants.

Solid wastes that are deposited in the landfill will tend to mix and degrade chemically and biologically to produce a variety of gaseous (methane), liquid (leachate), and solid (metal oxides) products. The amount of moisture, oxygen, temperature, and contents of the fill effect the level of decomposition and the products produced. The major products of concern include leachate and methane. Leachate is usually a dense, dark liquid produced when surface water and groundwater infiltrate through the solid waste and pick up solid matter, chemicals, and waste products of microbial decomposition, which serve as a potential serious source of contamination to surface and groundwater (Figure 11.11). Methane gas is produced when organisms anaerobically decompose organic waste such as a food waste and sewage sludge disposed in the landfill. Methane gas can migrate laterally to enter the cellars of nearby structures and produce potentially explosive concentrations.[11] Although biodegradation does occur in a limited amount in landfills, the marketing of biodegradable products to the public may be overstated. Archeologist William Rathje has been excavating and researching landfills for years. His discovery of 25-year-old hot dogs, guacamole, steaks, and even older intact newspapers has Rathje theorizing that very little biodegradation takes place in a landfill. He concludes that most garbage is "mummified" and takes up space indefinitely since microbes have insufficient moisture and air to decompose the bulk of landfill ingredients.[13]

The critical elements of a landfill include (1) a bottom liner or double bottom liner, (2) a leachate collection system, (3) a cover, (4) the natural hydrogeologic setting, and (5) monitoring systems for detecting methane and leachate plumes (Figure 11.12). The natural hydrogeologic setting should be away from surface water and aquifer with impervious material below (fractured bedrock should be avoided) while also having geology simple enough to permit the installation of wells to intercept escaping pollutants. The bottom liners may be composed of one or more layers of clay or a synthetic flexible membrane(s) (or some combination of both). The synthetic liners are often constructed of HDPE. However, other materials may be used for plastic liners (referred to as Flexible Membrane Liners or FMLs). Leachate seeps to the bottom of the sloped landfill and is collected by a system

> Leachate—formed when rain water filters
> through wastes placed in a landfill. When
> this liquid comes in contact with buried
> wastes, it leaches, or draws out,
> chemicals or constituents from those
> wastes.

FIGURE 11.11
Definition of leachate. (From USEPA.gov.)

FIGURE 11.12

Typical construction of a landfill and potential leachate routes.

of pipes that are laid along the bottom. The leachate should be periodically pumped and transported to a wastewater treatment plant where solids are removed and liquids treated. The cover cap of the landfill is designed to restrict the entry of water from precipitation and usually is constructed of several layers sloped away from the center to promote the shedding of water. The first layer(s) are usually membrane covered by a layer of permeable sand or gravel, and then a layer of topsoil where vegetation may root.

Landfills Are Not Secure

Landfills are vulnerable to leakage from several different sources. Landfill liners, even constructed of HDPE, can be expected to leak more than 187.08 Litres per Hectare (L/ha/ volume area) per day through tiny holes created during manufacture and those created when the seams are welded together during installation.[14] Furthermore, such liners are also permeable to a variety of dilute solutions including xylenes, toluene, trichloroethylene (TCE), and methylene chloride that pass through the solid matrix of 100 mil thick HDPE liners in less than 2 weeks. Such liners also become brittle and will fracture or crack from stress and chemical assault in under 2 years.[15] Even benign chemicals such as margarine, shoe polish, vinegar, and peppermint oil may attack and degrade plastic liners.[13]

Leachate collection systems can fail as the polyvinyl chloride pipes fail from chemical assault; are crushed by the weight of the landfill; or become clogged from silt, mud, microbial growth, or precipitation of minerals. Once clogged, leachate accumulates, causing increased pressure, which may lead to leaks or overflows.[12]

Landfill caps or covers are meant to be impermeable but are disturbed by (1) burrowing and soil-dwelling mammals, insects, worms, and reptiles; (2) erosion by precipitation, freeze-thaw cycles, and wind; (3) roots of vegetation that may penetrate the cover; (4) uneven settling that causes rips and tears on the membrane or cracks in the clay surface; (5) upward migration of chemicals or piercing objects; and (6) sunlight on exposed surfaces. Caps that are not maintained will permit precipitation to accumulate causing leachate overflow or penetration through the bottom liner.

Leachate collection systems, bottom liners, and caps are ultimately expected to fail and cause the contamination of groundwater, which is likely to remain contaminated for thousands of years.[15] Since groundwater is a diminishing resource, alternative methods of waste management must be employed. The combination of increased costs for landfill construction, the difficulty in siting landfills, and the moratorium on landfill development and expansion by some states have resulted in a significant decline in the number of land-fills since a peak of 8,000 community landfills in 1988.[8,16] There are fewer than 2,400 MSW landfills in the United States today. However, the total landfill capacity has not declined, as newly constructed landfills are much larger. The Northeast United States has the fewest landfills (208) and the fewest remaining years of landfill capacity.[3]

Source Reduction

Activities that reduce the total amount of waste prior to entering the waste stream are referred to as source reduction. Source reduction activities may include:

1. *Products package reuse*: The use of glass bottles has declined in favor of plastic containers, but glass beer or soft drink bottles represent an excellent example of containers that can be reused for their original purpose. Many chain grocery stores offer reusable cloth shopping bags. Clothing, furniture, and appliances are often informally reused among family and friends, sold at yard sales, or donated to agencies that will redistribute them for further use.

2. *Package or product redesign that reduces either material volume or toxicity*: Containers may be redesigned to use fewer materials, such as aluminum beverage cans that are nearly 20 percent lighter than they were in the 1970s or reducing the thickness of plastic liners used on cereal boxes.[5] Even the caps of milk cartons have been redesigned to be thinner and lighter. In addition, consumers may choose to purchase less toxic materials such as citrate-based cleaners or non-toxic pesticides. Manufacturers may use less toxic packaging by avoiding the use of highly bleached white corrugated material in favor of brown corrugated boxes printed with non-toxic inks.

3. *Reducing use by modifying practices*: Consumers have wide latitude to reduce material use. Many grocery stores now offer the opportunity to purchase foods or other materials in bulk thereby minimizing packaging. Paper can be used for copying on both sides, while electronic mail (E-mail) reduces the need for hardcopy mail. Many products are now offered in concentrated form, including juices, detergents, and fabric softeners that can be reconstituted in a larger and reusable container.

Trends in the use of non-durable papers (newspapers, office papers, etc.) and packaging containers (excluding corrugated boxes) have been used as indicators of overall source reduction efforts. For a few decades, packaging waste, by weight, had increased annually in the United States since the 1960s. There were strong concerns that generated packaging waste would continue to increase dramatically. However, exactly the opposite happened. Between the early 1990s and 2014 packaging waste as a percentage of the total weight of generated MSW declined from 36 percent to 30 percent.[17] Further, as the rate of gener-ated packaging waste nearly plateaued, its recycling rate continued at a steep rate, slowing

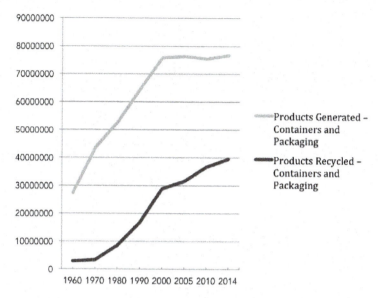

FIGURE 11.13

Packaging generation and recovery 1960–2014. (Adapted from United States Environmental Protection Agency, USEPA, Office of Land and Emergency Management, *2014 Advancing Sustainable Materials Management,* Washington, DC, November 2016.)

only slightly in the early 2000s (Figure 11.13). This indicated that a larger percentage of the packaging waste was being recycled, so the percentage of total packaging waste sent to landfills fell from 66.5 percent in 1994 to less 50 percent in 2014. Packaging waste *has* proven to be a sound indicator of source reduction. Since 2000, there has been an overall reduction in per capita waste generation, from a high of 2.15 kilograms per capita in 2000 to 1.99 kilograms per capita in 2014.[18]

A central challenge in municipal waste management in some smaller countries is a lack of space for landfills. Not surprisingly then, in countries where space is limited, such as Japan, Singapore, Taiwan, and Hong Kong, there is the promotion of source reduction through reduce, reuse, and recycle strategies (the 3Rs) to reduce the amount of waste. Encouraging responsibility for one's own trash can yield great benefits at a low cost and only minor inconvenience.[19]

Recovery for Recycling (Including Composting)

The act of removing materials from MSW for a productive use is referred to as recycling or resource recovery. It can be measured by determining the amounts of purchases of post-consumer recovered materials including the net exports of such materials. The recovery of materials for recycling is one of the most preferred waste management alternatives and continues to be among the most effective as well.[3,7] Citizens perceive recycling as desirable and are willing to participate in programs that are minimally demanding on time and effort. Recycling is perceived to (1) conserve resources by reducing the need for virgin and nonrenewable materials; (2) reduce the amount of pollution by using secondary materials that require less energy to process (using recycling aluminum cuts pollution and energy use by more than 90 percent compared to the use of virgin materials); and (3) save energy by using recycled materials, since less is required for processing. The USEPA considers the

FIGURE 11.14
Separating MSW for Pay As You Throw. (From EPA, US 'Pay As You Throw Workbook,' https://archive.epa.
gov/wastes/conserve/tools/payt/web/html/toolkit.html.)

recycling of materials in MSW among the more important waste management strategies, and many states have adopted recycling into their integrated waste management plans. These states have set recycling goals, including the percent of MSW to be recycled and the dates by which this will be accomplished.

Recyclable materials must be collected before they can be processed. Collection normally involves curbside collection, drop-off programs, container deposit programs, and buy-back operations. Curbside collection requires residents to separate their refuse into recyclables and discarded waste, and to place the separated recyclables in separate collection containers at the curb. Some states have adopted unit-based pricing systems, also known as "pay as you throw" programs (Figure 11.14). A typical program would require a resident to purchase plastic trash bags or stickers at selected community vendors bearing the name of the town. The cost varies from $1.00 to $2.00 per 113-liter bag or bag sticker. In some regions there is a tiered variable pricing system, where the first 75–151-liter waste can will cost $9–15 and any additional cans require payment of another fee.[20] Discarded material would go into the bags, and the resident would have a financial incentive to recycle and avoid filling the bags quickly. Such programs have been shown to be the greatest single stimulus to increasing residential recycling.[21] Nationally, slightly more than half of the U.S. population

TABLE 11.4

Countries with Bottle Bill Laws

Australia	Iceland
Austria	Israel
Barbados	Kiribati
Belgium	Lithuania
Canada	Micronesia
Croatia	Netherlands
Denmark	Norway
Estonia	Palau
Fiji	Scotland
Finland	Sweden
Germany	Switzerland
Guam	Turks and Caicos
	USA

Source: http://www.bottlebill.org /legislation/world.htm.

(about 135 million people) have access to one of the 9,000 curbside collection programs, although most are not based on "pay as you throw" programs.[3] There are in excess of 10,000 locations where citizens may drop off materials for recycling. There are currently 10 states that have container deposit systems in place. These are sometimes referred to as "bottle bill" states, which require consumers to pay a refundable deposit on soft drink and beer containers. Such programs have documented a marked increase in recycling of these containers, with over 35 percent of all recycled beverage containers occurring in these 10 states. The state of Oregon passed the first bottle deposit law in 1971, followed by Vermont in 1973. Most recently, Hawaii became a bottle bill state in 2002.[22] Around the world, 25 countries have some kind of bottle bill law to encourage recycling[23] (Table 11.4). While this is only a small percentage of the total of 195 countries, it is a start.

Once recyclable materials are collected, they must be further processed and separated to create large amounts of homogeneous secondary materials that are attractive to manufacturers as alternatives to virgin materials. The facilities that perform the function of preparing recyclables for marketing are referred to as materials recycling facilities (MRFs) (pronounced "murfs"). In a typical MRF, recyclables are separated in a sequential process (Figure 11.15). First, they are dumped on a cement floor (Figure 11.16), picked up by a front-end bucket loader, and placed in conveyor belts that carry the material past magnetic separators to remove steel cans and steel objects, and then past a line of people who remove recyclable materials manually from the stream (Figure 11.17). The materials often require further processing. Glass bottles are separated into clear, brown, and green glass and ground into small smooth pellets called cullet. An example of green glass cullet is shown in Figure 11.18. There are more than 480 MRFs operating in the United States, processing nearly 27,215 tons per day of material.[24] Some of these MRFs employ more sophisticated technologies to sort recyclables including magnetic pulleys, optical sensors, eddy currents, and air classifiers.

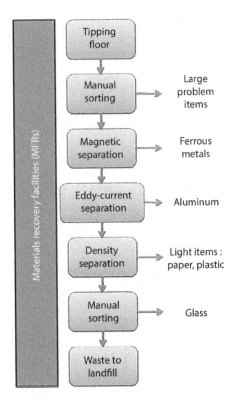

FIGURE 11.15
Step-by-Step Processes at a MRF. (From Report on current waste management systems in Europe, 30/06/2011, ec.europa.eu/environment/life/project/Projects/.)

FIGURE 11.16
Recyclables awaiting processing at a typical MRF.

FIGURE 11.17
Manual separation of recyclables at a typical MRF.

FIGURE 11.18
Green glass cullet as a recyclable product after processing.

There are two types of MRFs: clean and dirty. A clean MRF accepts co-mingled blue bin material, that is, separated recyclables. A clean MRF may be either single stream or dual stream, accepting either combined or separated recyclables. A dirty MRF handles household or commercial trash that has not separated out recyclables from trash. Dirty MRFs can recover recyclables placed in with trash. However, dirty MRFs require more manual processing.[25]

Dirty MRFs process mixed waste that would normally go to a landfill complete with food scraps, dirty diapers, and mustard jars. There are only 58 such facilities in the country with most of them being in the West. Dirty MRFs tend to be popular in areas of the country where administrators have opted not to involve citizens in recycling efforts. Theoretically, in these areas the percent of material recycled would be very high because all municipal refuse would be delivered to such facilities.

Trends in Resource Recovery

Materials recovered from MSW have been measured by the purchases of post-consumer recovered materials plus net exports of the materials. Since 1960, there has been a consistent increase in the weight of MSW recovered, and in the percentage of total MSW recovered (Figure 11.19). Notice that, between 1985 and 1995, there was a recycling inflection point, where the percentage of total MSW that was recycled/composted increased sharply, by more than 150 percent, from 10.1 percent to 25.7 percent.[7] These days, the percentage of recycled MSW continues to increase year on year, albeit more slowly. Even small percentage increases are significant, given the total weight of MSW. Specific 2014 MSW outcomes show that 60.24 million megagrams were recycled, 20.9 million megagrams were composted, 30.02 million megagrams were used to generate energy, called waste to energy, and the remaining 123.38 million megagrams of MSW went to a landfill.

Recycling rates tracked by product category reveal that lead-acid batteries, corrugated boxes, and steel cans lead the way, all recycled at higher than 70 percent of the generated amounts[7] (Figure 11.20). In 2014, recycling of nondurable goods—goods that last 3 years or less—was 33 percent. The newspapers and other paper products category of nondurable goods was recycled at 68 percent, the highest for nondurables. Interestingly, one product group that has seen increased recycling rates is select consumer electronics. In 2014 the recycling rate for these products was 41.7 percent or 1.4 million tons. This is a significant contrast to 2000, when the recycling rate for selected consumer electronics was only 10.0 percent. One ongoing driver of increased recycling is that certain federal agencies and many state governments mandate a minimum recycled content for targeted products

FIGURE 11.19
MSW recycling rates 1960–2014. (Adapted from United States Environmental Protection Agency, USEPA, Office of Land and Emergency Management, *2014 Advancing Sustainable Materials Management*, Washington, DC, November 2016.)

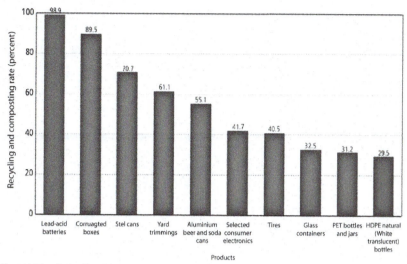

Does not include combustion with energy recovery.

FIGURE 11.20

Products recycled in 2014 in the United States. ('Superfund Liability', USEPA, Enforcement, www.epa.gov/enforecment/superfund-liability, accesssed July 15, 2018.)

such as newspapers or for products they purchase such as office paper. In addition, due to consumer preference, many manufacturers now use a percentage of recycled material in their product packaging and state that on the product.

Composting

There are approximately 2,200 municipal composting programs nationally directed at yard trimmings. Most of these programs are located in the Northeast and Midwest and handle more than 22680 megagrams per day.[7] There are 14 mixed waste composting facilities (eight in the Midwest) that remove large items and metal from the mixed waste and compost the combined paper, food waste, yard trimmings, and wood. Composting is a controlled process of degrading organic matter by microorganisms into a humus-like material. The composted material is often low in plant nutrients but is useful for conditioning soil by improving soil porosity and aeration and increasing water retention. Some communities are experimenting with adding domestic sludge to the mix and co-composting the sludge and biodegradable refuse together. Unfortunately, public acceptance of this co-composted material is not universal because of the real or perceived dangers of pathogens and/or toxic agents surviving in compost. The composting process is normally performed by removing inorganic materials such as glass, metal, and tires and shredding the organic material. In some sophisticated operations, the entire process from shredding to aeration, curing, and finishing is performed in an in-vessel composting system (Figure 11.21). Most composting uses a windrow process in which rows of material are placed next to each other outdoors and periodically mixed to incorporate air into the mix (Figure 11.22). This permits the organisms to aerobically digest the materials, increasing the heat of the compost pile to temperatures sufficient to kill most pathogens and insect eggs. When the materials have been left long enough to cure so that cellulose and lignin are broken down, the material is normally screened to size and either packaged or used in bulk for soil enhancement.

FIGURE 11.21
In-vessel composting system. This system is used to grind mixed food and plant wastes, aerobically digest them, and provide a finished compost product in days to a few weeks.

FIGURE 11.22
Compost piles arranged in rows at different stages of digestion and curing.

Compost that is consistently high in quality and blended to meet the needs of the end user is being successfully marketed. Compost is being used for wetlands mitigation, land reclamation, storm filtrates, soil amendments, mulches, and low-grade fertilizers.[24] The majority of composted material is used in landscaping, agriculture, and for landfill cover. Many of the markets have seen little penetration. However, as the quality and consistency of compost improve, there is likely to be higher consumer acceptance when combined with education on the benefits of compost use.

Combustion

The treatment of MSW by combustion includes the combustion of mixed MSW, the preparation of refuse-derived fuel (RDF), or a separated component of MSW (such as rubber tires) with or without the recovery of energy.[8] Generally, combustion with the production of energy is called waste-to-energy (WTE) also known as energy from waste (EfW). Combustion of MSW without energy recovery is called incineration. The burning of unsegregated wastes (mass burn) in incinerators was very popular prior to the 1970s before air quality regulations made the operational costs prohibitive. However, improved methods in pollution control and monitoring, combined with escalating costs of alternative refuse disposal, have encouraged the growth of the refuse-combustion industry. It is estimated that 17 percent or 31.75 million megagrams per year of MSW are combusted. Most facilities that combust MSW also incorporate recovery of energy as steam or electricity, or both. The sale of the generated energy helps to defray operational costs. These WTE plants are generally mass-burn plants, although many incorporate recycling activities that remove noncombustible items such as metals and glass prior to burning. A state-of-the-art WTE facility integrates with a statewide master plan to combust only waste that cannot be recycled or composted. An example of such a facility is operated by Springfield Resource Recovery, Inc. (SRRI) in Springfield, Massachusetts. The facility is designed to burn waste in stages to minimize emissions. The first stage incinerates MSW by controlled combustion, which minimizes combustion pollution. The second stage recovers energy and converts it to useable steam and electricity, and the third stage minimizes or neutralizes pollutants coming from the waste (Figure 11.23). Independent testing firms use EPA-approved testing methods to evaluate the levels of emissions including dioxins, furans, lead, mercury, beryllium, trace metals (cadmium, chromium, copper, manganese, nickel, tin, zinc), inhalable particles (PM_{10}), and total suspended particulate matter (TSP). This has been made necessary because of increasing concerns about possible toxic air emissions from these facilities. The USEPA approved tighter emissions controls in 1991 for combustion facilities in accordance with the Clean Air Act Amendments. Systems now employ specialized reactors, bag houses, computerized combustion control systems, gas scrubbers, and continuous monitoring systems to control and evaluate emissions. The fly ash captured from the bag house may be mixed with the bottom ash to reduce the probability of it being designated as hazardous waste.

More than 2000 WTE facilities are in use across the globe.[26] In the United States, there are 90 WTE facilities, most of these located in the Northeast, mid-Atlantic, and southern states. In the EU, there are over 400 WTE sites. Many of the remaining WTE plants are located in Asia. In particular, China has seen a boom in WTE plants, with 226 plants and counting and a five-year plan goal to achieve 35 percent of energy from WTE plants.[27] Most of these plants are located in the eastern portion of the country.

In 2013, the U.S. sent 29.63 million megagrams of waste to energy facilities, including 26.76 million megagrams of MSW, 462,664 megagrams of wood, and 2.4 million megagrams of tires. This is a slight decrease from 30.84 million megagrams in 2000. Overall, combustion with energy recovery has increased from less than 2 percent of generated MSW in 1980 to 12.8 percent in 2014.[7]

Preseparated MSW is deposited to the tipping floor. MSW is swept to the incinerator feed hopper and burned in a multiple hearth furnace in stages for 3 hours until fully combusted. A lean atmosphere and controlled temperatures reduce the formation of CO, NOx, and dioxin emissions.

Hot gases from the incinerator pass through water tubes in the boiler which heat water to create steam. The steam expands through turbines generating electricity.

Economizers extract heat and recycle it for plant uses. Acid gases are neutralized in a dry absorption reactor. Particulates are removed by filters in the bag house. Fly ash and bottom ash are collected and deposited in a monofill. All systems are continuously monitored for CO, CO2, NOx, SO2, and opacity. (Opacity is an indirect measurement of smoke indicated by the degree of obstruction of a beam of light.)

Tipping floor

Control room

Baghouse and stack

FIGURE 11.23
Stages and operations of a modern WTE facility. This diagram is based on the design of the WTE facility in Springfield, MA.

Hazardous Wastes

Background

During the spring of 1977, heavy rains raised the groundwater level and turned the Love Canal region in Niagara Falls, New York into a muddy swamp mixed with some of the 17236 megagrams of toxic wastes buried in the abandoned canal by Hooker Chemical Company. The chemically laced ooze seeped into basements, onto playgrounds, and over lawns, creating a storm of protests, claims of severe health effects, costs of over $37 million to relocate families from the area, and another $30 million to aid families. More than $150 million and 12 years were spent on clean-up efforts. Love Canal became a major turning point in directing public attention to other chemical wastelands scattered across the country in abandoned warehouses, manufacturing facilities, landfills, and processing plants.[28]

The USEPA estimates that in excess of 245 million megagrams tons of hazardous waste is generated every year in the United States, with most of it produced by the petrochemical industry.[29] Discarded or released hazardous substances have contaminated surface and groundwater, soils, and air. Drinking water has been contaminated, food sources polluted, and wildlife habitats destroyed or compromised. Too often these hazardous substances are released or improperly stored in poorer communities. The idea of *environmental justice* developed from studies that indicated minorities were disproportionately subjected to environmental pollutants. In the United States, this prompted President Clinton to issue an executive order in 1994 that decreed agencies must address the notion of environmental justice when considering the disposal of hazardous wastes. Additionally, the order required minorities be given a voice in environmental regulation and that the cleanup of the worst toxic waste sites in minority areas be given priority.[30]

What Is a Hazardous Waste?

The definition of hazardous waste in the United States RCRA regulation includes any discarded material that may pose a substantial threat or potential danger to human health or the environment if improperly handled. The USEPA has established that a waste is to be considered hazardous if it meets either of following qualifications:

First, the substance should be listed as one of the 400 wastes or waste streams under Title 40, Parts 261, 31–33 of the Code of Federal Regulations (CFR). Such substances include wastes generated by chemical manufacturers, vehicle maintenance shops, the paper industries, metal manufacturing, cleaning agents and cosmetics manufacturers, the construction industry, the printing industry, and many others. These are called listed wastes.

Second, wastes are considered hazardous if they possess certain properties or characteristics. These characteristic wastes are subject to regulation if they are toxic, ignitable, corrosive, or reactive according to the specifications outlined in RCRA. A substance is considered toxic if it is potentially poisonous to humans and can cause acute or chronic health problems. Wastes are considered toxic if they produce sufficient amounts under the Toxic Characteristic Leachate Procedure (TCLP) described by the USEPA. This process requires acidified liquid to be percolated through waste in a laboratory-created landfill condition for 24 hours. Samples of the leachate are tested for the amounts of the target substance released. Substances are ignitable if they have a flash point below 60°C or can cause fires spontaneously by friction, moisture, or chemical change. Certain plastics, oils, solvents, and paints fit this category. Wastes are corrosive if they have a pH below 2.0 or above 12.5, and include such wastes as battery acid, drain cleaners, and other compounds that will dissolve living tissue, metals, and other materials. Reactive wastes may include discarded metallic sodium or wastes from the manufacture of certain explosives that may react vigorously with air or water to generate toxic gases or even explode.

In the United States, the proliferation of hazardous wastes, industrial sources, and community outcry prompted Congress to pass the RCRA of 1976. This was the first federal regulatory program targeted at controlling present and future hazardous waste. This Act was intended to prevent the improper disposal of hazardous waste using a manifest tracking system that attempts to follow the waste from its point of generation to its final treatment and disposal. The process is called the "cradle-to-grave" system and is meant to protect public health by (1) defining what wastes are hazardous, (2) tracking wastes to the point of disposal, (3) assuring that treatment, storage, and disposal (TSD) facilities meet minimum national standards, and (4) making certain TSDs are properly maintained after closure, and that facility owner/operators are financially responsible for hazardous waste releases that may occur at their facility.

Everyone who handles the waste, including the generator, the shipper, and the manager of the TSD facility, must sign a multi-part manifest with copies going to the state and federal regulatory agencies and back to the generator after various signatures are applied. The final disposal of wastes has become expensive, and persons wishing to profit by dumping the waste along roadways during rainstorms, down storm sewers, or on privately owned land may realize significant profits by failing to report these activities at the outset. However, the penalties are severe and include criminal prosecution and hefty fines.

Hazardous Waste Regulations

The regulation of hazardous waste has changed significantly across the past 50 years. In the late 1970s and early 1980s, it was generally accepted that such wastes were inevitable and that acceptable methods of treatment and disposal were the answer to protecting the public health. However, the rapidly increasing generation of hazardous wastes, the discovery of tens of thousands of hazardous waste sites, and growing public concern demanded an alternative to this "end of pipe" approach. Consequently, regulations became more stringent. In the United States, amendments were made to RCRA in 1984 that significantly expanded its regulatory powers. The amendments are known as the Hazardous and Solid Waste Amendments (HSWA), and they are focused on protecting groundwater by (1) restricting the TSD of hazardous wastes in land management facilities; (2) mandating stricter requirements for landfills accepting hazardous waste; (3) requiring a schedule for determining if the landfilling of untreated hazardous waste should be phased out; (4) creating a new program for the detection, control, and management of hazardous liquids (primarily petroleum) in underground storage tanks; and (5) increasing the numbers of people who fall under RCRA regulations by including small-quantity generators. Small-quantity generators would include those who produce 100 kilograms per month of hazardous waste or 1 kilogram per month of extremely hazardous wastes. Previously, only large-quantity generators who produced more than 1000 kilograms per month of hazardous waste were subject to RCRA provisions. Land disposal in landfills and surface impoundments were considered unsafe and potential sources for leaks.[31] The same HSWA amendment called for the reduction of hazardous wastes generated at the source and that such activities should have priority. This effort was further strengthened by the passage of the Pollution Prevention Act of 1990 that emphasized a federal policy of preventing the formation of hazardous wastes, followed in priority by the recycling or reuse of hazardous materials, and the last activity in the hierarchy would be treatment and disposal in an environmentally safe way. The USEPA strongly committed to this policy by making waste minimization its highest priority and requiring hazardous waste generators to certify their efforts at waste minimization on the hazardous waste shipping manifests.[32] Similar requirements exist for owners and operators of TSD facilities. In the early 1990s, the USEPA created The National Waste Minimization Program, a long-term national effort to reduce the quantity and toxicity of hazardous wastes, with a focus on 31 toxic chemicals, deemed priority chemicals (PC). The goals of program are to:

1. Achieve a 50 percent reduction in the most persistent, bioaccumulative, and toxic (PBT) chemicals in the nation's hazardous waste by the year 2005 as compared to the baseline year of 1991 Bioaccumulative (B) refers to chemicals that tend to concentrate in animal and plant tissues. Many chlorinated organics and heavy metals tend to fit in this category.
2. Prevent the transfer of chemical releases from one medium (air, water, land) to another and give priority to source reduction (reducing waste prior to its being generated) and environmentally sound recycling over waste treatment and disposal.[33]

To accomplish these goals, the plan focused on several innovative processes including:[34]

- Lean Manufacturing
- Energy Recovery
- Environmental Management Systems (EMS)
- Green Chemistry

Other countries and regions of the world were on a similar journey with regard to hazardous waste. From the mid-1970s, the European Union followed a similar path toward increased control over such wastes. Starting with a Waste Framework Directive in 1975 and later signing on to the Basel Convention, the European Union's view on hazardous waste evolved further with the development of the EU Waste Shipments Regulation from 1993 to 2006.[35] This regulation monitors hazardous waste shipments across EU country borders by reporting requirements. Further, the Basel convention prevents OECD countries, which includes 35 developed countries, including much of the EU, from shipping hazardous waste to non-OECD countries, thereby preventing dumping hazardous waste on poorer countries. For EU-member countries, this means that such wastes are shipped primarily between countries in the EU.

In lesser developed countries that lack the infrastructure and resources to properly and safely manage a growing hazardous waste problem, there is assistance from other sources. For instance, the United Nations has established a program called UNEP's Harmful Substances and Hazardous Waste sub-program to help developing countries better contain hazardous waste. This UN program has four key areas of country assistance, which are (1) scientific assessments, (2) legal instruments, (3) national implementation of hazardous waste management, and (4) monitoring and evaluation. This UN program lays the groundwork for developing countries to establish sound management processes for hazardous wastes.

Are you interested in learning more about solid and hazardous wastes? Visit <https://people.umass.edu/~envhl565/> and click on one or more of the following sites for Chapter 11.

Envirofacts Querying Capability

Scorecard Home

The Basics of Landfills

Hazardous Waste: Introduction

EPA: RCRA Online

EPA: Hazardous Waste

EPA Transforming Waste Tool

European Environment Agency

The Management of Hazardous Wastes

The three major options for managing hazardous waste include (1) reducing the production of waste by reducing the amount generated or recycling/reusing the hazardous material after its generation, (2) reducing the volume and/or hazard of the waste, and (3) long-term storage or disposal (Figure 11.24).[32]

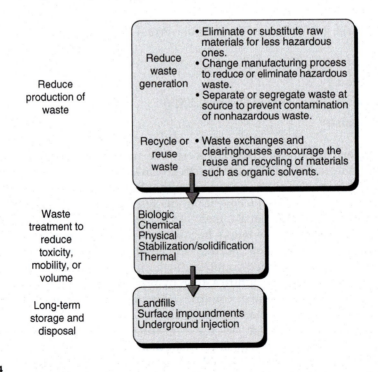

FIGURE 11.24
A desirable hierarchy for the management of hazardous wastes. (Adapted from Stephens, R.D., Hazardous wastes, in *Principles and Practice of Environmental Medicine*, Tarcher, A.B. (Ed.), Plenum Medical Book, New York, 1991.)

Reduction of Generation of Hazardous Waste

The highest priority item in the management of hazardous waste is to reduce the production of waste as the source. The generation of hazardous waste can be minimized by (1) eliminating or substituting raw materials for less hazardous ones, (2) changing the manufacturing process to reduce or eliminate hazardous waste, and (3) separating or segregating waste at the source to prevent the contamination of non-hazardous waste. Recycling and reusing hazardous materials before they enter the waste stream is also strongly supported by the USEPA and most state regulatory agencies. Waste clearinghouses and waste exchanges have been organized to facilitate the transfer of discard materials from one industry to another that may use it as raw material in their process. Organic solvents are among the most widely recycled and reused materials.

Technologies for Hazardous Waste Treatment

Biological

Microorganisms are used to break down or degrade organic compounds as they use waste materials for a food source and convert them into water, carbon dioxide, and simple inorganic and organic molecules. This process is known as bioremediation and has been used successfully to clean up oil spills and other organic wastes where soil permeability allows good oxygen penetration. Waste degradation may be enhanced by adding oxygen and/or nutrients to stimulate the waste-eating organisms. Treatment may also occur in aboveground facilities or reactors where conditions may be better controlled and higher levels of decontamination can be achieved on wastes fed through the system. Bioremediation is being extensively tested under USEPA or state oversight at more than 150 sites throughout the United States.

Chemical

Treatments are applied which neutralize acid or alkaline wastes, detoxify, dechlorinate, oxidize, coagulate, precipitate, or concentrate hazardous wastes. Chemical treatment is used most often to change the chemical structure of waste and render it nontoxic or less hazardous.[36]

Physical

Non-chemical methods that separate or concentrate components of the waste stream are termed physical methods. Such methods include activated carbon to remove organic materials from liquid waste, distillation, filtration, coagulation, centrifugation, reverse osmosis, and ion exchange.[37] Once separated, the hazardous components may be reused or more efficiently destroyed or disposed.

Solidification/Stabilization

Both physical and chemical methods are used to form a solid, inert substance that limits the migration of hazardous materials from a landfill. Compounds amenable to fixation include organic salts, heavy metals such as lead, mercury, and zinc, and many inorganic compounds. Binding agents may include organic polymers, silica, or cement-based materials that are used to encapsulate materials in a water-resistant matrix, which can then be more safely disposed in a landfill. Stabilization can also be achieved by inserting electrodes into the contaminated soil, which send huge amounts of electricity through it, melting and fusing the soil into a glass-like block. This process is called in-situ vitrification (ISV) and such stabilized blocks are often left in place since the danger of pollutant migration is largely diminished.

Thermal Treatment

Incineration accounts for less than two percent of the hazardous wastes managed although it offers the advantages of (1) completely destroying many forms of hazardous waste, (2) reducing the volume of waste, (3) and recapturing energy in the form of steam or electricity. Thermal treatment may include incineration or pyrolysis. Pyrolysis combusts materials in an oxygen-starved atmosphere and is not as widely used as incineration, which is conducted in an oxygen-rich environment. Incineration is normally conducted at temperatures of 425°C to 1650°C (800°F–3000°F) in a turbulent atmosphere of sufficient duration that will ensure good mixing and effective destruction of the waste approaching levels of the USEPA current standard of 99.99 percent.[36,37] Incinerators have been used successfully to treat waste paints, plastics, mineral oils, pesticides, solvents, sludge, resins, greases, and waxes. The threat of hazardous emissions, including acids, dioxin, and heavy metals, have led to strict regulations on emissions from incinerator facilities.[38] The materials generated in an incinerator include carbon dioxide, water, sulfur, and nitrogen oxides, acid gases such as hydrogen chloride, and ash. Gas streams must be monitored, and hazardous materials must be removed from the emissions by various scrubbers, filters, and electrostatic precipitators. This has prompted very high construction and maintenance costs but has become a useful alternative as the options for land disposal continue to decrease.

Hazardous Waste Disposal

Even after extensive treatment, some amount of hazardous wastes will remain in the form of ash, chemically treated residue, physically separated materials, or even in original waste form. Such waste maybe economically reusable or recyclable and is usually dispersed to landfills, surface impoundments, or underground injection wells. Nearly 80 percent of hazardous wastes are disposed to land with 96 percent of these wastes treated or disposed at the site where they are generated.[38,39] Historically, such wastes were deposited to unlined pits and lagoons (surface impoundments), which presented extensive environmental risk from overflows, leaching, and volatilization.

Landfills

A common method for hazardous waste disposal is the secure chemical landfill as designated in the 1984 HSWA of RCRA. Secure landfills require groundwater monitoring, double liners, a leachate collection system, and financial guarantees for post-closure activities (Figure 11.25). Just as with MSW landfills, current technologies almost guarantee the leakage of hazardous materials from commercial hazardous waste landfills.[27] Increased costs of construction, operation, and liability have curtailed the number of active commercial operations to less than 20 landfills. These landfills are further restricted by legislation to accept mostly pre-treated wastes that have been reduced in toxicity and/or treated to prevent leaching. The difficulty in siting for commercial hazardous waste landfills in New England and New Jersey because of public resistance and other factors has resulted in long-haul shipments at every high costs for businesses, schools, and commerce in these areas.

Deep-Well Injection

The process of deep-well injection involves pumping hazardous wastes through wells with double or triple casings into porous rock formations at depths of 304.8–3,048 meters below the surface and well below drinking water aquifers (Figure 11.25). New hazardous waste wells were banned in 1983. Most injection wells are used by petroleum refineries

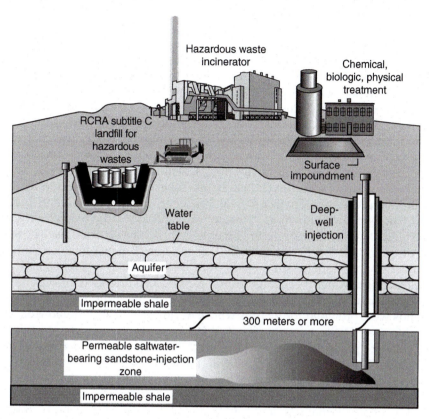

FIGURE 11.25
Typical methods for hazardous waste treatment and disposal.

and petrochemical plants located in the Midwest and Texas who pump nearly 27.22 million megagrams of hazardous waste into the ground each year. Despite legislation that prohibits the construction of injection wells within .4 kilometers of a drinking water source, there remain concerns about such wells. Detection of leaks and methods of correcting them are severely restricted by the depth. Further, the earth is a dynamic entity with continents drifting, tectonic plates shifting, and local geology under variations in pressure that combine to produce fractures and fissures that may release injected wastes to combine with groundwater at some future time. For more information see http://www.dep.state .fl.us/water/uic/index.htm.

Surface Impoundments

Surface impoundments are depressions in the ground or areas with dikes around them that contain the liquid wastes pumped into them. They vary in size and shape ranging from several hundred square meters to hundreds of hectares. Most of these pits, lagoons, or basins were historically unlined and located on-site at manufacturing facilities. Legislatures recognized the high risk of groundwater contamination and regulations now require such impounds to be equipped with double liners, leachate collection systems, and groundwater monitoring wells. However, just as with landfills, such impoundments will eventually leak and attention is focused on reducing hazardous waste through reduction, reuse, and recycling efforts.

Cleaning Up

Before the 1980s, it was common practice for industries and municipalities to haul the wastes to a depression in the ground, dump them, and cover them with dirt. Many local landfills weren't even covered with soil. In many cases, drums of toxic wastes were simply stored in piles on-site (Figure 11.26). Many of these sites were abandoned, forgotten, and even built on, as neighborhoods expanded with the pressures of a rapidly increasing population. Then the buried toxic sludge began to appear, as rusted 208-liter drums seeped their corrosive contents into the soil to contaminate groundwater or work their way to the surface where the true drama of poisonous mud and rusted containers could be seen. Driven by the horrors of Love Canal, Chem-Dyne, Swartz's Creek, and dozens of such hazardous waste sites the United States Congress acted in December of 1980 by passing the Comprehensive Environmental Response, Compensation, and Liability Act (CERCLA). This Act, also known as "Superfund," authorized the federal government to spend $1.6 billion over a five-year period for emergency clean-up activities. The Superfund program also provided for response to the release of hazardous substances from leaking underground storage tanks, spills from transportation accidents, and pesticide-contaminated groundwater.[40] The Superfund program also provided for activities to oversee long-term containment of hazardous sites. Despite the term "'compensation' in the law, no funds were provided for victims whose health may have suffered from exposure to toxic waste sites. The term "Superfund" is attributed to the fact that the bill created a trust fund financed primarily by excise taxes on chemicals and oil, and an environment tax on corporations. The USEPA was charged with the responsibility of identifying those hazardous waste sites in most need of cleaning up. Superfund requires that for each site, the USEPA identify potentially responsible parties (PRPs) who are considered to be any entity, however large or small, that has anything to do with the site. The liability rules applied include retroactive liability, joint and several liability, and strict liability (Figure 11.27). This may force individuals, companies, or organizations (entities) to pay for cleanups even

FIGURE 11.26
Stored (and discarded) drums of hazardous waste.

Strict liability: The government needs to prove only involvement at a waste site. Negligence or strict causation is not necessary to establish liability.

Joint and several liability: Any involved party can be legally responsible for cleaning up a site no matter what the degree of involvement. In practice, liability is often apportioned.

Retroactive liability: The PRPs can be held liable for releases resulting from actions before Congress enacted CERCLA in 1980.

FIGURE 11.27
Definitions of liability under Superfund. (Adapted from USEPA, 'Superfund Liability', www.epa. gov/enforcement/superfund-liability, accessed July 15, 2018.)

when they have had little or even no part in causing the damage. In practice, state or federal agency personnel will attempt to identify entities with the largest pockets (most financial resources) as PRPs. Such rules have been held responsible for 36–60 cents of every dollar spent in Superfund going to legal and other transactional costs.[41,42]

There are presently nearly 33,000 hazardous waste sites that have come under Superfund authority. In 2017, 1,336 sites were on the USEPA's National Priority List (NPL) (Figure 11.28). The majority of these sites (1,179) are private sites and only 157 are

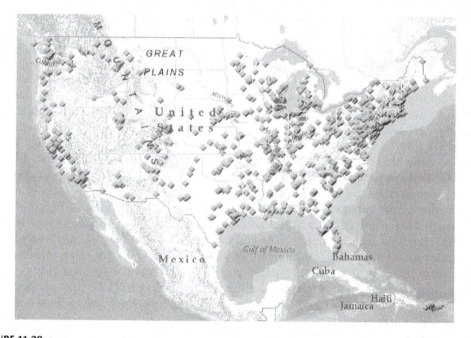

FIGURE 11.28
National priority list active Superfund sites. (From EPA Facility Resource https://www.epa.gov/enviro /frs-data-resources.)

Federal Facility sites. The NPL was established because hazardous waste sites vary in size, waste content, containment systems, and the degree of threat to the surrounding populations and environment. To make sense of this range of hazards, the USEPA established a hazard-ranking system (HRS) based on the estimated hazard potential of the hazardous waste site. The factors used to develop this estimate include waste characteristics (toxicity, quantity, solubility, reactivity); the distance to the local population; and the proximity to surface water, groundwater, and drinking water supplies. Other factors include the level of containment such as the condition of the containers (i.e., rusted drums), and whether there are leachate collection systems or liners. The NPL is updated annually, with some sites deleted, others cleaned up, and new sites added.

Buttressed by a large contingent of concerned citizens, environmental groups charged that RCRA was not being vigorously enforced. These actions caused Congress to respond by passing the Superfund Amendments and Reauthorization Act of 1986 (SARA), which increased the program's funding to $85 billion and provided new and stricter standards. A series of delays, lawsuits, and accusations followed demanding that the Superfund program be repaired. Superfund had spent billions of dollars of taxpayer money and private sector funds since 1980 to clean 550 sites.[43] This cost was expected to escalate to $42 billion in the next 70 years to clean up an estimated total of 2,300 NPL sites.[42]

At the end of 1995, the tax authorization for Superfund expired. The balance of some $6 billion in the Hazardous Substance Superfund Trust Fund was being supported by (1) taxes on domestically produced and imported oil; (2) a tax on feedstock chemicals; (3) a corporate environmental income tax; and (4) general revenues, reimbursements, penalties, and interest on the trust fund.[28] These fees generated $1.5–$2 billion annually for cleanups, and the trust fund reached a high of $3.6 billion in 1995. Unfortunately, in the same year, the Superfund fee system lapsed, and Congress has failed to reinstate it. The trust fund is diminishing and more of the cleanup burden is falling on taxpayers' shoulders, while the pace of cleanups has declined dramatically. The administration has failed to request reinstatement of Superfund's fee system for three years running. In its 2004 budget request, the administration requested $1.39 billion for the Superfund program, expecting taxpayers to shoulder approximately 80 percent of program costs. Taxpayers paid only 18 percent of program costs in the last year before the Superfund fee system expired. Industry has avoided paying about $4 million a day, totaling over $10 billion since the Superfund fees expired in 1995. More disturbing is that oil and gas companies are exempted from liability under the Superfund law.[44] In fiscal year 2005, the USEPA listed 1,303 sites including 59 proposed sites awaiting final agency action, 54 in the general Superfund section and five in the federal facilities section.[45] In fiscal year 2005, EPA's Superfund program obligated $524 million to perform construction and post-construction activities and to conduct and oversee emergency response actions. They conducted or oversaw 665 ongoing construction projects (by EPA, PRPs, and federal facilities) at 422 sites. Federal facilities accounted for 220 of these ongoing projects. The EPA Superfund program completed the construction phase of cleanup at 40 sites across the country for a total of 966 or 62 percent of the sites on the NPL. Further, underscoring EPA's commitment to the "polluter pays" principle, the Agency secured private party funding commitments of more than $1.1 billion in Fiscal Year 2005. Of this amount, PRPs agreed to conduct more than $857 million in future response work and to reimburse EPA for $248 million in past costs.[46]

In 2017, a new administration entered and the new administrator of the USEPA, E. Scott Pruitt, created a Superfund task force to address what he saw as a too-slow remediation of Superfund sites.[47] The focus areas of this Superfund task force are:

- Expedite Cleanup and Remediation
- Re-invigorate Responsible Party Cleanup and Reuse
- Encourage Private Investment
- Promote Redevelopment and Community Revitalization
- Engage Partners and Stakeholders

As it is early days for this new Superfund plan, implementation of which will begin in 2018, its execution and outcomes remain unknown. Of course, the hope is that environmental mistakes of the past are not repeated: Hazardous wastes improperly stored, cavalierly discarded, shoddily converted to green space, only to years or decades later contaminate the land and the water, damaging human health and the environment. Time will tell. As of 2015, 53 million people live within three miles of a superfund site, with potential exposure to lead, mercury and arsenic, the top superfund contaminants.[48] As American's tax dollars increasingly pay for superfund site cleanup, millions of Americans may pay the steep price of environmental disease due to proximity to these dangerous contaminants.

References

1. Van Lohuizen Noor, K., Drowning in garbage, *Washington Post*, Washington, DC, November 21, 2017.
2. Mills, C., *15 of the Worlds Largest Landfills with Photos and Statistics*, November 28, 2017. https://owlcation.com/stem/15-of-the-Worlds-Largest-Landfills (accessed April 3, 2018).
3. United States Environmental Protection Agency (USEPA), Characterization of Municipal Solid Waste in the United States: 1997 Update, Prepared for U.S. Environmental Protection Agency, Municipal and Industrial Solid Waste Division, Office of Solid Waste, by Franklin Associate, Prairie Village, KS, Report No., EPA 530-R-98-Oct, May 1998, Chap. 3.
4. Nadakavukaren, A., *Our Global Environment*, 4th ed., Waveland Press, Prospect Heights, IL, 1995, Chap. 16.
5. *Solid Waste Digest*, 11(12), National ed., Chartwell Information Publishers, Alexandria, VA, 2001.
6. Mian, M., Zeng, X., Nasry, A.A., and Al-Hamadani, S., Municipal solid waste management in China: A comparative analysis, *Journal of Material Cycles and Waste Management*, 19(3), 1127–1135, 2017.
7. United States Environmental Protection Agency, USEPA, Office of Land and Emergency Management, *2014 Advancing Sustainable Materials Management*, Washington, DC, November 2016.
8. Land of Waste: American Landfills and Waste Production, Save on Energy. www.saveonenergy.com (accessed July 15, 2018).
9. Zhou, B., Sun, C., and Hongtao, Y., Solid waste disposal in Chinese cities: An evaluation of local performance, *Sustainability*, 9(12), 2234, 2017.
10. Waste statistics, 2017. http://ec.europa.eu/eurostat/statistics-explained/index.php/Waste_statistics#Total_waste_generation.
11. Koren, H., and Bisesi, M., *Handbook of Environmental Health and Safety: Principles and Practices*, Vol. 11, 3rd ed., Lewis/CRC Publishers, New York, 1995, Chap. 2.
12. Pellerano, M.G., and Parkhurst, B. The basics of landfills, *Environmental Research Foundation*, 1995, http://www.enviroweb.org/enviroissues/landfills/index.html.
13. Anonymous, Garbologist goes where few scientists dare, *Los Angeles Times*, 1, July 24, 1993.

14. Bonoparte, R., and Gross, B.A., *Field Behavior of Double-Liner Systems, in Waste Containment Systems: Construction, Regulation, and Performance*, R. Bonaparte (Ed.), Geotechnical Special Publication, No. 26, American Society of Civil Engineers, New York, 1990, Chap. 2.

15. Lee, G.F., and Jones, A., *Municipal Solid Waste Management in Lined, "Dry Tomb" Landfills: A Technologically Flawed Approach for Protection of Groundwater Quality, EL Macero, California*: G. Fred Lee and Associates, March, 1992. Available from: G. Fred Lee and Associates, 27298 East El Macero, Drive, El Macero, CA, 67 p.

16. Commonwealth of Massachusetts, Department of Environmental Protection, *Executive Summary: 1997 Solid Waste Master Plan Update*, Executive Office of Environmental Affairs, Department of Environmental Protection, 1997. http://www.magnet.state.ma.us/dep/bwp /dswm/dswmpubs.htn#swmp.

17. Lilienfeld, B., *From Crisis to Myth: The Packaging Waste Problem, Use Less Stuff*, April 22, 2015. https://www.livescience.com/50581-packaging-no-longer-the-nightmare-some-claim.html (accessed April 3, 2018).

18. USEPA Municipal Solid Waste Generation, *Recycling, and Disposal in the United States: Facts and Figures for 2012*, Washington, DC, 2014.

19. Zhou, B., Yi, H., and Sun, C., Solid waste disposal in Chinese cities: An evaluation of local performance, *Sustainability*, 9(12), 2234.

20. EPA Archive, Pricing Systems, January 2017. https://archive.epa.gov/wastes/conserve/tools /payt/web/html/top13.html (accessed April 3, 2018).

21. United States Environmental Protection Agency (USEPA), *Characterization of Municipal Solid Waste in the United States: 1997 Update*, Prepared for US Environmental Protection Agency, Municipal and Industrial Solid Waste Division, Office of Solid Waste, by Franklin Associate, Prairie Village, KS., Report No., EPA 530-R-98 October–May 1998, Chapter 4.

22. Nash-Hoff M., *Why Are There So Few States with 'Bottle Bill' Laws?* September 16, 2015. http:// www.industryweek.com/regulations/why-are-there-so-few-states-bottle-bill-laws.

23. Bottle Bill Resource Guide, September 16, 2015, http://www.bottlebill.org/legislation/world .htm.

24. Yoon, R.-H., and Camicia, N.T, *Recycling of Municipal Solid Wastes in the United States*, Center for Advanced Separation Technologies, Industrial Waste Recycling R&D Center, 2001, Blacksburg, VA

25. LeBlanc, R., *What Is a Materials Recovery Facility (MRF) and How Does It Work?* November 03, 2016. https://www.thebalance.com/what-is-material-recovery-center-2877733 (accessed April 3, 2018).

26. Kalogirou, E., *Current Status of WTE Worldwide Potential for Developing Nations*, WTERT Council NOVISAD, September 19, 2016, ISWA, Novi Sad, Serbia, 2016.

27. Cord Homme, C., *Myths and Facts about WtE Plants*, CNIM, International Solid Waste Association, http://www.iswa.org/working-group-energy-recovery/.

28. Congressional Research Service, 'Foreword' Superfund reform: Cleaning up America's toxic waste sites, *Congressional Digest*, 7(3), 68, 1998.

29. Stephens, R.D., Hazardous wastes, In: *Principles and Practice of Environmental Medicine*, A.B. Tarcher, (Ed.), Plenum Medical Book, New York, 1991.

30. Nassiter, A., Villain is dioxin: Relocation is response, *NY Times*, October 20, 1996.

31. Council on Environmental Quality (CEQ), Environmental quality, *15th Annual Report, 1984*, U.S. Government Printing Office, Washington, DC, 1986.

32. United States Environmental Protection Agency, *Safer Disposal for Solid Waste: The Federal Regulations for Landfills*. Report EPA/530-SW-91-092, Washington, DC, 1993.

33. United States Environmental Protection Agency, *Waste Minimization National Plan: Reducing Toxics in Our Nation's Waste*, EPA Solid Waste and Emergency Response, EPA530-F-97-010, Washington, DC, 1997.

34. EPA Archive 2017. https://archive.epa.gov/epawaste/hazard/wastemin/web/html/index .html (accessed April 3, 2018).

35. Prevention of hazardous waste in Europe—the status in 2015, EEA Report No 35/2016, Copenhagen, Denmark.
36. Senkan, S.N., and Stauffer, N.W., What to do with hazardous waste, *Technology Review*, 84, 34, 1987.
37. Harris, R.H., English, C.W., and Highland, J.H., Hazardous waste disposal: Emerging technologies and public policies to reduce public health risks, *Annual Review of Public Health*, 6, 269, 1985.
38. Rose, J., Incineration regulations set to tighten, *Environmental Science and Technology*, 28(12), 512A, 1994.
39. U.S. Congress, Office of Technology Assessment, *Technologies and Management Strategies for Hazardous Waste Control*, OTA-M-197, U.S. Government Printing Office, Washington, DC, 1983.
40. U.S. Congress, Office of Technology Assessment (OTA), *Assessment Superfund Strategy*, OTA-ITE-282-, U.S. Government Printing Office, Washington, DC, 1985.
41. 'Superfund Liability', USEPA, Enforcement, www.epa.gov/enforecment/superfund-liability (accessed July 15, 2018).
42. The Congressional Institute, *Superfund*, July 17, 1, 1996. http://www.conginst.org/cidoc-l.nsf/e74d.,167f730.d9718525636coo79ao93?Open Document.
43. USEPA, Superfund Redevelopment Program, At A Glance, *Approximately 550 Superfund Sites are Ready for Reuse, or Have Already Been Returned to Productive Uses*, Update Marcy, 2006. http://www.epa.gov/superfund/programs/recycle/index2.htm.
44. Zdeb, S., Polluters must pay: Superfund reauthorization, *Green Scissors*. http://www.greenscissors.org/other/superfund.htm (accessed June 2006).
45. Smith, R., Superfund, *Cleaning up the Nation's Hazardous Waste Sites*, USEPA, Superfund, Washington, DC, April 18, 2006. http://www.epa.gov/superfund/.
46. USEPA, Superfund, *Superfund National Accomplishments Summary Fiscal Year 2005, as of November 22, 2005*, last updated on March 1, 2006. http://www.epa.gov/superfund/action/process/numbers05.htm.
47. USEPA Superfund Task Force Recommendations, July 25, 2017. https://www.epa.gov/superfund/superfund-task-force-recommendations (accessed April 3, 2018).
48. Anderson, B., Taxpayer dollars fund most oversight and clean-up costs at Superfund sites, *The Washington Post*, September 20, 2017. www.washingtonpost.com/national/taxpayer-dollars-fund-most-oversight-and-cleanup-costs-at-superfund sites/ (accessed July 12, 2018).

12

Assessing Human Risk

OBJECTIVES

A student reading this chapter will be able to:

1. Define risk and environmental risk

2. Describe several characteristics of risk

3. Define risk assessment and describe the tools used to perform risk assessment

4. Explain the concepts of dose, extrapolation, and acceptable daily intakes (ADI)

5. Outline and discuss the process of risk assessment, including hazard identification, dose-response assessment, exposure assessment, and risk characterization

6. Outline and discuss the major components of risk management and risk communication

---

Introduction

A farmer stands in his field, staring in horrified awe at the stream of black waste flowing onto his land from an adjacent river. This scene occurred on April 29, 1998, when a waste reservoir in Southern Spain burst, releasing toxic sludge into waterways and onto surrounding farmland. The reservoir served as a holding area for zinc, lead, iron, and cadmium wastes from a nearby Canadian-owned zinc mine. The dangerous waste posed both an immediate and long-term threat to the health and livelihoods of locals and to the environment. Crops were ruined and land contaminated by the toxic mess. Locals complained of burning eyes and throats. Nine megagrams of dead fish and shellfish were carted away from the acidic waters. With shock and dismay, people wondered at the odds of this unexpected event. A shift in the earth apparently sent the reservoir wall tumbling, releasing millions of cubic meters of metal waste into the surrounding region. This event typifies the multitude of environmental risks facing the world today. These types of environmental risks are uncertain and unexpected. The consequences can be severe, even deadly.[1]

Risk is the possibility or chance of an unwanted event or outcome.[2] Risk is an integral part of life, permeating every aspect, from eating, breathing, socializing, working, and playing. Some *professions*, such as insurance and finance, are based on risk. The word "risk" can describe a wide range of activities, situations, and concepts, from drinking a glass of red wine daily to skydiving, to unwanted chemical exposure. Risk can imply not only uncertain consequence and danger, but also opportunity and even adventure.[3] In everyday language, the word "risk" is used to describe types of people or situations. For example, the term "risky business" implies shady dealings and a "risk-taker" suggests a brave, perhaps foolhardy individual. "Risk-averse" describes someone who shies away from uncertainty. "Risk-free" describes an activity that contains no uncertainty or negative consequences. A comparison of several human risks is shown in Figure 12.1.[4]

Many of life's risks are voluntary, based on people's personal choices, such as diet and exercise, hobbies like rock climbing and sunbathing, and habits like smoking. We accept many risks in our lives, not only because we enjoy the risks themselves, but also because the risks (i.e., the likelihood) of an adverse event or outcome seem both unlikely and remote. As a consequence, we cross streets, take pharmaceuticals, drive cars, and fly in airplanes, with little thought or concern of safety. We are under the assumption that these activities are safe, or safe enough that we may do them without harm to ourselves.

However, even these everyday activities carry some risk and uncertainty of the outcome. But because these activities are familiar and routine, they do not appear threatening or dangerous.

Other risks involve a more obvious element of danger, but people partake in them regardless, as they believe the benefit outweighs the risk. For instance, people risk money in the stock market in the hopes of making a profit; people live in earthquake- and hurricane-prone areas in exchange for the lovely climate or scenery often found in those regions. Some people seek out extraordinarily high levels of risk, engaging in skydiving, bungee jumping, rock and ice climbing, and other extreme sports (Figure 12.2). People choose these activities because they recognize and even enjoy the element of danger and uncertainty, and the adrenaline "rush," from taking a chance. Their "acceptable" level of risk might be too high for most people but it is their choice.

This chapter examines how scientists and regulators identify the acceptable range of a variety of environmental risks, which are typically chemicals agents we experience in the world around us.

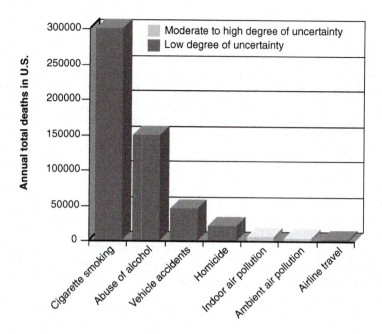

FIGURE 12.1
A comparison of some common risks. (Adapted from Moeller, D.W., *Environmental Health*, revised ed., Harvard University Press, Cambridge, MA, 1997, Chap. 16.[4])

FIGURE 12.2
Extreme sports are becoming popular because of the excitement associated with high risk. (From Corel sampler, limited edition CD-ROM.)

Environmental Risk

Environmental risks are the potential health threats in the world we live—in the air we breathe, the water we drink, the earth in which we grow food, including all the natural and synthetic chemicals that are introduced into these systems, from fertilizer on a field to chlorine in the drinking water to exhaust from vehicles. People are often unaware of the environmental risks they experience, as these risks are often in minute quantities. But even in small quantities, they may still pose health hazards. Therefore, acceptable levels of exposure are often determined by scientists, who quantify and characterize the potential dangers of environmental risks.

The study of environmental risk began humbly as a desire to understand and avoid obvious dangers to human health. Prior to the development of sophisticated scientific testing, knowledge of health effects was limited by what people could observe with the naked eye. For instance, poisons like arsenic were known to have visible and acute adverse effects on health.[5] As scientific tools improved, scientists were able to test for the presence of a specific chemical. Developed in 1836, the Marsh test could indicate the definitive presence of arsenic. This test, the first of its kind, launched the field of forensic toxicology. Over time scientists were able to examine a broader set of chemicals, to study how a chemical affected a living organism, with the intent to understand not only the immediate health effects but also longer-term consequences of exposure. This study developed into a discipline known today as risk assessment. In this chapter, we will look at the tools and components parts of risk assessment, as it is used to understand environmental risk.

Environmental risk is a bleak reality in today's world. There is a seemingly endless supply of chemicals to which we are exposed. Consider that on a daily basis, we may use soap, toothpaste, hair products, cosmetics, pharmaceuticals, and household cleaning products. Beyond those we choose to experience, many environmental risks we face are involuntary. Energy and manufacturing waste products may create new risks through exposure to harmful substances in our air and water. Environmental degradation may even expose us to environmental risks from far away. Across the world, environmental disasters threaten millions of people living in the vicinity of manufacturing or storage facilities but can also threaten people living thousands of kilometers away. The 2011 Fukushima Daiichi nuclear disaster in Japan is an example of an environmental disaster that acutely impacted Japan—150,000 residents had to evacuate and the threat of radioactive contamination persists.[6] But this catastrophic event has had negative environmental impacts across the globe. Most of the radioactive waste, more than 80 percent, was swept into the Pacific Ocean. In 2015, traces of this radioactivity, specifically two isotopes of Cesium, were detected along the western Canadian and United States coastlines.[7] The Fukushima disaster shares similarities with the 1986 Russian Chernobyl disaster. The effects of that nuclear accident are still being experienced today since the surrounding land remains contaminated.

As mentioned previously, a central factor of environmental risk is that it is often involuntary. People do not choose to ingest chemical pollutants such as pesticides, trace pharmaceuticals, or industrial solvents in their food and water; to undergo workplace exposures to dangerous chemicals; to breathe polluted, ozone-laden air; or to experience radiation exposure from nuclear fallout or faulty nuclear power plants. These broad environmental risks pose a unique problem to regulators charged with protecting the public's health. Limited information may be available on the health effects of some of these risks. Consequently, in an effort to protect the public's health, various government agencies study these potential hazards to determine the levels of risk they pose. This effort to understand these risks, and to quantify their impact on human health, is the work of risk assessment.

Risk Characteristics

The two characteristics that most describe risk are uncertainty and unknown. Earlier, we defined risk as the likelihood of an unwanted occurrence.[4] A broader definition is the likelihood of an unwanted occurrence coupled with an element of uncertainty about when the risk might occur.[8] Other definitions of risk exist, all varying slightly, with most focused on those concepts of the uncertainty of a risk and the unknown degree of danger inherent in that risk. Additional risk characteristics include the voluntary or involuntary exposure to a risk, such as lead in a town's drinking water, and the immediacy or latency of a risk's effects. For instance, radiation exposure can lead to immediate effects and/or those that persist through generations. There is the catastrophic potential of risk, as seen during a chemical disaster such as the Bhopal explosion in India, and, the threat to future generations, such as through persistent PCB contamination of soil and water. The field of environmental risk encompasses the ideas of morbidity and mortality to humans, that is, the possibility of sickness and death.[9] Certain risk characteristics, such as catastrophic potential, generate greater fear in people than risks with consequences that are dispersed over time. Consider that in the U.S., many more people die in automobile accidents each year than in plane crashes, but people generally are more afraid to fly than to ride in a car. Assessing human risk through quantitative risk assessment is one method to address fear and anxiety of a poorly understood environmental risk, by increasing knowledge of a risk and its potential health effects and devising methods to limit and manage risk exposure.

Development of Risk Assessment

Risk assessment is the process of reviewing information on a potential health hazard, such as a water contaminant in a factory's effluent or wastewater, to characterize that hazard's impact on human health. This process involves a review of scientific studies to develop an understanding of the properties of a risk, an assessment of levels of human exposure and dose, and a conclusion about the likelihood, impact, and extent of a risk. By employing these methods, risk assessment allows researchers to develop conclusions on the severity and consequences of environmental risks.

Risk assessment developed over the past 30 years in the United States out of a need to understand modern risks. It began with an increased interest in the environment, from the government, scientific, and public sectors. Rachel Carson's 1962 book *Silent Spring* spurred concern about environmental issues. Environmental crises such as Love Canal, Three Mile Island, and the chemical accidents in Seveso, Italy, and Bhopal, India, encouraged research into the effects of these environmental risks. The purpose was to assess the potential health impact of these hazards and to develop methods to cope with and manage these risks and exposures.

The field of risk assessment developed as a response to several changes in the modern world. First, the number of environmental risks society faces today has increased dramatically compared to the past. For example, over 70,000 synthetic chemicals are used today, in industrial and manufacturing processes and in many consumer products and household items. Little detailed risk information is available for most of these substances.[10] Second, technology has improved our ability to measure risk. We can quantify to the parts per billion and trillion the amount of a chemical agent in drinking water. We can measure the dose at which deleterious effects take place in the body and on what systems. Through the continued development of risk assessment, we can study the hazards of consumer goods and products that have been considered healthy or at least harmless. Third, the number of government agencies directly or indirectly monitoring environmental risk have increased, such as the Food and Drug Administration, the Environmental Protection Agency, and state-run Departments of Environmental Protection. Even towns and cities have agencies regulating local risks to public health. Fourth, the number of laws and regulations governing the environment and protecting the public health has increased dramatically over the past thirty years. Some of these laws drastically changed the management of risks and dissemination of information. The Superfund Amendments and Reauthorization Act of 1986 (SARA), also known as the Emergency Planning and Community Right-to-Know Act, ensured that communities were prepared for chemical emergencies and had access to information on chemicals at sites in the area. Other examples of regulations include the Food, Drug and Cosmetic Act, the Toxic Substances Control Act, and the Clean Water Act. Fifth, an increase in public interest in environmental risks has spurred research in quantifying risks. Special interest groups and grassroots movements have involved the public in environmental issues on both local and national levels.[11,12] The development of risk assessment has indeed increased knowledge of environmental hazards and risks associated with our lifestyles. Progress has been made on educating people on issues and outcomes of risk assessment, about food they eat, the air they breathe, the products they use, and even the cars they drive.

Environmental risk transcends geography, as these risks exist all over the world, not restricted by a country's borders, size, or location. Water contamination in the U.S., air pollution in China and Iran, and oil contamination in the Gulf of Mexico and the Niger Delta of Nigeria are all recent environmental crises with significant human health impacts. The lead crisis in Flint, Michigan is a notable example for its impact on children. When high levels of lead were found in Flint's municipal water supply, following the city's switch from Lake Huron to the Flint River as a water source, a public health crisis developed, as local families realized their children had been drinking lead-tainted water for over a year.[13] Lead exposure in children has a detrimental impact on cognitive development. While the Flint problem was isolated, some environmental risks are pervasive and ongoing. An example with broad geographic impact is Chinese air quality. In a 2016 study led by Mr. Song Xi Chen of Peking University, a team reviewed three years of air quality data from China's five largest cities, looking specifically at the levels of $PM_{2.5}$, a type of fine particulate matter linked to health problems.[14] Over 3 years, $PM_{2.5}$ concentrations fell in all five cities. However, although the $PM_{2.5}$ levels fell 20–25 micrograms, the levels remained above the World Health Organization's suggested limit of 35 micrograms per cubic meter. For instance, in Beijing, the 2014 $PM_{2.5}$ level was 99 micrograms per cubic meter and this dropped to 81 microgram per centimeter by 2016.

For better or worse, these types of environmental risks are part of the human experience. Risk assessment can help us understand and manage them. Through risk assessment, a qualitative and quantitative risk description can be made by gathering relevant study information, considering the important factors, and developing conclusions and recommendations about the risk, such as how much of a chemical agent is safe in the public drinking water supply. This risk description can be used in regulatory decision-making and the development of environmental and public health policy. For instance, risk assessment can help to estimate the likelihood of developing cancer from exposure to radiation, to ascertain whether a synthetic food product, such as olestra or saccharin, is safe for human consumption, or to determine if genetically-modified products pose a health risk. Risk assessment allows the public to make informed decisions and weigh the risks and benefits in their community. Further, the public is more aware of environmental issues and impacts because of the outputs of risk assessment.[15]

Tools of Risk Assessment

Risk assessment employs several scientific disciplines in its goal to characterize a risk (Figure 12.3). Significant amounts of information, both quantitative and qualitative, are needed to develop an adequate picture of a risk. Toxicology, epidemiology, clinical trials, and cellular studies provide some of the information for a risk assessment. The types of information reviewed include characteristics of the chemical, mechanism of action in the body and on what systems, type and length of exposure, dose, animal response to exposure at different exposure levels, and human response. These disciplines are discussed in greater detail in the following sections.

Toxicology

Toxicology is the study of chemicals to determine their physiological and health impacts on humans. Risk assessment relies heavily on toxicological studies and conclusions. A particular branch of toxicology, regulatory toxicology, aims at guarding the public from dangerous chemical exposures. For example, the health effects of nicotine, saccharin, and benzene, a gasoline additive, have been characterized through toxicology studies. Toxicological studies involve controlled laboratory studies of mammals. During the course of the study, the animals are exposed to a chemical agent and are evaluated for their responses to different doses of the chemical. These controlled research conditions are crucial for building a health effects profile of the chemical. Prior to a toxicology study,

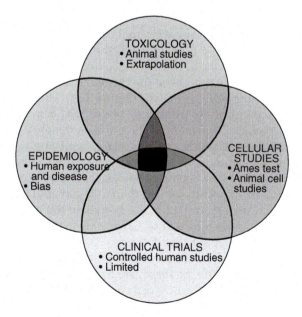

FIGURE 12.3
Tools of risk assessment.

little may be known about the health effects of a chemical, such as what dose level causes minimal adverse health effects, substantial adverse health effects, or even death. In a toxicology study, scientists aim to learn about a substance's activity in a living organism, in terms of carcinogenicity, mutagenicity, reproductive effects, and toxicokinetics of the substance. An initial study will help delineate the lower and upper limits of a chemical's toxic potency, that is, the amount that causes no effect in any animal and the amount at which animals die. After these limits are established, scientists can work within this range to discover additional information about a chemical's health effects. As a chemical may have a myriad of health effects, the researchers will focus their interests on certain health responses. These specific physiological changes are called *endpoints*. Some common endpoints studied include the effective dose of a chemical, named the No Observable Effect Level (NOEL), the No Observable Adverse Effect Level (NOAEL), the Lowest Observable Adverse Effect Level (LOAEL), and death. All of these endpoints provide valuable information on exposure levels and effects of a chemical. This information can be incorporated into a risk assessment.[10,12] Do you want to learn more about risk assessment? Point your browser to <https://people.umass.edu/~envhl565> and click on any of the Chapter 12 Web links associated with risk assessment including:

- Welcome to Center for Applied Environmental Public Health
- USEPA Superfund Risk Assessment
- International Society of Exposure Analysis
- The CRESP Homepage
- The Risk Assessment Information System
- Society for Risk Analysis: Risk Assessment, Characterization, Management, Communication, and Policy

Dose-Response Assessment

Toxicologists use different doses to assess animal responses to chemical exposure. The scientist will ascertain the threshold dose, which is that dose at which no effects are seen in the study animals.

As the administered dose increases above the threshold amount, animals will begin to show adverse effects and some animals will die (Figure 12.4). The dose at which 50 percent of all the test animals die is called the lethal dose or the LD_{50}. Similarly, the ED_{50}, or effective dose, is the dose at which 50 percent of the animals demonstrate a response to the chemical. A very toxic chemical will have a low LD_{50}, signifying a low dose is sufficient to kill 50 percent of the test animals. A less toxic chemical will have a higher LD_{50}. The Maximum Tolerated Dose (MTD) is a common level of chemical exposure in animal studies. At this dose, researchers expect to see less than a 10 percent loss in weight, no death, and the absence of clinical toxicity or pathologic lesions in the animal population. Some critics of the MTD believe it is too high a dose. Chemicals that cause cancer at the MTD in test animals may not cause cancer at lower doses.[15]

In an effort to improve study efficiency and precision, alternative methods were developed to identify the LD_{50}. One of these used globally is the up-and-down (UDP) method.[16] The UDP aims to identify the LD_{50} of a chemical agent using fewer animals than previous toxicology study protocols. In UDP, groups of test animals, between three to five mice or rats are sequentially given an increasing dosage of the chemical. When a dosage kills a certain number of test animals in a dose group, the next test group receives a lower

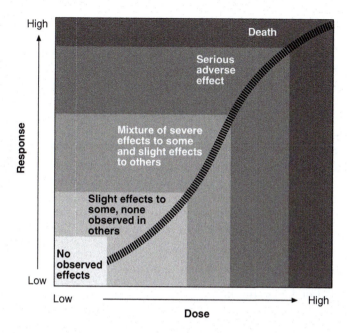

FIGURE 12.4
The dose response curve. (Adapted from Covello, V.T., and Merkhofer, M.W., *Risk Assessment Methods*, Plenum Press, New York, 1993, Chap. 1.[8])

dose. Hence, the name "up-and-down." Statistical simulation then expands the data set from the individual animals' responses to a multitude of hypothetical responses, yielding an LD_{50}. The use of powerful statistical programs allows for using fewer animal endpoints while successfully locating the LD_{50}.

Two other sequential testing methods are the fixed dose method and the acute toxic class method.[16] Using the fixed dose method, groups of five animals are dosed in a sequential fashion from five fixed doses. The endpoint is signs of toxicity. The acute toxic class method also uses the same five fixed doses, but tests groups of three or six animals at each dose. The acute toxic class method uses death as the endpoint. Unlike the UDP, these methods do not indicate the LD_{50}.

Animal studies occur under controlled conditions within a laboratory, ensuring that the test animals receive exact doses of the chemical of interest. By controlling the amount, the timing, and the duration of exposure, scientists can develop a broad understanding of a chemical's health effects. Large doses may cause one type of response; small doses, another. Timing of a dose may also be critical, if exposure is timed to occur during a "window of vulnerability." For example, in a fetus, minuscule amounts of an agent may cause severe damage. Outside this "window," this same amount would be harmless to the fetus. In addition, this control allows for variability in study design. Toxicology studies range in duration from 2 weeks to 2 years and are called Acute, Sub-chronic, and Chronic studies, depending on a study's length.[15]

Short-term, or acute exposure, studies occur over two weeks and involve high doses of the chemical substance. Medium-length, or sub-chronic, studies use lower chemical doses over a longer time period, from 28 to 90 days. Long-term or chronic studies may last up to 2 years and give much smaller chemical doses to the study animal over this longer time period. This range of study duration improves understanding of a chemical's activity over different lengths of time and dosage.[15] To ensure that the changes seen are caused by the chemical, animal studies include a control group not exposed to the chemical agent under consideration. This group provides the researchers a level of certainty that the unique physiological responses seen after exposure to the chemical are in fact due to that exposure.[17,18]

A new technology that places tissue from human organs on a chip about the size of a USB memory stick is a promising new way to simulate an in vivo–like environment and therefore produce a more realistic mock-up of human organs of interest. Microfluidic tubes, each less than a millimeter in diameter, are lined with human cells taken from the organ of interest inside the chip.[19] When nutrients, blood, and test-compounds such as experimental drugs are pumped through the tubes, the cells reproduce some responses similar to a living organ. This methodology could potentially provide models for studying environmental toxicology that provide more realistic responses to toxic chemical exposure. It may even soon be possible to possible to construct an interconnected network of multiple-organs-on-a-chip toward a realistic human-on-a-chip system.[20]

Extrapolation from Animal Studies to Human Exposure

Extrapolation is defined as using known information to conclude that an observed data trend or outcome will continue in the same way.[21] In environmental risk assessment, the term "extrapolation" refers to the use of animal study data to predict human response to exposure. Extrapolation is widely used in toxicology, to deduce the risks humans face from exposure to a chemical agent. Extrapolation is a vital tool of toxicology, since human exposure studies are not widely accepted, for ethical reasons. The most common laboratory animal used in toxicology studies is the rat. During studies, animals generally receive high doses of the chemical of interest for a measured span of time. Laboratory animals have shorter life spans than humans, allowing studies to document a full lifetime of exposure, from infancy to maturity to death. The results from these high-dose studies are used to extrapolate human response to the longer-term, lower-level exposures we generally receive. The difference in dose level and duration of exposure may generate some doubt about the validity of extrapolation. Furthermore, biological and metabolic differences between the test animal and humans may undermine the validity of extrapolations. For instance, the test animal may have one method for detoxifying a substance, whereas a human may use a different metabolic pathway. Alternatively, a test animal may lack an adequate detoxification pathway, whereas humans may be very effective at detoxifying a substance. Chocolate is a well-known example: safe for humans but toxic for dogs. Because of these differences, a substance that causes cancer in one animal may not cause cancer in another animal, or vice versa. This fact limits the accuracy in extrapolating from animal data to human response.[17] Despite these drawbacks, extrapolation is a crucial method in studying potential human outcomes based on the results of animal studies.

In toxicology, methods of extrapolation have changed over time, with new technology and desire to improve on older methods. One criticism of the NOAEL method is that it does not consider the *slope* of the dose-response curve, which reflects the increasing rate of a physiological response at increasing doses of the chemical agent.[22] The steeper the slope of the dose-response curve, the smaller amounts of a chemical it takes to achieve a response at each upward point on the curve. To better reflect the dose-response slope, a method called the Benchmark Dose Method (BMD) is frequently used to extrapolate from animal data to human exposure. This method relies on statistical modeling, which considers the slope of the dose-response curve.

FIGURE 12.5
Extrapolation using the benchmark dose. (Adapted from EPA, *Benchmark Dose Technical Guidance*, 2014.)

Calculating Acceptable Daily Intakes

The purpose of toxicology studies is to establish an acceptable level of human exposure or dose of a substance that is considered "safe." This level, which poses little risk, is termed the acceptable daily intake (ADI). In extrapolating from animals to human studies, scientists incorporate safety factors when determining acceptable exposure (dose) levels. The NOEL established by the threshold dose in test animals is translated into the ADI or the reference dose for humans.[23] This level includes safety factors or uncertainty factors from 10 to 10,000, depending on the protected group. The safety factor is incorporated by dividing the threshold dose by some factor of ten, thereby reducing the acceptable level of exposure and risk from exposure and increasing safety (Figure 12.5). Again, which factor of 10 is used depends on the level of protection sought, the uncertainty inherent in the study data, and who is being protected. Lower safety factors are used when high-quality data is available. Higher safety factors, up to 10,000, are used as the uncertainty level goes up and the quality of the data diminishes. Although the ADIs are based on scientific studies and include a margin of safety, these numbers are not foolproof. As more data becomes available about a substance, the ADIs may change, either lower or higher. For example, as we have gained knowledge about the dangers of exposure to lead, the minimum level of exposure considered harmful to human health has decreased, as adverse health effects are seen at extremely low concentrations of lead, especially in children.[8] The World Health Organization has developed 10 guiding principles to address levels of uncertainty in scientific studies across a broad range of chemicals.[24]

Epidemiology

What Is It?

Epidemiology is "the study of the distribution and determinants of disease frequency in the human population."[14] In an attempt to determine causality between an exposure and a disease, epidemiologists study human response to various environmental agents, such as tobacco smoke, pharmaceuticals, exercise, and even hair dye. These are not controlled laboratory studies, but rather, occur in the outside world, subject to fluctuating conditions.

When an increase in the incidence of a disease is reported, epidemiologists try and account for as many factors that can be related to its occurrence. This was true when high incidences of morbidity were discovered among inner-city children with asthma and it was hypothesized that this could be related to the exposure to cockroach allergen.[25] Children from eight major inner-city areas were studied. It was concluded that this variable alone could not account for this situation. Many variables are introduced into these studies, such as personal choice, socioeconomic factors, age, demographics, and other factors. These factors, called confounders, can bias or skew the results of the study either towards more of an association than actually exists or towards less of an association than actually exists, depending on the confounding factor. Consequently, in designing a study, scientists will try to "control" for these confounding factors.[26]

Study Types

Several different types of epidemiological studies exist, including cross-sectional, cohort, and case-control studies. These study types differ in design by the methods categorization of study subjects and the timing of information gathering. Each type of study has its own strengths and weaknesses in finding an association between an exposure and a disease. A cross-sectional study is a type of descriptive study design in which both a subject's exposure and disease are determined at the same time. This type of study provides general information on the disease, such as who and where. Cohort studies and case-control studies are types of analytic studies, focused more on answering the question of why a particular disease develops. Cohort studies group people by exposure and then follow them over time for disease development. This type of prospective study is useful for studying rare exposures. Case-control studies group people by disease and then look retrospectively for a history of exposure. This type of study is useful for studying rare diseases. All of these study types provide valuable information on disease occurrence in human populations.[26,27]

Bias

Epidemiology involves the study of humans. Consequently, it avoids the extrapolation problem seen in animal studies. However, it is not without drawbacks. A significant problem in epidemiology is bias. Bias is a type of error introduced into epidemiological studies through differential treatment of the cases and controls in a study, either during the selection process or information gathering. Selection bias, information bias, and non-differential misclassification bias may lead to inaccurate study results. Researchers can address the problem of bias through careful study design. Despite this and other limitations, epidemiology remains an important tool for evaluating the link between exposure and disease.[26,27]

Clinical Trials

Clinical trials provide a direct means to evaluate human response to a substance under more controlled conditions than epidemiology studies. This type of risk assessment provides detailed information on dose-response, toxicity, efficacy, and side effects. In clinical drug trials, individuals volunteer to be exposed to a substance or to ingest a drug and are assessed for their health response. The types of health responses measured are generally minor, short-term, and reversible, since testing human subjects involves ethical consideration and the trials are limited in duration. Volunteers must be informed of the risks, must give their consent, and may receive monetary compensation for their participation. Despite the volunteer status of subjects in human clinical trials, dissent exists about the ethics of human studies. Past studies on prisoners were halted due to ethical considerations. Recent controversy erupted when several pesticide manufacturers announced human testing of pesticides to determine safety. The companies resorted to human trials to demonstrate the safety of their products in the face of strict food quality legislation passed in 1996. Opponents felt the companies were taking advantage of volunteers, enticing them with significant monetary compensation, downplaying the inherent uncertainty of the exposure and misleading them by calling pesticides "drugs." Clinical trials are used widely by pharmaceutical companies prior to releasing drugs to the public. Although they are limited in scope, depth, and duration, clinical human trials yield crucial information on humans.[28,29] Clinical trials provide us with information on the safety and efficacy of drugs. Human trials can provide great risk to participants but they also offer a great service to humanity. The promise of drugs that may eradicate the flu, the AIDS virus, or the common cold cause a stir at the mere thought of the possibilities and lives that may be saved. The National Cancer Institute investigated the drug Tamoxifen in 1992 as a possible cure for previously treated patients and preventative agent for breast cancer. The drug was conducted as a placebo-controlled study on 13,388 healthy women.[30] Half of the women were given tamoxifen and the other half were given the placebo, with varying results. The clinical trial found that there were very satisfactory results from the study with fewer cases of breast cancer than the placebo group of women; however, the women who took the drug had two to three times the incidence rate of uterine cancer. Only through clinical trials can we ascertain the health hazards associated with drugs that may be helpful and harmful. Their importance is crucial in the discovery of new drugs and their effect on the human body, both long and short term.[31]

Cellular Testing

Cellular studies are used to test chemicals for mutagenicity and carcinogenicity. These tests can be done on bacteria or animal cells. A widely-known bacterial test is the Ames test, which uses *Salmonella* bacteria lacking the ability to produce histidine, a vital growth chemical. These bacteria are exposed to the chemical in question and assessed after two to three days for signs of growth. Indications of growth suggest a mutation has occurred in the bacteria's genetic information, conferring the ability to produce histidine. Although relatively inexpensive and quick, cellular studies are not always accurate in identifying carcinogens. This problem relates to the difficulty in using single-celled organisms to predict human response to a substance. Other cellular studies include animal cellular studies, which are used to identify potential carcinogens. In these tests, the test chemical is applied to rodent or hamster cells. After a one-to-three-week period, these cells are assessed for changes in growth. Unlike the Ames test, this type of study is fairly accurate in predicting carcinogenesis.[15]

The Process of Risk Assessment

With this arsenal of tools, the process of risk assessment can begin to identify and enumerate the risk of a particular hazard. This process has four steps: hazard identification, dose-response evaluation, exposure assessment, and risk characterization (Figure 12.6). These four steps create a picture of the risk through qualitative and quantitative evaluations.

FIGURE 12.6
Schematic of risk assessment process.

Hazard Identification

The initial step in risk assessment, hazard identification, involves identifying chemicals that present a risk to human health. This step entails performing a qualitative assessment of a chemical's potential for negative health impacts on humans. A thorough evaluation of scientific research about the risk determines the likelihood of adverse effects on humans. A chemical's activity in living organisms is researched through a review of animal studies, cellular studies, epidemiological studies, and human trials. In addition, to ascertain the possibility of human exposure, a substance's boiling point, volatility, and solubility are considered. If a chemical has been identified as a potential threat to human health, the process of risk assessment continues.[4,32]

Dose-Response Evaluation

The next step, dose-response evaluation, provides a quantitative view of the risk from a chemical agent. This step also involves a review of scientific studies and data. In this case, as we learned in the toxicology discussion, the magnitude of health response is correlated with the dose. The health response is reviewed at each dose to note the differences in response. This correlation provides valuable information about levels of risk at different levels of exposure. In addition, available epidemiological studies are assessed to determine the agent's strength, or potency, and capacity to produce negative health effects in humans. Although human studies are preferable for this evaluation, such data is not always available. In such an instance animal data is used.[4,32]

Exposure Assessment

After the dose-response assessment, an exposure assessment is performed. The purpose of the exposure assessment is to measure or estimate people's level of exposure. To accomplish this, the length of exposure, the amount of exposure, the route of exposure, the duration of exposure, and the number of people exposed are considered. Exposure is different from dose in that exposure refers to the amount of a substance in the environment, while dose is the level of a substance actually taken in by an organism. Dose can be influenced by many factors, such as how the substance enters the body, whether absorbed through the skin, ingested with food, or inhaled (Figure 12.7). This discrepancy between the amount of a substance measured in the environment and the actual dose is included in an exposure assessment. To perform an exposure assessment, specific information must be considered. The properties of the substance, such as its volatility, water-solubility, and ability to become airborne or migrate through soil will influence its relationship with the

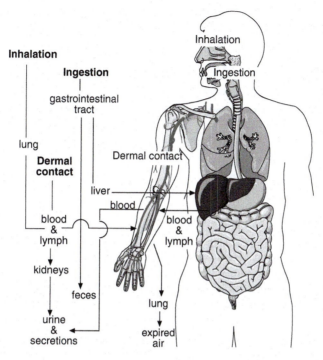

FIGURE 12.7
Routes of chemical exposure. (Adapted from Cohrssen, J.J. and Covello, V.T., *Risk Analysis: A Guide to Principles and Methods for Analyzing Health and Environmental Risks*, The Council on Environmental Quality, Washington, DC, 1989, Chap. 3.[17])

environment and affect its routes to human exposure. Besides considering the substance's properties, the assessment should specify the group exposed, with special attention given to those particularly vulnerable, such as pregnant women, children, the elderly, and the health-compromised. Finally, the type of exposure must be considered, whether continuous, intermittent, short term, long term, or chronic.[4,12]

Risk Characterization

The final step of risk assessment is risk characterization. Risk characterization is the development of a comprehensive picture of the risk that includes its severity, likelihood of exposure, and consequences. All of the previous information from the hazard identification, the dose-response evaluation, and the exposure assessment are incorporated into the risk characterization. The risk characterization includes an estimate of the negative effects to exposed individuals, such as the number of expected cases of cancer or deaths per 100,000 people (Figure 12.8). Other types of risk estimates include an individual lifetime risk, such as a one-in-a-million chance of cancer, a loss of life expectancy, as in days or years lost from exposure to a substance, and relative risk, that is, the risk of the exposed population versus the unexposed. Probabilities of these effects are also provided. Any drawbacks, limitations, and uncertainties inherent in the estimate are incorporated into the risk characterization.[4,15]

Deaths per million people in a selected population (nation).

Deaths per million people within a certain radius of the released pollutant.

Deaths per weight of the toxic substance(s) released.

Loss of life expectancy associated with exposure to the toxic material(s).

FIGURE 12.8
Types of risk estimates. (Adapted from Stern, P.C. and Fineberg, H.V., *Understanding Risk*, National Academy Press, Washington, DC., p. 50, 1996.[33])

Limitations of Risk Assessment

The conclusions drawn from a risk assessment are not foolproof. Limitations include uncertainty, variability, and effect of multiple exposures. Uncertainty exists in all estimates and recommendations about environmental risks. Generally, "zero risk" is not a possibility, due to high degrees of uncertainty involved in risk assessment. Although people want to be assured about a safe level of exposure to a chemical, sometimes too little is known to about the substance to provide any real assurance. A frightening example is represented by pharmaceuticals pulled from the market. Although a drug may undergo years of laboratory testing and even human trials, sometimes, rare side effects are not seen until the drug is approved and subsequently sold to the public. It is then generally used by a much larger number of people than in the clinical trials. An increased number of users broadens the exposed population, increasing the likelihood of seeing rare side effects within that larger and more varied population. This scenario occurred with the popular diet drug, fen-phen, and the fast-acting analgesic, Duract. In the case of both drugs, dangerous and even deadly side effects appeared in human users shortly after the drugs were available by prescription. Both drugs were subsequently taken off the market.[33] Even with rigorous laboratory testing and careful risk assessment, a degree of uncertainty remains. This uncertainty factor can erode the purpose of recommended doses, as people may doubt that any safe exposure level exists. In addition, interpersonal variability may strongly affect an individual's risks. Some people in a population are simply more sensitive to exposures than others and may be more likely to experience adverse health effects from a "safe" dose. Therefore, the built-in safety factors may not be enough to protect all individuals from harm. In addition, risk assessment rarely considers the fact that people are exposed to multiple risks at the same time. Rather, risk assessment focuses on exposure to a single agent. Any synergistic or additive effects are not addressed, potentially providing an incomplete picture of the risk of exposure. Despite these limitations, risk assessment remains a valuable tool for exploring and understanding the risks of the world.[10]

Risk assessment is rarely a static process. Rather, it may be iterative as research continues, improved technology yields new insight, or perhaps the regulation of a risk changes. Further, the output of one risk assessment may suggest the need for further study or modifications as new health risk data comes to light. Therefore, it may be helpful to view risk assessment as part of a larger and ongoing problem-solving process, which starts with planning and leads to an informed decision about a risk (Figure 12.9).

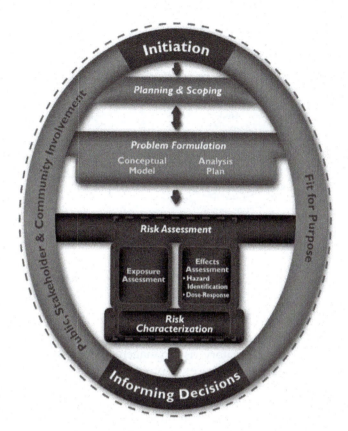

FIGURE 12.9

Health risk assessment. *Framework for Human Health Risk Assessment to Inform Decision Making*, EPA, April 5, 2014.

Risk Management and Communication

Risk Management

Risk assessment provides us with information about risks from chemical agents. The next step is to share that information. This is accomplished through risk management and risk communication. Risk management involves merging the results of risk assessment with various social factors, such as socioeconomic conditions, political pressures, and economic concerns (Figure 12.10). Three types of risk management are educational, economic, and regulatory.[34] These are not mutually exclusive but can be used together to manage a risk. Educational risk management may include use of the media to inform the public about a risk, such as the strategic placement of warnings, as on drugs and other chemical products. Economic risk management is the use of pollution taxes and permits that support risk reduction through monetary incentives or disincentives. Regulatory risk management, also called "command and control," involves the use of the law to decrease risk, for example the numerous laws governing potential environmental and health hazards in the

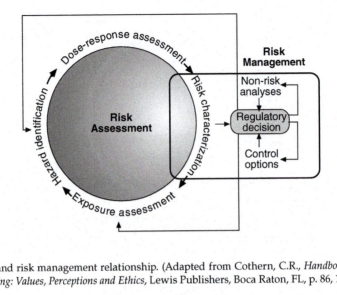

FIGURE 12.10
Risk assessment and risk management relationship. (Adapted from Cothern, C.R., *Handbook for Environmental Risk Decision Making: Values, Perceptions and Ethics*, Lewis Publishers, Boca Raton, FL, p. 86, 1996.[35])

United States. The Clean Water Act, the Safe Drinking Water Act, and the Toxic Substances Control Act are examples of regulatory risk management, where laws restrict and limit pollution, chemical exposures, and releases.[35] Another example of risk management is the use of warning labels, symbols, and images on products that may carry a human health risk, such as warnings seen on cigarettes and on strongly alkaline and acidic consumer products, including sodium hypochlorite, ammonia, and hydrochloric acid.

Risk management generally involves weighing benefits and costs, comparing the cost of reducing the risk with the benefit gained from risk reduction (Figure 12.11). Several analytical methods include cost-benefit analysis, risk-benefit analysis, and contingent analysis. These methods allow regulators and agencies to determine the "best" course of action for managing or controlling a risk, given available funding and the degree of a risk. Economics are a critical factor in risk management. Consequently, the "best" course of action is not always the one that reduces the most risk, but rather, that which is the most economically feasible option, which reduces the greatest amount of risk per dollar spent.

Risk management decisions do not solve all problems surrounding a risk or satisfy everyone's concerns: a decision that benefits one group may be unsatisfactory to another. An example is the process of selecting the location of an ash dump. A risk management decision is made to site the dump in a sparsely populated, rural area. This decision solves the problem of where to dispose of incinerated waste. However, most residents of the rural community receive little or no benefit from the dump. Instead, they face the new risk of land and water pollution from the ash, a decrease in aesthetic value of their land from the existence of the dump, and the noise of waste trucks through a previously low-traffic area.[36]

Risk management also encompasses the idea of an "acceptable" level of risk and an acceptable number of cases. We cannot eliminate all risk, but we can hopefully reduce risk to some acceptable level. This idea of an "acceptable number" of cancer cases or an "acceptable number" of deaths is an uncomfortable but inherent aspect of risk management. Clearly, risk management is a complex undertaking that requires careful consideration of the risk characterization, political, economic, and social factors.[37,38] Unfortunately, sometimes the multi-level process of risk management may result in indecision, delayed action, or no

Risk-reducing regulations

FIGURE 12.11
Comparison costs of risk management. (Adapted from Viscusi, W.K., *Rational Risk Policy*, Clarendon Press, Oxford, UK, p. 98, 1998.[36])

action due to the absence of funding or stakeholder agreement. This has been the case in the Niger Delta oil spill, where cleanup has only now begun after a 10-year delay.[39]

Risk Communication

Risk communication is a dialogue or information-sharing between any two or more parties involved in a risk. These parties may be federal, state, or local government agencies, industry stakeholders, and citizen groups. Methods of risk communication include public hearings, websites, emergency telephone hotlines, and information pamphlets. Internationally agreed-upon symbols may communicate environmental risk with images, rather than words, to overcome language barriers (Figure 12.12). The goal of risk communication is to effectively relay risk information developed through risk assessment to various interested groups. Risk communication can be challenging, as it requires addressing people's different risk perceptions, biases, scientific knowledge, educational backgrounds, even race and gender. Many of these areas require significant sensitivity and tact. Not all risk communication is effective. Certain factors are required for effective risk communication. Most importantly, trust and credibility are needed. As trust and credibility are influenced by the related factors of openness and honesty, knowledge and expertise, and concern and care, these factors also influence the success of risk communication.[40] The public often mistrusts industry, so industry must act to earn the public trust. On the other hand, industry often perceives the public as lacking scientific knowledge, so the public must strive for a clear understanding of scientific principles. In addition to requiring trust, risk communication must often address differences in perception and meanings. Successfully addressing these differences allows for open, effective communication among parties subject to the risk.[34]

FIGURE 12.12
Corrosive Substance Warning Label. *A Guide to The Globally Harmonized System of Classification and Labeling of Chemicals (GHS)*, UNECE, 2003.

Other obstacles that hamper effective risk communication are the scientific terms and expressions used to discuss and describe risks. The general public's knowledge and an expert's understanding of a risk may differ significantly, impacting each person's interpretation.[41] For example, the phrase "one in a million" can be interpreted in two opposite ways. The expert states the statistic to address society's risk as a whole. However, the individual, interested in his own chance of developing cancer, applies the probability to himself and attempts to determine his chance of being that "one in a million." In addition, scientific language with which the expert is comfortable may be foreign and meaningless to a layperson. A term such as "parts per billion" is difficult to interpret and may render any risk information useless to someone who cannot relate to the expression. Translating terms such as "parts per billion" into comprehensible and applicable measurements, such as 1 second in 6.4 months, or 3.96 centimeters in 40,233 kilometers or one-half drop of liquid in 64,000 liters, can facilitate comprehension of scientific language and increase risk communication (Figure 12.13). Risk communication involves building bridges among opposing groups, so that all parties can

TIME: One second in 6.4 months.

LENGTH: 3.96 centimeters in 40,233 kilometers (the circumference of the earth at the equator).

AREA: One sq. yard in 839 square kilometers.

VOLUME: Approximately a half drop of water in a 64,000-liter pool.

WEiGHT: A particle of dust (<0.01 grams) on a 907-kilogram automobile).

FIGURE 12.13
Parts per billion translated into comprehensible terminology. (Adapted from Peters, P.G. et al., *Risk Anal.*, 17, 43, 1997.[40])

understand the relevant concepts and issues, although they may not agree on them. Risk communication is an essential aspect to managing environmental risks today. In 2017, the World Health Organization released its first-ever guidance document on risk communication in public health emergencies.[42]

References

1. Simons, M., Big sludge spill poisons land in southern Spain, *The New York Times,* May 2, 1998, p. A3.
2. Oxford Living Dictionary, Oxford University Press, 2018, www.en.oxforddictionaries.com (accessed January 8, 2018).
3. Jardine, C.G. and Hrudey, S.E., Mixed messages in risk communication, *Risk Analysis,* 17, 1977, Chap. 1.
4. Moeller, D.W., *Environmental Health,* revised ed., Harvard University Press, Cambridge, MA, 1997, Chap. 16.
5. Lane, R.L., The Wissenschaften of toxicology: Harming and helping through time, A.W. Hayes and C.L. Kruger (Eds.), Hayes' *Principles and Methods of Toxicology,* 6th ed., Taylor & Francis Group, Raven Press: New YorK, 2014, Chap. 1.
6. O'Conner, T., *Fukushima Nuclear Disaster: Radioactive Water May Have Been Leaking from Reactors for Months,* Newsweek, September 29, 2017.
7. Fukushima Site Still Leaking After Five Years, Research Shows, Woods Hole Oceanographic Institute News Release, March 7 ,2016, Woods Hole, MA, www.whoi.edu/news-release/fukushima-site-still-leaking (accessed January 10, 2018).
8. Covello, V.T., and Merkhofer, M.W. *Risk Assessment Methods,* Plenum Press, New York, 1993, Chap. 1.
9. Derby, S.L. and Keeney, R.-L., Understanding "How safe is safe enough?" *Risk Analysis,* 1, 217, 1981.
10. Beck, B.X., Rudel, R., and Calabrese, E., *The Use of Toxicology in the Regulatory Process,* Principles and Methods of Toxicology, A.W. Hayes (Ed.), Raven Press, Sacramento, CA, 1994, Chap. 1.
11. U.S. Department of Health and Human Services, Task Force on Health Assessment, *Determining Risks to Health,* Auburn House Publishing, Dover, MA, 1986, pp. 3–4.
12. Halbeib, W.T., *Emergency Planning and Community Right-to-Know Act,* Environmental Law Handbook, 14th ed., T.E.R. Sullivan (Ed.), Government Institutes, Rockville, MD, 1997, p. 481.
13. Milman, O., Flint water crisis: EPA concerned with 'inadequate' state and city response, *The Guardian,* February 20, 2016.
14. Tatlow, D.K., China air quality study has good news and bad news, *New York Times,* March 30, 2016.
15. Malachowski, M.J., *Health Effects of Toxic Substances,* Government Institutes, Rockville, MD, 1995, Chap. 4.
16. Rispin, A., Farrar, D., Margosches, E. et al. Alternative methods for the median lethal dose (LD50) test: The up-and-down procedure for acute oral toxicity, *ILAR Journal,* 43(4), 233–243, 2002.
17. Cohrssen, J.J., and Covello, V.T., *Risk Analysis: A Guide to Principles and Methods for Analyzing Health and Environmental Risks,* The Council on Environmental Quality, Washington, DC, 1989, Chap. 3.
18. Parasuraman, S., Toxicological screening, *Journal of Pharmacology and Pharmacotherapeutics,* 2(2), 74–79, 2011.
19. Al-Rodhan, N. Organs-on-chips allow new views of human biology, *Scientific American,* June 23, 2016.

20. Soohee, C., and Yoon, J.-Y., Organ-on-a-chip for assessing environmental toxicants, *Current Opinion in Biotechnology*, 45, 34–42, 2017.
21. *The Merriam Webster Dictionary*, Merriam-Webster, Springfield, MA, 1995, p. 183.
22. Davis, J.A., Gift, J.S., and Zhao, Q.J., Introduction to benchmark dose methods and U.S. EPA's benchmark dose software (BMDS) version 2.1.1. *Toxicology and Applied Pharmacology*, 254(2), 181–191, 2011.
23. National Research Council (US) Subcommittee on Chronic Reference Doses for Selected Chemical Warfare Agents, *Review of the U.S. Army's Health Risk Assessments for Oral Exposure to Six Chemical-Warfare Agents*, National Academies Press, Washington, DC, 1999.
24. World Health Organization, *Guidance Document on Characterizing and Communicating Uncertainty in Exposure Assessment*, Geneva, Switzerland, 2008.
25. Rosenstreich, D.L., Eggleston, P., Kattan, M. et al., The role of cockroach allergy and exposure to cockroach allergen in causing morbidity among inner-city children with asthma, *The New England Journal of Medicine*, 336, 1356–1363, 1997.
26. Hennekens, C.H., and Buring, J.E., *Epidemiology in Medicine*, Little Brown and Company, Boston, MA, 1987, Chap. 2.
27. Schettler, T., Solomon, G., Burns, P., and Valenti, M., *Generations at Risk: How Environmental Toxins May Affect Reproductive Health in Massachusetts*, MASSPIRG, Martin/VanderLoop Associates, Cambridge, MA, 1996, Chap. 3.
28. Hennekens, C.H., and Buring, J.E., *Epidemiology in Medicine*, Little Brown and Company, Boston, MA, 1987, Chap. 1.
29. Stecidow, S., New food quality act has pesticide makers doing human testing, *Wall Street Journal*, 28, p. Al, 1998.
30. Pappert, A. "*TamoxiSpin*" 39–42. October/November 1999.
31. U.S. Environmental Protection Agency, *Risk Assessment*, Government Printing Office, Washington, DC, 1992, p. 3.
32. Stern, P.C., and Fineberg, H.X., *Understanding Risk*, National Academy Press, Washington, DC, 1996, p. 50.
33. Sharpe, R., 20 Painkillers: How a drug approved by the FDA turned into a lethal failure, *Wall Street Journal*, 1998, p. Al.
34. Cothern, C.R., *Handbook for Environmental Risk Decision Making: Values, Perceptions and Ethics*, Lewis Publishers, New York, 1996, p. 86.
35. Somers, E., Perspectives on risk management, *Risk Analysis*, 15, 677, 1995.
36. Viscusi, W.K., *Rational Risk Policy*, Clarendon Press, Oxford, UK, 1998, p. 98.
37. Morgan, M.T., *Environmental Health*, Revised ed., Morton Publishing Co., Englewood, CO, 1997, p. 279.
38. Hallenbeck, W.H., *Quantitative Risk Assessment for Environmental and Occupational Health*, 2nd ed., Lewis Publishers, Ann Arbor, MI, 1993, p. 2.
39. Idris, A., *Nigeria oil spill: Shell begins clean-up after n10-year delay*, Al-Jazeera.com, October 31, 2017 (accessed January 10, 2018).
40. Peters, P.-G., Covello, V.T., and McCallum, D.B., The determinants of trust and credibility in environmental risk communication: An empirical study, *Risk Analysis*, 17, 43, 1997.
41. DiNardi, S.P., *Calculation Methods for Industrial Hygiene*, Van Nostrand Reinhold, New York, 1995, p. 76.
42. WHO, *Communicating Risk in Public Health Emergencies, A WHO Guideline for Emergency Risk Communication (ERC) policy and practice*, World Health Organization, Geneva, Switzerland, January 10, 2017.

13

Environmental Laws and Compliance

OBJECTIVES

A student reading this chapter will be able to:

1. Discuss how a law is made and describe the system of environmental laws.
2. List the major components of the major federal environmental laws including: RCRA, CERCLA, EPCRA, SARA Title IIII, Pollution Prevention Act, CAA, CWA, SDWA, stormwater regulations, pesticide regulations, and underground storage regulations.
3. Describe the major components of environmental compliance.

Introduction

Environmental laws are a maze of confusing and overlapping regulations that now cover several thousand pages of federal documents. It is not within the purpose or capability of this brief chapter to clarify or detail all of the environmental regulations, but to outline some of the major environmental legal initiatives and compliance issues. In pursuing this objective, the following areas will be covered:

- Environmental laws–some fundamentals
- Resource Conservation and Recovery Act (RCRA)
- Comprehensive Environmental Response, Compensation, and Liability Act (CERCLA)

- Emergency Planning and Community Right-to-Know (SARA Title 111), transportation of hazardous materials and wastes
- Pollution Prevention and Improved Waste Management Programs, regulation of underground storage tanks
- Pesticide Regulation
- Air quality control
- Water Quality Control and Stormwater Management, Safe Drinking Water Acts

Environmental Laws—Some Fundamentals

The Making of a Law

It may take several years for an environmental law to be passed at the federal level. Every federal law starts as a bill and passes through a complex system of checks and balances in the Congress. Checks and balances were originally created by the framers of the Constitution as a way of dispersing power and preventing tyrants from abusing that power. The bill is first introduced into the House. If it is passed, it goes to the Senate. These bills are then forwarded by the leadership to a standing committee (subcommittee) for review and support. Nearly 90 percent of bills never make it through this process. Recommended bills are brought forward through hearings for comments and opinions. The committee then meet to mark up (discuss and amend) the bill and to vote on it. If the subcommittee and the parent committee in the Senate vote favorably for the bill, it is sent (reported) to the full chamber for debate and a vote. Before a bill is sent to the House, it goes first to the Rules Committee where a time limit is set for debate and it also indicates whether floor amendments are allowed. This is not so in the Senate where riders are often attached to popular bills that may have no relationship to the original bill, but often are passed along with it, forcing the president to accept it or veto the entire bill.[1]

If both the Senate and the House pass a bill with some differences (as is usually the case), then they go to a conference committee to resolve these differences. Should these differences be resolved, and the same version approved by both House and Senate, the bill moves forward to the president who may sign it into law or veto it. Although Congress may override a veto by a two-thirds majority vote in each chamber, it is very difficult to do so.

Environmental Laws Are Part of a System

Environmental law is more than a collection of unrelated regulations; it is a system of statutes, regulations, executive orders, factual conclusions, case-specific interpretations, and accepted environmental principles that have evolved into "Environmental Law."[2] Understanding the unifying principles of this law system is both a challenge and the key to understanding its complexity. Environmental law is a system that uses all of the laws in our legal system to minimize, prevent, punish, or remedy the consequences of actions that damage or threaten the environment and public health and safety. Environmental law encompasses all the environmental protections that originate from (1) the United States Constitution and state constitutions; (2) federal, state, and local statutes; (3) regulations published by federal, state, and local agencies; (4) presidential executive orders; (5) court decisions interpreting these laws and regulations; and (6) the common law.[2]

Virtually all of the environmental laws require some form of compliance that falls within one or more of the following areas:[2]

1. Notification requirements
2. Discharge or waste controls
3. Process controls and pollution prevention
4. Product controls
5. Regulation of activities
6. Safe transportation requirements
7. Response and remediation laws
8. Compensation requirements

Notification requirements are meant to deal with any intended or accidental releases of pollutants or hazardous wastes that need to be immediately reported to local and state and/ or federal authorities. Failure to do so can significantly impact the environment, endanger public health, and cause significant liability to the persons associated with the release. Discharge or waste controls are regulations designed to prevent or minimize the discharge, release, or disposal of wastes and pollutants into the environment. *Process controls and pollution prevention* are regulations that promote minimizing waste generation, reducing the quantities produced and released and preventing the release or discharge of pollutants into the environment. *Product controls* can reduce the generation and disposal of solid waste by altering the design or packaging of a product, along with the process by which it may be produced. Some regulations promote the use of less hazardous materials and reduced packaging to achieve these goals. The *regulation of activities* that threaten natural habitat, fragile ecosystems, and vital resources may be necessary to provide protection for resources, endangered species, or habitat. Some of those activities include mining, harvesting of lumber, oil refining, and construction. *Safe transportation requirements* recognize that accidents and spills along our highways occur at an enormous rate, and laws have been established to reduce the risks associated with the transport of hazardous materials. These laws provide for special packaging, labeling, placarding, operator training, and methods of response. *Response and remediation laws* have been promulgated that regulate the cleanup of pollutants and hazardous wastes released to the environment. Laws such as CERCLA have specific provisions for the payment of cleanup costs. These same laws provide for *compensation requirements* requiring responsible parties to pay for cleanup costs for damages done to the health or environment of the private sector. Compensation for damages done to public assets may be recovered by representatives of the "public interest."[1,2]

The system of environmental law includes federal statutes, executive orders, state laws, tax laws, local and municipal laws, business and regulatory law, environmental law and judicial decisions, common law, and torts.

Federal statutes include such regulations as the Federal Clean Air Act (CAA) or Clean Water Act (CWA) that establish federal and state regulatory programs that allow states to enact and enforce laws that meet federal minimum standards and to achieve the regulatory objectives established by Congress. Most states have taken over regulatory programs as allowed under federal regulations and are often more stringent than the federal counterpart. These states are subject to federal intervention only when they fail to enforce the regulations. *Executive orders* are issued by the president and require federal facilities to provide leadership in protecting the environment and to comply with federal statutes and policies regarding the protection of the environment. They were first used in 1970 and include more than 18 executive orders including such areas as federal compliance with right-to-know laws and pollution prevention requirements, to federal actions that address environmental justice in minority populations and low-income populations.[3]

Tax laws are being used to create incentives for environmentally safe products and activities and disincentives against environmentally detrimental products and activities. Among the approaches that have been adopted or seriously considered are recycling tax credits, taxes on use of virgin materials, taxes on hazardous waste generation, and excise taxes on various products. Many state attorney generals use their business regulatory authorities to control environmental claims made for products. Full disclosure of environmental liabilities in statements and reports has been required by the Securities and Exchange Commission for some time. Local and municipal laws include zoning and noise control ordinances, nuisance laws, air emission requirements, landfill restrictions or closures, local emergency planning, and product recycling incentives. These are not trivial ordinances and regulations and must be incorporated into any compliance program. Such laws have been effectively used to control the operation and location of facilities.[2] Court interpretations of these laws and regulations are dynamic and changing, and therefore, judicial decisions have important impacts on the manner in which environmental laws are applied. The environmental system of laws also incorporates *common law*, which is a body of rules and principles that pertain to the government and the security of persons and property. These basic rules were originally developed in England and were then brought to the American colonies. After the American Revolution, these basic rules were formally adopted and enforced by the states.[2] Civil suits, in which the plaintiff seeks to remedy a violation of his rights, are common law actions. Under common law, *"tort"* is a private wrong or wrongful act for which the injured party can bring forth a civil action. There is a general legal duty and responsibility to avoid causing harm to others by acts of omission or commission. When a person's rights have been violated by acts of carelessness, the injured party may seek compensation or restitution by means of a lawsuit. The 1990s have been called the era of "toxic torts" due to the filing of tens of thousands of lawsuits involving asbestos and toxic chemical litigation cases. The three most common types of torts encountered in environmental law include:

1. *Nuisance*: Generally, persons may use their land or personal property as they see fit with the limitation that the owners use their property in a reasonable manner. Whenever persons use their property to cause material injury or annoyance to their neighbor, their actions constitute a nuisance.

2. *Trespass*: An invasion of another's rights is the general definition of trespass. A more limiting definition of trespass is an injury to the person, property, or rights of another immediately resulting from an unlawful act. Unlawful intent is not necessary to constitute trespass.

3. *Negligence*: This is the portion of the tort law that deals with acts not intended to inflict injury. A case becomes one of criminal law if an intent to inflict injury exists. The degree of care and caution that would be taken or used by an ordinary person in similar circumstances is defined as the standard of care required by law in negligence cases. A defendant is liable if his actions cause injury and without those actions the end result would not have happened. This involves a natural and continuous sequence of actions, unbroken by intervention. Persons harmed by the careless and improper handling or disposal of hazardous wastes can file suit to recover damages for their losses even if there has been full compliance with all government regulations and permit conditions.[2]

Do you want to learn more about environmental laws and compliance? Point your browser to <https://people.umass.edu/~envhl565/> and click on any of the Chapter 13 Web links associated with this topic:

- EPA Introduction to Laws and Regulation
- USEPA Compliance and Enforcement
- EPA regulations
- The Environmental Law Institute

Federal Environmental Laws

Managing Hazardous Waste

Hazardous waste management has become a crucial issue. Human health has been threatened and water supplies have been polluted by the improper handling, release, and disposal of hazardous waste. Most hazardous waste is banned from community landfills. Just a few commercial hazardous waste disposal facilities exist, and their total capacity is limited. Liability and costs of waste disposal are increasing. Federal regulations require a cradle-to-grave approach for managing hazardous waste. These regulations establish a system to determine which wastes are hazardous, track wastes from generation to disposal, operate disposal and temporary storage facilities, and set up state hazardous waste programs.[4] An environmentally sound and economic approach to meeting legal requirements for hazardous waste management is reducing the amount of hazardous waste generated by the facility. Hazardous waste minimization is now the law. An integrated hazardous waste management program must be implemented with initiatives for reducing the use of hazardous waste in order to protect the facility and its employees.[5]

Resource Conservation and Recovery Act

The federal hazardous waste laws evolved after a series of amendments to the Solid Waste Disposal Act (SWDA) of 1965. The RCRA was passed as an amendment in 1976 to the SWDA originally promulgated in 1965. General guidance in meeting hazardous waste requirements was provided to USEPA by RCRA. However, USEPA did not issue key hazardous waste regulations until May 1980. The discovery of hazardous waste problems at Love Canal, New York in 1979/80, and at many other locations thereafter, increased pressures for faster regulatory action.[6] The foundation for RCRA waste management regulations can be divided into three distinct programs or subchapters.

1. Subchapter C establishes a system for controlling hazardous waste from the point of generation to the ultimate disposal.
2. Subchapter D establishes systems for controlling solid or non-hazardous waste, such as household waste.
3. Subchapter I regulates underground tank storage of toxic substances and petroleum products.

Subchapter C of the RCRA is concerned with the everyday environmental management of hazardous wastes and involves a "cradle-to-grave" system that tracks hazardous waste from its origin through its ultimate disposal. Facilities must comply with regulations if they generate, transport, store, treat, and dispose of hazardous waste in amounts specified under RCRA. Hazardous waste generators can be separated into three principal categories. State regulations may provide somewhat different designations and responsibilities to the three groups:

1. Large quantity generators (LQG) produce 1,000 kilograms or more of hazardous waste or more than 1 kilogram of acutely hazardous waste (AHW) in a calendar month.

2. Small quantity generators (SQG) produce 100–1,000 kilograms of hazardous waste and less than 1 kilogram of AHW in a calendar month.

3. SQGs that are conditionally exempt include those that generate no more than 100 kilograms per calendar month of hazardous waste, or no more than 1 kilogram per month of AHW. These generators may be exempted from RCRA requirements providing their wastes are properly managed (e.g., sent to a solid waste disposal facility).

4. A very small quantity generator (VSQG) is a newly defined term in USEPA's final Hazardous Waste Generator Improvements Rule ("Rule") that replaces the term "Conditionally Exempt Small Quantity Generator." A VSQG is a generator that generates less than or equal to the following amounts of the listed substances in a calendar month: (1) 100 kilograms of non-acute hazardous waste; (2) 1 kilogram of acute hazardous waste; and (3) 100 kilograms of any residue or contaminated soil, water, or other debris resulting from the cleanup of a spill, into or on any land or water, of any acute hazardous waste. The VSQG is included in the "Rule," which went into effect on May 30, 2017.[7] The Improvements Rule contains approximately 60 revisions to the hazardous waste generator regulatory program and provides significant reorganization of the hazardous waste generator regulations,

The administration of RCRA is done by USEPA, and state programs must be consistent with federal RCRA regulations. Comprehensive cradle-to-grave regulation of hazardous wastes is provided for in most state laws. The lead agency in implementing RCRA is usually the state environmental regulatory agency. In general, most state hazardous waste regulations closely parallel federal provisions of RCRA.[5] Within some states, special problems are addressed by the addition of more stringent requirements. Hazardous waste management has been changed in many ways. There is a ban on land disposal of certain wastes, SQGs are under tighter controls, and regulation of used oil as a hazardous waste has gone into effect in some states.

Identifying a Hazardous Waste

There are two types of hazardous wastes: listed and characteristic. Listed wastes include hundreds of hazardous wastes including wastes from nonspecific sources (F Wastes), manufacturing processes (K Wastes), and discarded chemical products (P and U Wastes) that are listed in the RCRA regulations. The Code of Federal Regulations, Title 40, Part 261, Subpart D, contains these lists of hazardous wastes. Products that are not listed may be characteristically hazardous. Under the direction of RCRA, USEPA developed generalized criteria for evaluating the potential hazard produced by a waste. The four criteria include:

1. *Ignitability*: Easily catches fire, with a flash point of less than 60°C

2. *Corrosivity*: Easily corrodes materials or human tissue, very acidic or alkaline (pH less than 2 or greater than 12.5).

3. *Reactivity*: Explosive, reacts with water or acid, unstable.

4. *Toxicity*: Causes local or systemic damages that result in adverse health effects in an organism—that is, a poison (such as oil), asphyxiant (such as carbon monoxide, CO), mutagen (a chemical that alters DNA), teratogen (one that causes birth defects), or carcinogen (one that causes cancer). Toxicity is determined by a test known as the Toxicity Characteristics Leachate Procedure (TCLP).

USEPA has developed lists of hundreds of hazardous wastes using these criteria. Generators of non-listed waste must analyze the waste to determine if it exceeds USEPA standards. If the waste exceeds any of USEPA's four criteria, it is considered a "characteristic hazardous waste."

Tracking Hazardous Waste

All SQGs and LQGs, transporters, and treatment, storage, and disposal (TSD) facilities must obtain a USEPA identification number as required by most state environmental agencies and RCRA. These ID numbers are used by state agencies and USEPA to track the flow or whereabouts of hazardous waste. This tracking is accomplished by using a paper trail created by shipping manifests. RCRA assures uniform record keeping for all shipments of hazardous wastes. A "Uniform Hazardous Waste Manifest" must be used for all shipments of RCRA-regulated hazardous wastes. State manifest forms, adapted from USEPA form, supplement this documentation in a number of states (Figure 13.1). Each shipment of hazardous waste requires a USEPA manifest form.[6]

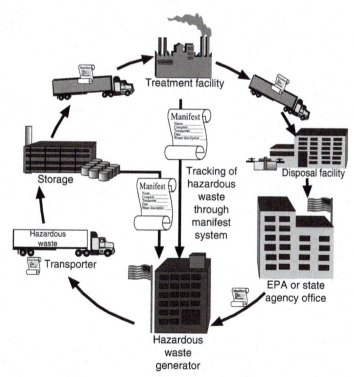

FIGURE 13.1

The RCRA "cradle-to-grave" system tracks hazardous waste from its origin through its ultimate disposal by a manifest system.

Other Requirements Under RCRA

Accurate record keeping is essential to a successful hazardous waste management program. Complete and up-to-date records are necessary to ensure compliance with environmental regulations. Shipments of hazardous waste need to be properly labeled, marked, and placarded according to the U.S. Department of Transportation (DOT) requirements. The facility managers remain responsible for the proper management of waste after it leaves the property, and thus, great care must be taken in selecting a waste hauler and a TSD facility.[7] The Land Disposal Program Flexibility Act of 1996 allowed some flexibility in the procedures for land disposal of certain wastes. For example, a waste is not subject to land disposal restrictions if it is sent to an industrial wastewater treatment facility, a municipal sewage treatment plant, or is treated in a "zero discharge" facility.[8]

Comprehensive Environmental Response, Compensation and Liability Act

The cleanup of inactive and abandoned hazardous waste sites is provided for by CERCLA (Superfund). CERCLA and RCRA have the same basic objectives. However, the purpose of RCRA is to prevent hazardous waste problems at active sites, whereas CERCLA is designed to provide clean-up of existing problems at inactive sites. CERCLA also provides a mechanism of establishing financial liability to any or all parties responsible for the release of hazardous substances. These potentially responsible parties (PRPs) are liable for the costs of cleaning up those substances. These strict joint and liability provisions of CERCLA can result in costly investigative and cleanup activities at waste disposal sites. According to USEPA, the average cost per Superfund site is $35 million, and it is therefore important to manage wastes in a way that reduces the potential for exposure to these costs.[6]

The Superfund Amendments and Reauthorization Act (SARA) of 1986 amended and strengthened the Comprehensive Environmental Response, Compensation, and Liability Act of 1980 (CERCLA), the basic Superfund law. The Regional USEPA is the state's lead agency for implementation of the Superfund. Although crude oil, petroleum products, and natural gas products are excluded from this list, oil spills are addressed by the National Oil and Hazardous Substances Pollution Contingency Plan.[6]

Steps in Superfund: Find, Prioritize, and Clean

The first step in Superfund was to find the hazardous waste sites. The identification of hazardous waste sites was/is accomplished in three major ways. First, USEPA required all owners and operators (including former owners and operators) of hazardous substance TSD facilities to report the existence of these facilities by June 11, 1981. Second, USEPA already regulated TSD facilities under RCRA, and these reported facilities were added to the list. Third, state and local governments, as well as the public, have identified many additional hazardous waste dump sites and facilities. The CERCLA Information System (CERCLIS) database incorporates information for approximately 30,000 sites. Additionally, USEPA established specific reportable quantities (RQs) for the release of hazardous substances. Under the provisions of the Superfund, any releases of hazardous substances larger than these RQs must be reported to the National Response Center by the owners and operators of the facilities or vessels that release hazardous substances (except under federal permit). The Center then notifies other appropriate agencies and begins any necessary emergency response or cleanup actions. Once sites are identified, USEPA is required to collect data and make a preliminary assessment report. The sites are then ranked using a risk assessment procedure for their likelihood of causing human health impacts or environmental damage. The sites with the highest ranking (Hazard Ranking System or HRS) are placed on the National Priorities List (NPL) and are targeted for cleanup with monies coming from PRPs or from Superfund monies derived from taxes on chemical feedstocks and crude oil supplies. However, the $4 billion in Superfund as of 1995 were depleted by 2003 when congress failed to reauthorize collection of the funds.[9] Therefore, funds for cleanup were then provided by taxpayers. During the Obama Administration the number of Superfund sites cleaned continued to decrease, and since then, under the direction of USEPA Administrator Gina McCarthy, Superfund cleanups decreased even more from 20 in 2009 to only 8 in 2014.[10] The preliminary 2018 budget under the Trump administration 2018 would cut the Superfund program by $330 million out of its nearly $1.1 billion budget, nearly a one-third reduction to the U.S. Environmental Protection Agency (USEPA) program.[11]

Historically, a next step in the Superfund process would be the identification of those parties responsible for site contamination. These "potentially responsible parties" and "responsible parties" identified by USEPA and state agencies can be required to finance cleanup activities, either directly or through reimbursement of federal Superfund expenditures. The final steps involve the cleanup of contaminated sites. The program includes extensive provisions for site investigations, selection of cleanup sites, and establishing the levels of cleanup to be attained.[6] USEPA and state agencies use the HRS to calculate a site score (ranging from 0 to 100) based on the actual or potential release of hazardous substances from a site. A score of 28.5 places a site on the NPL, eligible for long-term remedial action (i.e., cleanup) under the Superfund program. As of August 9, 2016, there were 1,328 sites listed, an additional 391 had been delisted, and 55 new sites have been proposed.

Emergency Planning and Community Right-to-Know (SARA Title 111)

The possibility that mishaps at facilities handling hazardous chemicals could harm neighboring communities was brought to worldwide attention with the December 1984 toxic gas disaster in Bhopal, India. In response to this disaster, U.S. federal, state, and local governments created a variety of laws to improve accident prevention and emergency response planning activities by chemical-handling facilities and local governments. "Right-to-Know" laws increased public access to information about the storage and use of hazardous chemicals.

Title III of SARA was the first congressional response to the Bhopal accident. A federal program subtitled the Emergency Planning and Community Right-to-Know Act of 1986 (EPCRA) was mandated by SARA Title III. Title III was one of those laws that was appended to the SARA legislation as a way of getting it passed, and it really has very little to do with the Superfund program. SARA Tile III is focused on worker safety regulations. The Occupational Safety and Health Administration (OSHA) took the lead in developing the regulations because these new requirements were extensions of OSHA's traditional work safety roles. Identical requirements have been adopted by USEPA and are applied to state and local government employees (over whom OSHA has no jurisdiction). The purpose of Title III was to assure the public and emergency response agencies that information regarding hazardous chemicals would be made available to them, thereby lessening the likelihood of Bhopal-like disasters in the United States. USEPA and the individual states are responsible for these efforts. USEPA enforces SARA's requirements for compliance. Any facility that stores or handles extremely hazardous substances (EHS) at or in excess of USEPA-prescribed threshold planning quantities (TPQs) must comply with

SARA's release, planning, and release reporting requirements. This also covers unauthorized discharges (leaks or spills) of any EHS, or of any substance already subject to emergency notification requirements under section 103(a) of CERCLA. Any discharge of RQ must be immediately reported to appropriate local emergency response agencies (police, fire department), to national and state emergency response agencies, and to state and district Title III agencies.

Title III reporting requirements (SARA Sections 311–312) and the federal Occupational Safety and Health Act of 1970 require all facility owners or operators to prepare or have available material safety data sheets (MSDSs) (Figure 13.2). Businesses are required to provide either an MSDS for each on-site chemical in excess of its threshold quantity or a list of all such chemicals to state and district Title III agencies and to the local fire department.

The reporting requirements in Title III (Section 313) require selected businesses to file annual reports with USEPA and the state on USEPA "Form R." These reports must include the following information for the current and previous calendar years:

- Estimated quantities of both routine and accidental releases of listed "toxic chemicals"
- The maximum amount of listed on-site chemicals during the calendar year
- The amount of listed chemicals contained in wastes transferred offsite

These reports estimating releases during the previous calendar year are due each July 1.

In October of 1999 President Clinton declared that beginning January 1, 2000 USEPA would require industries to release fuller reports on 27 additional chemicals, including the endocrine disruptor, dioxin, for the first time. These new regulations require industries to report if they use 45 kilograms per year. In the case of dioxin, the industry must report if one-tenth gram is released.

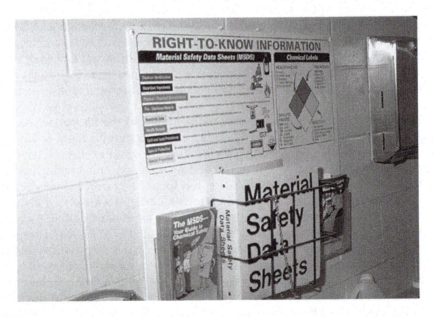

FIGURE 13.2
Material safety data sheets (MSDSs) on hazardous chemicals in the workplace must be made available to employees in a convenient place.

Transportation of Hazardous Materials

The DOT regulates the transportation of hazardous wastes and materials and it is within this department that the Hazardous Materials Transportation Act (HMTA) is administered. States are allowed to adopt standards that are more stringent than federal regulations. State provisions are often similar to the federal standards. The regulations of the transportation of hazardous wastes can differ from state to state.[6]

What are hazardous materials under HMTA? As defined by HMTA, hazardous materials are those that are capable of creating an "unreasonable risk" to health, safety, or property while being transported. Any hazardous substance listed in DOT regulations (49 Code of Federal Regulations sections 172.101, 172.102) that is transported in a package containing a quantity that equals or exceeds the listed RQs is considered a regulated material. Each hazardous material is designated by DOT and depending on its risk to health, safety, or property, is assigned to classes and divisions. Each class has specific packaging, handling, labeling, and registration requirements.

Packaging, labeling, and construction requirements must now meet United Nations (UN) recommendations. There are three packaging groups under the UN-based regulations: Group I for the most hazardous to Group III for the least hazardous. Shipping papers must accompany every shipment. Basic information must be provided within the papers including name of shipper, shipping names of materials, hazard class, standard materials ID numbers under either the UN or NA (North America) system, packaging group (if there is one), quantity of each material being shipped, and the emergency response telephone number.

All containers being shipped must be appropriately marked and labeled. The labels must include the proper shipping name, ID number, and the consignee's name and address. The labels are distinctively diamond shaped, which must clearly identify the hazard class. Even the size, color, and text of each label is specifically laid out by DOT (Figure 13.3). Transport

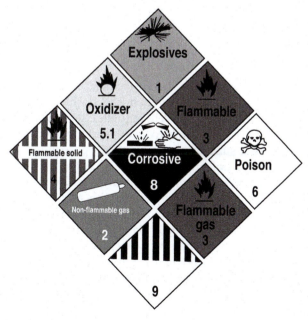

FIGURE 13.3
Examples of several distinctive diamond-shaped shipping labels as required under DOT.

vehicles must also display placard(s) in a specified pattern. Emergency response information must also accompany each hazardous material being transported. The transporting company must also provide a monitored 24-hour emergency response telephone number.[7]

Pollution Prevention and Improved Waste Management Programs

The Pollution Prevention Act of 1990 has become integrated into today's environmental regulations at both the state and federal levels. Pollution prevention, also referred to as "source reduction," is defined as the deliberate decrease in the amounts of hazardous substances that enter the environment via recycling, treatment, or disposal.[11] The emphasis of the Pollution Prevention Act focuses on remediation and treatment, rather than the prevention of pollution. Any pollution that cannot be prevented should be recycled or treated and disposed of in a way that is not harmful to the environment. Mandatory requirements as well as voluntary programs to minimize pollution were established. The Act provides financial assistance to states for their role in prevention. Businesses are encouraged to share and transfer source-reduction technology.[12]

For example, in response to the environmental problems posed by methyl tertiary butyl ether in California, a new "biorefinery" is beginning to convert rice straw, orchard prunings, and other agricultural waste into ethanol. This is being used as an additive in gasoline in the place of methyl tertiary butyl ether. The plan is to phase out MTBE and replace it with the cleaner alternative, ethanol. However, worldwide production of MTBE has been constant owing to growth in Asian markets.[13]

Source reduction must be reported by facilities generating over a certain amount of toxic emissions during the previous calendar year. The form, referred to as "Form R," must be filled out for each toxic chemical to include information such as the amount of toxic chemical recycled and treated. Also included is the amount of toxic chemical that entered a waste stream, percent change from the previous year, and the amount anticipated for the following 2 years. Recycling efforts as well as source reductions must be reported for the previous calendar year.[4]

Toxic Substances Control Act

There are over 70,000 chemicals listed in the Toxic Substances Control Act (TSCA) inventory. TSCA mandates that manufacturers of chemicals develop safety and health data on chemicals and mixtures and requires USEPA to regulate chemical substances and mixtures that present an "unreasonable risk of injury to health and the environment." USEPA is also directed to act when chemical substances and mixtures pose "imminent hazards" and is authorized to restrict the production and use of such hazardous materials. USEPA has done this by banning the use of polychlorinated biphenyls (PCBs) and asbestos-containing products. Congress intended that USEPA not abuse this authority and shouldn't unnecessarily impede or create economic barriers to technological innovation.

There are a number of reporting and record-keeping activities and data collection procedures under TSCA. Additionally, there are mechanisms for public information and "citizen suits" to compel USEPA action when TSCA is violated. The basis of the reporting system is the TSCA inventory originally compiled during 1977 and consisting of two lists: a main inventory and a confidential inventory for all regulated substances that are protected as trade secrets.

Since 1977, each new toxic substance falling under TSCA regulations needs to be added to the list. Before new chemical substances may be introduced into commerce, TSCA requires pre-manufacture notification, as well as notification of significant new uses of existing chemicals. TSCA regulations stipulate guidelines and standards for chemical testing and testing laboratories, rules for the reporting of chemical information, and the use of health and safety data.[6]

TSCA gives USEPA the power to regulate, restrict, and even prohibit the manufacture, importation, and use of certain toxic chemicals. Under this mandate, USEPA has enacted strict regulations for the phase-out, removal, and disposal of the following:

- PCBs
- Asbestos and asbestos-containing materials
- Fully halogenated chlorofluorocarbons (CFCs; once used as aerosol propellants, responsible for ozone depletion in the upper atmosphere)
- Dibenzodioxins and dibenzofurans (contaminants present at Times Beach, Missouri[2])

On June 22, 2016, President Obama signed into law the Frank R. Lautenberg Chemical Safety for the Twenty-first Century Act (Lautenberg), which produced marked changes to the TSCA. It would be the first revision of our domestic chemical management law in 40 years.[14] The new law includes a number of important improvements, consistent with the Administration's 2009 Principles for TSCA Reform. These changes include:

- For the first time, requiring USEPA to evaluate the safety of existing chemicals in commerce, starting with those most likely to cause risks
- Requiring USEPA to evaluate new and existing chemicals against a new risk-based safety standard that includes explicit considerations for vulnerable populations
- Empowering USEPA to require the development of chemical information necessary to support these evaluations
- Establishing clear and enforceable deadlines that ensure both timely review of prioritized chemicals and timely action on identified risks
- Increasing the public transparency of chemical information by limiting unwarranted claims of confidentiality and allowing for the appropriate sharing of confidential information with states and health and environmental professionals
- Providing a source of funding for USEPA to carry out these significant new responsibilities

EPA views the law as a major victory for chemical safety, public health, and the environment—particularly the mandatory duty to evaluate chemicals and the new risk-based safety standard.[15]

Regulation of Underground Storage Tanks

In the late 1970s and early 1980s, lawmakers became aware that thousands of underground storage tanks (USTs) containing petroleum products such as gasoline were beginning to leak, or were in danger of leaking, and contaminating groundwater. The lawmakers developed standards for storage tank construction, installation, and removal and defined a UST as any tank or underground piping connected to the tank that has a minimum of 10 percent of its volume located underground. The federal UST law is a component of the

Hazardous and Solid Waste Amendments of 1984 (HSWA) under the RCRA. This is known as Subtitle I, and this subtitle deals with USTs that store petroleum and other hazardous substances, but not hazardous wastes, which are regulated by other sections of RCRA. USEPA issued its final technical requirements for USTs on September 23, 1988, These regulations include detailed technical standards for new USTs installed after December 22, 1988 and retrofit requirements for tanks already in service on that date. These particular regulations only applied to the storage of petroleum or certain hazardous chemicals. The hazardous chemicals are the same as those listed under CERCLA, except for those chemicals listed as hazardous wastes. All of the UST regulations also applied fully to the underground piping associated with the tank(s) in question. These standards also included tank registration and monitoring. Regulations were initially enforced secondary to tanks that were leaking, thereby causing contamination of groundwater. Aging storage tanks and steel associated with gas piping and oil become a hazard when these pipelines begin to leak and contaminate soil and water supplies.

There are several types of tanks and piping systems that are not included by UST regulations, including septic tanks, tanks for heating oil for onsite use, and farm and residential fuel tanks under 4,163 liters.

Leak detection devices had to be installed in all USTs per federal mandates by December 1993. At the same time, federal and state regulations enforced the use of overfill protection or spill protection. Two categories of record keeping must be maintained including records generated at the point of old tank removal and new tank installation and records of proof of maintenance of monitoring devices, inventory reconciliation, and registration and permit renewal.

All tanks installed after December 22, 1988, are required to have the following construction standards: new underground piping must be corrosion resistant and new USTs must have both spill and overfill protection, as well as leak detection provisions. Also included in the standards are regulations for new petroleum USTs. These USTs may continue to be single-lined, whereas all other USTs must now be double-lined for both primary and secondary containment (Figure 13.4).[4]

In the July 15, 2015, *Federal Register* (PDF) (119 pp, 1.5 MB), USEPA published the 2015 UST regulation and the 2015 state program approval regulation. The revisions strengthen the 1988 federal UST regulations by increasing emphasis on properly operating and maintaining UST equipment. These revisions are designed to prevent and also detect UST releases and ensure all USTs in the country meet the same provisions. USEPA added new operation and maintenance requirements and addressed UST systems deferred in the 1988 UST regulations. The changes include:[16]

- Added secondary containment requirements for new and replaced tanks and piping
- Added operator training requirements
- Added periodic operation and maintenance requirements for UST systems
- Added requirements to ensure UST system compatibility before storing certain biofuel blends
- Removed past deferrals for emergency generator tanks, field-constructed tanks, and airport hydrant systems
- Updated codes of practice
- Made editorial and technical corrections

FIGURE 13.4
A typical underground storage tank installation meeting current regulatory requirements.

Pesticide Regulation

USEPA is responsible for regulating pesticides under the Federal Insecticide, Fungicide, and Rodenticide Act (FIFRA) and the Food Quality Protection Act of 1996 (FQPA).[17] Studies are performed to determine the conditions in which the material is safe to use and the effectiveness against the intended pest(s).[18] USEPA regulates pesticides to reduce the risk of adverse effects to humans or the environment. Registered pesticides are reviewed every 15 years to ensure they meet the proper standards. It is during the registration process that a label is created with safety use directions. USEPA regulates pesticides under two main acts, both of which were amended by the FQPA. In addition to USEPA, the United States Department of Agriculture (USDA) and the United States Food and Drug Administration (FDA) set standards for the level of pesticide residue that is allowed on or in crops. USEPA looks at what potential human health and environmental effects might be associated with the use of the pesticide.[19]

Additionally, USEPA uses the National Research Council's four-step process for human health risk assessment: (1) Hazard Identification, (2) Dose-Response Assessment, (3) Exposure Assessment, and (4) Risk Characterization.

The basic national framework for pesticide control is provided by FIFRA, which is administered by USEPA in cooperation with state and local agencies. Initially enacted in 1947, FIFRA has been amended several times. Pesticides are chemicals specifically formulated to be toxic to living things and are therefore subject to regulation. Pesticide regulations are intended to protect public health, livelihoods, and the environment by ensuring that workers are trained in proper application techniques; pesticides are properly handled and stored; the location or content of chemicals is made known; and pesticides are known to emergency response units, medical personnel, and workers who may be exposed to the chemicals.[6]

Applicators of restricted-use pesticides must be certified based on demonstrated competency in the use of pesticides. Regulations also establish worker safety precautions, including requirements to warn field workers before pesticide applications and to prevent exposure to workers not actually involved in pest control activities. USEPA uses the authority under FIFRA to collect information necessary to register and control the "active ingredients" in pesticides, while state and local agencies control the registration and actual use of the pesticides themselves. Congress has directed USEPA to develop regulations for the design of pesticide containers to promote safe storage and disposal of pesticides, and standards for the removal of pesticides from containers prior to disposal. States, which have been delegated primary FIFRA enforcement, have the responsibility to incorporate and enforce these new requirements within 2 years.

Air Quality Control

A comprehensive national framework for protecting air quality was created by the CAA. In 1990, amendments were enacted to measure compliance. There are 11 sections, referred to as "titles." The first seven titles are most pertinent to most facilities and are briefly described in the following section. The CAA was first enacted in 1963, but its scope was significantly expanded in 1970 and 1977. The 1990 CAA amendments include the following titles:[20]

Title I: Provisions for Attainment and Maintenance of the NAAQS

USEPA monitors six indicator pollutants for the primary protection of human health: ozone, CO, particulate matter (PM), sulfur oxides (SO_x), nitrogen oxides (NO_x), and lead (Pb). Areas where pollutant concentrations exceed NAAQS are referred to as non-attainment areas. These areas are categorized according to the severity of their air pollution problems. Deadlines by which NAAQS must be reached for these areas are seen in Table 13.1. Later deadlines have been assigned to those areas with the most serious pollution problems. Non-attainment areas that move too slowly toward attainment, or that fail to develop acceptable attainment plans, face USEPA sanctions. Currently, there are

TABLE 13.1

Anticipated NAAQS Implementation Milestones (September 2017)

Pollutant	Infrastructure SIP Due	Attainment SIP Due	Attainment Date
$PM_{2.5}$	October 2009	December 2014	December 2019
Pb	October 2011	June 2012–2013	December 2015–2019
$PM_{2.5}$	December 2015	October 2016	December 2021 (Mod) December 2025 (Ser)
NO_2	January 2013	N/A	N/A
SO_2	June 2013	April 2015, March 2018 (2019, 2022)	October 2018, September 2021 (2023, 2026)
Ozone (2008)	March 2011	Mid 2015–2016	Mid 2015–2032
Ozone (2015)	October 2018	TBD	TBD

roughly 100 non-attainment areas, home to some 150 million people. Five classes of ozone violations (marginal, moderate, serious, severe, and extreme) and two categories (moderate and serious) for CO and particulates are established by these amendments. Compliance deadlines and control requirements exist for each category. Severely polluted areas are given more time to comply, but they must adopt more stringent controls for ozone, CO, and PM_{10}. The CAA directs USEPA to set air quality and emissions standards at levels that protect public health and the environment.

Other Related Programs Established Under Previous Air Quality Regulations

USEPA gives states with approved *state implementation plans (SIPs)* as required under the 1970 CAA, the authority to run federal clean air programs. The air quality in designated regions, basic strategies to achieve or maintain NAAQS, and enforcement strategies are described in a SIP and every state must submit an SIP to USEPA. In various states, SIPs are the blueprints for achieving air quality goals. Standards more stringent than those required by CAA may be imposed by the states. Once a state submits an SEP or SIP amendment, USEPA either approves it or orders the state to revise it. If a state fails (or refuses) to submit an acceptable SEP, USEPA may either draft the SEP for that state or write (and implement) a federal implementation plan (FIP). SIPs are reviewed and revised continually. Table 13.1 lists the anticipated NAAQS implementation milestones for State Implementation Plans (SIPs) for NAAQS as of September 2017.[21]

Each state must identify in its SIP, emission controls and other measures that prevent significant deterioration (PSD) of air quality with respect to two criteria pollutants: SO_2 and PM_{10}. *The prevention of significant deterioration (PSD)* of air quality in pristine areas is the focus of this CAA program. Special attention is directed to national parks, national wilderness areas, national monuments, national seashores, and other areas of special natural, recreational, scenic, or historic value by the PSD program. A region's classification (Class I, II or III) determines the allowable levels Of SO_2 and particulates for that region. The most severe restrictions are imposed on Class I regions; further construction of major stationary sources in these regions without a permit issued after PSD review is generally prohibited.

Emissions standards for new and modified sources of air contaminants are set by USEPA. New and modified sources are forced to achieve the greatest emission reductions that are technologically and economically feasible by these *new source performance standards (NSPSs)*. States are required by CAA to develop emissions standards for existing sources. If a state fails to submit a satisfactory SIP, USEPA may set these standards. New sources are typically subject to stricter emissions standards than existing sources because new sources are considered to have greater flexibility to incorporate innovative emission control technologies. The 1990 CAA amendments add important new controls on stationary source.

USEPA announced new NAAQS for ground-level ozone and particulates on July 17, 1997. Ground-level ozone is a major component of smog that is photochemically produced as a secondary pollutant of the stratosphere from the interaction of sunlight, NO_x, and hydrocarbons. The previous ozone standard was last revised in 1979 and set at 0.12 ppm for one hour. The previous standard did not adequately protect the public health, and USEPA replaced it with an 8-hour standard set at 0.08 ppm. Similarly, USEPA announced a new standard for PM under the national ambient air quality standards (NAAQS) based on the review of hundreds of scientific studies. USEPA has added a new annual $PM_{2.5}$ standard set at 15 micrograms per cubic meter ($\mu g/m^3$) and a new 24-hour $PM_{2.5}$ standard set at 65 micrograms per cubic meter. USEPA is retaining the current annual PM_{10} standard set at 50 micrograms per cubic meter.

Clear Skies Program

There has been a great deal of controversy since President Bush first announced the Clear Skies Act on February 14, 2002. Clear Skies legislation was first introduced in both Houses of Congress in July 2002 and reintroduced in February 2003. On January 24, 2005, Senators Inhofe and Voinovich introduced their version of the Clear Skies Act (S 131). The Senate Environment and Public Works Committee held hearings on this legislation on January 26 and February 2 of 2005. The Bill never moved out of committee when members were deadlocked and the Bush Administration moved to implement key measures, such as NO_x, SO_2, and mercury trading provisions of the Bill administratively through USEPA. It is unclear how court challenges may eventually impact these administrative changes. According to the USEPA and Bush administration officials, the Clear Skies Act would:

- Establish federally enforceable emissions limits (or "caps") for three pollutants including sulfur dioxide (SO_2), nitrogen oxides (NO_x), and mercury. Advocates claim that the emissions reductions from Clear Skies would help to alleviate our nation's major air pollution–related health and environmental problems by targeting fossil-fueled electric generators and reducing fine particles, ozone, mercury, acid rain, nitrogen deposition, and visibility impairment. Mercury requirements would affect only those power plants that are coal-fired.

- Use a regulatory approach known as emission caps and trading that provides power plants with some flexibility in reducing emissions in the least costly way.

- Maintain the authority of state and local government to set source-specific emissions limits for sources within their borders to ensure that ambient air quality standards are met.

Administrators claim that Clear Skies would:

- Provide immediate benefits to human health and the environment when it is passed. USEPA claims human health benefits would save $110 billion per year by 2020 in health-related costs, lost work, and other health considerations. This would be substantially greater than the annual costs of $6.3 billion associated with the costs of meeting the Clear Skies requirements.

- Help to avoid from 8,400 to 14,000 premature deaths by 2020, saving at least $21 billion in health benefits.

- Reduce hospital and emergency room visits by 30,000 and decrease the days with respiratory symptoms (including work loss days and school absences) by 12.5 million.

- Increase the number of people that breathe air meeting the national ozone and fine particle standards by 18 million.[23]

Environmental groups such as the Sierra Club and the National Resources Defense Council (NRDC) claim that the Clear Skies program would actually result in greater pollution than if we enforced current air pollution laws. The Sierra Club claims that by the 15th year of the Clear Skies program there would be 408233 more megagrams of NO_x, 0.907 million more megagrams of SO_2, and 8.62 more megagrams of mercury than if there were strong enforcement of existing CAA programs.[24,25]

Title II: Provisions Relating to Mobile Sources

Pollution-reduction requirements for motor vehicles are defined by Title II. These requirements are of great significance to the facility and its large vehicle fleet. In areas with CO levels above 16 parts per million (ppm) or in areas designated as seriously, severely, or extremely above the limit for ozone, Title II requires centrally fueled fleets to burn only clean fuels.

Vehicle emissions requirements: The 1990 CAA pollution standards are stricter than the 1970 act for automobile and truck emissions. Tail pipe emissions of hydrocarbons, CO, and NO are expected to be reduced by these standards. New vehicles purchased by the facility (cars and light-duty trucks up to 2,721 kilograms) have stricter emission controls. The maintenance of appropriate emissions standards for non-methane hydrocarbons, CO, and NO_x will affect fleet management.

Inspection/maintenance requirements: As mandated by the CAA, the Inspection and Maintenance (I&M) program requires periodic inspections of vehicles to ensure that emissions of specified pollutants are not exceeding established limitations. All ozone and CO non-attainment areas in the United States are required to have an I&M program. Annual inspection programs are required unless a state can demonstrate that a biennial program is just as effective.

Reformulated gasoline requirement: In 1995, reformulated gasoline with a 2 percent minimum oxygen content will be required during the winter months in non-attainment areas for carbon monoxide. Presently being sold as gasoline, this "oxygenated" fuel contains 10 percent ethanol. MTBE, another fuel oxygenator, is no longer used in most areas of the country because of its widespread ability to contaminant groundwater and liability issues arising from this problem.

Title III: Hazardous Air Pollutants

Hazardous air pollutants are defined as those substances that are hazardous to human health or the environment but are not specifically covered under other portions of the CAA. A list of 189 toxic air pollutants for which emissions must be reduced is included in this law. A more comprehensive program for regulating HAP emissions was introduced by the 1990 amendments. All new and existing stationary sources that emit HAPs must comply with air toxics emissions standards to be adopted by USEPA.

Maximum achievable control technology (MACT) will have to be applied by stationary sources to control their HAP emissions. MACT is a standard that USEPA uses to define various categories of sources. USEPA was to produce a list of all categories of stationary sources that emit HAPs by November 1992. USEPA must adopt MACT standards for at least 40 categories. USEPA must then adopt MACT standards for 25 percent of the remaining categories by November 1994, with another 25 percent due by November 1997, and the balance by November 2000.[6] Area-wide control strategies must be formulated by USEPA by the year 2000. The strategy must identify at least 30 HAPs emitted by area sources presenting the greatest threat to public health in most areas and identify source categories emitting 90 percent or more of each of the 30 HAPs. The strategy must also show how USEPA intends to reduce by at least 75 percent, the cancer risk posed by the release of HAPs from stationary sources. Sources will have 9 years to comply with the strategy. Reports to Congress that identify high-risk metropolitan areas and describe actions being taken to reduce health risks posed by HAP emissions are due in 1998 and 2002.[6]

Title IV: Acid Deposition Control

Title IV regulates the sources of acid deposition. Most of the acid deposition in the United States comes from the burning of fossil fuels. The operation of facility boilers is affected by these Title IV regulations. Emissions of SO_2 and NO, from fossil fuel-fired electric utility plants are the leading cause of acid rain. An ambitious program to reduce and control these emissions is included in the 1990 amendments. However, the program does not affect existing units with a generating capacity of 25 megawatts or less, qualifying cogeneration units, or qualifying small power producers.

The reduction Of SO_2 and NO_x emissions is to be done in two phases, the first beginning in 1995 and the second in 2000. Each SO_2 source will be given tradable allowances equal to the number of tons Of SO_2 it is permitted to emit. If emissions are reduced below this amount, a source can sell its extra allowances to another source. New SO_2 sources will have to purchase allowances from existing sources for their emissions. Certain boiler units in Phase I and others in Phase II must meet USEPA emissions standards to achieve NO_x reductions.

Plans for achieving the required emission reductions must be prepared and submitted to USEPA by units covered by the SO_2 and NO_x regulations. These same units must obtain air emissions permits (see earlier), install and operate continuous emission monitoring systems, and submit detailed records of their SO_2 and NO_x emissions. Clean coal technology demonstration projects are exempt from NSPSs and non-attainment requirements, therefore, SO_2 and NO_x sources are encouraged to switch to clean coal technologies.

Title V: Permits

Title V requires all major sources of air pollution to obtain permits and extends emissions controls to thousands of sources in many unregulated areas. For the first time, a nationwide program of air emissions permits had been established by the 1990 amendments. Following USEPA guidelines, the states were encouraged to set up their own permit programs, which are subject to USEPA oversight. Each state proposed a permit program, and states faced sanctions if they failed to submit acceptable permit programs covering non-attainment areas.

Permit holders must certify that the facility complies with the terms of its permit on an annual basis. Under certain conditions, compliance with the permit terms will be considered compliance with CAA. Monitoring progress reports must be submitted at least once every 6 months by permit holders.[6]

Title VI: Stratospheric Ozone Protection

A complete production phase-out of ozone-depleting chemicals (especially CFCs and halons) is required by Title VI. A new national program designed to minimize human impact on the stratospheric ozone layer by phasing out the use of CFCs was introduced by Title VI of the 1990 CAA amendments. Congress originally directed that parts of the phase-out be completed by the year 2000. The Bush Administration accelerated the phase-out schedule for certain listed CFCs to 1995, based on new evidence concerning the deterioration of the ozone layer. The United States signed an international treaty on preventing ozone depletion in 1987 known as the Montreal Protocol. The accelerated phase-out schedule keeps the U.S. in terms with this ozone treaty. Ozone-depleting chemicals are divided into Class I and Class II substances and phase-out schedules for each class are prescribed by Title VI. The greatest threat to the ozone layer is Class I chemicals, which were not to be produced after December 31, 1995. Methyl chloroform is an exception, and its production may continue until January 1, 2002. In the interim period, specific annual percentage reductions are required. Beginning January 1, 2030, all production of Class II substances is banned by Title VI.

Title VII: Provisions Relating to Enforcement

The Amendments contain a broad array of provisions that brings the law up to date with the other major environmental statutes and thus makes the law more readily enforceable.

Water Quality Control

The basic framework of the Federal Water Pollution Control Act, passed by Congress over President Nixon's veto in 1972, remains today as a part of the modern CWA. Effluent limitation guidelines, water quality requirements, and the permit program are continually enforced. The goals of the CWA are to eliminate the discharge of pollutants into surface waters and to achieve water quality that "provides for the protection and propagation of fish, shellfish and wildlife," and "for recreation in and on the water."[10] The Act also incorporates the prohibition of the "discharge of toxic pollutants in toxic amounts."

The CWA only applies to surface waters and the groundwater that is hydrostatically connected to surface water. The Safe Drinking Water Act provides federal protection of groundwater sources. One priority of the CWA is the discharge of mainly industrial pollutants into our nation's waters, referred to as "point sources." These controls are also known as technology-based effluent limitations. The CWA ensures that the water quality standards for the waters receiving the discharge are met, and that the discharger applies the required technology-based standards to the facility through application of a comprehensive permitting scheme known as the National Pollutant Discharge Elimination System (NPDES).[6]

The second major focus is that of water quality-based controls. The CWA describes three categories of pollutants: (1) conventional pollutants (biochemical oxygen demand, total suspended solids, pH, fecal coliform, oil, and grease); (2) toxic pollutants, including the 65 pollutants that are identified by USEPA; and (3) non-conventional pollutants (all pollutants not categorized as either toxic or conventional).[4]

The CWA allows states to apply to USEPA for authorization to administer various aspects of the federal NPDES program. Generally, for a state to obtain USEPA authorization to administer all or part of the permitting program, it must match the requirements mandated by CWA or apply more stringent ones. The owner/operator of a facility must obtain a permit prior to discharging any pollutant into the waters of the state from any point source under the NPDES program requirements. Both CWA and most State Acts broadly define these terms to regulate almost all activities that result in the release of pollutants from a discrete point (e.g., a pipe).[6]

Stormwater

Water quality may be degraded and pollutant concentrations may exceed water quality standards as a result of polluted stormwater. Pollutants may dissolve in the stormwater, become suspended, or float on the surface. Pollutants that could potentially occur in stormwater draining from sites include sediments, solids, nutrients, oxygen-demanding

FIGURE 13.5
Stormwater runoff is a significant contributor to water pollution and is regulated under amendments to the CWA.

materials, bacteria, hydrocarbons, heavy metals, herbicides, pesticides, and other organic substances. The runoff and pollutants enter storm drains (Figure 13.5) and are then discharged into receiving waters, such as streams and lakes. During dry weather periods pollutants can also enter the storm drain system as a result of improper connections or illegal dumping.[26]

In 1987, Congress amended the CWA requiring USEPA to establish phased NPDES requirements for stormwater discharges. NPDES permits are required for the discharge of pollutants from any point source into waters of the United States, which includes almost any surface water or wetland. These permits have provisions for monitoring the discharge of pollutants to U.S. waters and for establishing source controls where necessary. On November 16, 1990, USEPA published a final rule in the Federal Register (55 CFR 47990) that contains permit requirements for stormwater discharges. Facilities regulated under the new federal stormwater program must comply with the state and/or federal stormwater regulations. There are three distinct types of stormwater permits—individual, general, and group—which are established by the federal regulations.

Oil and Hazardous Substance Spill and Reporting Requirements

The CWA contains specific provisions regulating the handling of oil and hazardous substances. Specific penalties and rules of liability for unauthorized releases of these materials are established by CWA. Similar to the requirements outlined under Superfund, these provisions focus on reporting unauthorized leaks, spills, and discharges to water (discharges without a permit). Under the CWA, any person in charge of a vessel or an onshore or offshore facility who has knowledge of an unauthorized release of oil or any hazardous substance must immediately report the release in a manner consistent with the National Oil and Hazardous Substances Pollution Contingency Plan (commonly called the National Contingency Plan).[27]

The Oil Pollution Act of 1990 was created in response to the Exxon Valdez oil spill in 1989, which dumped 41.6 million liters of oil into the waters off south-central Alaska. Media coverage of black oily beaches and the resulting deaths of thousands of birds, sea otters, and whales, among other wildlife, created public outrage. This, in addition to the spill size and location, led to OPA 90. This act provides tougher penalties and liabilities for oil spills, allocates more funds and other resources for dealing with spill cleanup, and places more responsibility on the Executive Branch to respond to these incidents.

Compliance Strategies

Since the 1970s, employers have been given increased responsibilities. Managers are becoming more responsible for environmental health and employee safety. Various agencies have implemented regulations that are enforced to further protect the employee and the environment. Employers have developed worker safety programs, and the method in which these programs are enforced can directly impact worker health and safety, liability and insurance premiums, worker's compensation, remediation costs and fines, and costs for meeting regulatory requirements. However, it is not uncommon that such programs result in a negative company image due to the inadequacy of the implemented programs. Unfortunately, these programs are often not given enough attention and support. Additionally, supervisor and worker training programs are frequently insufficient.

There are criminal penalties for non-compliance with environmental health and safety programs. There has been a tremendous increase in liability for both the federal agencies and for individuals from the civil and criminal provisions of recently re-authorized environmental legislation. Even though the thousands of pages of regulations are difficult to interpret, they still apply. Some of the regulatory initiatives include pollution prevention and waste minimization in facilities. The regulations are enforceable throughout all facets of a facility's operation including air, water, and solid waste disposal. The Community-Right-to-Know Act plays a large role in this process, requiring facilities to publicly report their waste production for hazardous wastes, while the Pollution Prevention Act ensures that waste is reduced through raw material substitution, process modification, and recycling. The RCRA targets recycling programs.

After gaining a basic understanding of the systems of environmental laws, employers need to apply their knowledge in order to remain in compliance. The development of such strategies for compliance is based on the premise that the most effective protection against aggressive enforcement and other efforts to assess liability is aggressive compliance.

The Government Institutes has published a variety of books that discuss and detail the elements of appropriate compliance organization. There are several basic principles with which one should become familiar. The first principle is that environmental law compliance is the responsibility of everyone. The second principle describes the need to demonstrate a concern for compliance by providing appropriate education and training. The third principle focuses on the prevention of violations and minimization of liability through aggressive implementations of environmental objectives. The fourth principle recommends periodic "audits" to verify compliance and identify areas that can be improved upon. These audits will be conducted to address environmental issues and controls, such as asbestos, indoor air quality, and UST regulations.[2]

Trends in Regulatory Compliance

The states are responsible for nearly 90 percent of regulatory inspections involving air, water, and hazardous wastes, with USEPA conducting the remainder of the inspections. Routine inspections usually consist of sampling analysis and record keeping. Inspections may also include investigations of operation and maintenance programs. USEPA has developed target strategies, based on the quantity and severity of toxic materials discharged, emitted, or disposed of by a facility as reported under the Community-Right-to-Know Act. This data is compared to other self-reported data required by other Acts, and with information stored in USEPA files such as manifest reports and permit applications. USEPA will likely focus on repeated violations and evidence of suspicious self-reporting.

Traditionally, regulators have focused largely on the use of standard enforcement tools, such as sanctions, to persuade industries found to be in violation of environmental laws to comply. However, another approach is being explored by some environmental experts. It is being suggested that this approach be combined with an effort to educate industry officials regarding environmental laws in hopes of inducing increases in the standard of current practice and promoting a healthier environment for the future.

A situation that has recently arisen is the question of whether U.S. companies may be sued for environmentally damaging acts committed in other countries. In San Carlos, Ecuador, carcinogenic by-products of extraction methods were poured into open pits and left to seep into the ground. They also contributed significantly to air pollution when the pits were set fire to burn off their contents. As a result of this issue 30,000 Ecuadorians filed a class action suit against Texaco. It is being said that Texaco executives violated human rights and should be held accountable for their actions.

On the other side of the world, China has begun to deal with the issue of environmental destruction that is resulting from its own development. Over the past two decades China has implemented several pollution control programs to curb the amount of pollution produced by its growing industries. These include such steps as enacting emissions taxes and setting a general goal of stopping environmental deterioration.

References

1. Burns, I.A., Peltason, J., Cronin, T.E., and Magleby, D.B., *Government by the People*, 15th ed., Prentice Hall Publishers, Englewood Cliffs, NJ, 1993, Chap. 14.
2. Sullivan, T.E.P., *Fundamentals of Environmental Law*, Government Institutes, Rockville, MD, 1995, Chap. 1.
3. United States Postal Service, *A Proposal to Establish Environmental Coordinators in District Offices of the Northeast Area*, Northeast Area Environmental Compliance Staff, Northeast Area Processing and Distribution, 6 Griffin Road North, Windsor, CT 06006-7030, February, 25, 1993.
4. Vance, B., Weinsoff, D.J., et al., *Toxics Program Commentary: Massachusetts*, prepared by Touchstone Environmental Inc., and published by Specialty Technical Publishers, North Vancouver, B.C., Canada, V7M IA5, 1992.
5. United States Postal Service, *Waste Reduction Guide*, Handbook AS-552, Northeast Area Environmental Compliance Staff, Northeast Area Processing and Distribution, 6 Griffin Road North, Windsor, CT, 06006-7030, February, 1992.
6. Vance, B., Weinsoff, D.J., Henderson, M.A., and Elliot, J.E., *Toxics Program Commentary: Massachusetts*, prepared by Touchstone Environmental, Inc., and published by Specialty Technical Publishers, North Vancouver, BC, Canada V7M I A5, 1992.

7. Greenberg Glusker Fields Claman & Machtinger LLP, RCRA Update: New Hazardous Waste Generator Rule Goes into Effect in May 2017, April 2017, https://www.jdsupra.com/legalnews /rcra-update-new-hazardous-waste-80701/.

8. Land Disposal Program Flexibility Act of 1996, P.L. 104-119, 110 Stat. 830, 42 U.S.C. § 6924, March 26, 1996.

9. Superfund Program: Updated Appropriation and Expenditure Data (PDF), U.S. Government Accountability Office, February 18, 2004. p. 6, Retrieved January 27, 2014.

10. Beins, K., and Lester, S., Superfund Polluters Pay So Children Can Play, 35th Anniversary Report, December 2015, 80 pp, The Center for Health, Environment & Justice.

11. King, L., President Trump's budget would cut Superfund toxic cleanup program by 30%, *USA Today*, April 7, 2017.

12. United States Postal Service, *Hazardous Waste Management Guide*, Handbook AS-553, Northeast Area Environmental Compliance Staff, Northeast Area Processing and Distribution, 6 Griffin Road North, Windsor, CT, 06006-7030, May, 1992.

13. Winterberg, M., Schulte-Korne, E., Peters, U., and Nierlich, F., "Methyl Tert-Butyl Ether" in *Ullmann's Encyclopedia of Industrial Chemistry*, 2010, Wiley-VCH, Weinheim, doi:10. 1002/14356007.a16_543.pub2.

14. Bergeson and Campbell, PC, TSCA reform news and information, 2018, http://www.lawbc .com/knowledge-resources/tsca-reform-news-info.

15. United States Environmental Protection Agency (EPA), Assessing and Managing Chemicals Under TSCA: The Frank R. Lautenberg Chemical Safety for the 21st Century Act: Frequent Questions, retrieved February 20, 2018, https://www.epa.gov/assessing-and-managing-chemicals-under-tsca/frank-r-lautenberg-chemical-safety-21st-century-act-0#briefings.

16. United States Environmental Protection Agency (EPA), Underground Storage Tanks: Revising Underground Storage Tank Regulations-Revisions to Existing Requirements and New Requirements for secondary Containment and Operator Training; Final Rule, retrieved February 20, 2018, https://www.epa.gov/ust/revising-underground-storage-tank-regulations -revisions-existing-requirements-and-new.

17. Pesticides and Public Health|Pesticides|US EPA, *Epa.gov*, Retrieved April 1, 2013, https:// www.epa.gov/pesticides#regulation.

18. Data Requirements for Pesticide Registration," *US EPA*, https://www.epa.gov/pesticides.

19. US Environmental Protection Agency (February 16, 2011), Pesticide Registration Program epa .gov. Retrieved on February 25, 2011, https://www.epa.gov/pesticides.

20. United States Postal Service, *Clean Air Act Compliance*, Handbook AS-553, Northeast Area Environmental Compliance Staff, Northeast Area Processing and Distribution, 6 Griffin Road North, Windsor, CT, 06006-7030, April, 1992.

21. Anne Marie Wood, Director, Air Quality Policy Division, USEPA, NAAQS and other Implementation Updates, September 2017, http://www.4cleanair.org/sites/default/files /Documents/Fall_2017 _Wood-and_Banister_NAAQS_and_Other_Implementation _Updates.pdf.

22. The Library of Congress, Thomas, S.313, http://thomas.loc.gov/cgibin/bdquery/z?d109:s . 00131: June, 2006.

23. United States Environmental Protection Agency, *Clear Skies*, last updated on Thursday, March 2, 2006, http://www.epa.gov/air/clearskies/legis.html.

24. Mund, N., *Clean Air: Facts about the Bush Administration's Plan to Weaken the Clean Air Act*, Sierra Club, Retrieved June, 2006, http://www.sierraclub.org/cleanair/clear_skies.asp.

25. National Resources Defense Council (NRDC), *Dirty Skies, The Bush Administration's Air Pollution Plan*, last revised October 5, 2003, http://www.nrdc.org/air/pollution/qbushplan.asp.

26. United States Postal Service, *NPDESI Stormwater Guide*, Handbook AS-554, Northeast Area Environmental Compliance Staff, Northeast Area Processing and Distribution, 6 Griffin Road North, Windsor, CT, 06006-7030, May, 1992.

27. United States Environmental Protection Agency (USEPA), When are you required to report an oil spill and hazardous substance release? Retrieved July 8, 2018. https://www.epa.gov/ emergency-response/when-are-you-required-report-oil-spill-and-hazardous-substance-release.

14

Emergency Preparedness

OBJECTIVES

A student reading this chapter will be able to:

1. Discuss the history and origins of the Incident Command System (ICS), the National Incident Management System (NIMS), and The National Response Plan (NRP)

2. List the major components of ICS, NIMS, and NRP

3. Describe the role of Environmental Health professionals in emergency/disaster response

---

Introduction

Pre-9/11

It is not terrorist bombs or the release of anthrax spores that spawned the birth of modern-day emergency preparedness. It was not the 9/11 attack on the World Trade Center twin towers in New York that began this effort to prepare the United States for disaster response. It began with a wildfire in Southern California in 1970 that swept through several counties during a 13-day period, burning more than 200,000 hectares, destroying over 700 structures, and costing 16 lives. The awesome fury of this fire swept across county lines and created chaos among responding agencies that could not coordinate efforts because there was no clear person in charge, no common easily understood language among police, fire, and emergency response agencies, and no previously established coordination plan among these multiple agencies.[1]

Because of this tragedy, the 92nd Congress approved funding for the U.S. Forest Service Research group to design a system or process to significantly improve the capacity of the agencies in Southern California to provide protection against wildfires. Multiple state and federal agencies in Southern California joined to form a workable interagency system that could deliver resources and persons to a major fire in a coordinated fashion. This became known as Firefighting Resources of California Organized for Potential Emergencies (FIRESCOPE). The participating agencies determined that there were five major program components that required planning and attention. These included:

1. Coordination of the resources among many agencies during a large incident
2. Creating a standard and easily understood terminology by all participants to avoid the use of professional jargon that would differ among police, fire, and emergency medical responders
3. Developing systems and technical capacity that would permit communication among many participating agencies
4. Providing training to all members of the participating agencies in communication, terminology, and command structure
5. Developing better methods for predicting the behavior of fires

These objectives were ultimately condensed into two components of the FIRESOPE system designated as:

1. Incident Command System (ICS) that provided a specific command structure and coordinated incident management tool
2. Multi-Agency Coordination System (MACS) that improved coordination among multiple agencies for large incidents that demanded massive resources

In 1976, the agencies involved in FIRESCOPE agreed on common terminology and procedures for the ICS structure and successfully employed ICS on a number of wildland and urban fires in 1978. As the success of the ICS management of fire responses grew, the use of the system expanded to non-fire incidents. It became clear that the management structure of ICS was well adapted to numerous types of incidents to which fire agencies must respond, including hazardous materials accidents, earthquakes, floods, plane crashes, and more. The term the agencies began to apply to the management of these incidents was "all-hazard" as they employed the well-documented and proven command structure of ICS to a wide variety of incidents. The single incident management system of ICS was being successfully applied on an all-hazards basis.[1,2]

According to a report by the FIRESCOPE agencies, efforts were made to "extend these proven systems to other public safety services as the foundation of a uniform and comprehensive emergency management system." Some of the initial steps to extend this ICS management structure began in 1983 with the creation of the National Interagency Incident Management System (NIIMS). The agency made a few minor organizational and procedural modifications and small changes in terminology that allowed improved adaptation to a multi-agency, all-hazards environment.[2] In practice, these applications were generally limited to regional or multi-state applications involving responses to natural disasters. Natural disasters of national significance such as the flooding of New Orleans, or the terrorist attacks on the World Trade Center towers in New York had not yet occurred to test the system at the national level.

Post 9/11

The importance of an organized response to terrorist threats to our country was made very clear on September 11, 2001, when terrorists took control of several airplanes filled with passengers and used those planes as missiles. Two of the planes were hurled into the World Trade Center towers in New York causing their collapse. Another plane crashed into the Pentagon in Washington, D.C. and destroyed one side, while a fourth plane dived into a Pennsylvania field after brave passengers tried to retake control of the aircraft. More than 3000 people were killed by these brutal attacks. The Bush administration declared these to be "acts of war," and asked Congress for $20 billion to help rebuild and recover from these attacks.[3]

In swift reaction to these events, The Homeland Security Act of 2002 (H.R. 5005-8) was approved with the purpose of establishing a Department of Homeland Security (DHS) as an executive department of the United States. The mission of DHS is to prevent terrorist attacks within the United States, reduce the vulnerability of the U.S. to terrorism, and minimize damage and assist in recovery from those terrorist attacks that do occur in the United States. The DHS was officially established on January 23, 2003 causing a massive reorganization of the government with the intent of providing a unifying core for a vast national network of institutions and organizations that are important to national security and emergency response.

More than 20 federal agencies were directed to become part of the DHS and to be compartmented into one of four directorates: (1) the Border and Transportation Directorate, (2) The Emergency Preparedness and Response Directorate, (3) the Science and Technology Directorate, and (4) the Information Analysis and Infrastructure Protection Directorate.[4,5]

President Bush issued Homeland Security Presidential Directive 5 (HSPD-5) on February 28, 2003. HSPD-5 directed Secretary of Homeland Security Tom Ridge to develop and then administer a National Incident Management System (NIMS) in order to provide a consistent management tool for government, private, and nongovernmental organizations to function together during a major incident. NIMS represents a national approach to incident management that scales up and modifies ICS to be applicable to a broad range of possible natural and manufactured disasters of varying size and complexity.

ICS and NIMS

Among the most likely reasons for response failure in an emergency are confusion about who is in charge, unclear lines of authority, and lack of clear communication. The management model used successfully for more than 30 years to overcome these problems is ICS. This is a standard, on-scene, all-hazard management system that is highly flexible, interdisciplinary, and able to adjust to one or more incidents that vary in size and complexity.

TABLE 14.1

Titles for Supervisory Levels Under ICS

Organizational Level	Title
Incident Command	Incident Commander
Command Staff	Officer
General Staff	Chief
Branch	Director
Division/Group	Supervisor
Unit	Leader

The following is a brief description of ICS. Those wishing to learn more and/or take quali-fied courses may go to http://training.fema.gov/EMIWeb/IS/is100.asp.[6] ICS position titles are distinct and recognizable anywhere in the country and at any agency or organization using the ICS system. There is only one Incident Commander at each incident and only that person is given the title of Commander. Members of the command staff are called Officers, and members of the general staff are called Chief as described in Table 14.1 and Figure 14.1.

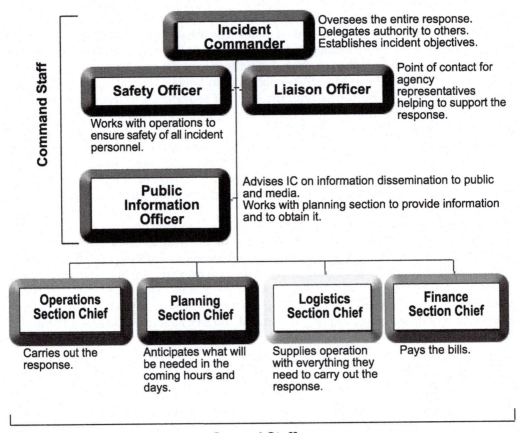

FIGURE 14.1
Typical ICS management structure and brief description of responsibilities.

ICS provides a very clear reporting structure so that each response person knows to whom to report and what they are expected to do. All response personnel answer to one immediate supervisor and can be assured that communication will be understandable and in plain English. Typical ICS management structure as shown in Figure 14.1 may be reduced to only an Incident Commander if the incident is very small while ICS structure may be greatly expanded if the incident grows or is already complex and large. As the incident grows there may be a need to add supervisory layers to the organizational structure. As an example, the Incident Commander can add branches, Divisions, Groups, Units, and Single resources to the management structure as needed (Figure 14.2).

Another important element of ICS is the use of a manageable span of control. Effective span of control refers to the number of people or elements answering to an individual such as a commander or officer. This number is usually maintained between three to seven reporting elements with a desired ratio of one supervisor to five reporting elements. When the numbers exceed this range, expansion of the organization should occur.

There are more elements to ICS, and the reader is referred to http://training.fema.gov for more complete information on ICS. It is understood that the uniformed services of

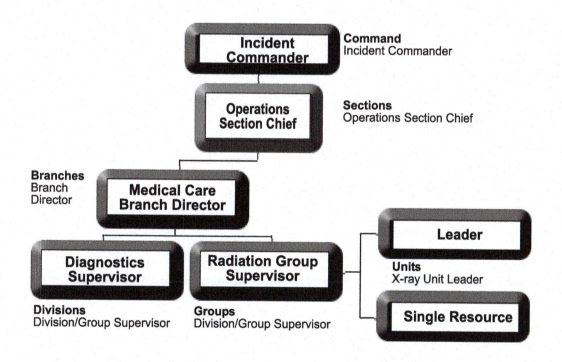

FIGURE 14.2

Example of options for expanding incidents that require added supervisory layers to the ICS organizational structure.

communities such as fire and police function under ICS on a daily basis and so they are intimately familiar with it as second nature. Public health doesn't normally function within an ICS command structure since most public health departments and agencies are quite small, personnel have defined roles, communication is usually simple, and there are very few immediately life-threatening incidents. However, the threat of larger, natural, and manufactured disasters requires local and state public health to coordinate with multiple agencies and so they must communicate within ICS and the overall NIMS framework.[7] Public Health may play a subordinate role such as surveillance of water to assure the public an incident such as a flood or hurricane has not contaminated the water or providing information to the public about preventative measures to protect one's health. Public Health may also play a major role in a multi-agency response such as a pandemic flu outbreak. Whether large or small incidents, the response of Public Health needs to be within the NIMS/ICS framework. In order to better understand how ICS might work within a Public Health structure, a detailed description of the Public Health Incident Command System is available online at http://www.Ualbanycphp.org /pinata/phics/guide/default.cfm.

How does the NIMS fit into this picture? NIMS recognizes that ICS will be used at the local level. However, until the appearance of NIMS, there have been no standards for domestic incident response that encompassed the governments and agencies at all state and federal levels. The purpose of NIMS is to provide a unified national response to incident management for all potential incidents regardless of size and complexity. A major objective of NIMS is also to improve coordination and cooperation among public and private agencies and organizations throughout a whole range of domestic incident management activities.

NIMS is composed of six related components that function together as a system to prepare for, prevent, respond to, and recover from domestic incidents.[8]

These include:

- Command and Management
- Preparedness
- Resource Management
- Communications and Information Management
- Supporting Technologies
- Ongoing Management and Maintenance

The *Command and Management* elements of NIMS during an incident is based on the use of ICS for establishing the operation and structure of incident management organizations; MACS, which defines the operation and structure of supporting entities; and Public Information Systems that establishes the processes, communication systems, and procedures for delivering timely and accurate information to the public.

NIMS compliance also requires participating entities to *be prepared* and to provide continuing pre-incident training, planning, exercises, personnel qualification checks, certification standards, and equipment acquisition and certification standards.

NIMS requires effective *resource management* so that there are standard methods for managing equipment, supplies, and personnel throughout the incident. Similarly, there needs to be a standardized system for communications information management including systems and technology procedures that work across all agencies and a mechanism of assuring that information flows smoothly through a commonly accepted architecture.[8]

This description of NIMS is incomplete in this text, and the interested student is directed to www.fema.org for more complete information. In October 2017 FEMA published a revision of the NIMS. The 2017 NIMS document replaces the 2008 NIMS.[9] Key changes to the 2017 NIMS include:

1. Strengthened guidance for effective incident management
2. Using lessons learned from exercises and real-world incidents and disasters to update policies
3. Building a foundation for developing a national qualification system by clarifying the processes and terminology for qualifying and credentialing incident personnel
4. Making clear that NIMS applies to all incident personnel from the Incident Command Post (ICP) to the National Response Coordination Center
5. Showing the application of NIMS to all five mission areas (Prevention, Protection, Mitigation, Response, and Recovery)
6. Adding guidance on the Intelligence and Investigations function
7. Providing expansion of Emergency Operations Centers (EOC) guidance based on input from local, state, tribal and territorial EOC leaders
8. Explaining the relationship among Incident Command Structure (ICS), EOCs, and senior leaders/policy groups (Multiagency Coordination Group)
9. Describing common functions and terminology for staff in EOCs allowing for flexibility in differing missions, authorities, and resources of EOCs nationally
10. Refining the model for the Operational Period Planning Cycle

A guiding principle of NIMS added in 2017 includes Unity of Effort. Unity of effort refers to coordinating activities among various organizations to accomplish common objectives. This unity of effort will enable organizations with specific jurisdictional responsibilities to support each other while maintaining their own authorities. NIMS 2008 defined five NIMS components: Preparedness, Communications and Information Management, Resource Management, Command and Management, and Ongoing Management and Maintenance. Three NIMS framework components are defined in NIMS 2017. This was to enable diverse organizations to integrate capabilities and achieve shared goals. These include Resource Management, Command and Coordination, and Communications and Information Management. Resource Management under NIMS 2017 describes standard mechanisms to systematically manage resources, including personnel, equipment, supplies, teams, and facilities, both before and during incidents in order to allow organizations to more effectively share resources when needed. Command and Coordination describes leadership roles and suggested organizational structures for incident management at both operational and incident-support levels. NIMS2017 further explains how these structures interact to manage incidents effectively and efficiently. Communications and Information Management describes systems and methods that help to ensure that incident personnel and other decision makers have the means and information they need to make and communicate decisions.[9]

There are differences between NIMS and ICS based on the level of response required. When there is more than one responding agency or there are incidents that cross jurisdictional boundaries, a Unified Command may be employed whereby command decisions

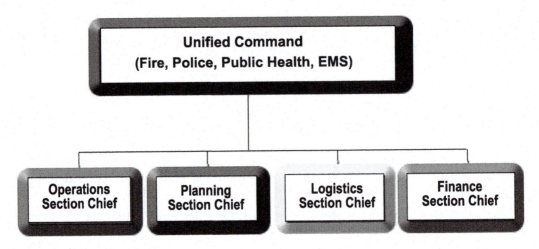

FIGURE 14.3
Example of unified command structure.

and planning are completed by senior representatives of the participating organizations who comprise the Unified Command (Figure 14.3).

An *Area Command* (1) may be employed if there are multiple incidents that are being managed by an ICS organization, (2) a large incident crosses jurisdictional boundaries, (3) no specific site can be identified as are many public health emergencies, (4) there is no identifiable cause, and (5) an incident is geographically dispersed and evolving over time such as a widespread foodborne outbreak. The Area Command is responsible for setting overall strategy and priorities, allocating critical resources, ensuring proper management of the incidents, and that objectives are met and strategies are followed. Incidents that fall within these categories require coordination at higher jurisdictional levels (Figure 14.4).

An Area Command is based on an ICS structure but without an Operations component since those functions are performed at each site (Figure 14.5).

When large or wide-scale emergencies require significant resources and complex responses, a MACS is usually employed. An example would be a statewide response to a natural disaster

FIGURE 14.4
Area Command is used for coordinating multiple incidents with different incident command posts (ICPs).

FIGURE 14.5
An Area Command structure does not require an Operations section since those functions are normally performed at each site.

such as a major hurricane. Typically, EOCs established by the emergency management agency at the state level performs this function. See the following link for information on your state emergency management association: <http://www.emergencymanagement.org/states/>. Such EOCs coordinate multiple agencies that provide a broad array of resources in response to an emergency or disaster. EOC staffing and organization may include personnel from multiple jurisdictions and functional disciplines including police, fire, public health, public works, environmental management, medical response units, and possibly many others. The numbers and disciplines of EOC staff will depend on the expected incident needs, size of the jurisdiction, and the resources available (Figure 14.6).

National Response Plan

The Homeland Security Presidential Directive 5 (HSPD-5) led to the creation of the National Response Plan (NRP) in establishing a "single, comprehensive approach to domestic incident management to prevent, prepare for, respond to, and recover from terrorist attacks, major disasters, and other emergencies."[10] The NRP is constructed around the previously described NIMS and is intended to be an all-hazards plan that provides the structure for incident management at all levels of federal, state, and local government and for any

FIGURE 14.6
An example of an EOC management structure.

major type of incident including a major public health emergency. The NRP is designed to coordinate activities among all levels of government, tribal, private-sector, and non-governmental agencies. The NRP is designed to be scalable and flexible to meet the needs of incident management ranging from multi-jurisdictional to *Incidents of National Significance* (INS) such as massive hurricane. The Secretary of Homeland Security will consult with other departments, agencies, and the White House before declaring INSs. The NRP serves as the core operational plan for national incident management and establishes the process, protocols, and structures for such responses. The NRP and NIMS are companion documents that function together to integrate the capabilities and resources at all government levels, non-government entities, tribal, and private-sector organizations into a coordinated national response for domestic incident management.

How NRP Should Work

When an incident occurs, it is managed at the lowest jurisdictional level possible. If it is a train derailment with the release of a toxic gas, as an example, it may be handled locally with the assistance of state agencies and response teams. The NRP does alter the typical ICS response of first responders to carry out their functions or responsibilities. Should the derailment incident expand to multiple jurisdictions or states with greater potential consequences, then more than one federal, state, or local agency will have jurisdiction. The Incident Commanders from each agency will form a Unified Command to coordinate resources and develop a common Incident Action Plan (IAP) with a unified set of objectives. An ICP will likely be designated and the Unified Command develops the NIMS/ICS command structure from the top-down according to the size and complexity of the incident. The ICP serves as the reporting center for responders to receive their assignments. EOCs will be notified and receive direction from the Unified Command. The EOCs will coordinate support functions and provide resources. Should the Secretary of Homeland Security or his/her designee determine there is a need to implement one or more components of the NRP, then one or more components of the NRP may be implemented. If the Secretary of DHS determines the incident to be of national significance then the Secretary will manage the federal response. Otherwise, the incident will be managed by other authorities in coordination with the NRP framework.

Independent study courses on NIMS (NIMS: An Introduction IS-700) and NRP (The NRP: An Introduction IS-800) are available through the Federal Emergency Management Agency's (FEMA) Emergency Managements Institute of DHS at http://www.training. fema.gov/emiweb/IS/crslist.asp. You may also obtain copies of these plans at http://www .dhs.gov/dhspublic/interapp/editorial/editorial_0566.xml, or by calling (800) 368-6498.

The Role of the Environmental Health Professional in Emergency/Disaster Response

Background

Environmental health (EH) professionals must have a large arsenal of skills and knowledge to competently perform their day-to-day activities. Much of this material is available through a Web-based course developed by the Twin Cities Metro Advanced Practice

Center and produced by the University of Minnesota, Center for Public Health Education and Outreach and available at the following site: <http://www.umncphp.umn.edu /preparedness/site/index.htm>.

This online course provides training on EH concepts at the awareness level, likely roles for EH staff during an emergency, some essential skills in responding to these emergencies, a guide to reducing health impacts, and some instruction on enhanced communication and collaboration with other responders.

The American Public Health Association (APHA) collaborated with EH practitioners to create three major EH competency areas now in use by the Centers for Disease Control and Prevention (CDC) and the Food and Drug Administration (FDA).[11] These three major EH competencies include:

1. *Assessment*: gathering information, data analysis, evaluation
2. *Management*: record-keeping, problem solving, project management
3. *Communication*: risk communication, conflict resolution, education

The EH professional will use these competencies in carrying out their daily duties as well as responding to emergencies. The role they ultimately pursue during an emergency will depend on the incident type and gravity and the emergency response plans developed by the health department and the community. EH personnel should participate in the development of the emergency response plans since the plans provide direction for EH professionals during an incident response. The EH professional will have responsibilities for identifying, monitoring, and reducing environmental hazards that may impact on the health of the community during a disaster. Conducting these activities requires EH personnel to work in partnership with other responding organizations in the emergency management structure including police, fire, public works, Red Cross, medical response teams, and so forth. This coordinated effort will focus on restoring secure sources of water, food, and shelter and providing sanitation and waste management.[11]

These are the roles that EH personnel normally pursue on a daily basis, but some activities may be more pronounced during an emergency response and must be done within the structure of ICS. Therefore, the EH person will report to a single person within the ICS and carry out assignments directed by that person according to the IAP. The EH personnel will be expected to perform their assigned duties often based on a predefined set of activities defined in a Job Action Sheet (JAS) created as part of the emergency response plan for their agency. A JAS describes the functional role of a position/person in an emergency response. Tasks on the JASs may be amended by the addition or deletion of tasks according to the needs of the incident. The modification of tasks would be performed by the Section Chief or Unit Leader who is issuing the JAS (Figure 14.7).

The EH professional is expected to function within the ICS during an emergency response to (1) assess hazards and assist in risk analysis, (2) act through providing services and applying regulations to reduce hazards, and (3) assure public awareness by working with others to communicate hazards and methods to minimize those hazards. As an example, a natural disaster such as a hurricane with significant flooding would require EH professionals to evaluate the safety of food and water. The EH professional could act by imposing regulations that close down unsafe food service establishments and closing water supplies and ordering the importing of safe water until adequate disinfection of water supplies can be reestablished. EH personnel would also be expected to help create accurate and timely information for the public on what the environmental issues are and what steps to take for

Reviewed: 6/06

AnyPlace Department of Health
Incident Command System (ICS)
Emergency Response
Job Action Sheet

Position Title

Operations

Reports to   Epidemiology Branch Director

Reports to: <u>Operations Section Chief</u>
Operations Command Center Location: _____ Telephone: _____
Name of Event:_____Date and Time ICS role instituted:____

◄─── Mission

Mission: Interpret pattern of disease, coordinate investigation, develop appropriate standardized tool to use in case investigations.

Immediate: Tasks must be completed first when reporting for duty

☐   Read entire Job Action Sheet
☐   Obtain briefing from Operations Section Chief
☐   Assign patient Tracking, Investigation, and/or Syndromic Surveillance Group

Intermediate: Tasks to be completed after immediate ones

☐   Update and/or tailor standardized questionnaire (or tool(s) as needed
☐   Relate information back to Medical Consultant and Operations Section Chief
☐   Establish and maintain ongoing contact with Planning Section Chief to coordinate data and analysis of information

Extended: Tasks to be completed later on or on a continuig basis during work shift

☐   Document all action decisions and intervention
☐   Observe all staff for signs of stress. Report issues to Operations Section Chief.
☐   Provide rest periods and relief for staff, as needed.

EPI Branch Director   1 of 1

FIGURE 14.7
An example of a typical job action sheet (JAS) in public health.

personal protection. The EH professional will be expected to function within their traditional role of inspection and regulatory enforcement to make certain that licensed facilities such as food service establishments, lodging, camps, public swimming pools, and mobile home parks meet the sanitation standards of local and state ordinances and regulations. They will also be expected to work as partners with multiple agencies to provide environmental information on general sanitation, food safety, drinking water distribution, waste disposal, inspection of temporary housing, and mass feeding centers.[11]

Immediate Actions

A set of immediate actions suggested by the Minnesota Department of Health to rapidly restore EH services during and after a disaster would include[11]:

1. Obtaining information on the movements of populations at or near the incident site; identifying the location of facilities used to house displaced persons, hospitals, and medical facilities; and determining where relief workers will be housed to determine if those living quarters meet sanitary and housing codes.

2. Determine what capacities remain for delivering basic EH services including inspection and regulation enforcement of food, water, and housing.

3. Quickly assess what damage has occurred to those food, water, housing, solid waste, and other environmental entities that fall under the Public Health regulations and ordinances. The EH professional should also attempt to discover what the immediate needs are for safe water, housing, food, and sanitation. These conditions should apply to any temporary settlements where displaced persons and relief volunteers may be housed. The major response and recovery elements likely to be present themselves during or after a disaster requiring prompt attention by environmental professionals include:[11,12]

- Water safety
- Food safety
- Liquid waste disposal (sewage)
- Solid waste disposal
- Housing/mass care shelters
- Vector control
- Hazardous materials and household hazardous waste
- Medical and infectious waste
- Radiological materials

The CDC National Center for Environmental Health has also produced a Web site that serves as a clearinghouse of information resources related to emergency and terrorism preparedness for EH practitioners found at <http://www.cdc.gov/nceh/ehs/ETP/default.htm>.

The CDC Web site provides resources for emergency and terrorism preparedness specifically for EH practitioners on the subjects of general preparedness, food protection, water security, hazardous materials, building environment, radiation, vector control, and biologic and infectious waste.[13] As an example, The CDC section related to food protection provides specific information on how practitioners can effectively prevent and respond to foodborne outbreaks and the site emphasizes food-security related issues. Examples of some of the information available at this site include:

- *CDC Public Health Emergency Preparedness and Response*: Offers information on food safety after disasters such as floods, hurricanes, or tornadoes.
  - Keep Food and Water Safe after a Natural Disaster or Power Outage
  - Hand Hygiene After a Disaster
- *CDC's Foodborne Outbreak Response and Surveillance*: Includes tools for investigating and responding to foodborne disease.
  - Outbreak Investigation Toolkit
  - Foodborne Diseases A–Z
  - Epidemiologic Case Studies
  - Epi Info Software
- *Emergency Handbook for Food Managers* (*PDF, 397 KB*): Provides food service establishments, food safety trainers and government EH staff with a quick-reference handbook that is easy to use for emergency preparedness and response.

- Handbook Companion Discussion Guide (PDF, 177 KB)
- Photo Lessons (PDF, 964 KB)
- Photo Lessons in Spanish (PDF, 992 KB)
- *Food Safety Self Inspection Checklist (PDF, 776 KB)*: Provides EH professionals and food managers with the tools for routine food security and safety evaluation by food service establishments. The two checklists address food security concerns such as inspecting products for signs of tampering, restricting access to records and food areas, and evaluating general food safety.
- *In an Emergency*: Equipment & Supplies List for Food Service Fact Sheet (PDF, 978 KB)
- *In an Emergency*: Discard or Salvage Fact Sheet (PDF, 40 KB)

Do you want to learn more about emergency preparedness? Point your browser to <https://people.umass.edu/~envhl565/> and click on any of the Chapter 14 Web links associated with this topic:

- FEMA training for IS 100
- Main FEMA site
- Public Health Incident Command System
- State Emergency Management Associations
- Independent Study Courses on NIMS and NRP
- Source for Disaster Field Manual for EH Professionals
- Training on Emergency Preparedness and Response Fundamentals for Environmental Health Professionals
- CDC Emergency and Terrorism Preparedness for Environmental Health Practitioners

References

1. FIRESCOPE (Firefighting Resources of California Organized for Potential Emergencies), Past, current and future directions: A progress report, October 1988, http://www.firescope.org.
2. Irwin, B., FIRESCOPE: A record of significant decisions, June 1981, http://www.firescope.org.
3. Sept 11 News, Today's 9/11 news headlines and archives of the September 11, 2001 attack on America, http://www. September11News.com.
4. Department of Homeland Security (DHS), DHS organization, http://www.dhs.gov/dhspublic /interapp/editorial/editorial_0413.xml.
5. United States Environmental Protection Agency (USEPA), Homeland security: Our history, http://www.epa.gov/homelandsecurity/htm/ohs-history.htm, last updated February 27, 2006.
6. Department of Homeland Security, Federal Emergency Management Institute (FEMA), *FEMA IS Course Material Download: IS-100 Introduction to Incident Command System, IS-100*, http:// training.fema.gov/EMIWeb/IS/is100.asp.
7. Qureshi, K., Gebbie, K., and Gebbie, E.N., *A Guide for the Management of Emergencies or Other Unusual Incidents within Public Health Agencies, PHICS: Public Health Incident Command System (Volume 1)*, October 2005, http://www. Ualbanycphp.org/pinata/phics/guide/default.cfm.

8. Department of Homeland Security, Federal Emergency Management Agency, *FEMA Independent Study Program: IS-700 National Incident Management System (NIMS), An Introduction*, May 12, 2006, http://www.training.fema.gov/EMIWeb/IS/IS700.asp.

9. US Dept of Homeland Security, FEMA Emergency Management Institute, ICS Resource Center, National Incident Management System (NIMS) 2017 Learning Materials, https://training.fema.gov/nims/docs/nims.2017.instructor%20student%20learning%20materials.pdf.

10. The Department of Homeland Security (DHS), *Quick Reference Guide for the National Response Plan*, Version 4.0, May 22, 2006, http://www.dhs.gov/dhspublic/interapp/editorial/editorial_0566.xml.

11. Palchick, S., et al., *Emergency Preparedness and Response Fundamentals for Environmental Health Professionals*, Online Course, prepared by Twin Cities Metro Advanced Practice Center, Hennepin County, Minnesota, http://www.umncphp.umn.edu/preparedness/site/index.htm.

12. California Conference of Directors of Environmental Health (CCDEH) Disaster Preparedness Technical Advisory Committee, *Disaster Field Manual for Environmental Health Specialists*, California Association of Environmental Health Professionals, 1998.

13. CDC, National Center for Environmental Health, Environmental Health Services, *Emergency and Terrorism Preparedness Specifically for Environmental Health Practitioners*, July 24, 2006, http://www.cdc.gov/nceh.

Appendices

Acronyms and Abbreviations

AHERA	(Federal) Asbestos Hazard Emergency Response Act
ASHAA	(Federal) Asbestos School Hazard Abatement Act
AST	Above-ground storage tank
ATSDR	Agency for Toxic Substances and Disease Registry
BAT	Best available technology economically achievable
BOD	Biochemical oxygen demand
BPT	Best practicable control technology currently available
CAA	(Federal) Clean Air Act
CERCLA	(Federal) Comprehensive Environmental Response, Compensation, and Liability Act of 1980 (Superfund)
CERCLIS	CERCLA Information System
CFC	Chlorofluorocarbons
CFR	Code of Federal Regulations
CO	Carbon monoxide
CWA	(Federal) Clean Water Act
DEP	Department of Environmental Protection
DOT	(US) Department of Transportation
DPH	Department of Public Health
DWPL	Drinking Water Priority List
EHS	Extremely hazardous substances/Extraordinarily hazardous substances
EIS	Environmental impact statement
EOC	Emergency Operations Center
EPA (USEPA)	(US) Environmental Protection Agency
EPCRA	Emergency Planning and Community Response Act
FIFRA	Federal Insecticide, Fungicide, and Rodenticide Act
FWPCA	Federal Water Pollution Control Act
HAP	Hazardous air pollutant
HCFCs	Hydrochlorofluorocarbons
HIV	Human immunodeficiency virus
HMTA	(Federal) Hazardous Materials Transportation Act
HRS	Hazard ranking system
HSA	(Federal) Hazardous Substances Act
HSWA	(Federal) Hazardous and Solid Waste Amendments of 1984
HW	Hazardous waste
IAP	Incident action plan
ICP	Incident Command Post

ICS	Incident Command System
IPM	Integrated Pest Management
LAER	Lowest achievable emission rate
LEPC	Local emergency planning committee
LEPD	Local emergency planning district
LOAEL	Lowest observable adverse effect level
LQG	Large quantity generators
LUFT	Leaking underground fuel tank
LUST	Leaking underground storage tank
MACT	Maximum achievable control technology
MCL	Maximum contaminant level
MSDS	Material safety data sheet
NA	North America
NAAQS	National Ambient Air Quality Standards
NESHAPs	National Emission Standards for Hazardous Air Pollutants
NIMBY	Not in my backyard
NIMS	National Incident Management System
NIIMS	National Interagency Incident Management System
NO$_2$	Nitrogen dioxide
NOx	Nitrogen oxides
NOEL	No observable effect level
NOAEL	No observable adverse effect level
NPDES	National Pollution Discharge Elimination System
NPL	National Priorities List
NRP	National Response Plan
NSPS	New source performance standard
NSR	New source review
OEM	Original equipment market
OPP	Office of Pollution Prevention
OSHA	(US) Occupational Safety and Health Administration
PCBs	Polychlorinated biphenyls
PEL	Permissible exposure limit
PIO	Public information officer
PM$_{2.5}$	Particulates <2.5 microns in size
PM$_{10}$	Particulates <10 microns in size
POTW	Publicly owned treatment works
PPA	(Federal) Pollution Prevention Act
ppb	Parts per billion
ppm	Parts per million
PRP	Potentially responsible parties
PSD	Prevention of significant deterioration
PWS	Public water system
RCRA	(Federal) Resource Conservation and Recovery Act

R&D	Research and development
Reg	Regulations
ROD	Record of decision
RP	Responsible party
RQ	Reportable quantity
SDWA	(Federal) Safe Drinking Water Act
SIC	Standard Industrial Classification
SIP	State implementation plan
SITE	Superfund Innovative Technology Evaluation
SOP	Standard operating procedure
SO₂	Sulfur dioxide
SPCC	Spill Prevention Control and Countermeasure (plan)
SQG	Small quantity generator
Stats	Statutes
SWDA	(Federal) Solid Waste Disposal Act
TCLP	Toxicity characteristic leaching (leachate) procedure
TPQ	Threshold planning quantity
TPY	Tons per year
TSCA	(Federal) Toxic Substances Control Act
TSD	Treatment, storage, and disposal
UIC	Underground injection control
UN	United Nations
USA	United States of America
UST	Underground storage tank
VOC	Volatile organic compound
VSQG	Very small quantity generator

Glossary

abiotic: components of the ecosystem that include such things as war, air, sunlight, minerals, and their interaction to produce climate, salinity, turbulence, and other conditions that influence the physical conditions under which organisms survive.

abortion: refers to the medical means of terminating a pregnancy.

absorption: passing through the skin.

acceptable daily intake (ADI): an acceptable level of exposure or dose of a substance that is considered "safe" or which poses little risk.

acid: a material that has a pH of less than 7.0. Acids are corrosive to human tissue and some metals.

acidic deposition: characterized by emissions of nitrogen and sulfur oxides that are partially converted in the atmosphere to nitric and sulfuric acids that return to the earth in rain, snow, fog, and on dry particles.

active transport: a process that involves ATP in conjunction with special carrier proteins to move molecules through a membrane against a concentration gradient (i.e., high concentration to low).

acute effects: immediate effects from a high-intensity chemical exposure that results in a change in health usually within 24 hours. Many diseases such as measles are acute and have a rapid onset, are usually self-limiting, and are of relatively short duration.

acutely hazardous waste: any hazardous waste with an EPA Waste Code beginning with the letter "P" or any of the following F codes: F020, F021, F022, F023, F026, and F027. These wastes are subject to stringent quantity standards for accumulation and generation.

acute toxicity: the ability of a substance to cause poisonous effects resulting in severe biological harm or death soon after a single exposure or dose. Also, any severe poisonous effect resulting from a single short-term exposure to a toxic substance.

additive: occurs when the effects of two substances taken together are equal to the sum of the two substances taken separately.

adduct: carcinogenic residues bound to DNA.

adsorb: to collect the molecules of a chemical on the surface of another material.

age distribution: describes the percent of the population in various age categories from birth to death.

age-sex composition: sex ratio and age distribution of a population. The age-sex composition of the population has a profound effect on the birth and death rates of a country.

air exchange rate: the rate at which air is replaced in the structure by external air.

air pollutants: refers to substances that enter the atmosphere from human activities and have the capacity to cause human disease and welfare effects resulting in disturbances to plant life and physical structures.

albedo: the combined reflective ability of cloud cover and ground surfaces.

alkali: a material that has a pH of greater than 7.0. Generally corrosive to human tissue.

allowance: defined under 1990 CAAA as the right to emit 0.907 megagrams of sulfur dioxide.

anaerobically: in the absence of oxygen.

anaphylactic: a type I hypersensitivity that is associated with hay fever, asthma, and anaphylactic shock from insect bites, certain foods such as shellfish, and a variety of chemicals. The reactions are usually immediate, occurring within moments in persons previously sensitized by exposure to the offending antigen.

angiogenesis: a process whereby malignant cells grow into a small mass that develops nutrient-bearing blood vessels.

antagonistic: a combined effect of two or more substances that is lesser than the sum of those substances acting independently.

anthropogenic: generated or created by humans.

anticodon: the recognition site for nucleotides on the sRNA molecule.

apoenzyme: protein component of an enzyme.

apoptosis: a process of cellular self-destruction.

area command: an organization established to (1) oversee the management of multiple incidents each of which has its own operating Incident Command System organization or (2) oversee the management of a very large incident that has multiple Incident Management Teams assigned to it. Area Command sets the unifying strategy and priorities, assigns critical resources based on priorities, makes certain the incident(s) are properly managed, and ensures that objectives are being met and that strategies are followed.

ash: the residue that remains after burning.

asphyxiant: a substance that causes suffocation.

atoms: the basic units of elements that consist of a small dense center called a nucleus surrounded by a cloud of negatively charged electrons.

autotrophic: refers to organisms that synthesize their own food from inorganic substances and obtain energy from the sun. Most plants are autotrophs.

base: the physical location where primary logistics functions for an incident are coordinated and administered. There is only one Base per incident. The Incident Command Post may be at the same place as the Base.

best management practices: schedules of activities, prohibitions of practices, maintenance procedures, and other management practices to prevent or reduce the pollution of U.S. waters, including treatment requirements, recycling, reduction, reuse, operating procedures, and practices to control facility site runoff, spillage or leaks, sludge or waste disposal, or drainage from raw material storage.

bias: a type of error introduced into epidemiological studies through differential treatment of the cases and controls in a study, either during the selection process or information gathering.

bioaccumulation: refers to the storage of substances in the body at levels above what would be normally expected.

biodegradable material: waste material that is capable of being broken down by microorganisms into simple stable compounds such as carbon dioxide and water.

biogeochemical cycling: refers to the recycling of nutrients.

biological oxygen demand (BOD): a standard test to determine the amount of dissolved oxygen in water used by microorganisms that feed on organic matter in the water. The test is performed over a 5-day period under controlled conditions.

biological transmission: a process in which the disease agent goes through some reproductive stage in the vector and is transmitted by a bite, fecal matter, or vomit of the vector in an active process.

biological vector: refers to vectors (usually arthropods) in which a microbial disease agent goes through development or multiplication prior to becoming infective.

biomarkers: these are mutations occurring along a gene appearing at characteristic sites and with unique base sequences.

biomass: the accumulation of organic material in an ecosystem.

biomes: based on the dominant types of vegetation, which are strongly correlated with regional climate patterns.

biosphere: that part of the earth and its atmosphere that can support actively metabolizing life.

biotic: components of the ecosystem include living organisms and the products of these organisms including waste products and decay such as urine, feces, decaying leaves and twigs, bones, and flesh.

biotic potential: the unrestricted growth of populations resulting in the maximum growth rate for a particular population.

biotransformation: a process in which metabolic processes alter the structure and characteristics of a chemical.

birth rate: refers to the number of individuals added to a population through reproduction (live births) and is normally expressed as the number of live births per 1,000 population (counting the population at the midpoint of the year).

bogs: usually a stagnant, acidic, nutrient-poor ecosystem filled with peat and sphagnum moss. Its abundance of decaying material consumes enormous amounts of carbon dioxide.

boiling point: the temperature at which a liquid turns to a vapor at ambient pressure.

bottom ash: larger residue pieces that fall through the bottom of the boiler grate after burning is completed.

bottomland hardwood forests: these are deciduous forested wetlands consisting of gum and oak trees adapted to areas of seasonal flooding or covered with water most of the year and found along streams and rivers in the Southeast and Southcentral flood plains of the United States.

building-related illnesses (BRI): a phenomenon characterized by well-defined illnesses that occur in a building that can be traced to specific building problems.

buttress: a shallow root system formed by trees with bases that spread at the bottom of the tree. This growth provides support in thin soil and may enhance the ability of the tree to secure nutrients over a broader area.

bypass waste: waste that cannot be recycled, reused, or burned but must be buried at the landfill.

canopy: an almost unbroken sea of green produced by leaves of trees at 21–30 meters above the forest floor.

carcinogen: cancer-producing substance.

carnivores: animals that eat primarily animal flesh. These animals include lions, tigers, hyenas, frogs, or even ladybugs (they feed on aphids).

carriers: persons who exhibit no symptoms but harbor and transmit the disease organisms.

carrying capacity: the sum of these limiting factors that serve to control the numbers of a species that can survive in a defined area over time.

ceiling limit: the concentration of chemicals in air that should never be exceeded during any part of a working exposure.

chain of command: a succession of control, executive, command, or management positions in hierarchical order of authority.

chaparrals: areas of moderately dry climate and limited summer precipitation are further characterized by small (0.9–4.5 meter) shrubs with leathery leaves that contain aromatic and flammable substances.

chlorofluorocarbons (CFCs): a family of inert, nontoxic, and easily liquefied chemicals used in refrigeration, air conditioning, packaging, or insulation or as solvents or aerosol propellants. Because CFCs are not destroyed in the lower atmosphere, they drift into the upper atmosphere, where the chlorine is released and destroys ozone.

characteristic waste: a waste classified as hazardous because it is ignitable, corrosive, reactive, or toxic as determined by the toxicity characteristic leaching procedure. It has an EPA waste code in the range D001 to D043. Each of these four characteristics is defined in 40 CFR 261.20 Subpart C.

chemical formula: a listing of the atoms in a molecule.

chlorinated solvent: an organic solvent containing chlorine atoms, for example, methylene chloride and 1,1,1-trichloromethane, that is used in aerosol spray containers and in certain paints.

chronic effects: effects of repeated, low-dosage chemical exposure that may not show up for many years or may last for many years. Diseases having a slow onset and lasting for extended periods are said to be chronic and may include cancer, emphysema, and some forms of heart disease.

Clean Water Act (CWA): re-designated name for the Federal Water Pollution Control Act following the 1977 amendments.

climate: can be viewed as average weather within a geographical area over years or even centuries.

climax community: is identified as one that forms in an undisturbed environment and continues to grow and perpetuate itself in the absence of further disturbance.

clonal evolution: a process whereby a single cell develops a genetic mutation that enables it to divide even when normal cells don't replicate.

cloning vector: a fragment of self-replicating DNA molecule.

Code of Federal Regulations (CFR): the detailed regulations written by federal agencies to implement the provisions of laws passed by Congress. Regulations in the CFR have the force of federal law.

co-dominant: refers to genes that are partly expressed if present as a single allele (an allele is a pair of genes situated at the same site on paired chromosomes) and fully expressed if present on both.

codon: a group of three consecutive (triplet) mononucleotide units in the mRNA that specifies the position of one amino acid in a protein molecule.

command staff: the Command Staff in an incident management organization or incident command system (ICS) that includes the Incident Command and the special staff positions of Public Information Officer, Safety Officer, Liaison Officer, and other positions as required, who report directly to the Incident Commander. They may have an assistant or assistants, as needed.

commensals: a relationship in which two kinds of organisms live in or near each other with one obtaining benefits, and the other is not damaged or benefited.

complement: a system composed of several (at least 20) serum proteins that follow a sequence or cascade of actions beginning with the Ag-Ab binding. This sequence of actions enhances phagocytosis, inflammation, and cell lysis.

composting: a controlled process of degrading organic matter by microorganisms into a humus-like material.

conditionally exempt generator: a hazardous waste generator that meets the following criteria: (a) in every single month of the year, the site generates no more than 100 kilograms of hazardous waste, no more than 1 kilogram of acute hazardous waste, and no more than 100 kilograms of material from the cleanup of spillage of acute hazardous waste.

congenital defects: the defects occurring during the embryonic stages of development.

contaminant: any physical, chemical, biological, or radiological substance or matter that has an adverse effect on air, water, or soil.

corrosives: chemicals that are acidic or basic in pH and destructive to human tissue and many metals.

crossbreeding: this involves the selection of genetic and morphologic variants that possess one or more of the desired characteristics and then crossing the plants/animals in the hope that the most desired characteristics of the different species will appear in the new crossbred organisms.

cytogenetic defects: abnormalities in the number or structure of chromosomes.

death rate: refers to the number of individuals removed from a population through death and is normally expressed as the number of deaths per 1,000 population (counting the population at the midpoint of the year).

decomposers: insects, bacteria, fungi, and protozoans that sequentially break down complex organic materials into low-energy mineral nutrients that once again may be reabsorbed and used by plants.

definitive host: the host in which the parasite reaches sexual maturity.

deforestation: the permanent decline in crown cover of trees to a level that is less than 10 percent of the original cover.

demographic transition: the simultaneous decline of death rates and birth rates in a country resulting in a diminishing difference between birth rates and death rates and a very low rate of natural increase, permitting a stable population with very long doubling times.

denitrifying bacteria: bacteria that live in the mud and sediment of lakes, ponds, streams, and estuaries and recycle organic nitrogen back to atmospheric nitrogen.

director: the ICS title for individuals responsible for supervision of a Branch.

deserts: areas defined by their climates that average less than 25 centimeters of precipitation a year and where evaporation tends to exceed this precipitation causing a negative annual water budget.

developmental disease: a disease that occurs when faults or mistakes occur within the genes (or chromosomes) or stages in development of the fetus are disturbed.

diffusion: a passive process that occurs when molecules move from areas of high concentration to one of low concentration.

direct contact transmission: occurs when the disease is passed directly from a source to the host by physical contact without an intervening object.

discharge: the discharge of a pollutant when used without qualification.

disposal: final placement or destruction of toxic, radioactive, or other wastes; surplus or banned pesticides or other chemicals; polluted soils; and drums containing hazardous materials from removal actions or accidental releases. Disposal may be accomplished through use of approved secure landfills, surface impoundments, land farming, deep well injection, ocean dumping, or incineration.

division: divisions are used to separate an incident into geographical areas of operation and organizationally are located within the ICS organization between the Branch and the Task Force/Strike Team.

DOT: the United States Department of Transportation.

doubling time: the length of time required for a population to double its size.

droplet infections: a process whereby the organisms are contained within mucous droplets that are expelled to the air in sneezing, coughing, or even talking and travel short distances to be inhaled or ingested by persons very nearby (usually less than a meter).

dry adiabatic lapse rate: a theoretical condition in which a parcel of warm, dry air will cool at the rate of 10 degree Celsius per kilogram of altitude.

dynamic equilibrium: processes that tend to balance each other so that there is little net loss over long periods of time.

ecosystems: refers to identifiable areas within nature where the organisms interact among themselves and their physical environment and exchange nutrients largely within that system.

effective demand: the "demand level" corresponding to purchasing power without regard to food requirements.

effective dose: the amount of a chemical that reaches the general circulation and ultimately the location where a particular effect is seen or felt.

effluent: the water and the quantities, rates, and concentrations of chemical, physical, biological, and other constituents that are discharged from a point source.

emergency operations center (EOC): the physical location where coordination of information and resources to support domestic incident management activities usually occurs. An EOC may be a temporary facility or may be located in a more central or permanently established facility.

emergent trees: taller trees that penetrate the lower canopy to form an umbrella-like canopy over the trees below. These trees reach heights of 39.6 meters.

emigration: this is a process by which some species disperse or migrate out of the area.

emissions: gases and particulates that reach the air from the combustion or industrial processes.

endemic: refers to diseases that already exist in a community where it is maintained in a low but constant incidence.

endocrine disruptors: synthetic chemicals that have been referred to as xenoestrogens, estrogenic, hormone-mimicking, and endocrine-disrupting chemicals that have the ability to mimic or block hormones, cause unwelcome changes in hormone receptor sites, cause changes in the baseline production of hormones, or replace hormones on the carrier proteins. This makes some proportion of the population of native functional hormone unavailable and produces adverse health consequences that may be barely noticeable to very severe.

endpoints: the final or targeted result of a health event such as death or recovery. Researchers will focus their interests on certain health responses or endpoints from exposure to biological, physical, or chemical agents.

energy recovery: the generation of power in the form of steam or electricity from the burning of garbage.

enforcement: efforts made by official agencies to ensure agreement with environmental laws and regulations.

environment: in its broadest definition, the term refers to personal and cultural behavior including smoking, diet, alcohol consumption, sexual and reproductive patterns, workplace, infections, food additives, and pollution along with the strictly physical environment.

environmental disease: refers to any pathologic process having a characteristic set of signs and symptoms that are detrimental to the well-being of the individual and are the consequence of external factors including exposure to physical or chemical agents, poor nutrition, and social or cultural behaviors.

environmental lapse rate: the actual change of temperature with altitude.

environmental resistance: refers to those pressures that limit population and may include such factors as disease, wars, predatory behavior, toxic waste accumulation, or species competition.

environmental tobacco smoke (ETS): the tobacco smoke consisting of side-stream smoke that originates from the burning end of the cigarette, and exhaled smoke.

EPA (also USEPA): the United States Environmental Protection Agency. Responsible for making and enforcing the laws and regulations regarding protection of the environment.

EPA identification number: a 12-character number assigned by either EPA or the authorized state to each hazardous waste generator; transporter; and treatment, disposal, or storage facility.

epidemic: a situation in which there is a marked increase in the incidence of disease within a limited area affecting ever-increasing numbers of people.

epiphytes: air-plants that grow on branches high in the trees. These plants are supported by the tree limbs and extract moisture from the air, trapping leaf-fall and wind-blown dust for nutrients.

estuaries: inland marine waterways found worldwide. They are dominated by grass or grass-like plants and lie mostly between barrier islands and beaches.

exoskeleton: a hard outer skin that protects internal organs of arthropods and serves as an attachment site for muscles to provide movement.

exponential growth phase: that part of the growth curve where the population increases at a rate proportional to its size, or exponentially.

extrapolation: refers to the use of animal data to predict human response to chemical exposure.

extremely hazardous substances (EHS): any of 406 chemicals identified by EPA on the basis of toxicity and listed under SARA Title III. The list is subject to revision.

facilitated diffusion: a process in which molecules such as amino acids and sugars require specialized carrier proteins to be transported across a membrane.

fallowing: an agricultural process whereby fields remain unplanted every few years, and this resting permits moisture and nutrients to return.

family planning: programs that are directed at assisting couples in having the number of children they desire regardless of how many.

flammable limits: the percent concentrations of gas or vapor in air that will support combustion.

flashpoint: the temperature at which a flammable liquid will flash when provided with an ignition source.

fly ash: very small solid particles of ash and soot generated by burning and carried in gases that fly up into the environmental control equipment.

foodborne pathogens: agents that include parasites, viruses, fungi, and bacteria that have the potential to contaminate a food supply.

food security: a situation that occurs when all people have physical and economic access to the basic food they need to work and function normally.

food web: the complex and interrelated patterns of consumption among many trophic levels.

form R: Toxic Chemical Release Inventory (TRI) Report Form. This form is required by Section 313 of the Emergency Planning and Community Right-to-Know Act (Title III of the Superfund Amendments and Reauthorization Act of 1986), Public Law 99-99. The information on this form was last collected for calendar year 1990 with the report due by July 1, 1991.

general staff: the command personnel reporting to the Incident Commander. The General Staff consists of:
Operations Section Chief
Planning Section Chief
Logistics Section Chief
Finance/Administration Section Chief

generation: refers to the weight of materials and products as they enter the waste management system from residential, commercial, institutional, and industrial sources and before materials are recovered or combusted.

generator: a person, company, site, or mobile source that produces solid or liquid waste.

gene transfer: the transfer of genetic information from one organism to another by artificial means in order to create new organisms with desired characteristics.

greenhouse effect: a process in which heat energy (infrared) reflected from the earth is absorbed by infrared-absorbing gases such as carbon dioxide, chlorofluorocarbons (CFCs), methane, and water vapor thereby trapping the warmth.

halon: bromine-containing compounds with long atmospheric lifetimes whose break-down in the stratosphere causes depletion of ozone. Halons are used in firefighting.

hazardous material: a substance or compound that is capable of producing adverse health and/or safety effects.

hazardous waste: a by-product of society that, because of its quantity; concentration; or physical, chemical, or infectious characteristics may cause, or significantly contribute to, an increase in serious, irreversible illness or pose a substantial present or potential hazard to human health, safety, or welfare or to the environment when improperly treated, stored, transported, used, disposed of, or otherwise managed. Possesses at least one of four characteristics (ignitability, corrosivity, reactivity, or toxicity) or appears on special EPA lists.

herbivores: animals that eat only plants.

holoenzyme: refers to a complete and active enzyme consisting of an apoenzyme and its cofactor.

humus: organic litter that decays further to a rich organic mat.

hypersensitivity: an exaggerated immune response to the presence of an antigen.

immigrate: the process in which species settle in a new area.

incident action plan: an oral or written plan containing general objectives clarifying the overall strategy for managing an incident. It may include the identification of operational resources and assignments. When written, the Plan may have a number of forms as attachments (e.g., traffic plan, safety plan, communications plan, map, etc.).

incident commander: the individual responsible for the management of all incident activities, including the development of strategies and tactics and the ordering and release of resources.

incident command post (ICP): the physical place where the primary command functions are carried out. The ICP may be co-located with the incident base or other incident facilities.

incident command system (ICS): a standardized on-scene emergency management construct specifically designed to provide for the adoption of an integrated organizational structure that reflects the complexity and demands of single or multiple incidents, without being hindered by jurisdictional boundaries. ICS is the combination of facilities, equipment, personnel, procedures, and communications operating within a common organizational structure, designed to aid in the management of resources during incidents. It is used for all kinds of emergencies and is applicable to small as well as large and complex incidents. ICS is used by various jurisdictions and functional agencies, both public and private, to organize field-level incident management operations.

incident management team: the Incident Commander and appropriate Command and General Staff personnel assigned to an incident.

incineration: refers to the burning of certain types of solid, liquid, or gaseous materials; or a treatment technology involving destruction of waste by controlled burning at high temperatures.

incomplete demographic transition: a decrease in deaths, but without the time for adjustment in birth rates. The number of births remains high, and the combination of high birth rates and lowered death rates results in very high rates of natural increase.

incomplete metamorphosis: refers to insects such as roaches, body lice, and grasshoppers that go through three developmental stages including egg, nymph, and adult stages.

indirect contact transmission: a process in which the disease agent is transmitted from a reservoir to a host by a nonliving object.

infectious diseases: result from the pathologic process occurring when a microbial agent invades the body.

information officer: a Command Staff member who is responsible for interfacing with the public and media or with other agencies requiring information directly from the incident. The Information Officer may have assistants, but there is only one Information Officer per incident.

ingestion: swallowing.

integrated solid waste management: the practice of using several alternative waste management techniques to manage and dispose of specific components of the postal waste stream. Waste management alternatives include source reduction, recycling, composting, energy recovery, and landfilling.

intermediate host: host in which a parasite passes through its larval or asexual stages of development.

inversion: a meteorological phenomenon in which a stable, slow-moving air mass results in the formation of a dense, cool layer of air near the earth with warmer air above. The warming of air with increasing elevation is a reversal of the normal tropospheric cooling process.

ionizing radiation: the energy resulting from radiant energy that interacts with matter to form charged particles and may include electromagnetic radiation (gamma and x-radiation) or particle radiation (alpha, beta, and neutron).

isotopes: atoms of the same element that have different mass numbers.

K-strategy: a reproductive strategy in which large organisms with relatively long life spans have only a few offspring but devote their energies to protecting and nurturing the offspring to enhance their individual survival until they can reproduce.

kwashiorkor: a disease of dietary insufficiency usually seen in children 1 to 3 years old.

lag phase: the initial portion of the growth curve in which the organisms show no increase in growth rate but are undergoing changes that ready them for rapid proliferation.

landfill: a huge basin ranging in size from 8 to 28 hectares wide scooped out of the earth. The sides and bottom are lined with layers of clay and geotextiles. Inside are also 30-to-45-centimeter layers of sand in which is buried a piping system that collects the leachate that runs through the garbage.

large quantity generator (LQG): a generator of hazardous waste that meets any of the following criteria: (a) the site generates in one or more months during a year 1,000 kilograms or more of RCRA hazardous waste; (b) the site generates in one or more months during the year, or accumulates at any time, 1 kilogram of RCRA acute hazardous waste; or (c) the site generates or accumulates at any time more than 100 kilograms of spill cleanup material contaminated with RCRA acute hazardous waste.

LD_{50}: defined as the dose of a chemical that will cause death in 50 percent of a defined animal population.

leachate: the liquid formed as moisture and rainfall pass through refuse in landfills.

lianas: woody vines that grow quickly up the tree trunks when a temporary gap opens in the canopy. Here they flower and fruit in the tree tops.

liaison officer: a Command Staff member who is responsible for coordinating with representatives from cooperating agencies.

lichens: mixtures of fungi and algae that grow together.

listed wastes: wastes specifically named in 40 CFR 261.3. These wastes are listed as hazardous under RCRA but have not been subjected to the toxic characteristics listing process because the dangers they present are considered self-evident. They bear EPA Waste Codes beginning with the letters F, P, U, or K.

loam: soils best suited to agriculture that consist of sand, silt, and some clay in a homogeneous mixture.

logistics section: the ICS Section responsible for providing services, facilities, and materials for an incident.

macronutrients: nearly all life consists of large percentages of sulfur, phosphorous, carbon, oxygen, hydrogen, and nitrogen. These are chemical elements necessary for life in amounts usually larger than 1.0 part per million.

marasmus: nutritional marasmus may appear in children 1 year of age who lack sufficient food and show signs of wasting, failure to grow, and complications of infections with multiple vitamin deficiencies.

material recovery facility (MRF): a type of solid waste recycling facility that separates out all usable products and usable parts from individual products.

material safety data sheet (MSDS): an information sheet, required under the OSHA Communication Standard, that provides information on a specific chemical health and physical hazards, exposure limits, and precautions. Section 311 of SARA requires facilities to submit MSDSs under certain circumstances.

mass number: the total number of protons and neutrons in an atomic nucleus.

materials recycling facilities (MRFs): facilities that perform the function of preparing recyclables for marketing.

maximum mixing depth (MMD): refers to the height a column of air can rise. Generally, the higher the column of rising air, the better the mixing or dispersion of pollutants.

mechanical injury: injury that results from a transfer of a damaging excess of kinetic injury to tissues resulting in abrasions, lacerations, punctures, contusions, broken bones, and projectile wounds.

mechanical transmission: a process of passive transmission by which organisms such as arthropods transmit a disease by simply carrying the disease organism(s) on the feet or other body part to a person's food or an open sore on the person.

metastasis: a process in which malignancies travel, producing tumors at distant sites.

monoculture: the planting of single crops.

monitoring: the measurement, sometimes continuous, of air or water quality.

mucociliary streaming: a process whereby particles such as dust and pollen of 10 grams or larger are removed by the constant streaming of mucous propelled from the bronchial and tracheal passages by the cilia beating at over 1300 times per minute.

Multi-Agency Coordination System (MACS): the system that integrates personnel, equipment, procedures, facilities, and communications. When activated, MACS coordinates the resources of assisting agencies in a multi-agency or multijurisdictional environment.

mutagens: agents in the environment including chemicals or radiation that promote mutations.

mutation: a change in the nucleotide base sequence of DNA.

National Ambient Air Quality Standards (NAAQS): refers to standards that have been set under the Federal Clean Air and its Amendments for criteria pollutants thereby establishing maximum concentrations allowed in the ambient or outdoor air environment.

National Incident Management System: a system mandated by HSPD-5 designed to provide a unified national response to incident management for all potential incidents regardless of size and complexity. A major objective of NIMS is to improve coordination and cooperation among public and private agencies and organizations throughout a whole range of domestic incident management activities.

National Interagency Incident Management System (NIIMS): a program developed by NWCG that includes five major subsystems providing a unified systems approach to all-hazard incident management. The subsystems include The Incident Command System, Qualifications and Certification, Training, Supporting Technologies, and Publications Management.

National Pollutant Discharge Elimination System (NPDES): a provision of the Clean Water Act that prohibits discharge of pollutants into waters of the United States unless a special permit is issued by EPA, a state, or (where delegated) a tribal government on an Indian Reservation.

necrosis: a process in which lesions are produced that are local and often result in killing areas of tissue.

neoplasia: new and uncontrolled growth of abnormal tissue from the transformation of normal body cells (tumors).

NIOSH: National Institute for Occupational Safety and Health.

non-attainment areas: those Air Quality Control Regions (AQCRs) as designated by the USEPA that exceed the levels of one or more of the criteria pollutants during the year.

non-point source: refers to pollutants entering the environment from a broad area and may include scattered sources.

normal lapse rate: a characteristic loss of temperature as air rises. Air will lose heat from such mechanisms as radiative cooling at an average of −6.50 degree Celsius per kilometer.

off-site facility: a hazardous waste treatment, storage, or disposal area that is located at a place away from the generating site.

omnivores: animals that eat plants and animals, and include rats, bears, humans, hogs, foxes, and many others.

oncogenes: cancer-causing genes.

on-site facility: a hazardous waste treatment, storage, or disposal area that is located on the generating site.

OSHA: Occupational Safety and Health Administration. Designed to monitor the working conditions of individuals and ensure that they are safe.

oxidant: a substance that readily gives up an oxygen atom or removes hydrogen from a compound.

packaging: a product container or product cushion or wrapping around the product or container. Packaging may also refer to pre-consumer shipping and storage containers as well as packaging on the shelf.

pandemic: an epidemic that spreads throughout the world, as in the case of various outbreaks of influenza.

PCBs: Polychlorinated biphenyls. A known carcinogen.

permafrost: lies as a permanent layer of ice or frozen soil beneath a thin layer of topsoil.

permanent threshold shift (PTS): a permanent hearing loss resulting from damage and loss of hair cells in the hearing apparatus.

permissible exposure limit (PEL): an exposure limit to a chemical, as determined by OSHA, below which most individuals can work day after day with no harmful effects. This limit is legally enforceable.

pesticide: substance or mixture of substances intended for preventing, destroying, repelling, or mitigating any pest. Also, any substance or mixture of substances intended for use as a plant regulator, defoliant, or desiccant. A pesticide can accumulate in the food chain or contaminate the environment.

pH: a unit for measuring hydrogen ion concentrations. A pH of 7 indicates a neutral water or solution. At a pH lower than 7, a solution is acidic; at a pH higher than 7, a solution is alkaline.

phagocytosis: this process occurs when specialized cells such as phagocytes extend out streams of cytoplasm much like an amoeba and engulf solid particles to form a sac around the particle(s).

photochemical: refers to the initiation of chemical reactions by sunlight.

picocurie: the equivalent of one trillionth of a curie representing slightly more than 2.0 radioactive disintegrations per minute.

pinocytosis: a process in which cells absorb nutrients and liquids. A small drop adheres to the cell membrane surface where an indentation forms to surround the drop and form a vesicle, which then detaches from the membrane.

pioneer plants: these are early plants that establish the beginning stages of secondary succession and may include wildflowers, followed by tall grasses and compact woody bushes.

point mutations: single-gene defects.

point source: refers to pollutants entering the environment from a specific point such as a pipe or a specific source such as a factory or treatment plant.

pollution: refers to the presence of a foreign substance—organic, inorganic, radiological, or biological—that tends to degrade the quality of the environment so as to create a health hazard.

pollution prevention: the reduction of multimedia pollutants at the source and by the use of environmentally sound recycling. This prevention includes all regulated toxic and nontoxic substances.

polymorphisms: these are genetic differences reflecting the existence of several variations of the same gene.

population: considered to be the breeding group for an organism. Different species not only have differing physical attributes, but they also differ in the population characteristics.

population control: government-directed programs that set a policy for establishing an optimum population size.

population momentum: a situation in which a population has grown so much that even smaller growth rates lead to additions of larger numbers of people to the population.

prairie potholes: glacially created freshwater marshes found primarily in the upper Midwestern United States.

pulmonary: having to do with lungs and breathing.

radioisotope: a radioactive isotope.

rate of natural increase: determined by subtracting the death rate from the birth rate.

recovered materials: waste material and by-products that have been recovered or diverted from solid waste, excluding materials and by-products generated from, and commonly used within, an original manufacturing process.

recruitment failure: refers to the failure of eggs, larvae, and juvenile fish to thrive at pH levels of water below 5.5. Fish populations decline with the most sensitive, juvenile fish first disappearing.

recyclable materials: materials that still have useful physical or chemical properties after serving their original purpose and which can be reused or remanufactured into new products.

recycled materials: materials, which otherwise would have been destined for disposal, that have been collected, reprocessed, or remanufactured and made available for reuse.

refuse-derived fuel (RDF): a pelletized fuel derived from MSW that can be mixed with coal and burned in a standard boiler.

refuse reclamation: conversion of solid waste into useful products (for example, composting organic wastes to make soil conditions or separating aluminum and other metals for melting and recycling).

replacement level: defined as a total fertility rate (TFR) that corresponds to a population exactly replacing itself.

reservoir: living organisms or inanimate objects that provide the conditions where organisms may survive, multiply, and also provide the conditions necessary for transmission.

Resource Conservation and Recovery Act (RCRA): the federal statute that regulates the generation, treatment, storage, disposal, or recycling of solid and hazardous waste.

resource recovery: the act of removing materials from MSW for a productive use.

resource recovery facility: any place, equipment, or facility designed and/or operated to separate or process solid waste into usable materials or to incinerate solid waste to create heat and energy.

respiration: a process in which the respiratory system supplies oxygen to the body's cells and expels carbon dioxide from the body.

restriction enzymes: these are microbial enzymes that are used to cut DNA into several predictable and reportable pieces.

retrogression: a process resulting in a new ecosystem state with fewer overall species.

replication: a process whereby dividing cells provide each daughter cell with one complementary strand of the parent and then synthesize a complementary strand to produce double-stranded DNA that is a duplicate of the parent DNA.

rotation: agricultural practices that traditionally have included the planting of nitrogen-using (corn) and nitrogen-fixing (beans) crops in alternative years in the same space or planting them in alternating bands in the same year.

R-strategy: a reproductive strategy characterized by populations that are typically small, short-lived organisms, which produce large numbers of offspring and receive little or no parental care.

safety officer: a Command Staff member who is responsible for assessing and monitoring safety hazards or unsafe situations, and for developing measures that protect the ICS personnel.

saltwater intrusion: a condition in which saltwater can seep into an aquifer as freshwater is pumped out of the aquifer, thereby polluting the freshwater.

sanitary sewer: a channel or conduit that carries household, commercial, and industrial wastewater from the source to a treatment plant or receiving stream.

saprophytic: organisms that feed primarily on dead organic matter. As these organisms process components of the dead tissues, they leave behind undigested materials, which are in turn processed by the next group of decomposers until finally there is nothing left but energy-poor mineral nutrients.

scavengers: organisms that prefer to feed upon the dead remains of animals.

scurvy: develops from a lack of vitamin C (ascorbic acid) in the diet and consequently the synthesis of collagen and mucopolysaccharides is disturbed.

sedimentary: nutrients originating from soil or rocks.

sedimentary cycle: a process by which substances are gradually leached from sedimentary rock by the actions of rain or erosion from water.

sick building syndrome (SBS): a phenomenon in which occupants of a building display acute symptoms without a particular pattern and the varied symptoms cannot be associated with a particular source.

sewer: a channel or conduit that carries wastewater and stormwater runoff from the source to a treatment plant or receiving stream. Sanitary sewers carry household, industrial, and commercial wastes. Storm sewers carry runoff from rain or snow. Combined sewers are used for both purposes.

sex ratio: refers to the number of males relative to the number of females in the population.

sludge: a semi-solid residue from any number of air or water treatment processes. Sludge can be a hazardous waste.

small quantity generator (SQG): a hazardous waste generator that meets all the following criteria: in one or more months of the year the site generates more than 100 kilograms of hazardous waste, but in no month (1) generates 1,000 kilograms or more of acute hazardous waste, (2) generates 1 kilogram or more of acute hazardous waste, or (3) generates 100 kilograms or more of material from the cleanup of a spillage of acute hazardous waste.

solid waste: non-liquid, non-soluble materials ranging from municipal garbage to industrial wastes that contain complex, and sometimes hazardous, substances. Solid wastes also include sewage sludge, agricultural refuse, construction and demolition wastes, and mining residues. Technically, solid waste also refers to liquids and gases in containers.

solid waste management: supervised handling of waste materials from their source through recovery processes to disposal.

solvent: substance (usually liquid) capable of dissolving or dispersing one or more other substances. Solvents include, but are not limited to, the non-spent materials listed by the USEPA.

source reduction: any action taken before waste is generated that reduces its volume and toxicity.

span of control: the supervisory ratio under ICS management that ideally should be three-to-seven individuals, with five-to-one being established as optimum.

spermicides: chemicals that are designed to kill sperm.

species: normally considered to be a group of organisms that can breed together with the production of a viable and fertile offspring.

stable ecosystems: ones in which materials are constantly recycled within the system through growth, consumption, and decomposition.

storage: temporary holding of waste pending treatment or disposal. Storage methods include containers, tanks, waste piles, and surface impoundments.

stratosphere: a region of the atmosphere extending from the tropopause to about 50 kilometers above the earth. This region contains the ozone layer extending from 24 to 40 kilometers.

succession: refers to the predictable and gradual progressive changes of biotic communities toward the establishment of a climax community.

succulents: plants that store water in their tissues to be used during dry spells. The water may be stored in different parts of the plant including stem, leaf, fruit, or roots. The familiar American cactus is a stem succulent. The aloe plant is an African leaf succulent.

Superfund: the program operated under the legislative authority of the Comprehensive Environmental Response, Compensation, and Liability Act (CERCLA) and Superfund Amendment Reauthorization Act (SARA) that fund and carry out the EPA solid waste emergency and long-term removal remedial activities. These activities include establishing the National Priorities List, investigating sites for inclusion on the list, determining their priority level on the list, and conducting and/or supervising the ultimately determined cleanup and other remedial actions.

survivorship curve: a mathematical plot in which the fate of young individuals is followed to the point of death in order to describe mortality at different ages.

swamps: often transitional environments between large open water and drier land. They are home to species from both extremes and are marked by flood-tolerant trees and shrubs.

synergistic: the combined effect of two or more substances is much greater than the sum of those substances acting alone.

taiga: boreal forest that is an almost continuous band of coniferous (cone-bearing) trees extending in a giant arc from Alaska, through Canada, Europe, and Siberia.

temperate broadleaf deciduous forests (TBDFs): forests that are located in western and central Europe, eastern Asia, and eastern North America. These are climax deciduous forests including oak, maple, hickory, chestnut, elm, linden, and walnut trees.

temperate evergreen forests: areas where soil is poor and droughts and fires are frequent, the predominant species tend to be evergreens including coniferous, needle-leafed trees with some broad-leafed evergreens. In such areas, spruce, fir, and ponderosa pine proliferate.

temperate rainforests: in cool coastal climates where there is considerable rainfall or frequent heavy fogs, one is likely to see forests of giant trees such as redwoods that characteristically reach heights of 60 meters to 90 meters. Such areas exist along the Pacific coast of North America.

temporary threshold shift (TTS): a reversible hearing loss normally lasting from moments to hours, although longer periods of temporary hearing loss are possible.

teratologic defects: defects that arise during the embryonic period of development.

terracing: an agricultural process where farmers must create level fields along the contour of the slope when they are driven to plant on hillsides.

thermal injury: injury that results when tissues are exposed to extremely low or extremely high temperatures.

threshold limit value (TLV): the maximum concentration of a chemical that a worker can be exposed to day after day as determined by ACGIH. This value is not legally enforceable.

tipping fee: the cost for a truck to unload its waste at a landfill.

total fertility rates (TFR): represents the number of children a woman in a given population is likely to bear during her reproductive lifetime providing that birth rates remain constant for at least a generation.

toxic: harmful to living organisms.

toxic characteristic leachate procedure (TCLP): this is a process described by the USEPA that requires acidified liquid to be percolated through waste in a laboratory-created landfill condition for 24 hours. Samples of the leachate are tested for the amounts of the target substance released.

toxic substance: those that (1) can produce reversible or irreversible bodily injury; (2) have the capacity to cause tumors, neoplastic effects, or cancer; (3) can cause reproductive errors including mutations and teratogenic effects; (4) produce irritation or sensitization of mucous membranes; (5) cause a reduction in motivation, mental alertness, or capability; or (6) alter behavior or cause death of the organism.

toxicity test: the means to determine the toxicity of a chemical or an effluent using living organisms. A toxicity test measures the degree of response of an exposed test organism to a specific chemical or effluent.

trace elements: elements required in tiny amounts for life such as zinc, manganese, chlorine, iron, and copper.

transporter: a person engaged in the off-site transportation of hazardous waste by air, rail, road, or water.

tragedy of the commons: a behavior in which society will likely pass on the consequences of their destructive actions if they will benefit in the short term and receive little or no negative consequences from that action.

transstadial: a form of transmission in which disease organisms are passed from infected adult ticks through the egg, larval nymphal, and adult stages.

treatment: any method, technique, or process, including neutralization, designed to change the physical, chemical, or biological character of composition of any hazardous waste so as to neutralize such wastes, to recover energy or material resources from the waste, or to render such waste nonhazardous or less hazardous; safer to transport, store, or dispose; or amenable to recovery, storage, or reduction in volume.

treatment, storage, and disposal facility (TSD): site where a hazardous substance is treated, stored, or disposed. TSD facilities are regulated by USEPA and states under RCRA.

trophic level: refers to each stage of a food-web through which energy travels, consisting of producers, primary consumers (herbivores), and secondary consumers (carnivores). It is a system of categorizing organisms by the way they obtain food.

tropical rainforest: the tropical rainforest is complex in terms of species diversity and structure. The combination of constant warmth, with average monthly temperatures above 17.8 degree Celsius and precipitation greater than 254 centimeters per year, encourages rapid plant growth throughout the year.

troposphere: a region of the atmosphere extending from the earth's surface to approximately 10 kilometers. The troposphere is greater over the equator (12 kilometers) and lesser over the poles (8 kilometers).

TSDR: treatment, storage, disposal, or recycling.

tubal ligation: a surgical procedure requiring small incisions near the navel and just above the pubic area in which the fallopian tubes are tied off or cauterized. This procedure blocks the entry of eggs into the uterus.

tundra: a biome limited to the upper latitudes of the northern hemisphere that forms a belt around the Arctic Ocean.

understory: area on the floor of the rainforest where there is sparse vegetation in the inner sections of the undisturbed rain forest, and these sections have to compete intensely in an area with little sunlight and poor soil.

unified command: Unified Command is a management system under ICS that promotes a common response, set of objectives, and incident action plan among all the participating agencies with responsibility for the incident without giving up agency authority, responsibility, or accountability.

unified area command: a Unified Area Command is established when incidents under an Area Command are multijurisdictional.

unity of command: the practice within an organization whereby each person reports to one and only one designated person.

uniform hazardous waste manifest: the shipping document (EPA Form 8700-22 or 8700-22a) that pertains to hazardous waste and is duly signed by the generator.

urbanization: the mass migration of people to the cities.

vapor density: the weight of a gas or vapor compared to an equal volume of air. Gases with a vapor density of greater than one (1) settle to the ground; gases that have a vapor density less than one (1) rise in air.

vapor pressure: the pressure of vapor above a liquid.

vasectomy: male sterilization accomplished by making an incision on either side of the scrotum and snipping out a piece of the vas deferens.

vectors: refers to animals that carry pathogens from one host to another, either from another human that is infected or from an infected animal.

vehicle transmission: the transmission of disease agents by various media such as air, water, food, intravenous fluids or blood, and drugs.

very small quantity generator (VSQG): a conditionally exempt generator of hazardous waste.

volume reduction: the processing of waste materials so as to decrease the amount of space the materials occupy, usually by compacting or shredding (mechanical), incineration (thermal), or composting.

waste: any material discarded as worthless, defective, or of no further use that, when disposed of, may pose a threat to human health or the environment.

waste management costs: the costs to procure, handle, treat, and dispose of wastes; a key factor in deciding to introduce source-reduction techniques.

waste minimization: the reduction, to the extent feasible, of hazardous waste that is generated or subsequently treated, stored, or disposed. It includes any source-reduction or recycling activity undertaken by a generator that results in: (1) the reduction of total volume or quantity of hazardous waste; (2) the reduction of toxicity of hazardous waste; or (3) both, as long as the reduction is consistent with the goal of minimizing present and future threats to human health and the environment.

waste reduction: any change in a process, operation, or activity that results in the economically efficient reduction in waste material per unit of production without reducing the value output of the process, operation, or activity, taking into account the health and environmental consequences of such change.

waste stream: a specific type of waste leaving a facility or operation.

waste-to-energy (WTE): combustion of refuse with the production of energy.

water stress: a term defined as the ratio of water withdrawal to water availability.

wetlands: those areas of land where water saturation is the major factor influencing the nature of soil development and the communities of plants and animals that live in the soil and on the surface.

windrow: a composting process in which rows of material are placed next to each other outdoors and periodically mixed to incorporate air into the mix.

zoonoses: diseases that are spread mostly from animals to humans.

Index

Note: Page numbers in italic and bold refer to figures and tables, respectively